T0199462

STOCHASTIC ANALYSIS AND APPLICATIONS

ADVANCES IN PROBABILITY and Related Topics

Editor: Peter Ney

Department of Mathematics
University of Wisconsin-Madison
Madison, Wisconsin

Vols. 1 and 2	Advances in Probability and Related Topics, edited by Peter Ney
Vol. 3	Advances in Probability and Related Topics, edited by Peter Ney ar Sidney Port
Vol. 4	Probability on Banach Spaces, edited by James Kuelbs
Vol. 5	Branching Processes, edited by Anatole Joffe and Peter Ney
Vol. 6	Multicomponent Random Systems, edited by R. L. Dobrushin and Ya. G. Sinai (in preparation)
Vol. 7	Stochastic Analysis and Applications, edited by Mark A. Pinsky

Other volumes in preparation

STOCHASTIC ANALYSIS AND APPLICATIONS

Edited by Mark A. Pinsky
Northwestern University
Evanston, Illinois

CRC Press
Taylor & Francis Group
Boca Raton London New York

CRC Press is an imprint of the
Taylor & Francis Group, an **informa** business

First published 1984 by Marcel Dekker , Inc.

Published 2019 by CRC Press
Taylor & Francis Group
6000 Broken Sound Parkway NW, Suite 300
Boca Raton, FL 33487-2742

First issued in paperback 2019

ISBN-13: 978-0-367-45178-3 (pbk)
ISBN-13: 978-0-8247-1906-7 (hbk)

Visit the Taylor & Francis Web site at
http://www.taylorandfrancis.com

and the CRC Press Web site at
http://www.crcpress.com

Library of Congress Cataloging in Publication Data

Main entry under title:

Stochastic analysis and applications.

(Advances in probability and related topics ; v. 7)
Includes index.
1. Stochastic analysis--Addresses, essays, lectures.
I. Pinsky, Mark A., [date]. II. Series.
QA273.A1A4 vol.7 [QA274.2] 519.2s [519.2] 83-27279
ISBN 0-8247-1906-9

Preface

Like many other branches of mathematics, probability theory has grown and developed with inspiration from other areas of science. This is especially clear in stochastic analysis, where ideas from theoretical physics and electrical engineering, for example, have exerted a noteworthy influence in the twentieth century.

The mathematical theory of Brownian motion began in the 1920s with the work of Norbert Wiener, who had earlier considered problems in turbulence. This theory was developed further by Paul Lévy and Kiyosi Itô, who rigorously formulated the theory of stochastic differential equations in the early 1940s. This gives a "path integral representation" for the parabolic differential equations that are familiar from the classical theory of heat flow. Recently the "stochastic calculus of variations" of Paul Malliavin has been developed to obtain probabilistic proofs of the differentiability theorems for the solution of second-order parabolic equations. This theory is naturally adapted to differentiable manifolds.

In 1947 the physicist Richard Feynman proposed a theory of "path integration" to solve the differential equations of quantum mechanics. This further stimulated the development of stochastic analysis and led to new connections between probability theory and differential equations, due largely to Mark Kac. The theory of one-dimensional diffusions, developed at an analytical level by William Feller, was given a thorough probabilistic treatment by Kiyosi Itô and Henry McKean. Many of these theories were developed independently by the Soviet school, under the leadership of Eugene Dynkin.

The mathematical theory of time series also provided an important source of problems in stochastic analysis. This subject was first treated by Wiener in the context of second-order stationary processes, without the benefit of Itô integration theory. Kalman and Bucy gave a complete analysis of linear filtering problems using stochastic integration, which was subsequently developed by several authors in the areas of nonlinear filtering, control, and prediction theory.

The present volume attempts to exhibit current research in four major areas: (i) stochastic integration, (ii) stochastic differential equations, (iii) stochastic optimization, and (iv) stochastic problems in physics and biology. Without pretending to give justice to the depth or technical sophistication of any of these works, we shall briefly indicate the topic area of the contributions to this volume. In particular, many of the papers cited below do not fit neatly into the above fourfold classification.

Albeverio and Høegh-Krohn have shown that the theory of Dirichlet forms may be used to investigate a class of symmetric random fields which arise in quantum mechanics. Kallianpur and Bromley, following earlier work of these authors, have developed a new theory of Feynman integration based on abstract Wiener spaces and methods from several complex variables. Freidlin considers a multidimensional nonlinear diffusion equation with periodic coefficients which he treats by the method of "large deviation estimates" to study the wave front of a solution. Nishioka has obtained new results on the stochastic solution of Schrödinger's equation. In a similar spirit, Motoo has studied the stochastic solution of a fourth-order parabolic equation.

The paper of Bensoussan and Menaldi is a continuation of the authors' previous work. They have obtained important new results on the optimal control of diffusion processes in a bounded domain, showing that one may use the class of non-anticipative controls to solve the minimization problem. Franke has shown that locally convex optimization techniques can be applied to stochastic control problems. Kushner has further developed the theory of weak convergence and other approximation methods while demonstrating their wide applicability in control and communication theory. Omatu has developed estimation theory for discrete parameter stationary processes, together with a number of concrete applications. Hitsuda considers a new problem in communication theory, using a Wiener-like integral with respect to a gaussian process to construct the conditional expectation of the signal process given the past of the observed process.

Ikeda and S. Watanabe have provided a report on recent results on stochastic flows, a powerful extension of the notion of diffusion process. Under suitable conditions, a stochastic differential equation on a differentiable manifold naturally defines a flow on the diffeomorphism group of the manifold and distinct flows can have the same "one point motions." An illuminating example of this possibility is provided by a certain stochastic flow on the torus. The paper of Kunita gives a sufficient condition for a stochastic differential equation to generate a flow of homeomorphisms, namely, that the adjoint stochastic equation have no explosions. It is further shown that under quite general conditions, any stochastic flow can be obtained as the solution of a suitable stochastic differential equation.

Jacod has provided a modern form of the differentiability theorems for stochastic differential equations indexed by a multidimensional parameter. Results of this type have acquired great importance in recent years, in connection with the "stochastic calculus of variations," mentioned above. Ogura has continued his previous work on the strict inequality in comparison theorems for solutions of stochastic differential equations. This can be measured either by the minimum distance or by the sojourn time, both of which lead to concrete estimates in terms of the modulus of continuity of the common diffusion coefficient of the two stochastic differential equations. Tanaka considers diffusion processes in a half space with periodic coefficients, where the drift coefficients for the interior process and the boundary process both satisfy a suitable centering condition. By inserting a small parameter one obtains "rapidly oscillating coefficients" and it is shown that the process converges in the sense of probability law to a limiting diffusion process in the half space. H. Watanabe has obtained results on approximating certain diffusion processes by discrete-time, nonmarkovian processes which arise in selection models of population genetics. Results of this type have been of great interest recently in the case of continuous time.

Finally, a token of appreciation to Vicki Davis, who retyped the entire manuscript. Her patience and dedication to this task is deeply appreciated.

Mark A. Pinsky

Contents

Contributors

SERGIO ALBEVERIO Institute of Mathematics, Ruhr University, Bochum, Federal Republic of Germany

ALAIN BENSOUSSAN University of Paris-Dauphine, and INRIA, Paris, France

C. BROMLEY University of North Carolina, Chapel Hill, North Carolina

JURGEN FRANKE* University of Heidelberg, Heidelberg, Federal Republic of Germany

MARK I. FREIDLIN Moscow, Union of Soviet Socialist Republics

MASUYUKI HITSUDA† Faculty of Integrated Arts and Science, Hiroshima University, Hiroshima, Japan

RAPHAEL HØEGH-KROHN Institute of Mathematics, University of Oslo, Oslo, Norway; Theoretical Physics Center, CNRS, Marseille, France, and University of Provence, Marseille, France

NOBUYUKI IKEDA Faculty of Science, Osaka University, Toyonaka, Osaka, Japan

JEAN JACOD‡ Mathematical Research Institute, University of Rennes, Rennes, France

Current affiliations:

*University of Frankfurt, Frankfurt am Main, Federal Republic of Germany
†Faculty of Science, Kumamoto University, Kumamoto, Japan
‡University of Paris, Paris, France

GOPINATH KALLIANPUR University of North Carolina, Chapel Hill, North Carolina

HIROSHI KUNITA Faculty of Engineering, Kyushu University, Fukuoka, Japan

HAROLD J. KUSHNER Division of Applied Mathematics, Brown University, Providence, Rhode Island

JOSÉ LUIS MENALDI* University of Paris-Dauphine, Paris, France

MINORU MOTOO Faculty of Science, Tokyo Institute of Technology, Tokyo, Japan

KUNIO NISHIOKA Tokyo Metropolitan University, Tokyo, Japan

YUKIO OGURA Faculty of Science and Engineering, Saga University, Saga, Japan

SHIGERU OMATU Faculty of Engineering, University of Tokushima, Tokushima, Japan

HIROSHI TANAKA Faculty of Science and Technology, Keio University, Yokohama, Japan

HISAO WATANABE Faculty of Engineering, Kuyushu University, Fukuoka, Japan

SHINZO WATANABE Faculty of Science, Kyoto University, Kyoto, Japan

---

Current affiliation:

*Wayne State University, Detroit, Michigan

1

Diffusion Fields, Quantum Fields, and Fields with Values in Lie Groups

SERGIO ALBEVERIO/Institute of Mathematics, Ruhr University, Bochum, Federal Republic of Germany

RAPHAEL HØEGH-KROHN/Institute of Mathematics, University of Oslo, Oslo, Norway; Center for Theoretical Physics, CNRS, Marseille, France, and University of Provence, Marseille, France

1. INTRODUCTION

In this chapter we would like to point to some aspects of the theory of Markov processes and Markov fields that have received great attention in the last few years. A common ground for these aspects is the theory of Dirichlet forms in finite and infinite dimensions. This theory gives the possibility of formulating in a unified way the basic analytic quantities for the description of Markov processes taking values in finite or infinite dimensional spaces, in particular those of diffusion type, which are the main ones occurring in the applications we shall discuss.

Markov processes taking values in a space of (generalized) functions arise naturally in the study of (generalized) Markov random fields, and a suitable description of the latter can thus be given using the theory of Dirichlet forms in infinite dimensions. In this chapter we shall be concerned mainly with Dirichlet forms giving rise to diffusion processes and diffusion fields. The study of such forms has its roots in classical potential theory, the Dirichlet integrals $\int_\Omega (\nabla f)^2\, dx$ (Ω a domain of $\mathbb{R}^d$, f a smooth function in the interior of Ω, with gradient ∇f) being the prototype diffusion form. The well-known relation between classical potential theory, the Brownian motion process, and the heat equation belongs to the great successes of the interaction between analysis and probability, and has its origins in the work of Bachelier, Einstein, and Smoluchowski on the description of the physical process of diffusion on one hand, and on the other hand, in Wiener's work on functional integration. This relation was extended later, in work by Kac, Dynkin, and many others to more general elliptic and parabolic equations and the corresponding Markov processes. Both the analytic and probabilistic approaches then developed in more abstract and general directions for which there are now systematic expositions.

Probabilistically, one can look at diffusions in[1] several ways: as continuous strong Markov processes [1], as solutions of Itô's stochastic differential equations [2], as solutions of a martingale problem [3], and as continuous processes associated with local Dirichlet forms [4]. The latter point of view allows for the consideration of diffusions which, in terms of the associated stochastic equation, have singular drift coefficients (which can also be generalized functions). This is one of the reasons for its particular interest in quantum mechanics for the description of singular interactions, another being the fact that Dirichlet forms produce naturally symmetric processes and semigroups, and these are precisely the ones of interest in quantum mechanics (the L^2-generator should essentially be the self-adjoint Hamiltonian of the quantum mechanical system). The theory of Dirichlet forms itself was initiated as an extension of potential theory to a L^2-framework by Beurling and Deny and pushed forward enormously by Fukushima and

Silverstein [4a,b,5] (for some new developments, see also, e.g., Refs. 4c, 6, and 30h). In this work the basic space is locally compact with countable basis for the topology. Extensions to the case of infinite dimensional linear spaces have been given by ourselves [7,6o-q] (see also Ref. 30h), Kusuoka [8], and Paclet [9]. As mentioned above, such extensions are useful for the description of the symmetric Markov processes associated with Markov random fields.

In Sec. 2 of this chapter, we give briefly some basic definitions and results for the theory of (local) Dirichlet forms in finite and infinite dimensions, in view of the applications in the subsequent sections. In Sec. 3 we discuss finite dimensional Dirichlet forms and the associated Markov processes and their connections with quantum mechanics. These connections have been discovered and exploited both from a foundational and a technical point of view. Of the former type is the relation with the canonical formalism [10,7,6n-r], (particularly stimulated by the corresponding point of view in quantum field theory [10,7]) and stochastic quantization [11]; of the latter type are problems involving singular potentials [30i-o,6a,j,m,n] and probabilistic estimates of certain quantities (eigenvalues, eigenfunctions, etc. [12]). We describe briefly some recent progress on these themes, concentrating particularly on aspects which are not already covered by other recent surveys; we refer to the latter for complementary aspects.

In Sec. 4 we discuss infinite dimensional Dirichlet forms, the associated Markov processes, and Markov fields and their connections with quantum fields. The point of this section is to present a theory of random fields over $\mathbb{R}^d$ which are homogeneous (with respect to the Euclidean group in $\mathbb{R}^d$) and Markov. We are particularly interested in those fields $\xi(x)$, $x \in \mathbb{R}^d$, which are of diffusion type, inasmuch as they give rise to families of Markov diffusions when looked upon as indexed by one of the coordinates with state space (generalized) functions of the remaining coordinates. Such fields arise, for example, in the theory of stochastic partial differential equations [13,30h], in certain models discussed in connection with statistics [14], in the theory of nonlinear filtering [15], in the theory of quantum fields [16,12], and in statistical mechanics [17]. Historically, the mathematical theory of random fields has several sources. One source is certainly the work on functional integration in connection with the construction of quantum fields, initiated by Friedrichs, Segal, Symanzik, Gross, Gelfand, and Minlos and followed up by many others, especially since the advent of constructive field theory (see, e.g., Ref. 6p). A second one is mathematical work concerned with extending the concept of random process to generalized random processes and fields (see Ref. 18). Another source is in problems of statistical mechanics and information theory, where a theory of discrete Markov random fields has been developed intensively (see, e.g., Ref. 19). A fourth source is in applied problems in different domain (see, e.g., Refs. 20 and 14).

In Sec. 5 we discuss briefly extensions of the study of diffusion fields to the case where the underlying index space is a manifold, rather than $\mathbb{R}^d$. Such extensions are motivated on the one hand as extensions of the theory of diffusions on manifolds (see, e.g., Ref. 21), and on the other hand by applications such as to the study of quantum fields on a curved space-time such as the ones occurring in the quantization of general relativity theory (see, e.g., Ref. 22).

In Sec. 6 we discuss an extension of the concept of Markov random field to the case of fields with values in a

Lie group. This is a point of view we have been pursuing since 1978 [23] and is related to attempts to create a non-commutative distribution theory [24] (representations of Sobolev-Lie groups) as well as with certain physical [25] and algebraic problems [26].

2. FINITE AND INFINITE DIMENSIONAL DIRICHLET FORMS AND SYMMETRIC MARKOV PROCESSES

2.1 Finite Dimensional Dirichlet Forms

The theory of Dirichlet forms on $\mathbb{R}^d$ or more generally on a locally compact space has its roots in classical potential theory, in connection with the theory of Dirichlet integrals (see, e.g., the historical surveys in Refs. 27 and 28). Work by H. Cartan on potentials of finite energy and of Aronszajn and Smith on reproducing kernels was extended by Beurling and Deny [29,28] to a systematic study of Dirichlet forms as positive bilinear forms characterized by a contraction property, giving rise to a general potential theory. The relations of this potential theory with symmetric Markov semigroups and with Hunt and diffusion processes was worked out systematically by Fukushima and Silverstein (see Refs. 4a, 5, and 6c and a sample of recent additional references 4b,c, and 6; for a new approach using nonstandard analysis, see Ref. 30h). Extensions to the nonsymmetric case have been given particularly by Carrillo Menendez and Le Jan (see the references in [4a] and [6c]). Applications to quantum mechanics have been given particularly in Refs. 30, 6a-c,j, m-r,t, 10d,e, 36-38. In this section we describe briefly the parts of the basic theory of Dirichlet forms which are connected with diffusion processes and quantum mechanical applications, reserving the description of recent developments and applications for Sec. 3.

Let (X,Σ,μ) be a measure space, with μ a positive measure. Denote by $L^2(\mu)$ the corresponding L^2-space of real-valued functions[2] f on X which are measurable and such that $\int f^2 \, d\mu < \infty$. A bounded linear operator T in $L^2(\mu)$ such that for any $f \geq 0$, $f \in L^2(\mu)$, one has $Tf \geq 0$ is said to be positivity preserving [and positivity improving if $Tf > 0$ (μ a.e.)].[3] T is said to be Markov[4] (or to have the Markov property) if for any $f \in L^2(\mu)$ such that $0 \leq f \leq 1$ one has $0 \leq Tf \leq 1$. T is Markov and conservative if $T1 = 1$ (this can only be the case for μ finite). Let N be an arbitrary function from $\mathbb{R}$ into $\mathbb{R}$ such that $N(0) = 0$ and $|N(x) - N(y)| \leq |x - y|$ for all $x,y \in \mathbb{R}$. We denote by Lip_1 the family of all such uniformly Lipschitz functions. For any $N \in Lip_1$ and any function $f \in L^2(\mu)$ we call $N(f)$ a normalized contraction of f. A bilinear form on a real Hilbert space H [e.g., $L^2(\mu)$] is a map E from a linear (dense) subset D(E) of $H \times H$ into $\mathbb{C}$ such that $E(f,g)$, $f,g \in H$, is $\mathbb{R}$-linear in both arguments. D(E) is called the domain of E. E(f,f) is then the associated quadratic form. The bilinear form is called symmetric if $E(f,g) = \overline{E}(g,f)$ where the overbar indicates a complex conjugate. One has E symmetric iff the associated quadratic form is real. If the latter is nonnegative (respectively, strictly positive) we say that the quadratic form and the bilinear form E are positive (respectively, strictly positive). We say that a symmetric positive bilinear form E on $L^2(\mu)$ "contracts under an $N \in Lip_1$" if for any $f \in D(E)$ one has $N(f) \in D(E)$ and $E(N(f),N(f)) \leq E(f,f)$. If E contracts under all $N \in Lip_1$, then we say that "E contracts under Lip_1." It is easily verified that the function

$\varphi_0(x) \equiv (x \wedge 1) \vee 0 \equiv \max(\min(x,1),0)$ is in Lip_1, hence if E contracts under Lip_1 then E contracts under "the unit contraction" φ_0. In turn, if E is closed[5], then if it contracts under φ_0, it also contracts under Lip_1, so that for closed symmetric positive bilinear forms E on $L^2(\mu)$ contracting under φ_0[6] and contracting under the whole Lip_1 are equivalent statements. The proof of this is given in Ref. 4a, sec. 1.4. Now the important point is that contracting properties of E are equivalent to the Markov property of the associated semigroup. We use here some elementary results and definitions of Hilbert space theory (see, e.g., Ref. 31, p. 322). Let E be a positive bilinear form in a Hilbert space H. To it there is associated a unique self-adjoint positive operator A_E such that $D(A_E) \subset D(E)$ and $E(f,f) = (A_E f,f)$ for all $f \in D(A_E)$, and one has $E(f,f) = (A_E^{1/2} f, A_E^{1/2} f)$, for all $f \in D(E)$. Then e^{-tA_E} is a contraction in H for all $t \geq 0$, and $(e^{-tA_E}, t \geq 0)$ is a strongly continuous contraction semigroup. A semigroup e^{-tA_E} of this type with the additional property that e^{-tA_E} is Markov for all $t \geq 0$ will be called a (symmetric) Markov semigroup. We can now state the main theorem relating forms and Markov semigroups.

THEOREM 2.1 If E is a closed symmetric positive bilinear form and if E contracts under Lip_1, then $(e^{-tH_E}, t \geq 0)$ is a Markov semigroup, where H_E is the self-adjoint positive operator uniquely associated with the closed positive form E. And vice versa, to any Markov semigroup $P_t = e^{-tH}$ there is associated a unique closed symmetric positive bilinear form $E(f,f) = (H^{1/2}f, H^{1/2}f)$.

For the proof, see Ref. 4a, theorems 1.4.1, 2.1.1. One

proves in fact that if E contracts under an $N \in Lip_1$ and for some $f \in L^2(\mu)$ one has $Nf = f$, then $N \circ \alpha G_\alpha f = \alpha G_\alpha f$, with $G_\alpha \equiv (\alpha + H_E)^{-1}$, the (Markov) resolvent of H_E, $\alpha > 0$.

Remark 1. That e^{-tH_E} is positivity preserving holds iff the closed symmetric positive bilinear form E contracts under taking absolute values [i.e., under $N(x) \equiv |x|$].

The proof is easily deduced from Ref. 4a.

A closed symmetric positive bilinear form which contracts under Lip_1 (or equivalently under φ_0) is called a <u>Dirichlet form</u>. Theorem 2.1 gives a one-to-one correspondence between the family of all Dirichlet forms on $L^2(\mu)$ and the family of all symmetric Markov semigroups on $L^2(\mu)$.

An additional important property of Dirichlet forms is the following. If a symmetric positive bilinear form which is closable (i.e., admits a closed extension) is <u>Markov</u> in the sense that for each $\varepsilon > 0$ there exists a real function $\varphi_\varepsilon(t)$, $t \in \mathbb{R}$ such that $\varphi_\varepsilon(t) = t$ for all $t \in [0,1]$, $-\varepsilon \leq \varphi_\varepsilon(t) \leq 1 + \varepsilon$, $0 \leq \varphi_\varepsilon(t') - \varphi_\varepsilon(t) \leq t' - t$, for all $t < t'$ and such that E (on its domain) contracts under φ_ε, then the closure of E is a Dirichlet form. For the proof, see Ref. 4a, theorems 1.4.1 and 2.1.1.

Since the property of contracting under φ_ε need only be verified on the domain of the not necessarily closed form E, this is often a more convenient criterion than the ones we had before, where contracting properties on the whole domain of a closed form had to be verified! There are representation theorems on Dirichlet forms on locally compact spaces X (with countable basis for the topology). Such theorems were just stated by Beurling and Deny; proofs appeared later (see, e.g., Refs. 4a and 5). In particular one has that for X a domain D of $\mathbb{R}^d$ any closable Markov symmetric positive bilinear form on $L^2(X,\mu)$ (with μ Radon and supp μ = X) and with $D(E) = C_0^\infty(D)$ is given by the sum of three forms: a "diffusion form," a "jumping form," and a "Killing form." In this chapter we restrict our attention to cases where the jumping part is absent. We are now going to characterize this situation. The structure theorem also holds for certain X locally compact and E regular in the sense that $D(E) \cap C_0(X)$ is dense in D(E) both in the sup-norm and in the E_1-norm $\|f\|_1 \equiv [E(f,f) + \|f\|^2]^{1/2}$ [$C_0(X)$ denotes the continuous functions on X of compact support]. We are interested here in so-called local forms. E is local if $f,g \in D(E)$ and supp f $\cap$ supp g = $\emptyset$, supp f, supp g compact, implies that $E(f,g) = 0$.

The structure theorem says that any regular local Dirichlet form E can be expressed uniquely for $f,g \in D(E) \cap C_0(X)$ by

$$E(f,g) = E^{(c)}(f,g) + \int_X f(x)g(x)k(dx)$$

for some positive Radon measure k on X, where $E^{(c)}(f,g) = 0$ for $f \in D(E^{(c)})$, $g \in D(E^{(c)})$, and such that g is constant on a neighborhood of supp f. $E^{(c)}$ is called the diffusion part of the Dirichlet form E. We are interested in the case $k \equiv 0$, in which case we call E a diffusion form. In the particular case X = D, a domain in $\mathbb{R}^d$, one has if $D(E) = C_0^\infty(D)$, $E^{(c)}(f,g) = \sum_{i,j=1}^{d} \int_D \frac{\partial}{\partial x_i} f(x) \frac{\partial}{\partial x_j} g(x)$ $d\nu_{ij}(x)$ for $f,g \in C_0^\infty(D)$ for some (not necessarily positive) Radon measures ν_{ij} such that for any $\lambda = (\lambda_1,\ldots,\lambda_d) \in \mathbb{R}^d$ and any compact $K \subset D$: $\sum_{i,j=1}^{d} \lambda_i\lambda_j\nu_{ij}(K) \geq 0$, $\nu_{ij}(K) = \nu_{ji}(K)$. Such Dirichlet forms are discussed in Sec. 3.

The basic result above gives the relation between

(symmetric) Dirichlet forms and (symmetric) Markov semigroups. What about the associated processes? In the special case where the Dirichlet form reads on $C_0^\infty(\mathbb{R}^d)$-functions

$$E(f,f) = (f, H_E f)$$

with

$$H_E f(x) = \sum_{i,j=1}^{d} a_{ij}(x) \frac{\partial}{\partial x_i} \frac{\partial}{\partial x_j} f(x) + \sum_{i=1}^{d} \beta_i(x) \frac{\partial}{\partial x_i} f(x)$$

with $a \equiv ((a_{ij}))$ a (strictly) positive definite matrix over $\mathbb{R}$, with smooth bounded $a_{ij}(x)$, $\beta_i(x)$, it is well known that an associated process ξ_t, $t \geq 0$, with values in $\mathbb{R}^d$ can be constructed by solving the associated Itô stochastic differential equation $d\xi_t = \beta(\xi_t)\, dt + \sigma\, dw_t$, with $a \equiv (1/2)\sigma\sigma^*$ and initial condition $\xi_0 = x$, where w_t is the Brownian motion on $\mathbb{R}^d$ (this is the "Itô-Levy approach" [2]). Another way is by taking the kernel of the associated semigroup e^{-tH_E}, H_E being now the self-adjoint operator associated with the Dirichlet form (the kernel existing and being smooth in $\mathbb{R}^d$, due to the assumptions on the coefficients), and constructing the associated process (e.g., by Kolmogorov's techniques [1]). A third way is to solve the associated martingale problem of showing that $f(\xi_t) - \int_0^t H_E f(\xi_s)\, ds$ is a martingale, for all $f \in C_0^\infty(\mathbb{R}^d)$ [3].

These various ways can be extended in many directions. However, the case where σ_{ij}, β_i are singular (in some cases not even Lebesgue measurable) functions is best covered by the L^2-methods (rather than the previously mentioned "C-methods") of the theory of Dirichlet forms developed by Fukushima and Silverstein (see also Ref. 32). As remarked by Fukushima [6b], the point is to get away from analysis on $\mathbb{R}^d$ by means of <u>functional</u> analysis (through Dirichlet forms).

How does the theory of Dirichlet forms proceed to

construct a process out of a Dirichlet form? First we observe that the one-to-one correspondence between Dirichlet forms and Markov semigroups extends to a one-to-one correspondence between these and μ-symmetric Markov transition functions. A family $(p_t, t \geq 0)$ of Markovian kernels on (X,B) [with $p_t(x,\cdot)$ a positive measure on B and $p_t(\cdot,A)$ a positive B-measurable function for each fixed $A \in B$] is a <u>μ-symmetric transition function</u> iff $p_t p_s f = p_{t+s} f$ for any bounded B-measurable function, $\int f p_t g \, d\mu = \int (p_t f) g \, d\mu$ for all $f,g \geq 0$, B measurable, and $\lim_{t\downarrow 0} p_t f(x) = f(x)$, μ a.e. $x \in X$, for all f which are continuous and of compact support. Now, if T is a symmetric Markov semigroup on $L^2(X,\mu)$, then $(p_t \chi_A)(x) \equiv p_t(x,A)$ is for any $x \in X$ and measurable A a μ-symmetric Markov transition function. Vice versa, any such transition function defines by $(p_t f)(x) \equiv \int p_t(x,dy) f(y)$ a symmetric Markov semigroup.

To any μ-symmetric transition function p_t on X one can construct a <u>Markov process</u> ξ_t <u>governed by</u> p_t (and hence by the associated Dirichlet form E); namely, there is a collection $M = (\Omega, A, \xi_t, P^x)$, $x \in X_\Delta$, with state space X_Δ (the one-point compactification of X) such that for all each $x \in X$, $P^x(\xi_{t+s} \in A | A_t) = p_s(\xi_t, A)$, P^x a.s. (<u>the Markov property</u>), $P^x(\xi_0 = x) = 1$, where A is any element of the σ-algebra B of X and A_t denotes the sub-σ-algebra of A generated by the ξ_s, $s \leq t$. It has been shown by Fukushima and Silverstein that if E is regular, then ξ_t (or more precisely M) is a (μ-symmetric in the sense that the associated p_t is μ-symmetric) <u>Hunt process</u> in the sense that M has the <u>strong Markov property</u> [$P^x(\xi_{\tau+s} \in A | A_\tau) = p_s(\xi_\tau, A)$ P^x a.s., for some right continuous σ-fields $(A_t, t > 0)$ and any A_t' stopping time τ] and almost every sample path is left continuous along any increasing sequence of stopping times. M is uniquely associated with E [in the sense that $p_t f$ is a quasi-continuous version of $p_t f \equiv e^{-tH_E} f$ for all $t > 0$ and all $f \geq 0$, Borel, in $L^2(\mu)$; see below]. M is a diffusion [in the sense that ξ_t has continuous sample paths up to the death time $\zeta(\omega) \equiv \zeta(\omega) \equiv \inf\{t > 0, \xi_t(\omega) \in \Delta\}$ P^x a.s., $x \in X$] iff E is local.

The above uniqueness is understood in the sense that one identifies μ-symmetric Hunt processes for which one has equality of transition functions except possibly on a Borel subset N of X of μ-measure zero and such that its complement is invariant. Thus to the local regular Dirichlet form there is associated a diffusion process. This applies in particular to the local regular Dirichlet forms E obtained by closed extensions of the form E_0 on $C_0^\infty(D)$, D a domain of $\mathbb{R}^d$:

$$E_0(f,f) = \int (\nabla f)^2 \, d\mu, \qquad \nabla f(x) \equiv \left(\frac{\partial f}{\partial x^1}, \ldots, \frac{\partial f}{\partial x^d}\right)$$

$$x = (x^1, \ldots, x^d) \in D$$

in $L^2(X,\mu)$. (In Sec. 3 we discuss assumptions on μ such that such closed extensions exist.) Dirichlet forms obtained from E_0 will be called diffusion forms or energy forms given by μ, and the associated processes will be called distorted Brownian motions [33,30a,6a].

Now above we mentioned the notion of "quasi-continuous." This notion is connected with the potential theory associated with the Dirichlet form E, of which we shall now say a few words. Given a regular Dirichlet form E on $L^2(X,\mu)$ the associated L^2-capacity Cap (A) for any open set $A \subset X$ is defined as Cap (A) = $\inf E_1(f,f)$, the infimum being taken over all $f \in D(E)$ such that $f \geq 1$ μ a.e. on A, where $E_1(f,f) \equiv E(f,f) + \|f\|^2$. Cap A is then extended to all $A \subset X$ as the outer capacity, yielding a Choquet capacity. Quasi-everywhere (q.e.) means then "except on a set of zero capacity" and a function f defined q.e. on X is said to be quasi-continuous if for any $\varepsilon > 0$ there exists an open set A with Cap A $< \varepsilon$ such that the restriction of f to X - A is continuous (for an alternative hyperfinite version of the construction of the above process, see Ref. 30h).

At this point we leave the discussion of the general theory of Dirichlet forms on $\mathbb{R}^d$ or a locally compact space. Having introduced the above concepts and structures, practically all results of the C-theory of Markov processes and the associated potential theory can be worked out in the L^2-setting proper to the theory of Dirichlet forms, and moreover many new or more general results can be obtained (e.g., boundary theory, classification of Markov extensions, equilibrium potentials, hitting times, additive functionals, random time changes, transiency, recurrency, asymptotic behavior, etc.). We take up some particular aspects and new developments in Sec. 3; for the rest we refer the reader to the rich literature [4-6].

2.2 Infinite Dimensional Dirichlet Forms

We shall now outline briefly the extension of the above theory to the case where the space X is an infinite dimensional space, not locally compact, typically a Hilbert space. We do this in much the same spirit as above, concentrating on basic results and leaving the discussion of newer developments for Sec. 4. Dirichlet forms on infinite dimensional spaces appeared at various places in the literature (e.g., Ref. 10c,d) and were studied systematically by ourselves in Ref. 7. Later this study was developed especially by Kusuoka [8] and Paclet [9]. We shall first introduce our framework. In this framework the role of the underlying space $\mathbb{R}^d$ (and its dual) is played by a triple $Q \subset \mathcal{H} \subset Q'$, where $\mathcal{H}$ is a real separable Hilbert space, Q is a dense nuclear subspace of H such that for $\varphi \in Q$, $h \in H$, $\varphi \to (\varphi, h)$ is weakly dense in the dual space Q' of Q, where $(\cdot,\cdot)$ is the inner product in H. The nuclear rigging is understood in the sense that there exists a scale of Hilbert spaces H_m, $m \in \mathbb{N}$, such that $H_m \subset H_{m+1}$, with the injection $H_m \hookrightarrow H_{m+1}$ of trace class, and with $Q = \bigcap_m H_m$, $Q' = \bigcup_m H_m^*$ as sets but also as locally convex linear spaces (see, e.g., Ref. 18). In particular we have that the topology in Q is given by the family of Hilbert norms $\|\cdot\|_m$, $m \in \mathbb{N}$, $\|\cdot\|_m$ being the norm in H_m.

Let μ be a Radon probability measure on Q' (this measure is going to play the role of the basic measure μ in the above theory of Dirichlet forms on locally compact spaces). Let $c(\varphi) \equiv \int_{Q'} e^{i\langle\varphi,\xi\rangle} d\mu(\xi)$, $\varphi \in Q$, be the characteristic function corresponding to the measure μ. $\langle\,,\rangle$ is the dualization between Q and Q'. $c(\varphi)$ is a continuous positive definite function on Q such that $c(0) = 1$. By a theorem of Bochner-Minlos the continuous positive definite functions on Q which are 1 at zero are in one-to-one correspondence with probability measures on Q'. Let $e_1,\ldots,e_n,\ldots$ be a complete orthonormal base in H of elements in Q. By FC^k we shall understand "cylinder functions which are C^k on their base," that is, functions $f(\xi)$ on Q' of the form $\tilde{f}(\langle e_1,\xi\rangle,\ldots,\langle e_n,\xi\rangle)$ for some $\tilde{f} \in C_b^k(\mathbb{R}^n)$ (bounded C^k functions). We have that FC^∞ is dense in $L^2(Q',\mu)$ and there is a countably generated algebra, (e.g., generated by arctan $\langle e_i,\xi\rangle$), which is uniformly dense in FC^∞. Hence the uniform closure $\overline{FC^\infty}$ of FC^∞ is, by Gelfand-Mazur representation theory, isomorphic to C(X), where X is a compact Hausdorff space with the topology determined by the linear functions $e_i \to \langle e_i,\xi\rangle$ embedded in X as a subset Q' of X. In particular X has a countable base for its topology. By the Riesz theorem μ extends to X with

$\mu(X) = 1$ and μ cannot vanish on an open subset of X if μ is quasi-invariant [in the sense that $\mu(\cdot + \varphi) \ll \mu$ for all $\varphi \in Q$] and the topology is determined by the above linear functions. Thus μ is a Radon measure on the compact Hausdorff space X with countable basis for the topology. Let E be a Dirichlet form on $L^2(X,\mu)$. By looking at functions $f \in L^2(Q',\mu)$ as embedded in $L^2(X,\mu)$ we can look upon E as a symmetric positive bilinear form on $L^2(Q',\mu)$, which we call a Dirichlet form on $L^2(Q',\mu)$. If E is regular in $L^2(X,\mu)$, we call E regular in $L^2(Q',\mu)$. We understand by canonical (with respect to the rigging $Q \subset H \subset Q'$) diffusion form or energy form given by a quasi-invariant measure μ on Q' a regular Dirichlet form E on $L^2(X,\mu)$ with core[7] FC^∞ and such that

$$E(f,f) \equiv \int_{Q'} (\nabla f)^2 \, d\mu$$

for all $f \in FC^\infty$, where

$$(\nabla f)^2(\xi) \equiv \sum_{i=1}^{n} \left(\frac{\partial \tilde{f}}{\partial x_i}\right)^2 (\langle e_1,\xi\rangle,\ldots,\langle e_n,\xi\rangle)$$

if

$$f(\xi) \equiv \tilde{f}(\langle e_1,\xi\rangle,\ldots,\langle e_n,\xi\rangle), \qquad x_i \equiv \langle e_i,\xi\rangle$$

$$(\nabla f)(\xi) \equiv \sum_{i=j}^{n} \frac{\partial \tilde{f}}{\partial x_i} (\langle e_i,\xi\rangle,\ldots,\langle e_n,\xi\rangle)$$

if

$$f(\xi) \equiv \tilde{f}(\langle e_n,\xi\rangle,\ldots,\langle e_n,\xi\rangle), \qquad x_i \equiv \langle e_i,\xi\rangle$$

Under which assumptions on μ is this form closable (i.e., has a closed extension?) Similarly as in the case of the finite dimensional theory (see Sec. 3) conditions are known [7,8] under which this is true. A sufficient condition is, e.g., the following:

THEOREM 2.2 Let μ be a probability measure on Q' (not necessarily quasi-invariant). If for all basis elements $e_i \in Q$ and all $f \in FC^\infty$, one has that $f \to \int e_i \cdot \nabla f \, d\mu = \int \frac{\partial \tilde{f}}{\partial x_i} \, d\mu$ is continuous in the $L^2(\mu)$-norm [i.e., the function identically 1 is in the domain of the adjoint of the derivative $(e_i \cdot \nabla)$ in the direction e_i for all i, i.e., $\beta_i(\xi) \equiv -(e_i \cdot \nabla)^* 1(\xi) \in L^2(\mu)$], then the canonical diffusion form defined on FC^∞ is closable.

Let us now suppose that μ is such that E is closable, and let H be the positive self-adjoint operator associated with the closure of E. Then H is the infinitesimal generator of a Markov semigroup e^{-tH}, $t \geq 0$. We observe that if $\langle e_i, \xi \rangle \in D(H)$, then $1 \in D((e_i \cdot \nabla)^*)$ and $FC^\infty \subset D(H)$. Moreover, $Hf = -\Delta f - \beta \cdot \nabla f$, with

$$(\beta \cdot \nabla) f(\xi) \equiv \Sigma_i \, \beta_i(\xi) \frac{\partial}{\partial x_i} \tilde{f}(\langle e_1, \xi \rangle, \ldots, \langle e_n, \xi \rangle),$$

$$\Delta f(\xi) = \Sigma_i \frac{\partial^2}{\partial x_i^2} \tilde{f}, \text{ for } f(\xi) = \tilde{f}(\langle e_1, \xi \rangle, \ldots, \langle e_n, \xi \rangle)$$

In this sense H is a diffusion operator, an infinite dimensional version of the diffusion operator $-\Delta_d - \beta \cdot \nabla_d$, with $\beta \in \mathbb{R}^d$, in $L^2(\mathbb{R}^d, \mu)$, the operator associated with a diffusion form $\int (\nabla f)^2 \, d\mu$ on $L^2(\mathbb{R}^d, \mu)$. If μ is Q quasi-invariant, we have that $L^2(Q', \mu)$ carries a representation of the canonical commutation relations, i.e., two strongly continuous unitary representations U,V of Q in $L^2(Q', \mu)$ such that

$$V(\varphi) U(\varphi') = e^{i\langle \varphi, \varphi' \rangle} U(\varphi') V(\varphi)$$

where

$$U(\varphi) f(\xi) \equiv e^{i\langle \varphi, \xi \rangle} f(\xi)$$

$$V(\varphi) f(\xi) \equiv \left[\frac{d\mu(\xi + \varphi)}{d\mu(\xi)}\right]^{1/2} f(\xi + \varphi)$$

That $\varphi \to V(\varphi)$ is a strongly continuous from Q into $L^2(Q', \mu)$ follows whenever, e.g., Q is Fréchet [66]. We now write $\varphi \cdot \beta \equiv \Sigma_i \varphi_i \beta_i$, where $\varphi_i \equiv \langle e_i, \varphi \rangle$. From Ref. 7a we have that $\varphi \to \varphi \cdot \beta$ is continuous from Q into $L^2(Q', \mu)$ in their respective topologies. Let $\pi(\varphi)$ be the self-adjoint infinitesimal generator of the unitary group $t \to V(t\varphi)$ and remark that $\langle \varphi, \xi \rangle$ is the corresponding generator for $U(t\varphi)$. Then one has easily that on FC^∞

$$\pi(\varphi) = \frac{1}{i}(2\varphi\cdot\nabla + \varphi\cdot\beta)$$

where $\varphi\cdot\nabla \equiv \Sigma_k \varphi_k \frac{\partial}{\partial x_k}$. In particular, $\varphi\cdot\beta = i\pi(\varphi)1$, so that $1 \in D(\pi(\varphi))$. From now on we shall suppose μ quasi-invariant and such that $1 \in D(e_i\cdot\nabla)^*$. The canonical diffusion form $E(f,f)$ is called analytic if for any $\varphi \in Q$ we have $\pi(\varphi)^n 1 \in D(E)$, for all $n \geq 0$, with $E_1(\pi(\varphi)^n 1, \pi(\varphi)^n 1) \leq n!\,(C_\varphi)^n$, for some constant C_φ independent of n. If zero is an isolated eigenvalue for the above H it was shown in Ref. 7b that a sufficient condition for E to be analytic is $\mathrm{ad}^n\,\pi(\varphi)(H) \leq (C_\varphi)^n(H + 1)$.[8] Let for any $m > 0$, Q_m' be the closed linear subspace of Q' given by $\langle e_i,\xi\rangle = 0$, $1 \leq i \leq m$. Writing $Q' = \mathbb{R}^m \times Q_m'$, and accordingly $\mu(\xi) = \mu(x,\eta)$, $\xi \in Q$, $x \in \mathbb{R}^m$, $\eta \in Q_m'$ and $\mu_m = \mu \upharpoonright Q_m' = \int_{\mathbb{R}^m} d\mu(x,\cdot)$, we have, by the conuclearity of Q' and the Q-quasi-invariance of μ: $d\mu(x,\eta) = \rho_m(x|\eta)\, d\mu_m(\eta)\, dx$, with $\rho_m(x|\eta)$ the conditional probability density and dx the Lebesgue measure on $\mathbb{R}^m$. We say that $d\mu$ is strictly positive if for any m and μ_m almost all η, $\rho_m(x|\eta)$ is bounded below by a strictly positive constant on any compact subset of $\mathbb{R}^m$. Moreover, μ is called analytic if for any m, $\rho_m(x|\eta)$ is a real analytic function of $x \in \mathbb{R}^m$ for μ_m a.e. η. Then one has (see Ref. 7b) that if E is analytic, then μ is strictly positive and analytic.

The point about these concepts is that all assumptions are verified in a large class of examples, as discussed in Ref. 7, and as we shall discuss further in Sec. 4. The study can be developed then in the direction of potential theory and ergodic theory (ergodic decompositions of μ, E, H and the associated quantities, under Q-translations and time translations), and this has been done in Ref. 7. But what about the associated processes? Since we realized the closure $\bar{E}$ of the form E in $L^2(X,\mu)$ as a regular local Dirichlet form associated with the locally compact Hausdorff space X with countable basis, we have automatically, by the theory of Dirichlet forms on such spaces, that there is a Hunt process associated with E in $L^2(X,\mu)$. There is also another point of view, namely to look at the form $\overline{E}$ in $L^2(Q',\mu)$ itself, defined as the closure of $\int (\nabla f)^2 \, d\mu$ from FC^∞ in $L^2(Q',\mu)$. By the Markov character of the latter form, we have immediately the Markov property of $\overline{E}$, and the regularity and locality of the form on $L^2(Q',\mu)$ have been studied in Ref. 7d. It follows from this work that there is a Hunt process associated with the form $\overline{E}$ which is continuous in suitable Banach norms and is thus in this sense a diffusion. Finally, it has been shown in Ref. 7a,b that the homogeneous process $\langle \varphi, \xi_t \rangle$, $\varphi \in Q$, $t \in \mathbb{R}$, associated (e.g., by the Kolmogorov construction) with the initial measure μ and the Markov semigroup e^{-tH} satisfies the stochastic equation $d\langle \varphi, \xi_t \rangle = \beta(\varphi)(\xi_t)\, dt + d\langle \varphi, w_t \rangle$, $w_t/\sqrt{2}$ being the standard Brownian motion associated with the rigging $Q \subset \mathcal{H} \subset Q'$ ([7],[10]f).

Kusuoka [8] has developed another point of view in the theory of infinite dimensional Dirichlet forms. He considers forms E associated with a Banach rigging $K \subset \mathcal{H} \subset B$, K a dense vector subspace of H, B a Banach space, H a real separable Hilbert space, μ a K-quasi-invariant measure on B. He defines E by

$$E(f,g) \equiv \int_B (A(\xi)Df(\xi), Dg(\xi))_{\mathcal{H}} \, d\mu(\xi)$$

D being a suitable derivative.

$(\cdot,\cdot)_H$ is the scalar product in H and $A(\xi)$ is a strictly positive symmetric bounded operator on H for μ a.e. $\xi \in B$. He shows that E is a Dirichlet form (in the sense of being Markov closed symmetric positive bilinear) if μ is strictly positive. For A = 1 one has a realization of the above canonical diffusion forms as the one defining a Sobolev space of order one (over B). Let E be a regular Dirichlet form on $L^2(X,\mu)$, X a locally compact separable metric space in which B is embedded such that X - B is of zero capacity. Then there is a diffusion process associated with E with state space the original space B and continuous sample paths (in the original topology of B). Recently, a new version of the theory of infinite dimensional Dirichlet forms and processes has been given using nonstandard analysis [30h]. In Sec. 4 we discuss applications and some further extensions of the theory of infinite dimensional Dirichlet forms.

3. $\mathbb{R}^d$-VALUED PROCESSES, QUANTUM MECHANICS

In quantum mechanics the states are represented by (equivalence classes of) vectors in a Hilbert space $\mathcal{H}$. The time evolution is given by a unitary group ($U_t = e^{-itH}$, $t \in \mathbb{R}$), where H is a self-adjoint operator, the Hamiltonian. When the system consists of N particles moving in $\mathbb{R}^3$ one can take $\mathcal{H}$ to be $L^2(\mathbb{R}^d, dx)$, with $d = 3N$ and dx Lebesgue measure on $\mathbb{R}^d$ (Schrödinger presentation). When the forces are "nice", H is set to be the sum (in a mathematical sense, e.g., operator sum, sum in the sense of quadratic forms, or limit of such) of a kinetic energy term H_0 and a potential term V, the potential term being given by multiplication by a function and the kinetic energy term being given (in suitable units and coordinates) by $-1/2\Delta$, where $\Delta \equiv \Sigma_{j=1}^d \partial^2/\partial x_j^2$ is the Laplacian acting in $L^2(\mathbb{R}^d, dx)$. This presentation does not work so well when a neat splitting is not possible, e.g., in the case of curved spaces, singular potentials, infinitely many degrees of freedom (i.e., formally $d = \infty$). In recent years an approach for handling these cases using the theory of Dirichlet forms and the associated symmetric Markov processes has been developed

[7a,30,6j-r]. More generally, stochastic processes and probabilistic methods have been used successfully in handling quantum mechanical problems (see, e.g., Ref. 12). In turn, new developments in the theory of stochastic processes have taken place under the stimulus of quantum mechanics (e.g., in the theory of stochastic equations with singular coefficients [7a,30a,4b,c,6], in the study of irreducibility questions for processes [6j,b,t], in the study of regularization of processes [6m], and in the development of a theory of Dirichlet forms and processes on infinite dimensional spaces [7-9]). In this section we discuss some of these problems, restricting ourselves to the case of processes with finite dimensional state space, the infinite dimensional case being treated in the next section.

In the case of bounded smooth potentials V it is well known, and will appear as a particular case below, that the Hamiltonian H can be described modulo unitary equivalence as the self-adjoint operator uniquely associated with the closure of the minimal form on $C_0^\infty(\mathbb{R}^d)$, $E(f,g) \equiv 1/2 \int \nabla f \nabla g \, d\nu^9$, with $d\nu = \varphi^2 \, dx$ and φ is the strictly positive smooth eigenfunction in $L^2(\mathbb{R}^d, dx)$ corresponding to the infimum of the spectrum of H. Hence we are led to study such sesquilinear forms, and we shall begin by reviewing some results about these forms and the associated processes (see Sec. 2). The associated processes are symmetric Markov diffusion processes and thus we start with a description of these. Let $\xi \equiv (\xi_t, t \geq 0)$ be a diffusion process on $\mathbb{R}^d$, with initial laws P^x, $x \in \mathbb{R}^d$ [i.e., Prob $(\xi_0 \in B) = P^x(B)$ for all Borel subsets B of $\mathbb{R}^d$]. Suppose ξ is symmetric with respect to some Radon measure ν on $\mathbb{R}^d$, strictly positive on each nonvoid open subset of $\mathbb{R}^d$, in the sense that the transition semigroup P_t^ν, $t \geq 0$, defined on nonnegative Borel functions f, g satisfies $\int (P_t^\nu f)g\, d\nu = \int f(P_t^\nu g)\, d\nu$. Then P_t^ν gives a strongly continuous semigroup of symmetric Markov operators in $L^2(\mathbb{R}^d,\nu)$, denoted by the same symbol. Let H_ν be minus the infinitesimal generator of P_t^ν so that $P_t^\nu = e^{-tH_\nu}$. The Dirichlet form of the process ξ is by definition the closed symmetric form E on $L^2(\mathbb{R}^d,\nu)$ defined by $E(f,g) = (H_\nu^{1/2}f, H_\nu^{1/2}g)$ with domain $D(E) = E(H_\nu^{1/2})$, the one of the operator $H_\nu^{1/2}$, and with $(\cdot,\cdot)$ the scalar product in $L^2(\mathbb{R}^d,\nu)$. If one assumes that the diffusion ξ_t admits no killing inside $\mathbb{R}^d$ and that $C_0^1(\mathbb{R}^d)$ is a core for E, then E has the representation for all f, g $\in C_0^1(\mathbb{R}^d)$ [4b]

$$E(f,g) = \frac{1}{2} \sum_{i=1}^{d} \int_{\mathbb{R}^d} \frac{\partial f}{\partial x_i} \frac{\partial g}{\partial x_j}\, d\nu_{ij} \tag{3.1}$$

with $\partial/\partial x_i$ the partial derivatives with respect to the ith coordinate x_i and with ν_{ij} a positive definite matrix of Radon measures on $\mathbb{R}^d$, i.e., such that $\Sigma_{i,j}\lambda_i\lambda_j\nu_{ij}(K) \geq 0$ for all compact K and all $\lambda_i \in \mathbb{R}$, $\nu_{ij}(K) = \nu_{ji}(K)$. It is also known that the sample path $\xi_t = (\xi_t^i,\ i = 1,\dots,d)$ of the diffusion ξ admits a unique decomposition

$$\xi_t^i = \xi_0^i + M_t^i + N_t^i \qquad i = 1,\dots,d \tag{3.2}$$

where M_t^i is a martingale additive functional of ξ locally of finite energy and N_t^i is a continuous additive functional of ξ, locally of zero energy. Moreover, there is a one-to-one correspondence between the family of continuous additive functionals of ξ of bounded variation and a family of signed Borel Revuz measures, and the above ν_{ij} is the Revuz measure of the quadratic variation $\langle M^i, M^i\rangle_t$ of the martingale part of ξ_t (Ref. 4a, theorem 5, 4.4). $M_t \equiv (M_t^i,\ i = 1,\dots,d)$ is the standard Brownian motion iff $\nu_{ij} = \delta_{ij}\nu$, ν being the Revuz measure of the (sure) additive functional $A_t = t$.

We can look upon the decomposition (3.2), and the corresponding one on functions of the process ξ, as an extension to the present situation [more general in the sense of allowing for processes which are not semimartingales, more special (for $d > 1$) in the sense of using symmetry] of Doob-Meyer's decomposition of supermartingales, Kunita-Watanabe's Itô formula for semimartingales, and Stroock-Varadhan's study of the martingale problem. The decomposition (3.2) is due to Fukushima [4,6a]. It extends a previous result by L. Streit and ourselves [30a] (see also [7a,6n]) that in the case where M_t is the standard Brownian motion w_t and $d\nu = \rho\, dx$ with $\rho \geq 0$, $\rho \in L^1(\mathbb{R}^d)$ and such that the corresponding Markov form

$$E(f,g) = \frac{1}{2} \int \nabla f\, \nabla g\, d\nu \tag{3.3}$$

[with $\nabla f \equiv ((\partial f/\partial x_i),\ i = 1,\ldots,d)$ the gradient of f] is closable from $C_0^1(\mathbb{R}^d)$ in $L^2(\mathbb{R}^d,\nu)$ and the distributional derivatives $(\partial/\partial x_i)\rho^{1/2}$ are in $L^2(\mathbb{R}^d,dx)$ for all $i = 1,\ldots,d$, one has that the Markov process ξ_t can be looked upon as a (weak) solution of the stochastic equation

$$d\xi_t = \beta(\xi_t)\, dt + dw_t \tag{3.4}$$

with $\beta \equiv (\beta_i,\ i = 1,\ldots,d)$, $\beta_i(x) \equiv (\partial/\partial x_i)\ell n\, \rho^{1/2}(x)$ (in the sense of distributions). Moreover, ξ_t is the Markov process associated with the Dirichlet form E in the sense of Fukushima and Silverstein [4,5]. Note that neither smoothness of β [except for weak assumptions on ρ assuring the closability of E (see below)] nor bounds on the growth of β for $|x| \to \infty$ are required. The need for such results for quantum mechanics had been stressed by Ezawa, Klauder, and Shepp [33]. Let us also remark that the assumption that β be the gradient of some function $\rho^{1/2}$ in (3.4) is no restriction for dimension $d = 1$. It is, however, a restriction in dimension $d > 1$; in fact, it is equivalent [34] to the symmetry of the Markov semigroup P_t^ν in $L^2(\mathbb{R}^d,\nu)$ associated with the solution process ξ_t or, equivalently, with the time reversibility of the process. Let us finally remark that there are also uniqueness results on the solutions of (3.4) under assumptions on β which are much weaker than the usual ones in the theory of Itô equations or in the martingale approach. In fact, it suffices that in addition to the above assumptions on ρ assuring closability one has, e.g., $\rho > 0$ a.e. on compacts, $\Delta(\rho^{1/2}) \in L^2_{loc}(\mathbb{R}^d)$, $V \equiv (1/2)\Delta(\rho^{1/2})/(\rho^{1/2})$ such that $-(1/2)\Delta + V$ is essentially self-adjoint on $C_0^\infty(\mathbb{R}^d)$ (see Ref. 30a, theorem 2.6).

REMARK 1. A recent paper by Fukushima and Stroock [6h] further clarifies the relation between the approach to diffusion processes by Dirichlet forms, of which the above are only particular results, and Stroock-Varadhan's approach by the martingale problem. As we saw above in the case where M_t is the standard Brownian motion, the Dirichlet form E associated with the process ξ is given by (3.3) on $C_0^1(\mathbb{R}^d)$ and is the closure of (3.3). Fukushima [6a] calls ξ, following Refs. 33 and 30a, a <u>distorted Brownian motion</u>, a nomenclature we shall use henceforth. Fukushima has recently discussed the question of when the distribution of ξ_t is absolutely continuous with respect to Wiener measure (extension of Girsanov's formula) [6a-c].

REMARK 2. In the above discussion, two different points of view were implicit. On the one hand, the process ξ is given; there is then an associated (symmetric Markov) semigroup and a Dirichlet form and Fukushima's decomposition (3.2) is then a property of ξ. On the other hand, as in the quoted result on the singular stochastic equation (3.4), the analytic data are given [β in equation (3.4)]; these determine a Markov form, from which it is required to construct a Dirichlet form, a symmetric Markov semigroup, and finally an associated process. These two points of view are typically in productive interaction in the whole theory of symmetric Markov processes and Dirichlet forms.

We recall from Sec. 2 that to any symmetric Markov semigroup P_t^ν in a $L^2(X,\nu)$, X locally compact with countable base, ν Radon measure strictly positive on nonvoid open sets, there is associated uniquely a Dirichlet form E in $L^2(X,\nu)$, and vice versa, the correspondence being given by $E(f,g) \equiv (H_\nu^{1/2}f, H_\nu^{1/2}g)$, if $P_t^\nu = e^{-tH_\nu}$. To any regular Dirichlet form there is associated a Hunt process which is actually a diffusion (continuous sample paths with probability 1) iff the Dirichlet form is local. Thus there is a complete, beautiful, and very fruitful correspondence between potential theoretic, analytic, and probabilistic notions. We shall mention here only those aspects that relate most directly to applications to quantum theory. What are in fact the relations to quantum theory? Let us consider a system with d degrees of freedom in $\mathbb{R}^d$. If its Hamiltonian (Schrödinger operator) in $L^2(\mathbb{R}^d,dx)$ is of the form $H = -(1/2)\Delta + V$ on $C_0^\infty(\mathbb{R}^d)$, with some real function V, and there exists a constant $E_0 > -\infty$ and a function $\varphi \in L^2_{loc}$, $\varphi > 0$ a.e., such that φ^{-1}, $\varphi^{-1}\nabla\varphi$, $\nabla\varphi$, $\varphi^{-1}\ \Delta\varphi \in L^2_{loc}(\mathbb{R}^d)$ and that $((-1/2)\Delta + V)\varphi = E_0\varphi$, then we can form $d\nu = \varphi^2\ dx$ and consider the Dirichlet form $E(f,g) \equiv (1/2)\int \overline{\nabla} f\ \overline{\nabla} g\ d\nu$, for all f,g in the operator domain of the closure $\overline{\nabla}$ of the gradient operator ∇, as an operator from $C_0^\infty(\mathbb{R}^d)$ into $\mathbb{R}^d \otimes L^2(\mathbb{R}^d,\nu) = L^2(\mathbb{R}^d,\nu;\mathbb{R}^d)$, i.e., the $\mathbb{R}^d$-valued function on $\mathbb{R}^d$ with values in $L^2(\mathbb{R}^d;\nu)$. The associated Markov semigroup is $P_t^\nu = e^{-tH_\nu}$, $t \geq 0$, so that $E(f,g) = (H_\nu^{1/2}f, H_\nu^{1/2}g)$, $(\cdot,\cdot)$ being the scalar product in $L^2(\mathbb{R}^d,\nu)$. We have $H_\nu = (1/2)\overline{\nabla}^*\overline{\nabla}$, $\overline{\nabla}^*$ being the adjoint of $\overline{\nabla}$, and H_ν is the Friedrichs extension of $-(1/2)\Delta - \beta\cdot\nabla$ on $C_0^\infty(\mathbb{R}^d)$, with $\beta = \varphi^{-1}\nabla\varphi$, $\beta\cdot\nabla = \Sigma_{i=1}^d \beta^i(x)(\partial/\partial x_i)$. An easy calculation, using $(H - E_0)\varphi = 0$, gives $\varphi^{-1}(H - E_0)\varphi = H_\nu$ on $C_0^\infty(\mathbb{R}^d)$. Then we have the relation $H - E_0 = U_\varphi H_\nu U_\varphi^{-1}$, where U_φ is the unitary operator from $L^2(\mathbb{R}^d, d\nu)$ onto $L^2(\mathbb{R}^d, dx)$ given by multiplication by φ; in particular, then, $H - E_0 \geq 0$. This unitary equivalence of $H - E_0$ and H_ν tells us, moreover, that the study of the unitary group e^{-itH} in quantum mechanics [the Schrödinger equation being $i(\partial/\partial t)\psi_t = H\psi_t$, hence being solved by $\psi_t = e^{-itH}\psi_0$] can be connected naturally with the study of the Markov semigroup e^{-tH_ν} [holomorphically extendable to e^{-itH_ν}, unitarily equivalent to $e^{-it(H-E_0)}$]. Vice versa, given any Dirichlet form looking like

$$E(f,g) = \frac{1}{2} \int_{\mathbb{R}^d} \nabla f \nabla g \, d\nu \tag{3.5}$$

on $C_0^\infty(\mathbb{R}^d)$ in $L^2(\mathbb{R}^d, \nu)$, for some Radon measure ν positive on all nonvoid open sets, and defined by closing then with domain that of the closed gradient operator $\overline{\nabla}$ (it is assumed that ν is such that this closure exists; see below), we get a positive self-adjoint operator H_ν in $L^2(\mathbb{R}^d, \nu)$, with $H_\nu = (1/2)\overline{\nabla}^*\overline{\nabla}$. In the case where $d\nu = \rho\, dx$, with $\rho \in L^1_{loc}(\mathbb{R}^d)$, $\rho > 0$ a.e. and $\nabla\rho^{1/2} \in L^2_{loc}(\mathbb{R}^d - N)$, with N some closed set of Lebesgue measure zero (e.g., $N = \emptyset$), we have $H_\nu = -(1/2)\Delta - \beta\cdot\nabla$ on $C_0^2(\mathbb{R}^d - N)$. If $\nabla\rho^{1/2}$, $\rho^{-1/2}\nabla\rho^{1/2} \in L^2_{loc}(\mathbb{R}^d)$, then $H \equiv U_\varphi H_\nu U_\varphi^{-1}$ is a self-adjoint operator in $L^2(\mathbb{R}^d, dx)$, which is $-(1/2)\Delta + V$, $V \equiv (1/2)\rho^{-1/2}\Delta\rho^{1/2}$ in the sense that its form restriction to $C_0^1(\mathbb{R}^d)$ is $(1/2)\int \nabla f\, \nabla g\, dx + \int fVg\, dx$, V being considered as an element in the dual of $C_0^1(\mathbb{R}^d)$. If, moreover, $V \in L^2_{loc}(\mathbb{R}^d)$, then $H = -(1/2)\Delta + V$ in the operator sense on $C_0^2(\mathbb{R}^d) \cup U_\varphi C_0(\mathbb{R}^d)$. Thus it is quite natural to call H, defined as $U_\varphi H_\nu U_\varphi^{-1}$ with the help of the operator H_ν associated with a Dirichlet form as (3.5), a (generalized) Schrödinger operator (for the above results, see Ref. 30a). By these considerations we see that the dynamical study of Markov semigroups and the associated Dirichlet forms and processes on the one hand, and Schrödinger operators and quantum mechanical quantities on the other, are closely related. We consider next a few problems where these relations have been useful.

3.1 Closability

When is the Markov form $(1/2)\int \nabla f \nabla g \, d\nu$ closable in $L^2(\mathbb{R}^d,\nu)$ on some minimal domain [like $C_0^\infty(\mathbb{R}^d)$], thus providing a self-adjoint operator H_ν? This study has been enhanced by the need to define singular Hamiltonians in quantum mechanics. Sufficient conditions for this to happen are known:

i. $d\nu = \rho\, dx$, $\rho > 0$ locally uniformly, $\rho \in L^1_{loc}(\mathbb{R}^d)$ [4a].[10]

ii. $\rho \in L^1_{loc}(\mathbb{R}^d)$, $(\partial/\partial x^i)\rho^{1/2} \in L^2_{loc}(\mathbb{R}^d - N)$, N some closed subset of $\mathbb{R}^d$ of Lebesgue measure zero, where $L^2_{loc}(\mathbb{R}^d-N)$, consists of generalized functions ψ such that $f\psi \in L^2(\mathbb{R},dx)$, for all $f \in C_0^\infty(\mathbb{R}^d - N)$ [30a,38].

iii. For d = 1 a necessary and sufficient condition is known: namely, $\rho \in L^1_{loc}(\mathbb{R})$, $\rho(x) = 0$ for a.e. x such that $\int_{x-\varepsilon}^{x+\varepsilon} (1/\rho(y))dy = \infty$ for all $\varepsilon > 0$. This is a result of Hamza [35] (see also Refs. 4a and 5) for the composite situation of the form
$\frac{1}{2}\int \frac{df}{dx}\frac{dg}{dx}\, d\nu$ in $L^2(\mathbb{R},dx)$

(Lebesgue measure instead of $d\nu$-measure!), which has been extended recently to the above case by Rullkötter and Spönemann [36].

3.2 Uniqueness

The Dirichlet form (3.5) is first given on a dense domain $\mathcal{D}$. The quantum dynamics requires a unitary group whose infinitesimal generator is a self-adjoint operator. We have seen that this corresponds to having a self-adjoint generator H_ν associated with the form (3.5) in the sense that $(H_\nu^{1/2}f,H_\nu^{1/2}g) = E(f,g)$ for all f,g in the domain of the closed form, which then coincides with the domain of $H_\nu^{1/2}$. So to the question of the possible quantum dynamics corresponds the question of which positive self-adjoint operators $\tilde{H}$ exist such that $(\tilde{H}^{1/2}f, \tilde{H}^{1/2}g) = E(f,g)$ for all $f,g \in \mathcal{D}$. The strongest uniqueness result would be available if $H_\nu \upharpoonright \mathcal{D}$ were already essentially self-adjoint on $\mathcal{D}$. Results of this type are known under some regularity assumptions on ρ:

1. $\rho \in L^1_{loc}(\mathbb{R}^d)$, $\rho > 0$, on compacts, $\nabla\rho^{1/2}$, $\Delta\rho^{1/2} \in L^2_{loc}(\mathbb{R}^d)$, $\rho^{-1/2}\nabla\rho^{1/2} \in L^2_{loc}(\mathbb{R}^d)$, $V \equiv (1/2)\rho^{-1/2}\Delta\rho^{1/2}$ such that $-(1/2)\Delta + V$ is essentially self-adjoint on $C_0^\infty(\mathbb{R}^d)$. Then H_ν is essentially self-adjoint on $\mathcal{D} = \varphi^{-1}C_0^\infty(\mathbb{R}^d)$ [30a].
2. $\rho \in L^1_{loc}(\mathbb{R}^d)$, $\rho > 0$, on compacts, ρ locally Lipschitz continuous. Then H_ν is essentially self-adjoint on $C_0^\infty(\mathbb{R}^d)$. This result has been obtained recently by N. Wielens [37]. A related result concerning uniqueness of Markov extensions had been obtained by Fukushima [6a,c]. For an extension to infinite dimensions, see Ref. 84.

The situation for d = 1 has been examined in detail (see Refs. 36-38, where the connection with Feller's theory is also explained). For general $d > 1$ the relations with Silverstein's boundary theory [5b] and Stroock-Varadhan's [3] theory should be investigated further.

3.3 Ergodic questions

The case d = 1 is best understood in relation with Feller's theory [4a, 36, 37]. In the general case $d > 1$ one has that ν is a probability measure iff 1 is an eigenfunction to the eigenvalue zero for H_ν and this is so, in the case $\rho > 0$ a.e., iff $\rho^{1/2}$ is an eigenfunction to the eigenvalue zero for H (the infimum of the spectrum of H, in this case). In this case one has for the kernel of P_t^ν : $P_t^\nu(x,\mathbb{R}^d) = 1$ for all $x \in \mathbb{R}^d$. The eigenvalue zero for H_ν is simple iff the same eigenvalue is simple for H and these properties are equivalent with the kernels of e^{-tH_ν} (respectively, e^{-tH}) being strictly positive a.e., if $\rho > 0$ a.e., i.e., with e^{-tH_ν} and e^{-tH} being ergodic, i.e., with the ergodicity of the associated process ξ_t (constructed by Fukushima and Silverstein, with invariant measure ν, continuous paths and state space $\mathbb{R}^d$ minus a Borel set of capacity zero) [30a, 7a]. One defines a Dirichlet form E as irreducible if any P_t^ν-invariant set B is trivial in the sense that either $\nu(B) = 0$ or $\nu(\mathbb{R}^d - B) = 0$ [5]. As proved by Fukushima [6b], E defined by the closure of (3.5) is irreducible if $\rho \in L^1_{loc}(\mathbb{R}^d)$, inf $\rho > 0$ on each compact set. For d = 1, inf p > 0 on each interval not containing zero, one has irreducibility iff $\int_{-b}^{b} dx/\rho(x) < \infty$ for $b > 0$. These questions are closely related to the next point.

3.4 Impenetrability, Capacity

In quantum mechanics one says that there is no tunneling through (there is impenetrability of) the boundary ∂G of a Borel set $G \subset \mathbb{R}^d$ if $L^2(\mathbb{R}^d) = L^2(G) \oplus L^2(\mathbb{R}^d - \bar{G})$, $\bar{G}$ being the closure of G, and e^{-itH} commutes with the orthogonal projections onto $L^2(G)$ and $L^2(\mathbb{R}^d - \bar{G})$. One proves, using the correspondence $H \leftrightarrow H_\nu$ described above, that no tunneling through G is equivalent with the existence of an increasing sequence of closed sets F_n such that, relative to the Dirichlet structure given by (3.5), $\text{Cap}(\mathbb{R}^d - F_n) \to 0$, and such that there exist modifications G' (respectively, G'') of G (respectively $\mathbb{R}^d - G$) such that $\nu(G' - G) = \nu(G'' - (\mathbb{R}^d - G)) = 0$ and G' and G'' are quasi-open and quasi-closed in the sense that $G' \cap F_n$, $G'' \cap F_n$ are both closed and open in F_n for each n [6e,t]. Necessary and sufficient conditions for no tunneling in terms of the measure ν have been found for all dimensions d [6j,b,t].

3.5 Approximation by Regularization of Dirichlet Forms and Processes

As we saw above in the formulation of quantum mechanics in

terms of Dirichlet forms, one can define the Hamiltonian H, and hence the time evolution, also in the case of very singular potentials V. For the study of some questions (e.g., in the scattering theory) it is useful to approximate H by Hamiltonians H_n constructed from smooth potentials V_n. For such Hamiltonians H_n, called "regularizations" of H, one has available many results proven using the fact that V_n are nice perturbations of the constant coefficient operator $-(1/2)\Delta$. Can one prove convergence in some precise sense of H_n to H and hence use the information about quantities associated with H_n to deduce information about the corresponding quantities for H?

In the other direction, there are situations, as we shall see below for the case of point interactions, where the Schrödinger operator H of a "singular potential V," representing a very idealized physical situation, allows for explicit computation of quantum mechanical quantities; these computations can then be used as an approximation to quantities associated to more realistic nonsingular potentials V_n, provided that one has suitable control over the convergence of H_n to H. For this general problem of convergence of "smooth" Schrödinger operators (H_n) to "singular" ones (H) one has available a class of results in the case where both V_n and V are in some L^p-spaces (see, e.g., Ref. [12].) The place where the theory of Dirichlet forms becomes very useful is where V is not necessarily a Lebesgue measurable function, e.g., $V(x) = \lambda\delta(x)$, $x \in \mathbb{R}^3$ (see Sec. 6.1). One can study approximations of the corresponding generalized Schrödinger operator H by the Schrödinger operators H_n to smooth potentials V_n, by using the Dirichlet forms E (respectively, E_n) associated with H (respectively, H_n). Results of this type have been obtained by Streit and ourselves [6m] (see also Ref. 38). For example, one has [6m,38]:

THEOREM 3.1 Let $d\nu_n = \rho_n dx$, $d\nu = \rho\, dx$, $\rho_n, \rho \in L^1_{loc}(\mathbb{R}^d)$ be such that the corresponding Dirichlet forms in $L^2(\mathbb{R}^d, \nu_n)$ [respectively, $L^2(\mathbb{R}^d, \nu)$] are closable. Suppose that $\rho_n > 0$ a.e., $\rho_n/\rho \uparrow 1$ (respectively, $\rho > 0$ a.e. $\rho_n/\rho \downarrow 1$) in $L^\infty(\mathbb{R}^d, dx)$, then $H_n \to H$ in the strong resolvent sense, the corresponding semigroups in $L^2(\mathbb{R}^d, dx)$ converge strongly, and the corresponding diffusion processes ξ_n converge to the diffusion process associated with H in the sense of convergence of all finite distributions.

As suggested by L. Streit, some new methods of T. Kurtz could help extend these results, and there is work in preparation along these lines [79].

3.6 Some Examples of the Uses of Dirichlet Forms in Quantum Mechanics and Statistical Mechanics

Point interactions

We shall look first at the description of the so-called point interactions by Dirichlet forms [7a,30a]. We consider the case of $\mathbb{R}^d$, d = 3, the cases d = 1, 2 being similar.

Let

$$\varphi(x) \equiv \frac{1}{2\pi} \frac{e^{\alpha|x|}}{|x|} \qquad \alpha \in \mathbb{R}$$

so that φ is the fundamental solution to $-(1/2)\Delta + (1/2)\alpha^2$. Since $\nabla\varphi \in L^2(\mathbb{R}^3 - \{0\})$ we can apply the result mentioned in Sec. 3.1 above to get that the closure E of the form $(1/2) \int \nabla f \nabla g \, d\nu$, $d\nu \equiv \varphi^2 dx$, $f \in C_0^1(\mathbb{R}^3)$ exists and is a local regular Dirichlet form in $L^2(\mathbb{R}^3, \nu)$. Since $\varphi > 0$ a.e. we have that $L^2(\mathbb{R}^3, \nu)$ is unitarily equivalent by U_{φ^2} to $L^2(\mathbb{R}^3, dx)$ and the unique self-adjoint operator in the $L^2(\mathbb{R}^3, dx)$, image of the self-adjoint operator H_ν associated with E, is $H = U_\varphi H_\nu U_\varphi^{-1}$. We have $H_\nu = (1/2)\bar{\nabla}^* \bar{\nabla}$ and $H_\nu = -(1/2)\Delta - \beta \cdot \nabla$ on $C_0^2(\mathbb{R}^3 - \{0\})$, with $\beta = \nabla \ln \varphi = -(-\alpha + 1/|x|)(x/|x|)$, and $H = -(1/2)\Delta + (1/2)\alpha^2$ on $C_0^2(\mathbb{R}^3 - \{0\})$. Since $\varphi \in L^2(\mathbb{R}^3, dx)$ for $\alpha < 0$, and $H\varphi = 0$ we have that, for such α, zero is an eigenvalue of H. Moreover, H is nonnegative and hence zero is the infimum of the spectrum of H. The resolvent of H is given explicitly for $z > 0$ or $\operatorname{Im} z \neq 0$ by

$$(H + z)^{-1} = (- \frac{1}{2}\Delta + \frac{1}{2}\alpha^2 + z)^{-1} + \frac{2\pi}{\sqrt{-\alpha^2 - 2z} + \alpha}(g_z, \cdot)g_z$$

with $g_z(x) \equiv (1/2\pi)|x|^{-1}e^{-\sqrt{-\alpha^2 - 2z}|x|}$ and $(g_z, \cdot)g_z$ the rank 1 operator such that $(g_z, \cdot)g_z f = (g_z, f)g_z$ for all $f \in L^2(dx)$, (g_z, f) being the scalar product in $L^2(dx)$. From this formula we also see that zero is a simple eigenvalue of H and H_ν. This is equivalent [7a] with each of the following conditions: (1) any bounded multiplication operator commuting with the symmetric Markov semigroup e^{-tH_ν} for all $t \geq 0$ is constant; (2) e^{-tH_ν} has ergodic kernel; (3) e^{-tH_ν} is positivity improving; and (4) the associated process is ergodic. The same properties hold with H_ν replaced by H and $L^2(\mathbb{R}^d, \nu)$ replaced by $L^2(\mathbb{R}^d, dx)$. The spectra of H and H_ν are purely absolutely continuous in $[\alpha^2/2, \infty)$. H can be characterized as the self-adjoint extension of $-(1/2)\Delta + \alpha^2/2 \restriction C_0^\infty(\mathbb{R}^3 - \{0\})$, with domain D(H) such that for $f \in D(H)$, $\alpha|x|f(x) - (x/|x|)\nabla(|x|f(x)) = 0$ at $x = 0$ (see, e.g., Ref. 30h-t). The symmetric diffusion process associated with H_ν satisfies, in the above sense, the stochastic equation $d\xi_t = -(-\alpha + 1/|\xi_t|)(\xi_t/|\xi_t|) + dw_t$. What is the capacity of points? It is zero for $x \neq 0$; it is > 0 for $x = 0$. By the relation with hitting times of the general theory of Dirichlet forms we see then that 0 is hit in finite time by the process. What about approximations of H, H_ν and the associated process?

A detailed study of the limit as $\varepsilon \downarrow 0$ in the strong resolvent sense of

$$H_\varepsilon = - \frac{1}{2}\Delta - (\frac{\pi^2}{8} - \alpha\varepsilon)\frac{1}{\varepsilon^2}\chi_1(\frac{x}{\varepsilon}) + \frac{1}{2}\alpha^2$$

with χ_1 the characteristic function of the unit ball in $\mathbb{R}^3$, to H has been made, partly inspired by methods of nonstandard analysis [39,30j,k]. This approximation of H by H_ε and in fact analytic expansions in ε for corresponding quantum mechanical quantities (scattering amplitudes, resolvent, eigenvalues, resonances) exhibits H as a point interaction Hamiltonian, formally $-(1/2)\Delta + \lambda\delta(x) + (1/2)\alpha^2$, with λ infinitesimal, as a limit of Schrödinger Hamiltonians with bounded potentials $V_\varepsilon = \lambda_\alpha(\varepsilon)\delta_\varepsilon(x) + (1/2)\alpha^2$, with δ_ε a δ-sequence and $\lambda_\alpha(\varepsilon) = -(\pi^2/8 - \alpha\varepsilon)(4/3)\pi\varepsilon$.

The results are independent of the particular local approximation; e.g., V_ε can be replaced by $\varepsilon^{-2}\lambda(\varepsilon)V(x/\varepsilon) + (1/2)\alpha^2$ with V such that, e.g., for some $a > 0$, $\iint e^{2a(|x|+|y|)}|V(x)||V(y)||x - y|^{-2}dx\,dy < \infty$, and such that -1 is an eigenvalue of $|V|^{1/2}(\text{sign } V)[-(1/2)\Delta]^{-1} \times |V|^{1/2}$ with eigenfunctions φ_j such that, for at least one j, $-(1/2)\Delta^{-1}|V|^{1/2}\varphi_j \notin L^2(\mathbb{R}^3)$; λ is such that $4\pi\lambda'(0)/\Sigma_{j=1}^N (|V|^{1/2},\varphi_j)^2 = \alpha$, when the φ_j are normalized so that $(\varphi_j, (\text{sign } V)\varphi_\ell) = -\delta_{j\ell}$, $j,\ell = 1,\ldots,N$. If these conditions on λ, V are not satisfied, one has convergence of H_ε to $-(1/2)\Delta + (1/2)\alpha^2$ [30j,k].

Recently, there have been extensions of the above results to the case where the free Hamiltonian $-(1/2)\Delta$ is replaced by the one, $-(1/2)\Delta + c/|x|$, of a Coulomb system in $\mathbb{R}^3$, the corresponding physical situation being then the one of, e.g., a mesic atom where in addition to the Coulomb force one also takes into account a point interaction coming from nuclear forces [30ℓ,s].

There have also been extensions to the case of operators given formally, instead of $-(1/2)\Delta + \lambda\delta(x) + \text{const}$ as above, by $-(1/2)\Delta + \Sigma_{x_i \in Y} \lambda_i\delta(x - x_i) + \text{const}$, with Y a fixed countable subset of $\mathbb{R}^3$ (e.g., $\mathbf{Z}^d$). These models are of importance in solid-state physics, nuclear physics (slow neutrons), and electromagnetism [30h].

Finally, we should mention that extensions of such models with point interaction to the case of random Schrödinger operators have been studied. These are models where the

Schrödinger operator $H(\omega)$ is defined as the limit as $\varepsilon \to 0$ of $H_\varepsilon(\omega) = -(1/2)\Delta + \Sigma_{x_i \in Y} \lambda_{\alpha_i(\omega)}(\varepsilon)\delta_\varepsilon(x - x_i)$, where the x_i are in $\mathbf{Z}^d$ and the $\alpha_i(\omega)$ are identically distributed uniformly bounded independent random variables with values in some fixed interval $[a,b]$. Kirsch and Martinelli have shown [30m,n], as a particular case of a much more general result, that the whole spectrum of $H(\omega)$ is a nonrandom set Σ, and the same holds for its essential, absolutely continuous, and pure point, singular continuous parts. Moreover, the discrete spectrum is almost surely void. The spectrum is the union of the spectra of the sure Hamiltonian operators $H(\omega_0)$, with ω_0 such that $\alpha_i(\omega_0) \in \text{supp}\, P_{\alpha_0}$, P_{α_0} being the distribution of α_0. From the deterministic case (see Ref. 30q) one knows that the spectrum of the operator H defined as $H(\omega_0)$, with $\alpha_i(\omega_0)$ replaced by α for all i, is absolutely continuous and for $\alpha < A$, $A < 0$, of the form $[E_0(\alpha), E_1(\alpha)] \cup [0,\infty)$, with $E_0(\alpha) < E_1(\alpha) < 0$. One gets then, for $\sup \text{supp}\, P_{\alpha_0} < A$, $(E_1(b),0) \cap \Sigma = \emptyset$ and if $E_0(b) \leq E_1(a)$, one has $\Sigma = \sigma(H_a) \cup \sigma(H_b) = [E_0(a),E_1(b)] \cup [0,\infty)$.

These results have recently been extended to the case of a random lattice where the x_i are also independent identically distributed (i.i.d.) random positions in some random sublattice $\Gamma(\omega)$ of $\mathbf{Z}^3$ (i.e., $\Gamma(\omega) = \{i \in \mathbb{Z}^3 | \xi_i(\omega) = 1\}$, the ξ_i being i.i.d. with values 0,1) [30o].

REMARK 3. All the above considerations on deterministic and random models with point interactions can also be made for cases of dimension d = 1, 2. For d = 1 the "deterministic" H is the unique self-adjoint operator unitarily equivalent by U_φ, $\varphi \equiv \sqrt{\alpha}\; e^{\alpha|x|}$, to the one associated with the closure of the form in $L^2(\mathbb{R}, |\varphi|^2 dx)$,

$$\frac{1}{2}\alpha \int \frac{df}{dx}\frac{dg}{dx}\; e^{2\alpha|x|}\; dx$$

first defined on $C_0^1(\mathbb{R})$. H can also be described as the self-adjoint operator associated with the closed form in $L^2(dx)$,

$$\frac{1}{2}\int \frac{df}{dx}\frac{dg}{dx}\,dx + \frac{1}{2}\alpha^2 + \alpha f(0)g(0)$$

The corresponding solid-state physics models with deterministic or random coefficients α_i and/or positions can also be studied [40; 30m,r; 83]. In the random case the spectrum Σ is equal to the known spectrum $\sigma(H_{\inf \operatorname{supp} P_{\alpha_0}})$ if $\inf \operatorname{supp} P_{\alpha_0} \geq 0$ which gives interesting results on the gaps in the spectrum of $H(\omega)$ [30m,r; 82; 83].

Application of the theory of Dirichlet forms to the statistical mechanics of polymers

Polymers are long chains of molecules consisting of repeated units (from a few hundred to a million) and forming building blocks (e.g., in biological systems). They are approximately described by self-avoiding random processes. There is interest (see, e.g., Ref. 42) in describing the case where the number of links N is very large (i.e., $N \to \infty$). For this purpose a justifiable approximation is given by Edward's model, in which the polymer chain of length t is taken to be a Brownian sample path followed up to time t. Then the formal density function (with respect to Wiener measure) for the location of the end point is $\exp[-J(b)]$ with $J(b) \equiv \lambda \int_0^t d\sigma \int_0^t d\sigma'\, \delta[b(\sigma') - b(\sigma)]$, δ being the Dirac measure on $\mathbb{R}$ and b a Brownian path starting at time 0 at 0, at time σ in $b(\sigma,\omega)$, $d\sigma$ being Lebesgue measure on $[0,t]$ and λ a (real) parameter. Thus the process is a Brownian motion forced to interact with itself by the interaction λJ. The model extends to one describing the interaction of m polymers described by densities of the above form with respect to the product of the Wiener measures for m independent Brownian motions b_i and with J given by $\Sigma\, \lambda_{ij} \int_0^{t_j} d\sigma \int_0^{t_i} d\sigma'\, \delta[b_i(\sigma') - b_j(\sigma)]$. Quantities of interest are, e.g., the correlations of the b_i for large times. The construction of these "polymer measures" has to do with control of the probability of self-intersection of Brownian paths, and this probability depends crucially on the dimension d. For d = 1, J is simply $\int T^2(x,b)\,dx$, $T(x,b)$ being the local time of Brownian motion. For d = 2 the construction has been done by Symanzik and Varadhan [41]. For d = 3 it has been achieved only recently in a very interesting work by Westwater [42].

A new approach that works partially also for d = 4 has been introduced recently using a combination of methods of nonstandard analysis [30h] and of the theory of Dirichlet forms. Using the Feynman-Kac formula for regularized versions of the above measures one reduces the problem of defining these measures to the one of defining for almost all sample paths of Brownian motion stochastic Hamiltonians of the form $H(\omega) = -(1/2)\Delta + \lambda \int_0^t \delta(x - b(\sigma,\omega))d\sigma$. By methods of nonstandard analysis similar to those already mentioned for point interactions, one can compute the kernel of the resolvent of the corresponding operator $H_\varepsilon(\omega)$, where λ is replaced by $\lambda(\varepsilon)$ and δ, by δ_ε, where ε is infinitesimal and δ_ε is a smooth approximation of the Dirac distribution. The kernel has a standard part [i.e., in standard terms, one has the resolvent of $H(\omega)$] for standard λ if d = 1, 2, 3, but for d = 4, λ has to be chosen infinitesimally negative. In this proof estimates on quantities like
$\int_0^t [-(1/2)\Delta - z]^{-1}[x - b(\sigma)]\,d\sigma$ and
$\int_0^t\int_0^t [-(1/2)\Delta - z]^{-1}[b(\sigma_1) - b(\sigma_2)]g(\sigma_1)g(\sigma_2)d\sigma_1\,d\sigma_2$,
$g \in C_0^\infty(\mathbb{R}^d)$, $\mathrm{Im}\, z \neq 0$, are needed, and here the properties of Dirichlet forms and of Brownian motion (and their hyperfinite counterparts [30h]) are used.

These results have also applications to the φ_d^4-model of quantum field theory, via Symanzik's representation of this model using the above polymer measures. At the time of writing the relations have been clarified for $d \leq 3$, the case $d = 4$ being only partially under control. In the case $d \geq 5$ a related discrete polymer representation has been used by Aizenmann and Fröhlich to prove the triviality of φ_d^4 (and some related models) for $d \geq 5$ when defined as a limit of the corresponding lattice versions [43] (for an early triviality result in other models see Refs. 62, 63).

4. MARKOV DIFFUSION FIELDS AND QUANTUM FIELDS

In recent years there has been an interest in developing a theory of random fields as an extension of the theory of stochastic processes to the case where the time parameter becomes multidimensional. This extension can be made in several directions. There is a general theory of processes where the index set ("time") need not be a linearly ordered subset of $\mathbb{R}$. Also, there is a theory of multiparameter martingales (e.g., Ref. 44). A well-studied domain is also the one where the index set is discrete, e.g., $\mathbf{Z}^d$ (this includes the lattice models of statistical mechanics [19, 17, 30h]. In this section we shall report particularly on some work done in recent years in the direction of extending the class of Markov processes of diffusion type to corresponding random fields. We shall stress particularly the case of stationary, isotropic stochastic Markov fields, as a multitime extension of the interesting class of stationary symmetric Markov processes. We begin with a consideration of the so-called free Markov field.

4.1 The Free Markov Field

The free Markov field is a probabilistic object that has arisen quite naturally in several contexts. In connection with quantum field theory it appears as such early in

Symanzik's work [41] (it was implicit also in earlier work by Schwinger and Nakano on "Euclidean quantum field theory") and has been fundamental in the approach by Symanzik, Nelson, and Guerra to the construction of quantum fields by "Euclidean methods." This work was then pushed forward by a large number of mathematicians and physicists (see, e.g., Refs. 12 and 60p).

In another context the free Markov field appeared in statistical work by Whittle and others [14] and in work on "Brownian harnesses" by D. Williams [45]. It was studied with a particular view toward its Markov property in work by Nelson, Molchan, Pitt, Wong, and many others (see, e.g., the references in [6p, 46, 47]). By definition, the free Markov field on $\mathbb{R}^d$ (of mass m) is the centered (i.e., zero mean) Gaussian (generalized) stochastic field $\xi(x) = (\xi(x)(\omega), x \in \mathbb{R}^d, \omega \in \Omega)$ with covariance operator $E(\xi(x)\xi(y)) = G_m(x - y)$, where $G_m(x - y)$ is the kernel, evaluated at x, y, of the operator $(-\Delta + m^2)^{-1}$ in $L^2(\mathbb{R}^d, dx)$, where m is a fixed parameter. For d = 1, 2 we suppose that $m > 0$; for $d \geq 3$ we suppose that $m \geq 0$ (the stricter assumption for d = 1, 2 is in order to have transience also in this case). By the very definition of generalized random fields this means that $\xi(x)(\omega)$ is a.s. a distribution, so that its evaluation $\langle \xi, \varphi \rangle(\omega)$ at $\varphi \in \mathcal{D}(\mathbb{R}^d) \equiv C_0^\infty(\mathbb{R}^d)$ can be looked upon as a real-valued linear process $\varphi \to \langle \xi, \varphi \rangle$ indexed by $\mathcal{D}(\mathbb{R}^d)$, with mean zero and covariance $E(\langle \xi, \varphi \rangle \langle \xi, \varphi' \rangle) = (\varphi, \varphi')_{-1}$, with $(\varphi, \varphi')_{-1} \equiv \int G_m(x - y)\varphi(x)\varphi(y)\, dx\, dy$, $\varphi, \varphi' \in \mathcal{D}(\mathbb{R}^d)$. Note that $G_m(x - y)$ being locally integrable for all d, this quantity is well defined. The process extends easily to a linear process indexed by $\mathcal{S}(\mathbb{R}^d)$ and by Minlos's theorem it then has a realization, again called ξ, in $\mathcal{S}'(\mathbb{R}^d)$.[11] Let μ_0 be the measure on $\mathcal{S}'(\mathbb{R}^d)$ giving the distribution of ξ, i.e., $\text{Prob}(\xi \in B) = \mu_0(B)$ for any Borel subset B of the (Suslin) space $S'(\mathbb{R}^d)\cdot\mu_0$ is the Gaussian measure on $S(\mathbb{R}^d)$ with Fourier transform $\int_{S'(\mathbb{R}^d)} e^{i\langle\xi,\varphi\rangle} d\mu_0(\xi) = e^{-(1/2)(\varphi,\varphi)_{-1}}$.

It is also useful to think of $d\mu_0$ formally as $e^{-(1/2)\int_{\mathbb{R}^d} \nabla\xi(x)^2 dx} \Pi_{x\in\mathbb{R}^d} d\xi(x)$ (this notation can actually be given a "direct" justification by methods of nonstandard analysis (see Ref. 30h)). Another possible description of (μ_0,ξ) is to realize μ_0 as the standard normal distribution associated with the real Sobolev Hilbert space $H_1(\mathbb{R}^d)$, the closure of $S(\mathbb{R}^d)$ in the norm $(\psi,\psi)_1 \equiv \int \psi(x)(-\Delta + m^2)(x - y)\psi(y)dx\, dy$. In fact, in this description $\int e^{i(\xi,\psi)_1} d\mu_0(\xi) = \int e^{i\langle\xi,\varphi\rangle} d\mu_0(\xi) = e^{-(1/2)(\varphi,\varphi)_{-1}} = e^{-(1/2)(\psi,\psi)_1}$ with $(-\Delta + m^2)\psi = \varphi$, and $E((\xi,\psi)_1(\xi,\psi')_1)) = E(\langle\xi,\varphi\rangle\langle\xi,\varphi'\rangle) = (\varphi,\varphi')_{-1} = (\psi,\psi')_1$, with $(-\Delta + m^2)\psi' = \varphi'$.

We also note that G_m is the potential of an exponentially killed transient Brownian motion with killing rate m^2, $G_m(x - y) = (4\pi)^{-d/2} \int_0^\infty e^{-m^2 t}(1/t^{d/2})e^{-|x-y|^2/4t}\, dt$. We have in particular

$$G_m(x - y) = \begin{cases} \dfrac{e^{-m|x-y|}}{2m} & \text{for } d = 1 \\[2ex] \dfrac{1}{2\pi} K_0(|x - y|)m & (K_0 \equiv \text{modified Bessel function}) \\ & \text{for } d = 2 \\[1ex] \dfrac{e^{-m|x-y|}}{4\pi|x-y|} & \text{for } d = 3 \end{cases}$$

The spectral density function of the stochastic random field is $\hat{G}_m(p) = (2\pi)^{-d}/(p^2 + m^2)$, so that

$$E(\xi(x)\xi(y)) = G_m(x - y) = (2\pi)^{-d} \int e^{ipx}\, \hat{G}_m(p)\, dp$$

The probabilistic study of the free Markov field ξ has different aspects:

1. The variance being infinite, $\xi(x)$ is a genuine generalized random field (for a more detailed study of

the properties of the support of μ, see the references in [60p; 56c]).

2. We also note that $\xi(x)$ satisfies the stochastic partial differential equation $(-\Delta + m^2)^{1/2}\xi(x) = \eta(x)$, where $\eta(x)$ is the white noise with values in $\mathcal{S}'(\mathbb{R}^d)$, in the sense that $E(\eta(x)\eta(y) = \delta(x - y)$ (η centered, Gaussian). (For a study of random fields as solutions of stochastic partial differential equations, see, e.g., Ref. 13; for stochastic analysis using white noise, see Ref. 76.)
3. There is a Wiener-chaos decomposition of the corresponding $L^2(d\mu)$-space, analogous to the classical one. This decomposition corresponds to the Fock-space realization of $L^2(d\mu)$. It is well known and we refer the reader to the references in [16, 12, 48]. We shall speak a little more about this decomposition below, in connection with other problems.
4. There is one property of the free Markov field that we want to discuss in more detail, since it is also central to our approach to the nonfree case, (i.e., interacting case) as discussed below. This property is the global Markov property. There are different versions of it, according to the degree of generality in the choice of regions and elements (functions, measures, distributions) in the indexing set. In all versions the aim is to extend the usual Markov property of Markov processes to the case where "time is multidimensional."

Let us begin with the case where the indexing set is the real Sobolev space $H_{-1}(\mathbb{R}^d)$, the closure of $C_0^\infty(\mathbb{R}^d)$ in the $(\cdot,\cdot)_{-1}$ norm. Let $(\Omega, \mathcal{A}, \mu_0)$ be the underlying probability space. To any open set $U \subset \mathbb{R}^d$ we can associate a σ-algebra $\mathcal{B}(U)$, which is by definition the σ-algebra of subsets of the underlying σ-algebra $\mathcal{A}$ generated by all random variables $\langle \xi, \varphi \rangle$ with $\varphi \in H_{-1}(\mathbb{R}^d)$, supp $\varphi \subset U$. For any Borel-measurable subset B of $\mathbb{R}^d$ one then defines $\mathcal{B}(B) = \bigcap_{U \supset B, U \text{ open}} \mathcal{B}(U)$. The global Markov property is then the property that for all open subsets $U \subset \mathbb{R}^d$:

$$E(u \mid \mathcal{B}(\tilde{U})) = E(u \mid \mathcal{B}(\partial U))$$

for all positive or integrable random variables u($\mathcal{B}(\overline{U})$-measurable), where $\tilde{U} \equiv \mathbb{R}^d - \overline{U}$, $\overline{U} \equiv$ closure of U, $\partial U \equiv$ boundary of U. This is equivalent to

$$E(uv \mid \mathcal{B}(\partial U)) = E(u \mid \mathcal{B}(\partial U))E(v \mid \mathcal{B}(\partial U))$$

for all positive or integrable $\mathcal{B}(\tilde{U})$-measurable u and all positive or integrable $\mathcal{B}(\overline{U})$-measurable v. The free Markov field ξ has this Markov property (see, e.g., the references in [46, 6p, 50]).

First one remarks that for any random field having the above global Markov property, one has, by taking U as the half-space in $\mathbb{R}^d$ given by $\{x \in \mathbb{R}^d \mid x^1 < 0\}$ so that $\partial U = \{x^1 = 0\}$, that the global Markov property yields the existence of a Markov semigroup P_t in $L^2(\Omega, A, P) \cap \mathcal{B}(\partial U)$ such that $E(F(\langle \xi, f_t \rangle) \mid \mathcal{B}(\partial U)) = (P_t F)(\langle \xi_0, \varphi \rangle)$, for all bounded continuous real-valued functions F on $\mathbb{R}$ and where $f_t(x) \equiv \delta_t(x^1)\varphi(x^2, \ldots, x^d)$, $t \geq 0$, $\varphi \in C_0(\mathbb{R}^{d-1})$, $\mathbb{R}^{d-1} \equiv \partial U$.

As a Markov semigroup (P_t, ν), with $\nu \equiv P \restriction \mathcal{B}(\partial U)$, gives a Markov process ξ_t with initial distribution ν and transition semigroup P_t, we say that the Markov process ξ_t is associated to the Markov field ξ. If the Markov field is invariant under the reflections $x = (x^1, \ldots, x^d) \to (-x^1, x^2, \ldots, x^d)$, then we can extend P_t to $t < 0$ by $P_{-t} = P_t$ and then P_t is a self-adjoint operator in $L^2(\nu)$ and the corresponding process ξ_t can be extended to a symmetric (time inversion invariant) process, denoted again by ξ_t, having ν as invariant measure.

By the theory of Dirichlet forms, (P_t, ν) is uniquely associated with a Dirichlet form $(H_\nu^{1/2} f, H_\nu^{1/2} g)$, with H_ν self-adjoint in $L^2(\nu)$, positive such that $P_t = e^{-tH_\nu}$, $t \geq 0$. The free Markov field has this invariance property since it is Gaussian, centered, and the covariance has this property. Hence it gives rise to a symmetric Markov process. In this case it can be shown that H_ν is essentially self-adjoint on smooth cylinder functions in $L^2(\nu)$ and coincides there with the operator $-(1/2)\Delta - \beta\cdot\nabla$ defined for cylinder functions $F = F(\langle\xi_0,\varphi_1\rangle,\ldots,\langle\xi_0,\varphi_n\rangle)$, $\varphi_i \in C_0^2(\mathbb{R}^{d-1})$ by

$$\left(-\frac{1}{2}\Delta - \beta\cdot\nabla\right)F = \sum_{j=1}^{d}\left(-\frac{1}{2}\frac{\partial^2}{\partial x_j^2}F - \beta_j\frac{\partial}{\partial x_j}F\right)(\langle\xi_0,\varphi_1\rangle,\ldots,\langle\xi_0,\varphi_n\rangle)$$

with
$$\left(\beta_j\frac{\partial}{\partial x_j}F\right)(\langle\xi_0,\varphi_1\rangle,\ldots,\langle\xi_0,\varphi_n\rangle) = \frac{\partial F}{\partial x_j}(\langle\xi_0,\varphi_1\rangle,\ldots,\langle\xi_0,(-\Delta+m^2)^{1/2}\varphi_j\rangle,\ldots,\langle\xi_0,\varphi_n\rangle)$$

Also, $H_\nu F$ is equal to $H_0 F$, where H_0 is the so-called free energy operator in the Fock space for (the time zero) quantum fields, when the Fock space is realized as $L^2(\nu)$. Thus $\langle\xi_0,f\rangle$ can be looked upon as the time-zero free quantum field. It follows, moreover, from the formula giving P_t that for $f = \delta_0\otimes\varphi$, $\varphi\in C_0^\infty(\mathbb{R}^{d-1})$ [such f are in $H_{-1}(\mathbb{R}^d)$!] one has that $\langle\xi,f_t\rangle = \langle\xi,\delta_t\otimes\varphi\rangle$ has the same law as $\langle\xi_t,\varphi\rangle$. Thus one can identify the stochastic random field $\xi(x)$, looked upon as an ordinary stochastic process $x^1 \to (x^1,x^2,\ldots,x^d)$ with time parameter $x^1\in\mathbb{R}$ and values in generalized functions of $(x^2,\ldots,x^d)$ with the stochastic process ξ_t, which also takes values in such generalized functions. In this picture the "euclidean field" $\xi(x)$ at $(x^1 = t, x^2,\ldots,x^d)$ appears as the time t stochastic process $\xi_t(x^2,\ldots,x^d)$.

The free Markov field has a further "homogeneity property." In general we understand by (euclidean) homogeneous stochastic field one which is stationary with respect to translations in $\mathbb{R}^d$ and isotropic with respect to rotations in $\mathbb{R}^d$ and moreover is symmetric with respect to reflections in $\mathbb{R}^d$, in short invariant under the full euclidean group $E(\mathbb{R}^d)$ in $\mathbb{R}^d$, in the sense that $E(\Pi_{i=1}^n\langle\xi,f_i\rangle) = \Pi_{i=1}^n E(\langle\xi,Tf_i\rangle)$, for all $T \in E(\mathbb{R}^d)$ and all $f_i \in H_{-1}(\mathbb{R}^d)$, where $(Tf_i)(x) \equiv f_i(T^{-1}(x))$.

It can be shown [51] that the covariance of a centered generalized random field which has the global Markov property is necessarily of the form

$$E(\langle\xi,f\rangle\langle\xi,g\rangle) = \int_0^\infty\int_{\mathbb{R}^d} \frac{\tilde{f}(p)\tilde{g}(p)}{p^2 + m^2}\, dp\, d\rho(m)$$

where $\tilde{f}$ means a Fourier transform of f and ρ is a positive measure which is at the same time a tempered distribution. This is the euclidean version of the "Källen-Lehman" representation [49].

In the literature on generalized stochastic fields the index set is often taken to be, instead as $H_1(\mathbb{R}^d)$ as above, a space of test functions $C_0^\infty(\mathbb{R}^d)$ (as in Ref. 18). In this case the Markov property is formulated differently since the σ-algebra $B(\partial U)$ for an open set U cannot be given a direct meaning. In this case $B(\partial U)$ and more generally $B(C)$ for closed sets C is defined as the intersection of the σ-algebras associated with all open sets covering C (all σ-algebras are understood to contain the σ-algebras of zero-measure sets). The free Markov field has this global Markov property (this has been discussed, e.g., in Ref. 50). On the other hand, global Markov fields in this sense are a much larger class than the ones in the previous sense (allowing for d = 1, "Markovicity of higher order"), more generally allowing as spectral densities entire functions of infra-exponential type (see, e.g., Ref. 51).

REMARKS

1. The global Markov property for fields indexed by $C_0^\infty(\mathbb{R}^d)$ have been discussed recently by D. Preiss and R. Kotecky [52] in relation to previous work. In particular they show that a generalized random field which is Markov in

the above sense does not in general have the property (P) that $E(v \mid \mathcal{B}(U)) = E(v \mid \mathcal{B}(U - C))$ for any $\mathcal{B}(\mathbb{R}^d - C)$-measurable bounded function v, $C \subset U \subset \mathbb{R}^d$, C closed, U open. They show explicitly that this property does not hold for the free Markov field for d = 1. By the martingale convergence theorem the converse is true, i.e., (P) implies the global Markov property. (P) is equivalent to the global Markov property, together with $\sigma(\mathcal{B}(C), \mathcal{B}(U - C)) = \mathcal{B}(U)$.

2. In the literature a notion of local Markov property appears [53, 54]. It is the Markov property in the above sense but is required only for all open bounded U. It is strictly weaker than the global Markov property. The local Markov property is the one one usually understands under the Markov property in classical statistical mechanics (in connection with the Gibbs fields) [19] (for concepts related to the Markov property, see also Refs. 58 and 30h).
3. Generalized random fields indexed by other function spaces have also been discussed in relation with the theory of Fourier hyperfunctions (see, e.g., Ref. 55). Moreover, we should mention work done using white noise fields (e.g., Ref. 76).
4. In the probabilistic literature as well as in the literature about quantum fields there are several results saying roughly that there are no "nice" fields having the invariance properties and "causality properties" (Markov property in the probabilistic case) one wants. For example, a Markov homogeneous Gaussian field with continuous covariance is by necessity a constant [51, 14]. A similar result in the non-Gaussian case with a restriction of ergodicity for the translations has been proven by Constantinescu and Thalheimer and Gudder [59]. In the

literature on quantum fields there is a related result [49].

We shall now discuss the case of stochastic fields indexed by spaces of measures. The usefulness of this study has been stressed by ourselves in Refs. 56 and 7 and by Dynkin [57]. Our study brought in methods of potential theory and such methods have now been extended considerably by Röckner [50]. Dynkin's methods are probabilistic. We shall briefly describe the potential theoretic methods first in the much restricted case of the free Markov field.

Let $m > 0$; then $\mathbb{R}^d$ with the harmonic sheaf of C^2 solutions h of $(-\Delta + m^2)h = 0$ is a self-adjoint harmonic space with Green function $G_m(x - y) = (-\Delta + m^2)^{-1}(x - y)$ in the sense of Constantinescu-Cornea-Maeda [78]. Introduce the space of signed measures ρ on $\mathbb{R}^d$ with finite energy, i.e., $|\rho| \in H_{-1}(\mathbb{R}^d)$, i.e., $(\rho,\rho)_{-1} \equiv \int d|\rho|(x)\, d|\rho|(y) G_m(x - y) < \infty$. It is a pre-Hilbert space in the $(\cdot,\cdot)_{-1}$ scalar product and its completion is a Hilbert space H. Let $\xi_\rho \equiv \langle \xi, \rho \rangle$, $\rho \in H$, be the corresponding linear process, which is the standard Gaussian process indexed by H, with mean 0 and covariance $E(\xi_\rho, \xi_\sigma) = \langle \rho, \sigma \rangle_{-1}$. ξ_ρ then has the Markov property with respect to arbitrary subsets $A \subset \mathbb{R}^d$, in the sense that $\mathcal{B}(\partial A)$ is a splitting σ-algebra for $\mathcal{B}(\bar{A})$ and $\mathcal{B}(\mathbb{R}^d - \bar{A})$, where for arbitrary closed C, $\mathcal{B}(C) = \sigma$-algebra generated by ξ_ρ, ρ any signed measure of finite energy with supp $\rho \subset C$ [A σ-algebra Σ is called a splitting σ-algebra for two σ-algebras F, G iff $E(uv \mid \Sigma) = E(u \mid \Sigma)E(v \mid \Sigma)$ a.s., for all bounded u, v which are F (respectively, G)-measurable.] This is called the global Markov property.

REMARK 5. This is a special case of a more general result which holds for the centered Gauss random fields ξ_ρ with covariance given by the Green function of some self-adjoint harmonic space.

The idea of the proof [56c,50] resides in an orthogonal decomposition of the process, for positive ρ of finite energy and such that for open U, $\rho(\tilde{\bar{U}}) = 0$; $\xi_\rho = \xi_{\rho^{\tilde{U}}} + \xi_{\rho-\rho^{\tilde{U}}}$, where $\rho^{\tilde{U}}$ is the measure ρ balayed on $\tilde{U}$; in particular, $\rho^{\tilde{U}}$ has support on ∂U. $\xi_{\rho-\rho^{\tilde{U}}}$ is orthogonal [in the sense that $E(\xi_{\rho-\rho^{\tilde{U}}}\xi_\sigma) = 0$] to all measures σ of finite energy which do not charge U, i.e., $\sigma(U) = 0$. In the proof the fact (proven in Ref. 50) that for any closed $A \subset \mathbb{R}^d$ the space of positive measures of finite energy and support in A is complete in the metric given by the energy is used (extension of a corresponding result by H. Cartan in the classic potential theoretical case). We shall come back later to the above important decomposition of ξ_ρ

REMARK 6. In Ref. 50 a "fine global Markov property" is introduced and proven for the free Markov field. It is formulated in the same way as the global Markov property for arbitrary subsets A of $\mathbb{R}^d$ and in it $\mathcal{B}(\partial U)$ [respectively $\mathcal{B}(\bar{A})$, respectively $\mathcal{B}(\bar{\tilde{A}})$] are replaced by $\mathcal{B}(\partial_f A)$ [respectively $\mathcal{B}(\bar{A}^f)$, respectively $\mathcal{B}(\bar{\tilde{A}}^f)$], where $\partial_f A$ is the so-called fine boundary of A, i.e., the boundary of A in the fine topology, and correspondingly for $\bar{A}^f$ and $\bar{\tilde{A}}^f$. One also shows that ξ_ρ has the property that for an arbitrary subset A of $\mathbb{R}^d$,

$$E(uv \mid \sigma_H(\partial A)) = E(u \mid \sigma_H(\partial A))E(v \mid \sigma_H(\partial A)) \qquad P \text{ a.s.}$$

for all bounded $\mathcal{B}(\bar{A})$-measurable u and all bounded $\mathcal{B}(\tilde{A})$-measurable v, with $\sigma_H(C) \equiv \sigma(\xi_T, T \in H_{-1};\ \text{supp}\, T \subset C)$ for any closed C. (All σ-algebras are always assumed complete, i.e., to contain all sets of measure zero.) The proof uses essentially the fact that $\sigma_H(C)$ is equal to the σ-algebra $\mathcal{B}(C)$ generated by the (ξ_ρ, $\rho \in H_{-1}$, supp $\rho \subset C$), by the fact that by an extension of a theorem of Deny the set of all signed measures of finite energy with support in C is $(\cdot,\cdot)_{-1}$ dense in $H(C) \equiv \{T \in H_{-1} \mid \text{supp } T \subset C\}$ (see Ref. 50). Finally, one shows that the free Markov field ξ as a process indexed by $C_0^\infty(\mathbb{R}^d)$ has the global Markov property in the sense that $\sigma_F(\partial A)$ is a splitting σ-algebra for $\sigma_F(\bar{A})$ and $\sigma_F(\bar{\bar{A}})$, for any $A \subset \mathbb{R}^d$, where $\sigma_F(A) = \sigma(\langle \xi, \varphi \rangle,\ \varphi \in C_0^\infty(\mathbb{R}^d),\ \text{supp } \varphi \subset A)$, and $\sigma_F(C) = \bigcap_{U \supset C} \sigma_F(U)$, for all open A and closed C (U open).

The utility of the above approach to the Markov property is most visible in the application to the proof of the global Markov property of certain non-Gaussian stochastic fields of interest in quantum field theory, since they lead to models of quantum fields with nontrivial interaction. Let us briefly describe these results.

To do so, we come back to the decomposition $\xi_\rho = \xi_{\rho^{\tilde{U}}} + \xi_{\rho - \rho^{\tilde{U}}}$ for any open U, for the free Markov field given above. This decomposition corresponds to the one given in Ref. 56c. We have that both ξ_{ρ^U} and $\xi_{\rho-\rho^{\tilde{U}}}$ are linear Gaussian-centered processes for all ρ of finite energy and compact support in U, and one has

$$E(\xi_{\rho^{\tilde{U}}}\, \xi_{\sigma^{\tilde{U}}}) = \iint K_{\partial U}(x,y)\, d\rho(x)\, d\sigma(y)$$

$$E(\xi_{\rho-\rho^{\tilde{U}}}\, \xi_{\sigma-\sigma^{\tilde{U}}}) = \iint G^{\partial U}(x,y)\, d\rho(x)\, d\sigma(y)$$

$$E(\xi_{\rho^{\tilde{U}}}\, \xi_{\sigma-\sigma^{\tilde{U}}}) = 0$$

for all ρ, σ of finite energy and compact support in U, with $G^{\partial U}(x,y)$ the Green's function on U, i.e., for U smooth, $G^{\partial U}(x,y) = (-\Delta_{\partial U} + m^2)^{-1}(x,y)$, with $\Delta_{\partial U}$ the Laplacian with Dirichlet boundary conditions on U. Moreover, $K_{\partial U}(x,y) \equiv G(x - y) - G^{\partial U}(x,y)$. Let $\xi^{\partial U}(x)$ be such that in the sense of generalized functions $\xi_{\rho-\rho^{\tilde{U}}} = \int \xi^{\partial U}(x)\, d\rho(x)$.

Then $\xi^{\partial U}$ has mean zero and covariance $G^{\partial U}$, and is the so-called free Markov field with Dirichlet boundary conditions on ∂U. Let P_x^U be the $\Delta - m^2$-harmonic measure of U, supported by the fine boundary of U, for all $x \in U$. It is shown in Ref. 56c and more generally in Ref. 50 that $\xi_{\rho^{\tilde{U}}} = \int \xi_{P_x^U}\, d\rho(x)$, where the integral is in the Bochner sense in $L^2(d\mu_0)$, after splitting $d\rho$ in the positive and negative parts. $x \to \xi_{P_x^U}$ is an (ordinary, nongeneralized) stochastic field with mean zero and covariance the continuous real function $K_{\partial U}(x,y)$. One has $\xi_{P_x^U} = \langle \xi, P_x^U \rangle$, in the sense of generalized processes.

We also remark that from $\xi_\rho = \xi_{\rho^{\tilde{U}}} + \xi_{\rho - \rho^{\tilde{U}}}$ we have, with

$$\xi_{P_x^U} \equiv \xi_{\partial U}(x),$$

$$\langle \xi, \rho \rangle = \langle \xi_{\partial U}, \rho \rangle + \langle \xi^{\partial U}, \rho \rangle$$

for any $\rho \in H_{-1}(\mathbb{R}^2)$. Now let f be an arbitrary positive μ_0-measurable or μ_0-integrable function and consider the conditional expectation $E(f \mid \mathcal{B}(\partial U))$ with respect to μ_0 and $\mathcal{B}(\partial U)$. Let $\mu_0^{\partial U}$ be the probability measure on $\mathcal{S}'(\mathbb{R}^2)$ with mean zero and covariance operator $(-\Delta_{\partial U} + m^2)^{-1}$. Then

$$E_{\mu_0^{\partial U}}(\xi^{\partial U}(x)\xi^{\partial U}(y)) = G^{\partial U}(x,y)$$

Since $\xi^{\partial U}$ and ξ are independent Gaussian and $\xi^{\partial U} + \xi_{\partial U} = \xi$, where ξ is Gaussian, we have for any bounded real-valued measurable f on $\mathbb{R}$ such that $f(\xi) \in L^1(\mu_0)$:

$$E_{\mu_0}(f \mid \mathcal{B}(\partial U)) = E_{\mu_0^{\partial U}}(f)$$

where on the right-hand side $f(\cdot)$ should be thought of as a function $f(\xi^{\partial U} + \xi_{\partial U})$ of the stochastic variables $\xi^{\partial U}$ and of the independent stochastic variables $\xi_{\partial U}$. This is a basic formula that we shall use in the next section for the computation of conditional expectation of the form $E_{\mu_0}(f \mid \mathcal{B}(\partial U))$.

REMARK 7. $\xi_{P^U_x}$ corresponds to the quantity denoted by $\psi^C_\xi(x)$, with $C \equiv \partial U$, in Ref. 56c. In those references $\psi^C_\xi(x)$ was looked upon as a solution of the Dirichlet problem in U with boundary condition $\xi \restriction C$ on C.

4.2 Regular Generalized Random Fields

It is possible to prove strong estimates on the fields $\xi_{\partial U}$ introduced in Sec. 4.1 [56c, 50]. In fact, these estimates hold also whenever one has a generalized random field $((\Omega, \mathcal{A}, P), \xi)$, not necessarily the free Markov field, which is regular in the following sense. There exists a constant $c > 0$ such that

$$E(\langle \xi, \varphi \rangle^2) \leq c(\varphi, \varphi)_{-1}$$

for all $\varphi \in C_0^\infty(\mathbb{R}^d)$. In the following ξ is always supposed to be regular. Theorem 4.1 gives estimates on the boundary behavior of $\xi_{P^U_x}$ as $x \to \partial U$.

THEOREM 4.1 For any $m > 0$ and all $U \subset \mathbb{R}^d$ open, $x \in U$, there exists $c_m^{(d)} > 0$ with

$$E(\xi^2_{P^U_x}) \leq c_m^{(d)} f_d(x,U) d(x, \partial U)^{-(d-1)/2} e^{-md(x,\partial U)}$$

where

$$f_d(x,U) \equiv [1 \vee d(x,\partial U)]^{-(d-2)} \qquad \text{for } d \geq 3$$

$$f_d(x,U) \equiv 1 \vee |\ln d(x,\partial U)| \qquad \text{for } d = 2$$

Moreover, for $K \subset U$ compact with $d(K,\partial U) \geq 1$, one has that for all $d \geq 3$ there exists $b_m^{(d)} > 0$ such that

$$E((\int_K |\xi_{P^U_x}| \, dx)^2) \leq b_m^{(d)^2} |K|^2 \, d(K,\partial U)^{-(d-1)/2} e^{-md(K,\partial U)}$$

$|K| \equiv$ volume of K

For d = 2 and any given $\alpha \in (0,1/2)$ there exists $b^{(2)}_{m,\alpha} > 0$ such that the same estimate holds, with $b^{(2)}_{m,\alpha}$ replacing $b^{(d)}_m$ and α replacing $(d - 1)/2$.

For the proof, see Refs. 56c and 50.

Using the harmonic property of $\int \varphi(x)\, dP^U_y(x)$ for $\varphi \in C^\infty_0(U)$ one then proves a mean value property for the stochastic field $\xi_{P^U_x}$ of the form $\xi_{P^U_x} = \int \xi_{P^U_y}\, dP^B_x(y)$, for any ball B with center $x \in U$, radius $r > 0$, and such that $\bar{B} \subset U$, the integral being a Bochner L^2-one. From this one then derives, using the Serrin inequality, the almost sure estimate

$$|\xi_{P^U_x}| \leq \frac{c}{\sigma_d(1) r^{d-1}} \int |\xi_{P^U_y}|\, d\sigma_{x,r}(y)$$

where $\sigma_{x,r}$ is the surface measure on ∂B, and $\sigma_d(1)$ is the surface of the sphere of radius 1 in $\mathbb{R}^d$, c being a constant. From these results one then arrives at the following asymptotic control on the solution $\xi_{\partial U}(x) = \xi_{P^U_x}$ of the Dirichlet problem:

THEOREM 4.2 Let ξ be a regular generalized random field over $\mathbb{R}^d$. Let $m > 0$ be a given number. Let U_n be a sequence of open subsets of $\mathbb{R}^d$ expanding to $\mathbb{R}^d$ in such a way that $d(0,\partial U_n) \to \infty$ as $n \to \infty$ For any n, let $K_n \subset U_n$ be compacts such that $|d(K^{(2)}_n, \partial U_n)|^{-\alpha/2} e^{-(\beta/2) d(K^{(2)}_n, \partial U_n)} \to 0$ as $n \to \infty$, for some $\alpha \in (0,1/2)$ for d = 2, and with $\alpha = (d - 1)/2$ for $d \geq 3$, and some $0 \leq \beta < m$. Here $K^{(2)}_n \equiv \{x \in \mathbb{R}^d \mid d(x,K_n) \leq 2\}$. Then there exists a subsequence n_k, $k \in \mathbf{N}$, $n_k \to \infty$, such that

$$e^{[(m - \beta)/2] d(K_{n_k}, \partial U_{n_k})} \sup_{x \in K_{n_k}} |\xi_{\partial U_{n_k}}(x)| \to 0$$

almost surely as $k \to \infty$.

COROLLARY 4.1 The solution $\xi_{\partial U}(x)$ tends to zero locally uniformly as $U \uparrow \mathbb{R}^2$ for almost every ξ.

For a complete proof, see Refs. 56c and 50.

Using similar methods one proves results corresponding to Theorems 4.1 and 4.2 for the field $\langle \xi, P_x^{U,U'} \rangle \equiv \xi_{U,U'}(x)$, with $P_x^{U,U'} \equiv P_x^U - P_x^{U'}$ for any open subsets U, $U' \equiv U \cap U_0$ for some fixed open U_0. Theorem 4.1 holds with ∂U replaced everywhere by U_0, and ξ replaced by $\langle \xi, P_x^{U,U'} \rangle$. Moreover, Theorem 4.2 holds with $K_n \subset U_n$ replaced by $K_n \subset U_0 \cup U_n$ and $\xi_{\partial U_{n_k}}$ replaced by $\langle \xi, P_x^{U_0 \cap U_{n_k}} \rangle - \langle \xi, P_x^{U_0} \rangle$. In particular, $\langle \xi, P_x^{U_0 \cap U_{n_k}} \rangle$ tends locally uniformly to $\langle \xi, P_x^{U_0} \rangle$ as $U \uparrow \mathbb{R}^d$, for a.e. ξ. Hence we get an asymptotic control on the solutions of the Dirichlet problem in U and $U \cap U_0$.

We think that these general results on regular generalized random fields on $\mathbb{R}^d$ might have an intrinsic interest. They certainly are basic tools in the study of potential theoretical questions for these generalized random fields. They have been used as a tool for solving the problem of the uniqueness of the Gibbs states associated with models of quantum fields for $d = 2$, as well as for the proof of the global Markov property for these fields [56c, 50]. These results also solve the problem of the construction of examples of non-Gaussian euclidean-homogeneous global Markov fields. We shall now briefly discuss these results. To do so we first describe the construction of locally Markov non-Gaussian homogeneous (generalized) random fields.

4.3 The Construction of Locally Markov Homogeneous Non-Gaussian Generalized Random Fields

Let $(\Omega, \mathcal{A}, \mu_0, \xi)$ be the free Markov field of mass m on $\mathbb{R}^d$. A <u>local additive functional</u> of ξ or, for short, an "interaction" is a family (Λ, U_Λ), where Λ runs over the bounded Borel subsets of $\mathbb{R}^d$ and U_Λ is $\mathcal{B}(\Lambda)$-measurable and such that $U_{\Lambda_1 \cup \Lambda_2} = U_{\Lambda_1} + U_{\Lambda_2}$ a.s. for all bounded rectangles Λ_1, Λ_2 with $\Lambda_1 \cap \Lambda_2 = \emptyset$ and $U_{T\Lambda}(T\cdot) = U_\Lambda(\cdot)$ for all Λ. We shall now consider local additive functionals constructed from "powers of the fields." Let $\alpha > 0$ and let $U^\alpha_{\Lambda,\varkappa}(\xi) \equiv e^{-(\alpha^2/2)G_\varkappa(0)} \int_\Lambda e^{\alpha\xi_\varkappa(x)} dx$, where $\xi \in \mathcal{S}'(\mathbb{R}^d)$, $\xi_\varkappa$ the convolution of ξ with a smooth function $\chi_\varkappa$ with Fourier transform having support in a ball with center at the origin and of radius $\varkappa$ and such that $\chi_\varkappa \to 1$ as $\varkappa \to \infty$. We have then $\exp(\alpha\,\xi_\varkappa) \in L^p(\mu_0)$ for all $1 \leq p < \infty$ and the factor $\exp[-(\alpha^2/2)G_\varkappa(0)]$ is such that $E(U^\alpha_{\Lambda,\varkappa}) = 1$. The following theorem holds:

THEOREM 4.3 $U^\alpha_{\Lambda,\varkappa}$ is a positive L^2-martingale with respect to the family of σ-algebras $\mathcal{B}_\varkappa$ generated by the $\xi_{\varkappa'}$, $\varkappa' \leq \varkappa \cdot e^{-U^\alpha_{\Lambda,\varkappa}}$ is a submartingale which converges a.s. and in $L^2(\mu_0)$ as $\varkappa \to \infty$. For $d = 1, 2$, $|\alpha| < \sqrt{4\pi}$, the limit is $e^{-U^\alpha_\Lambda}$, with U^α_Λ an additive functional which is for every bounded Borel Λ a nonnegative $L^2(\mu_0)$-function nonidentically constant given as the Radon-Nikodym derivative of positive measure $d\int_\Lambda \mu_0(\cdot - \alpha G_x)\, dx$ with respect to $d\mu_0$, where $G_x(y) \equiv G(x - y) = (-\Delta + m^2)^{-1}(x - y)$. For $d \geq 3$ or ($d = 2$, $|\alpha| > \sqrt{8\pi}$) one has $U^\alpha_\Lambda = 0$ a.s.

Proof: The proof for $d = 2$, $|\alpha| < \sqrt{4\pi}$, relies on the basic estimate [61] $E((U^\alpha_{\Lambda,\varkappa})^2) = \int_\Lambda\int_\Lambda e^{\alpha^2 G_\varkappa(x-y)}\, dx\, dy < \infty$, together with the positivity of $U^\alpha_{\Lambda,\varkappa}$. The proofs for $d \geq 3$ have been given in Ref. 62 (for $d \geq 4$) and Ref. 63. The case $d = 2$, $|\alpha|$ larger than some $\alpha_0 > 0$, is covered by Ref. 63, and that one can take $\alpha_0 = \sqrt{8\pi}$ has been shown in Ref. 64. A basic ingredient for this proof is the following lemma [63, 70a], of independent interest:

LEMMA 4.1 Let B be a bounded open cube in $\mathbb{R}^d$. For any $f,g \in C_0^\infty(B)$, let

$$(f,g)_p \equiv \int_B \sum_{i,j} p_{ij}(x) \frac{\partial}{\partial x_i} f \frac{\partial}{\partial x_j} g \, dx$$

where $P = ((p_{ij}))$ is a $d \times d$ matrix with real continuous coefficients $p_{ij}(x)$, with $p_{ij} = p_{ji}$ and such that $m1 \leq P(x) \leq M1$ for all $x \in B$. Let μ_P be the Gaussian measure on the dual of $C_0^\infty(B)$ with Fourier transform $\exp[-(1/2)(f,f)_p]$. Then for all $d \geq 3$ and all $\alpha \in \mathbb{R}$, or for $d = 2$ and $|\alpha| > 4\sqrt{2\pi}\, M^2/m^{3/2}$ there exists a measurable subset Q, independent of $\alpha \neq 0$ and $y \in B$ such that $\mu_P(Q) = 1$ and $\mu_P(Q + \alpha\delta_y(\cdot)) = 0$ for all $\alpha \neq 0$ and all $y \in B$.

For the proof of this lemma, see Refs. 63 and 70a. We note that the important fact is that Q is the same for all $y \in B$!

REMARK 8. Recently, there have been results in the related situation of trigonometric interactions (see below) concerned with the case left over by Theorem 4.3, namely $\sqrt{4\pi} \leq |\alpha| < \sqrt{8\pi}$ (see Ref. 65a). Also, an alternative definition of an interaction related to U has been given in Ref. 65b).

Now, let for $d = 2$: $:\xi^n:(\chi_\Lambda)$ be defined by

$$\frac{\partial^n}{\partial \alpha^n} U_\Lambda^\alpha \Big|_{\alpha=0}, \qquad \text{so that } \sum_{n=0}^\infty \frac{\alpha^n}{n!} :\xi^n:(\chi_\Lambda) = U_\Lambda^\alpha(\xi)$$

We easily prove that: $:\xi^n:(\chi_\Lambda) \in L^p(\mu_0)$ for all $1 \leq p < \infty$ and $E(:\xi^n:(\chi_\Lambda) :\xi^m:(\chi_\Lambda)) = 0$ whenever $n \neq m$.

One can extend by L^2-continuity the definition of $:\xi^n:(\chi_\Lambda)$ to $:\xi^n(f)$ for arbitrary $f \in S(\mathbb{R}^d)$. The generalized random field $:\xi^n:(f)$ is called the Wick-Itô power of the free Markov field ξ (see, e.g., Refs 6p, 12, 16, 48, 57). The family $(\Lambda, :\xi^n:(\chi_\Lambda))$ is a local additive functional of ξ.

Using the above facts on $:\xi^n:(f)$ and U_Λ^α it is possible to define other local additive functionals of ξ. Let $u(r)$ be any of the following real-valued functions of $r \in \mathbb{R}$:

1. $u(r) = \Sigma_0^N a_s r^s$, $a_{2N} > 0$.
2. $u(r) = \int e^{\alpha r}\, d\rho(\alpha)$.
3. $u(r) = \int \cos(\alpha r + \Theta)\, d\rho(\alpha)$.

In forms 2 and 3, ρ is any positive Borel measure with support in $(-\sqrt{4\pi}, \sqrt{4\pi})$ and $0 \le \Theta < 2\pi$.

4a. $u(r) = \int \alpha^{-4}(\cos \alpha r - 1 + \alpha^2/2\, r^2)\, d\rho_1(\alpha) + \int \alpha^{-2}(1 - \cos \alpha r)\, d\rho_2(\alpha)$ with positive measures ρ_i with $\operatorname{supp} \rho_i \in [0, \sqrt{4\pi})$.

b. $u(r) = \Sigma_{n=0}^\infty C_{2n} r^{2n}$ with $0 \le C_{2n} \le \exp(-n^{2n} e^{(\ell n)^\delta n})$ for some constants C and an arbitrary small $\delta > 0$.

c. $u(r) = \int_0^1 (\cos \alpha r - 1)\alpha^{-2-\varepsilon}\, d\nu(\alpha)$ for some $\varepsilon < 1$; and where ν is a positive Borel measure with $\int_0^\delta d\nu(\alpha) < C\delta$ for all $\delta \le 1$ and some constant C.

d. $u(r) = \int \alpha^{\varepsilon-2}(1 - \cos \alpha r)\, d\nu(\alpha)$, where ν is a real measure with support in $[0, \sqrt{4\pi})$.

e. $u(r) = \int \alpha^{-2}(\cosh \alpha r - 1)\, d\nu(\alpha)$, where ν is a positive measure with support in $[0, \sqrt{4\pi})$.

f. $u(r) = \int_0^1 \alpha^{-1} \sin \alpha r\, d\alpha$.

Now write $u(r)$ formally as a power series $\Sigma\, c_n r^n$, and define $:u(\xi):(x)$ formally by $\Sigma\, c_n :\xi^n:(x)$, with $:\xi^n:(x)$ defined above. Also define formally $U_\Lambda^u = \langle :u(\xi):, \chi_\Lambda \rangle$, for any bounded open $\Lambda \subset \mathbb{R}^2$. It has been shown (see below for references) that for any u of the above form one has that (Λ, U_Λ^u) is a local additive functional of μ_0. We shall call U_Λ^u the interaction in the region Λ given by u. Here are some names and references for the interactions U_Λ^u:

1. Polynomial, $P(\varphi)_2$-interactions: [12, 16]
2. Exponential, $\exp \alpha\varphi_2$-interactions: [61, 16, 6o,p, 7]
3. Trigonometric, Sine-Gordon interactions: [71, 56c]
4. $F(\varphi)_2$-interactions: [72]

It is easy to show, using the global Markov property of μ_0 that the probability measure on $S'(\mathbb{R}^d)$, $d\mu_\Lambda^u \equiv e^{-U_\Lambda^u} d\mu_0 / \int e^{-U_\Lambda^u} d\mu_0$ has the global Markov property and that the associated generalized random field ξ is again a global Markov field. Obviously, all $d\mu_\Lambda^u$ (for nontrivial choices of u) are non-Gaussian. However, none of them is homogeneous (because $\Lambda \neq \mathbb{R}^2$). The problem which has been solved by the construction of the μ_Λ^u is called in quantum field theory the "ultraviolet problem"; because of it it was necessary to introduce the renormalized powers $:\xi^n:(x)$ instead of simply $\xi(x)^n$ (which is a.s. infinite).

As is well discussed in the literature on constructive quantum field theory, in order to get homogeneous measures and from them relativistic quantum fields one has to take the limit ("thermodynamic or infrared limit") $\Lambda \uparrow \mathbb{R}^2$. Probabilistically, we have to show weak convergence of μ_Λ^u as $\Lambda \uparrow \mathbb{R}^2$ to some (nontrivial) probability measure μ^u on $S'(\mathbb{R}^2)$. This is the second part of the story in constructive quantum field theory and has been solved for all models above, under suitable restrictions on the coefficients and basically by two different refined techniques: the method of ferromagnetic inequalities (see, e.g., Refs. 16 and 12) and the method of cluster expansions (Ref. 12b). The result is that for a large class of interactions u, all moments of μ_Λ^u converge as $\Lambda \uparrow \mathbb{R}^2$ and μ_Λ^u converges weakly to a probability measure μ^u on $S'(\mathbb{R}^2)$. μ^u and the associated (generalized) random field ξ are homogeneous (stationary), in the sense that $\mu^u(T) = \mu^u$ for any transformation $T: S'(\mathbb{R}^2) \to S'(\mathbb{R}^2)$ induced in $S'(\mathbb{R}^2)$ by euclidean transformations τ in $\mathbb{R}^2$, i.e., $(T\xi)(x) = \xi(\tau^{-1}(x))$, τ in the group $E(\mathbb{R}^2)$ of transformations leaving the norm in $\mathbb{R}^2$ invariant. Several properties of the limit measure μ^u have been discussed in the literature. These are essentially estimates on the moments, including properties called Osterwalder-Schrader axioms for a euclidean field theory. These properties permit us, by a technique of analytic continuation, to get from the moments a set of relativistic, invariant, tempered distributions describing a relativistic local quantum field theory satisfying the basic Wightman axioms (cf. [49], [12], [16]).

The basic property allowing analytic continuation is a weak kind of Markovian property called Osterwalder-Schrader positivity. It holds for the above models given by μ^u. It is much weaker than the global Markov property we shall discuss below. The latter yields consequently stronger results but is much harder to prove; in fact, it has only been proven, at the moment of writing, for models with u of the forms (2) and (3) above [$\|\rho\|$ sufficiently small] (although only technicalities are involved in extending the proof to the other interactions, especially φ_2^4). So before going into the discussion of the global Markov property of homogeneous non-Gaussian fields (μ^u,ξ), let us describe briefly the Osterwalder-Schrader property and mention some related properties; in particular we shall examine its relations with the global Markov property. Write $\mathbb{R}^2 = \mathbb{R} \times \mathbb{R}$, $x = (x^0,x^1)$, $x^0 \equiv t$, $x^i \in \mathbb{R}$, $i = 0, 1$. Let $\pi_+ \equiv \{x \in \mathbb{R}^2 \mid x^0 \geq 0\}$, and let $\mathcal{B}(\pi_+)$ be the σ-algebra generated by the $\langle\xi,\rho\rangle$ with supp $\rho \subset \pi_+$, $\rho \in H_{-1}(\mathbb{R}^2)$. Let E_+ be the conditional expectation with respect to the probability measure μ^u and the σ-algebra $\mathcal{B}(\pi_+)$. E_+ can be identified with an orthogonal projection in $L^2(\mu^u)$. Let R be the unitary operator in $L^2(\mu^u)$ describing the reflection $(x^0,x^1) \to (-x^0,x^1)$, i.e., $R\,\xi(x^0,x^1) = \xi(-x^0,x^1)$. Then $H_0 \equiv E_+RE_+$ is a linear subspace in $L^2(\mu^u)$. Consider the sesquilinear form on H_0, $\langle E_+RE_+f, E_+RE_+g\rangle \equiv (Rf,g)$, $(\cdot,\cdot)$ being the scalar product in $L^2(\mu^u)$, for all $f,g \in H_0$.

We remark that $L^2(\mu^u \upharpoonright \mathcal{B}(\pi_0)) \subset \mathcal{H}_0$, where $\mathcal{B}(\pi_0)$ is the σ-algebra generated by the $\langle \xi, \rho \rangle$, supp $\rho \subset \{x^0 = 0\}$, $\rho \in \mathcal{H}_{-1}(\mathbb{R}^2)$. Moreover $\langle \cdot, \cdot \rangle \upharpoonright L^2(\mu^u \upharpoonright \mathcal{B}(\pi_0))$ coincides with $(\cdot, \cdot)$. Let E_0 (respectively, E_-) be the conditional expectation with respect to μ^u and $\mathcal{B}(\pi_0)$ [respectively, $\mathcal{B}(\pi_-)$], where $\mathcal{B}(\pi_-)$ is defined as $\mathcal{B}(\pi_+)$ with π_- replacing π_+, where $\pi_- \equiv \{x^0 \leq 0\}$. The global Markov property with respect to the half-plane π_0 can be formulated as the property that $E_+E_- = E_+E_0E_-$. The Osterwalder-Schrader positivity can be formulated as the property that $(Rf, f) \geq 0$ for all $f \in L^2(\mu^u \upharpoonright \mathcal{B}(\pi_+))$. In the case this property holds, $\mathcal{H}_0$ can be completed to a Hilbert space $\mathcal{H}$ and the family of operators defined by $P(t)E_+RE_+f \equiv E_+RE_+U(t)f$ for all $f \in L^2(\mu^u)$, with $U(t)f(\xi) = f(\xi_{-t})$, $\xi_{-t}(x^0, x^1) \equiv \xi(x^0 - t, x^1)$, is a strongly continuous self-adjoint contraction semigroup on $\mathcal{H}$ [58c,b,e]. As such, $P(t)$ has an analytic continuation to imaginary values of t and from this it then follows that the moments $E_{\mu^u}(\langle \xi, \delta_{t_1} \otimes \varphi_1 \rangle \langle \xi, \delta_{t_2} \otimes \varphi_2 \rangle)$, $t_1 \leq t_2$, have, for all $\varphi_1, \varphi_2 \in \mathcal{S}(\mathbb{R})$, analytic continuation to $\mathrm{Im}(t_1 - t_2) \geq 0$, this being then the so-called two-point, time-ordered Wightman function of a relativistic quantum model; and similarly for the higher moments (see, e.g., Ref. 12 for details).

Now, what is the difference with the global Markov property? The global Markov property with respect to the half-plane π_0 holds iff $E_+E_- = E_+E_0E_-$. In this case, and only in this case, we have $\mathcal{H}_0 = L^2(\mu^u \upharpoonright \mathcal{B}(\pi_0))$ and $P(t)$ is a Markov semigroup in $L^2(\mu^u \upharpoonright \mathcal{B}(\pi_0))$. What does this imply in the Gaussian case? For the case of centered ordinary processes, with values in $\mathbb{R}$ (and similar definitions for the relevant σ-algebras), this is discussed in Ref. 58b and d; Osterwalder-Schrader positivity is equivalent with the covariance being a completely monotonic function, i.e., of the form

$\int_0^\infty e^{-\alpha|t|}\, d\nu(\alpha)$ for some probability measure ν on $[0,\infty)$, (global) Markov (with respect to the origin $\{0\}$) corresponds to the covariance being the one, $e^{-\alpha|t|}$, of the Ornstein-Uhlenbeck velocity process. Similarly, one has that Gaussian generalized random fields μ on $S'(\mathbb{R}^d)$ with mean zero and covariance of the form $E_\mu(\langle\xi,f\rangle\langle\xi,g\rangle) = \iint f(x) \int_0^\infty \rho(m^2)(-\Delta + m^2)^{-1}(x - y)g(y)\, dx\, dy$ for some polynomially bounded measure $\rho(m^2)$ s.t. $\int_0^\infty (1/m)\, d\rho(m^2) < \infty$ [this condition making sure that $\langle\xi,\delta_t \otimes \varphi\rangle \in L^2(\mu)$ for all $\varphi \in S(\mathbb{R})$ ("existence of time zero fields")] are Osterwalder-Schrader positive. They are Markov iff $d\rho(m^2) = \delta(m^2 - m_0^2)$ for some m_0, which is exactly the case of the free Markov field.

Let us finally mention that an alternative construction of local relativistic fields starting from lattice models is possible using nonstandard analysis (see Ref. 30h). The semigroup of Osterwalder-Schrader's construction is here replaced by a hyperfinite semigroup [30h].

In the next section we discuss examples of non-Gaussian global Markov fields.

4.4 The Global Markov Property of Homogeneous (Non-Gaussian) Generalized Random Fields

Let u be of the form described in Sec. 4.3 and such that the weak limit of μ^u as $\Lambda \uparrow \mathbb{R}^2$ exists. At least in the case of interactions of forms 1, 2, and 3, and some cases of the interactions of the form 4 in Sec. 4.3, it is known that the limit measure μ^u is locally absolutely continuous with respect to the measure of the free Markov field, in the sense that, for any bounded open subset $\Lambda \subset \mathbb{R}^2$, $d\mu^u \restriction \mathcal{B}(\Lambda)$ is absolutely continuous with respect to $d\mu_0 \restriction \mathcal{B}(\Lambda)$ (see Refs. 53 and 54).

For such μ^u the local Markov property is easily proven.

Namely, let C be a simple closed smooth curve in the interior of Λ and let D be the region enclosed by C. Let f be a $\mathcal{B}(D)$-measurable function, nonnegative or μ_0-integrable. By the Markov property of μ_Λ^u, the properties of conditional expectations and the fact that U_Λ^u is an additive functional, we have

$$E_{\mu_\Lambda^0}(f|C) = \frac{E_{\mu_0}(e^{-U_\Lambda^u} f|C)}{E_{\mu_0}(e^{-U_\Lambda^u}|C)} = E_{\mu_D^u}(f|C)$$

Hence $E_{\mu_\Lambda^u}(f) = \int E_{\mu_D^u}(f|C)\,(d\mu_\Lambda^u \restriction \mathcal{B}(C))$. By taking the weak limit as $\Lambda \uparrow \mathbb{R}^2$ and using the local absolute continuity of μ_Λ^u with respect to μ_0 we get $E_{\mu^u}(f) = \int E_{\mu_D^u}(f|C)\Big(d\mu^u \restriction \mathcal{B}(C)\Big)$, from which it follows that

$$E_{\mu^u}(f|C) = E_{\mu_D^u}(f|C) \tag{4.1}$$

Now let C_0 be a simple closed smooth curve in the interior of D and let Λ_+, Λ_- be any two open sets in the interior of D such that Λ_- is in the interior of C_0 and Λ_+ in its exterior. Let $f_\pm$ be $\mathcal{B}(\Lambda_\pm)$ measurable. Using the Markov property of μ_D^u with respect to C_0 we get

$$E_{\mu_D^u}(f_+ f_-|C_0) = E_{\mu_D^u}(f_+|C_0)E_{\mu_D^u}(f_-|C_0)$$

and hence from (4.1),

$$E_{\mu^u}(f_+ f_-|C_0) = E_{\mu^u}(f_+|C_0)E_{\mu^u}(f_-|C_0)$$

which is the local Markov property of μ^u with respect to C_0. C_0 being eventually arbitrary, we have that μ^u has the local Markov property. In this sense μ^u is a Markov-Gibbs state in the sense of statistical mechanics. But what about the global Markov property? The first observation is that μ^u is a regular random field in the sense of Sec. 4.1. This is due to direct estimates on the second moment of μ^u and has been proven in the models above [56c]. We shall now describe briefly the method that has been used to prove the global Markov property for the regular generalized random fields associated with the trigonometric and exponential interactions.

As we saw in Sec. 4.1, we can compute conditional expectations of the form $E_{\mu_0}(\cdot|\mathcal{B}(\partial U))$ by computing corresponding expectations $E_{\mu_0^{\partial U}}(\cdot)$, where one uses the decomposition $\xi \to \xi^{\partial U} + \xi_{\partial U}$, with $\xi_{\partial U}(x)$ $(\equiv \xi_{P_x^{\partial U}})$ the solution of the Dirichlet problem with boundary condition $\xi \upharpoonright \partial U$ on ∂U. Let us now sketch briefly how this is used to prove the global Markov property. Let Λ be a bounded open subset of $\mathbb{R}^2$ and let C be a smooth curve dividing $\mathbb{R}^2$ - C into two disjoint subsets Ω_+^C, Ω_-^C.

1. Remark first that the equality $E_{\mu_\Lambda^u}(f|\mathcal{B}(\mathbb{R}^d - \Lambda) \cup C) =$ $E_{\mu_\Lambda^u}(f|\mathcal{B}(\partial\Lambda \cup C))$ follows from the fact that μ^u is a weak limit of μ_Λ^u as $\Lambda \uparrow \mathbb{R}^2$ and μ_Λ^u has the global Markov property.
2. It follows that it is enough to prove that $E_{\mu_\Lambda^u}(f|\mathcal{B}(\partial\Lambda \cup C))$ converges as $\Lambda \uparrow \mathbb{R}^2$ to $E_{\mu^u}(f|\mathcal{B}(C))$, because then the global Markov property $E_{\mu^u}(f_+f_-|\mathcal{B}(C)) = E_{\mu^u}(f_+|\mathcal{B}(C))\times$ $E_{\mu^u}(f_-|\mathcal{B}(C))$ for $\mathcal{B}(\Omega_\pm^C)$-measurable $f_\pm \geq 0$ follows from the corresponding property $E_{\mu_\Lambda^u}(f_+f_-|\mathcal{B}(\partial\Lambda \cup C)) =$ $E_{\mu_\Lambda^u}(f_+|\mathcal{B}(\partial\Lambda \cup C))E_{\mu_\Lambda^u}(f_-|\mathcal{B}(\partial\Lambda \cup C))$.
3. Express $E_{\mu_\Lambda^u}(f|\mathcal{B}(\partial\Lambda \cup C))$ by conditional expectations with respect to μ_0, using the definition of μ_Λ, and then by the expectation with respect to $\mu_0^{\partial\Lambda\cup C}$ using the above observation about splitting ξ into $\xi^{\partial U} + \xi_{\partial U}$. This yields that to prove the convergence in 2 it is enough to prove:

a. The convergence of $f(\xi^{\partial\Lambda} + \xi_{\partial\Lambda})$ as $\Lambda \uparrow \mathbb{R}^2$, for μ^u a.e. ξ.

b. The weak convergence [against bounded $\mathcal{B}(\Lambda_0)$-measurable functions, for some compact $\Lambda_0 \subset \mathbb{R}^2$] of $\exp[-U_\Lambda^u(\xi^{\partial\Lambda} + \xi_{\partial\Lambda\cup C})]\mu_0^{\partial\Lambda\cup C}(\xi^{\partial\Lambda\cup C})$, normalized to be a probability measure $\mu^u_{\Lambda,\partial\Lambda\cup C}$, to the conditional probability measure $\mu^u_{\xi_C}$ obtained by conditioning μ^u with respect to the σ-algebra generated by the fields $\langle\xi,\rho\rangle$ with supp $\rho \subset C$ and $\xi \in$ supp μ^u, i.e.,

$$E(f|\mathcal{B}_{\mu^u}(C))(\xi_C) \equiv \int_{S'(\mathbb{R}^2)} f \, d\mu^u_{\xi_C}$$

for $\mu^u \restriction \mathcal{B}_{\mu^u}(C)$ a.e. $\xi_C \in S'(\mathbb{R}^2) \restriction C$.

4. The weak convergence of $\mu^u_{\Lambda,\partial\Lambda\cup C}$ is proven by using two facts: By Theorem 4.2 we have

a. The convergence, locally uniformly in x, of $\xi_{C\cup C_n}$, $C_n \equiv \partial\Lambda_n - C$, $\Lambda_n \uparrow \mathbb{R}^2$ as $n \to \infty$, to $\xi_C(x)$, for $\mu^u \restriction \mathcal{B}_{\mu^u}(C)$ a.e. $\xi \restriction \mathcal{B}_{\mu^u}(C)$, where $\mathcal{B}_{\mu^u}(C)$ is the μ^u-completed σ-algebra associated with C. This yields the convergence in point (3a).

b. All considerations 1 to 4a above hold for all interactions u considered in Sec. 4.1. The remaining step to the global Markov property is the proof of the weak convergence of $\mu^u_{\Lambda_n,\partial\Lambda_n\cup C}$ as $\Lambda_n \uparrow \mathbb{R}^2$, $n \to \infty$. It is at this point that the proof differs for the different choices of u. In the case of trigonometric u, of the form 3 in Sec. 4.3, the basic observation is that one can use an adaptation of the "cluster expansion method." Roughly speaking $\mu^u_{\Lambda_n,\partial\Lambda_n\cup C}$ converges weakly as $\Lambda_n \uparrow \mathbb{R}^2$, since it is $\mu_0^{\partial\Lambda_n\cup C}$ perturbed by a multiplicative functional differing only in a localized way from the one,

$e^{-U_\Lambda(\xi^{\partial\Lambda_n \cup C})}$, that enters the usual interacting measure for a model with Dirichlet boundary conditions on C and trigonometric interaction. We refer to Ref. 56c for details in the case of trigonometric interactions.

Points 1 to 4 above then establish the following.

THEOREM 4.4 The homogeneous non-Gaussian generalized random field μ^0 given on $S'(\mathbb{R}^2)$ by the function $u(r) = \cos(ar + \theta)$ with $|\alpha| < \sqrt{2\pi}$ has the global Markov property with respect to any smooth bounded or unbounded curve, dividing the plane into two disjoint open sets.

REMARK 9. Practically the same method of proof yields a strong uniqueness result for the set of regular Gibbs states associated with the interaction u. Partial results of a similar type had been obtained before for other interactions (see the references in [56c]). We recall briefly that a Gibbs measure μ_G^u associated with an interaction (U_Λ^u) is by definition a probability measure absolutely continuous with respect to the free Markov field measure μ_0 and such that $E_{\mu_G^u}(f|\partial\Lambda) = E_{\mu_\Lambda^u}(f|\partial\Lambda)$, for all positive $\mathcal{B}(\Lambda)$-measurable f. The uniqueness result says that $E_{\mu_\Lambda^u}(f|\partial\Lambda)$ converges as $\Lambda \uparrow \mathbb{R}^2$ μ_0 a.s. to a constant, for all f which are $\mathcal{B}(\Lambda_0)$-measurable for some bounded measurable Λ_0, and this constant is equal to $E_{\mu_G^u}(f)$, for all Gibbs states μ_G^u which are regular in the sense that their covariance is bounded by the one of some free Markov field $\mu_0^{(m)}$. The uniqueness result is equivalent to the statement that the intersection of the convex set of Gibbs states with the regular ones consists of only one point, for the case of the interactions of

Theorem 4.4. The proof of this result relies essentially on Theorem 4.2, giving the local uniform convergence of $\xi_{\partial\Lambda}$ to 0 as $\Lambda \uparrow \mathbb{R}^2$, a.s. with respect to any regular random field measure.

Recently, the corresponding results in the case of exponential interactions (2) in Sec. 4.3 (with supp ρ sufficiently small) have been proven by R. Gielerak [60]. Gielerak's proof uses the above techniques coupled very ingeniously with "ferromagnetic inequalities" of the Fortuin-Kastelyn-Ginibre type. These inequalities permit us to show the weak convergence of the measures $\mu^u_{\Lambda_n, \Lambda_n \cup \partial C}$, as $\Lambda_n \uparrow \mathbb{R}^2$ and their strong clustering in the limit, replacing the role played in the case of trigonometric interactions by the cluster expansion. The result is the following:

THEOREM 4.5 The global Markov property holds in the same sense as in Theorem 4.4 for the exponential interaction μ^u given on $S'(\mathbb{R}^2)$ by the function $u(r) = \int e^{\alpha r}\, d\rho(r)$, with $d\rho(\alpha)$ a bounded positive measure with support on $(0,\alpha^*)$, for some sufficiently small $\alpha^* < \sqrt{4\pi}$. See Ref. 80 for extensions.

4.5 The Canonical Structure Associated with the Global Markov Homogeneous Random Fields

Using any homogeneous regular random field measure μ^u one can construct associated Markov processes in the same way as we did for the free Markov field. In the case where μ^u is global Markov the associated Markov process can be identified with the associated global Markov field in a way specified below. First one observes that since μ^u is regular, it has meaning to consider the σ-algebra $\mathcal{B}_{\mu^u}(C)$ associated with a curve C and the fields restricted to C. More precisely, if $\xi(x)$, $x \in \mathbb{R}^2$, is the random field distributed according to μ^u, $\xi \in S'(\mathbb{R}^2)$, then $\langle \xi, \rho \rangle$ for any measure ρ in $H_{-1}(\mathbb{R}^2)$ with support on C has a meaning as a random variable. In particular, for $C = \{(x^0,x^1) \in \mathbb{R}^2 \mid x^0 = 0\}$ we have with $d\rho = \delta_0(x^0)\varphi(x^1)dx^1, \varphi \in S(\mathbb{R})$: $\langle \xi, \delta_0(\cdot)\rangle \in L^2(S'(\mathbb{R}), d\mu^u \upharpoonright \mathcal{B}(C))$. We shall now write $t = x^0$, $y = x^1$ for the variables x^0, x^1.

Let $Q = S(\mathbb{R})$, $H = L^2(\mathbb{R})$, $Q' = S'(\mathbb{R})$, and $\nu_C \equiv \mu^u \upharpoonright \mathcal{B}(C)$. The structure (Q,H,Q',ν_C) satisfies all the postulates of our general theory of infinite dimensional Dirichlet forms which we recalled in Sec. 2. The verification of the postulates has been done by ourselves in Ref. 7b,e to which we refer. We shall here only use the result that from the structure (Q,H,Q',ν_C) one can construct a Markov semigroup $P_t^{C,F} \equiv e^{-tH_C^F}$, $t \geq 0$, acting in $L^2(\nu_C)$. [From now on we drop the index u on the different quantities; it is understood that $\nu_C = \mu^u \upharpoonright \mathcal{B}(C)$.]

The semigroup is conservative in as much as $P_t^{C,F} 1 = 1$, 1 being the function identically one in $L^2(\nu_C)$. The self-adjoint infinitesimal generator H_C^F associated with the closure of the local regular Dirichlet form

$(H_C^{F\,1/2}f, H_C^{F\,1/2}f) = (1/2) \int (\nabla f)^2 \, d\nu_C$ in $L^2(\nu_C)$, the gradient operator ∇ being defined in the obvious way on C-cylinder functions. We can also write, using variational (Gâteaux) derivatives

$$(H_C^{F\,1/2}f, H_C^{F\,1/2}f) = \frac{1}{2} \int_{S'(\mathbb{R})} \int_{\mathbb{R}} \left[\frac{\delta f(\eta)}{\delta \eta(y)}\right]^2 (y) \, dy \, d\nu_C(\eta)$$

H_C^F is the Friedrichs extension of the operator defined on C^2-cylinder functions f by

$$(H_C^F f)(\eta) = -\frac{1}{2} \int_{\mathbb{R}} \frac{\delta^2 f(\eta)}{\delta \eta(y)^2} \, dy - \int_{\mathbb{R}} \beta(\eta(y)) \frac{\delta f}{\delta \eta(y)} \, dy$$

where $\beta(\eta(y)) = -(1/2)[\delta/\delta\eta(y)]^* 1(\eta)$, i.e., for any $\varphi \in S(\mathbb{R})$, $\beta(\eta(\varphi)) = -(1/2)(\varphi\cdot\nabla)^* 1(\eta)$, $\varphi\cdot\nabla$ being the derivative in the direction φ and $*$ meaning adjoint in $L^2(\nu_C)$. The symmetric (i.e., time reversible) Markov process X_t^C associated with the self-adjoint Markov semigroup $P_t^{C,F}$ and the invariant measure ν_C, taken as the start measure, can be taken to take values in $\mathcal{S}'(\mathbb{R})$ and satisfies as a linear process the stochastic nonlinear differential equation, in the weak sense:

$$dX_t^C = \beta(X_t^C)\, dt + dw_t$$

where w_t is the standard Brownian motion associated with the rigging $\mathcal{S}(\mathbb{R}) \subset L^2(\mathbb{R}) \subset \mathcal{S}'(\mathbb{R})$. In this sense X_t^C is a diffusion process. Continuity properties of the sample paths of this process have been studied, as a diffusion in suitable compactifications of $\mathcal{S}'(\mathbb{R})$ as well as in Banach extensions of $L^2(\mathbb{R})$ [7b,d]. A closely related process has been studied by Kusuoka [8]. The latter is associated with a Dirichlet form which is an extension of the one above and coincides with the one above, e.g., in the case where $\nu_C \equiv \mu_0 \restriction B(C)$, μ_0 being the free Markov field (because in the latter case H_C^F restricted to smooth cylinder functions is already essentially self-adjoint, all self-adjoint extensions and all associated Markov semigroups coincide [7a]). Kusuoka's process runs in a Banach space extension of $L^2(\mathbb{R})$ with continuous sample paths there. Kusuoka's assumptions are satisfied in the present case of quantum fields, as follows from our results in Ref. 7b, the measure ν_C having all properties of a "ground-state measure" associated with the bottom of the spectrum of a uniformly elliptic operator [formally the eigenfunction of $H^{F,C}$ in $L^2(d\eta)$, i.e., formally equal to

$$-\frac{1}{2}\int \frac{\delta^2}{\delta\eta(y)^2}\, dy + m^2 \int \eta(y)^2\, dy + \int u(\eta(y))dy$$

In particular, ν_C is $\mathcal{S}(\mathbb{R})$-quasi invariant and is strictly positive in the sense of Sec. 2. Especially, for any $\varphi \in \mathcal{S}(\mathbb{R})$ there exists a measurable function f_φ on $\mathcal{S}'(\mathbb{R})$ such that $\nu_C = f_\varphi \sigma_\varphi$, ess $\inf_{|t|\leq T} f_\varphi(\xi + t\varphi) > 0$ for any

$\xi \in \mathcal{S}'(\mathbb{R})$, $T > 0$, with

$$\sigma_{\varphi}(E) = \int_{\mathbb{R}} ds \int \chi_E(\xi + s\varphi)\, d\nu_C(\xi)$$

for any Borel subset E of supp ν_C. This implies that ν_C conditioned with respect to any arbitrary finite dimensional subspace has strictly positive (on compacts) density with respect to the corresponding Lebesgue measure. Moreover, ν_C is smooth and even analytic in the sense of Sec. 2, implying that $\nu_C(\xi + z\varphi)$ is analytic in $z \in \mathbf{C}$ in the weak sense.

REMARK 10. It is expected that the semigroup considered by Kusuoka coincides with $P_t^{C,F}$. The analog of this has been proven in finite dimensions by M. Fukushima and N. Wielens (see Sec. 3).

Using ν_C and in particular its quasi-invariance one can construct a "Schrödinger representation of the canonical commutation relations," i.e., a pair of representations U_φ, V_φ of the abelian group $\mathcal{S}(\mathbb{R})$ by

$$(U_\varphi f)(\eta) = e^{i\langle\varphi,\eta\rangle} f(\eta) \qquad \eta \in \mathcal{S}'(\mathbb{R}),\ \varphi \in \mathcal{S}(\mathbb{R}),\ f \in L^2(\nu_C)$$

$$(V_\varphi f)(\eta) = \left(\frac{d\nu_C(\eta + \varphi)}{d\nu_C(\eta)}\right)^{1/2} f(\eta + \varphi)$$

These are strongly continuous [66, 7a,b].

We remark that calling $\pi(\varphi)$ the infinitesimal generator of the unitary group $t \to V_{t\varphi}$ for any fixed $\varphi, \psi \in \mathcal{S}(\mathbb{R})$, one has $[\pi(\varphi), \langle\psi,\eta\rangle] \subset -i(\varphi,\psi)]$ on a dense subset, where $[\cdot,\cdot]$ means commutator. Moreover, $i\pi(\varphi)1(\eta) = \beta(\eta(\varphi))$. From ν_C one can also construct an infinitesimal representation of the Poincaré group in $L^2(\nu_C)$, the Lorentz generator Λ being given on C^2 cylinder functions by

$$(\Lambda f)(\eta) = -\frac{1}{2}\int_{\mathbb{R}} y \frac{\delta^2}{\delta\eta(y)^2}\, dy - \int_{\mathbb{R}} y\beta(\eta(y)) \frac{\delta}{\delta\eta(y)}\, dy \quad [7e]$$

REMARK 11. It is not known whether in general this infinitesimal representation extends to one of the group. This is true in the free Markov field case [7a,b] and it is expected in the case of measures ν_C coming from global Markov ones, like the ones covered by Theorems 4.4 and 4.5. In such cases one then obtains a local relativistic quantum field theory.

We shall now discuss the case where $\nu_C = \mu^u \upharpoonright B(C)$, with μ^u global Markov. Also in this case one obtains a local relativistic quantum field theory, with Hilbert space $L^2(\nu_C)$. It is identifiable with the one in the remark above in the case where $H^{C,F}$ is essentially self-adjoint on C^2-cylinder functions in $L^2(\nu_C)$. In fact, if μ^u is global Markov we can introduce directly the associated Markov semigroup $E_C P_t E_C = U(t)$ described above, $U(t)$ being the shift operator on the paths $t \to \xi(t,y)$ of the euclidean random field associated with μ^u, and E_C being conditional expectation with respect to $B(C)$. Realizing ν_C as a measure on a suitable compactification of $S'(\mathbb{R})$ it is possible to look at P_t as a symmetric Markov semigroup in the sense of the theory of Dirichlet forms; hence there is an associated Dirichlet form $(H^{1/2}f, H^{1/2}f)$, where H is the infinitesimal generator of P_t. The form coincides on C^1-cylinder functions with the one given by $H^{C,F}$. Correspondingly, H and $H^{C,F}$ are two possibly different extensions of their restriction to C^2-cylinder functions.

Conjecture. $H = H^{C,F}$. The analogous statement for quantum mechanics has been proven [6a, 37]. Also, the statement is proven in the case $u \equiv 0$ (free Markov field) [7a,b]. Using (P_t, ν_C) and the associated Dirichlet form one can construct, independently of the conjecture, an associated symmetric Markov diffusion process. By construction it can be identified with the $S'(\mathbb{R})$-valued process $t \to \xi(t,\varphi)$, $t \in \mathbb{R}$, $\varphi \in S(\mathbb{R})$, where $\xi(t,y)$ is the (generalized) homogeneous global Markov field whose underlying probability measure is

μ. It is natural then to call $\xi(t,y)$, $t \in \mathbb{R}$, $y \in \mathbb{R}$, a <u>diffusion field</u>. To the canonical structure $(L^2(\nu_C), \xi(0,\varphi), \pi(\varphi), H)$ there is then associated, by techniques of Nelson, a structure of local relativistic quantum fields satisfying the Wightman axiom. Physically, the global Markov property, besides yielding the Markov semigroup P_t, implies that polynomials in the time-zero quantum fields generate the whole Hilbert space. The equation of motion can be written as

$$i[\pi(\varphi),H]f(\xi_0) = \langle\varphi, :u':(\xi_0)\rangle + \left\langle\left(-\frac{d^2}{dx^2} + m^2\right)\varphi, \xi_0\right\rangle$$

on C^2-cylinder functions, with $\xi_0(\varphi) \equiv \xi(0,\varphi)$. The above canonical structure with the stated properties realizes for these two-dimensional models all postulates of the most ambitious and natural program of field quantization [10, 6c,p, 7]. Probabilistically, what has been achieved is the construction of examples of homogeneous global Markov random fields over $\mathbb{R}^2$ of diffusion type. This is a first step toward carrying over to "higher time dimensions" the theory of stochastic (partial) differential equations. It also gives a partial answer to questions raised from the other points of view mentioned at the beginning of this section. Obviously, a serious limitation is the two-dimensionality. A challenging problem for the future is the study of similar constructions in the higher-dimensional case ($d > 2$).

For $d = 3$ the construction of some relativistic models has been achieved (see the references in [12b]); however, a probabilistic interpretation similar to the one discussed here has not been carried through (although construction of the models themselves at various stages uses refined probabilistic techniques). A more suitable concept of interaction and additive functionals seems to be required. We have mentioned some possibility in connection with polymer models (Sec. 3.6). It seems reasonable to hope that a whole new rich world of probability is to be found through the attempts to construct homogeneous random fields over $\mathbb{R}^4$.

5. DIFFUSION FIELDS ON MANIFOLDS

Interest in an extension of the concept of Markov fields of diffusion type to the case where the underlying space $\mathbb{R}^d$ is replaced by a Riemannian manifold X arises in several contexts, e.g., in connection with quantum fields in curved space-time (see, e.g., Ref. 22), and as we shall discuss in the next section, in connection with representations of groups of mappings. In Refs. 56a and 7c we introduced the concept of the free Markov field on a manifold. Let us describe briefly its definition and properties. Let $\mathcal{D}(X)$ be the C^∞ functions of compact support on X and $\mathcal{D}'(X)$ be its dual. The Dirichlet form $(\varphi,\varphi)_D \equiv \int_X (d\varphi(x),d\varphi(x))dx$ is defined on $\mathcal{D}(X)$, with dx the volume measure on X, and is regular in the sense of Fukushima [4a]. $(d\varphi(x),d\varphi(x))$ is the 1-form in the cotangent space at $x \in X$ given by the Riemann structure. We assume that the standard Brownian motion associated with the regular Dirichlet form $(\varphi,\varphi)_D$ on $L^2(X,dx)$ is transient, or equivalently, that the completion $H_D(X)$ of $\mathcal{D}(X)$ with respect to the Dirichlet norm $|\varphi|_D \equiv (\varphi,\varphi)_D^{1/2}$ is a Hilbert space, i.e., that the natural extension of $(\varphi,\varphi)_D$ to $H_D(X)$ is nondegenerate. In such a case we say simply that X is transient.

REMARK 1. X is transient, e.g., if it is R_+ or an open domain of $\mathbb{R}^d$, except for $X = \mathbb{R}, \mathbb{R}^2$. In the case $X = \mathbb{R}, \mathbb{R}^2$, or X a compact set, one has recurrence.

We shall assume from now on that X is transient. We define the standard diffusion field or Brownian field on X as the (generalized field associated with) the measure μ_0 on $\mathcal{D}'(X)$ such that

$$\int_{\mathcal{D}'(X)} e^{i\int (d\xi(x),d\varphi(x))dx}\, d\mu_0(\xi) \equiv e^{-(1/2)(\varphi,\varphi)_D}$$

Hence μ_0 is a realization in $\mathcal{D}'(X)$ of the standard normal distribution associated with $H_D(X)$. When $X = \mathbb{R}^d$, $d \geq 3$, then μ_0 can be identified with the free Markov field μ_0^0 of mass zero. In fact, the latter has, according to the definition in Sec. 3, the property that

$$\int_{S'(\mathbb{R}^d)} e^{i\langle \xi,\psi\rangle}\, d\mu_0^0(\xi) = e^{-(1/2)(\psi,G\psi)}$$

with $(\psi,G\psi) \equiv \iint \psi(x)G(x - y)\psi(y)\, dx\, dy$ for $\psi \in \mathcal{D}(\mathbb{R}^d)$, $G = (-\Delta)^{-1}$. Since by the transiency the range of $-\Delta$ is dense in $H_{-1}(\mathbb{R}^d)$ we can set $\psi = -\Delta\varphi$ for $\varphi \in \mathcal{D}(\mathbb{R}^d)$ and we get

$$\int_{S'(\mathbb{R}^d)} e^{i(\xi,\varphi)_D}\, d\mu_0^0(\xi) = e^{-(1/2)(\varphi,\varphi)_D}$$

where we used $(\xi,\varphi)_D = \langle \nabla\xi,\nabla\varphi\rangle = \langle \xi,\Delta\varphi\rangle$ in the distributional sense. This shows then that $\mu_0 = \mu_0^0$ in the case $X = \mathbb{R}^d$, $d \geq 3$.

We now want to define σ-algebras associated with subsets of X. We first describe the situation for $X = \mathbb{R}^d$. As shown in Ref. 50, the σ-algebra of μ_0^0-measurable sets is generated by the linear functions $\langle \xi,\rho\rangle$, where ρ runs over the bounded measures of bounded support and finite energy, since the linear span of the latter is dense in the Hilbert space $H_G(X)$ of distributions T for which $\iint T(x)T(y)G(x - y)\, dx\, dy < \infty$, where $G(x - y)$ is the fundamental solution to $-\Delta$, i.e., the kernel of $(-\Delta)^{-1}$, Δ being the self-adjoint Laplacian in $L^2(\mathbb{R}^d)$, $d \geq 3$. $H_G(X)$ is the dual space to $H_D(X)$.

For any set $\Lambda \subset X$ we define $B(\Lambda)$ as the smallest σ-algebra containing $B(K)$ for any compact $K \subset \Lambda$, where $B(K)$ is the σ-algebra generated by the linear functions $\langle \xi,\rho\rangle$, with ρ a positive measure in H_G and supported by K. In the realization of μ_0^0 as the standard normal distribution associated with $H_D(X)$ we have that $B(K)$ is also the σ-algebra generated by the linear functions $(\xi,\varphi)_D$ where φ is in the closed subspace H_D^K of functions in H_D which are orthogonal in the sense of H_D to all functions in $\mathcal{D}(X)$ with support in $X - K$. The latter definition can be extended to arbitrary transient Riemannian manifolds X, so that we define B(K) for any compact $K \subset X$ as the σ-algebra of μ_0-measurable subsets (μ_0 being the Brownian field measure) of $\mathcal{D}'(X)$ generated by the linear functions $(\xi,\varphi)_D$ with $\varphi \in H_D^K(X)$. Correspondingly, we define $B(\Lambda)$ for arbitrary Λ as above. In the same way that one proves the global Markov property of the free Markov field on $\mathbb{R}^d$ one proves the one for the Brownian field μ_0 on a transient manifold X.

REMARK 2. For $X = \mathbb{R}_+$ the standard Brownian field on X is just the Brownian motion on X, and the global Markov property is here just the Markov property of the Brownian motion.[12]

We now introduce the concept of exponential interaction on X. Let Δ be the Laplacian on $L^2(X,dx)$, i.e., the self-adjoint operator on $L^2(X,dx)$ associated with the Dirichlet form $(\varphi,\varphi)_D$, so that $e^{t\Delta/2}$ is the Markov semigroup associated with the standard Brownian motion on X. Then by transience the inverse G of $-\Delta/2$ exists and is given by $G = \int_0^\infty e^{t\Delta/2}\, dt$. Its kernel G(x,y), the potential kernel for the Brownian motion, is continuous for $x \neq y$, and one has $G(x,\cdot)$, $G(\cdot,y) \in L^1_{loc}(X,dx)$, so that $G(x,\cdot)$, $G(\cdot,y)$ can be looked upon as in $\mathcal{D}'(X)$. Hence for any $\alpha \in \mathbb{R}$ we have that $d\mu_0(\cdot - \alpha G_x)$, with $G_x(y) = G(x,y)$, is still a measure on $\mathcal{D}'(X)$ which depends continuously on $x \in X$. Hence for any Borel subset $\Lambda \subset X$ and any $g \geq 0$, $g \in L^1_{loc}(X,dx)$ such that $\int_\Lambda g(x)\, dx < \infty$ we have that

$$\mu_\Lambda(\xi) \equiv \int_\Lambda \mu_0(\xi - \alpha G_x) g(x)\, dx$$

is a positive measure on $\mathcal{D}'(X)$, such that its total variation $\|\mu_\Lambda\|$ satisfies $\|\mu_\Lambda\| \leq \int_\Lambda g(x)\, dx$. $\Lambda \to \mu_\Lambda$ is positive and σ-additive in Λ; i.e., μ_Λ is a measure-valued measure on X which is absolutely continuous with respect to dx. By a similar method used for exponential interactions [61] we prove that for dim X = 2, $|\alpha| < \sqrt{4\pi}$, $d\mu_\Lambda$ is absolutely continuous with respect to $d\mu_0$; hence $d\mu_\Lambda = a_\Lambda d\mu_0$, where $a_\Lambda(\xi)$ is a nonnegative, μ_0-measurable function. We shall now see that a_Λ is $B(\Lambda)$ measurable. We have that if $\Lambda = \Lambda_1 \cup \Lambda_2$, $\Lambda_1 \cap \Lambda_2 = \emptyset$, then $a_\Lambda(\xi) = a_{\Lambda_1}(\xi) + a_{\Lambda_2}(\xi)$ for μ_0 a.e. $\xi \in \mathcal{D}'(X)$. Moreover, if Λ is compact, then $a_\Lambda \in L^1(d\mu_0)$. $H_D(X - K)$ being the closure in $H_D(X)$ of $\mathcal{D}(X - K)$, we have $H_D(X) = H_D^K(X) \oplus H_D(X - K)$; hence the standard normal distribution μ_0 associated with $H_D(X)$ splits in a natural way as a product measure $\mu_0 = \mu_{H_D^K(X)} \times \mu_{H_D(\tilde{K})}$, where $\tilde{K} \equiv X - K$ and $\mu_{H_D^K(X)}$ [respectively, $\mu_{H_D(\tilde{K})}$] are the standard normal distributions associated with $H_D^K(X)$ and $H_D(X - K)$, respectively. Now for $\varphi \in \mathcal{D}(X)$ we have

$$\int e^{i(\xi,\varphi)_D} d\mu_\Lambda(\xi) = e^{-(1/2)(\varphi,\varphi)_D} \int e^{i\alpha(G_x,\varphi)_D} g(x)\,dx$$

$$= e^{-(1/2)(\varphi,\varphi)_D} \int e^{i\alpha\varphi(x)} g(x)\,dx$$

and if supp $\varphi \subset X - \Lambda$ this is equal to

$$e^{-(1/2)(\varphi,\varphi)_D} \int g(x)\,dx = e^{-(1/2)(\varphi,\varphi)_D} \|\mu_\Lambda\|;$$ hence for $\Lambda = K$ and any $\varphi \in \mathcal{D}(X - K)$

$$\int e^{i(\xi,\varphi)_D} a_k(\xi)\,d\mu_0(\xi) = e^{-(1/2)(\varphi,\varphi)_D} \int a_k d\mu_0$$

$$= \int e^{i(\xi_2,\varphi)_D} d\mu_{H_D(\tilde{K})}(\xi_2)$$

where we have used the above splitting of μ_0. But $\mathcal{D}(X - K)$ is dense in $H_D(X - K)$ and from this we conclude that $E(a_k|B(K)) = a_k$, hence a_k is $B(K)$ measurable. From this we conclude that for any measurable subset Λ of X, a_Λ is $B(\Lambda)$ measurable. From $\|a_\Lambda\|_{L^1(\mu_0)} \leq |\Lambda|$ we get that $\Lambda \to a_\Lambda$ is positive monotone. Analogously as in Sec. 4, we call a family (Λ, U_Λ) with U_Λ-$B(\Lambda)$ measurable and in $L^1(d\mu_0)$ and such that $U_{\Lambda_1 \cup \Lambda_2} = U_{\Lambda_1} + U_{\Lambda_2}$ when $\Lambda_1 \cap \Lambda_2 = \emptyset$, an additive functional of the standard Brownian field μ_0 on X. We have proven the following theorem (which first appeared in [56a]):

THEOREM 5.1 The family (Λ, a_Λ^g), Λ of finite volume, with $g \geq 0$, $g \in L^1(\Lambda, dx)$, $a_\Lambda^g \equiv d\mu_\Lambda^g / d\mu_0$, $d\mu_\Lambda^g \equiv \int_\Lambda \mu_0(\cdot - \alpha G_x) g(x)\, dx$, is for dim X = 2, X transient, $|\alpha| < 2\sqrt{\pi}$ a local additive functional of the standard Brownian field μ_0 on the Riemann manifold X. $\Lambda \to a_\Lambda^g$ is positive monotone and such that $\|a_\Lambda^g\|_{L^1(\mu_0)} \leq \int g(x)\, dx$.

There is an analogy to this construction for the case where X is replaced by $\mathbb{R}^2$ and μ_0 is replaced by the free Markov field of mass $m > 0$. In this case we have that $\int_\Lambda \mu_0(\cdot - \alpha G_x)\, dx / d\mu_0(\cdot)$ is the exponential interaction $U_\Lambda^{(\alpha)}(\xi) \equiv \int_\Lambda : e^{\alpha\xi} :(x)\, dx$ in the region Λ, as easily shown by the fact that μ_0 is a Gaussian formally given by a properly normalized measure $\exp[-(1/2)\int \nabla\xi(x)^2\, dx - m^2 \int \xi(x)^2\, dx]$. This is a reason for calling the above family $(\Lambda, U_\Lambda^{(\alpha)})$ with $U_\Lambda^{(\alpha)} = a_\Lambda^1$ the <u>exponential interaction</u> (of parameter α) for the Riemann manifold X.

Let $\rho = \Sigma c_i \delta_{\alpha_i}$ be any discrete measure on $\mathbb{R}$ and define $m_\Lambda^\lambda \equiv \int \exp(-\lambda U_\Lambda^{(\alpha)})\, d\rho(\alpha)$, $\lambda \geq 0$. $\Lambda \to U_\Lambda^{(\alpha)}$ is a positive additive functional of the standard Brownian field μ_0 on X; hence m_Λ^λ is a bounded positive multiplicative functional in the corresponding sense. Since μ_0 is a global Markov field, so is $\mu_\Lambda^\lambda \equiv m_\Lambda^\lambda \mu_0 / \|m_\Lambda^\lambda\|_{L^1(\mu_0)}$.

Let us now consider the positive definite function $f_\lambda(\varphi) \equiv \int \exp[i(\xi,\varphi)_D] m_\Lambda^\lambda(\xi)\, d\mu_0(\xi)$. We have

$$\frac{1}{\lambda}[f_0(\varphi) - f_\lambda(\varphi)] = \int e^{i(\xi,\varphi)_D} U_\Lambda^{(\alpha)}(\xi) [\frac{1}{\lambda} \int_0^\lambda m_\Lambda^\tau(\xi)\, d\tau]\, d\mu_0(\xi)$$

Now since $m_\Lambda^\lambda \geq 0$ we have $(1/\lambda) \int_0^\lambda m_\Lambda^\tau \, d\tau \leq 1$. Hence by Lebesgue's lemma on dominated convergence we get that the limit of $(1/\lambda)[f_0(\varphi) - f_\lambda(\varphi)]$ as $\lambda \downarrow 0$ exists and is given by

$$f'(\varphi) = -\int e^{i(\xi,\varphi)_D} m_\Lambda'(\xi)|_{\lambda=1} \, d\mu_0(\xi) = e^{-(1/2)(\varphi,\varphi)_D} \times \int \int_\Lambda e^{i\alpha\varphi(x)} \, d\rho(\alpha) \, dx$$

From this it follows that μ_Λ^λ is weakly differentiable at $\lambda = 0$ with derivative at zero given by the value at $\lambda = 0$ of $-[E_{\mu_0}(m_\Lambda') - m_\Lambda'(\xi)] \, d\mu_0(\xi)$. In particular, this implies that $d\mu_\Lambda(\xi)$ is non-Gaussian for $\lambda > 0$, $\rho \not\equiv 0$.

6. NONCOMMUTATIVE DIFFUSION FIELDS AND THE REPRESENTATION OF CURRENT GROUPS

Concepts related in one way or the other with extensions of the concept of (random) fields to (random) fields with values in a noncommutative group arise in several connections, e.g., in Gelfand's approach to noncommutative distributions (see, e.g., Ref. 24), in the theory of representation of current groups (see, e.g., Ref. 67), in the theory of gauge groups and fields (see, e.g., Ref. 25), and in an extension to multidimensional parameter of the theory of affine graded Lie algebras (see, e.g., Ref. 26).

We shall discuss here mainly an approach [23, 70] which leads to a construction of group-valued "diffusion fields" which are an extension of the (commutative) global Markov generalized random fields associated with Dirichlet diffusion forms in the sense of Sec. 3. In order to do so we recall the "algebraic side" of such global Markov fields.

Let μ^u be one of the global Markov measures of Sec. 4, e.g., the free Markov field. We can consider $L^2(\mu^u)$ as a representation space for the representation U of the Abelian group $S(\mathbb{R}^d)$ "by multiplication." The characters are given by $e^{i\langle \xi,\varphi\rangle}$, $\xi \in S'(\mathbb{R}^d)$, $\varphi \in S(\mathbb{R}^d)$. μ^u being $S(\mathbb{R}^d)$ quasi-variant, there is another unitary strong continuous representation V_φ of $S(\mathbb{R}^d)$ "by translations," namely

$$(V_\varphi f)(\xi) \equiv \left(\frac{d\mu^u(\xi + \varphi)}{d\mu^u(\xi)}\right)^{1/2} f(\xi + \varphi)$$

V_φ and U_φ are unitarily equivalent. We shall now seek non-commutative analogs of V_φ, U_φ.

Let X be a Riemann manifold, playing the role of $\mathbb{R}^d$, and G a compact semisimple Lie group, playing the role of the state space of the commutative processes and fields considered before. Let $\Omega \equiv \Omega(TX,g)$ be the smooth maps from the tangent bundle TX into the Lie algebra g, the maps being linear on each fiber and with compact support. We equip g with the euclidean structure given by the negative of the Killing form (here the assumption that G is semisimple compact enters of course). We equip the tangent space T_xX at $x \in X$ with the euclidean structure given by the Riemannian metric. Let ρ be a strictly positive smooth density with respect to the volume measure dx on X. Then $\langle \omega_1,\omega_2\rangle \equiv \int \mathrm{Tr}(\omega_1(x)\omega_2(x)^*)\rho(x)\, dx$, Tr being the trace in g and $*$ the adjoint mapping, is a scalar product on Ω making Ω into a pre-Hilbert space, and let H be the complex Hilbert space generated by Ω.

Let $C_0^\infty(X;G)$ be the group of C^∞ mappings from X into G of compact support in the sense that they are identically equal to the unit in G outside some individual compact subset of X. The group operation is defined as pointwise multiplication of the mappings. Using the Killing form the group $C_0^\infty(X;G)$ can be made into a metric group, called a Sobolev-Lie group $H_1(X;G)$. The metric is given by the distance $\delta(\varphi,\psi) \equiv \langle d(\varphi^{-1}\psi), d(\varphi^{-1}\psi)\rangle$ [see Ref. 23 for more details;

see also Ref. 68 for the structure of $H_1(X;G)$ as an infinite dimensional Lie group]. In the following we shall denote the group $H_1(X;G)$ simply by G^X.

A natural pointwise unitary representation V of G^X in H is given, for $\psi \in G^X$, $\omega \in H$, by

$$V(\psi)\omega(x) \equiv \mathrm{Ad}\ \psi(x)\omega(x)$$

where $\mathrm{Ad}\ \psi(x)\alpha \equiv \psi(x)\ (\exp \alpha)\psi(x)^{-1}$ for all $\alpha \in g$, exp being the exponential mapping. The unitarity of the representation is due to the invariance of the Killing form and hence of the scalar product in H under the adjoint representation. Let β be the Maurer-Cartan form on G, so that $\beta(h) \equiv dh\ h^{-1}$ for all $h \in G$. It is easily verified that $\psi(x) \to \beta(\psi(x))$ is a 1-cocycle for the group G^X and the representation V, in the sense that $\beta(\psi_1(x)\psi_2(x)) = \beta(\psi_1(x)) + V(\psi_1(x))\beta(\psi_2(x))$.

This cocycle is called the Maurer-Cartan cocycle [74b]. By a general procedure of Araki, Streater, Parthasarathy-Schmidt, Gelfand, Vershik and Graev, and Guichardet [69], given in general a unitary representation V and a 1-cocycle one can construct a new unitary representation U, the "exponential representation." As we shall see in our case, with our Hilbert space and the Maurer-Cartan cocycle, this representation is the extension of the representations U, V of $S(\mathbb{R})$ discussed above in the commutative case. It can be defined as a representation in a L^2-space, by introducing the canonical Gaussian measure associated with the real part of H. In this presentation U is the representation induced in $L^2(\mu)$ by the "translation" $\beta(\psi)$ and the "rotation" $V(\psi)$ in H, i.e., $(U(\psi)f)(\omega') \equiv \exp\ [i\langle\beta(\psi),\omega'\rangle]f(V^{-1}(\psi)\omega')$, $\omega' \in \mathrm{supp}\ \mu$, $f \in L^2(\mu)$, $\langle\cdot,\cdot\rangle$ being the natural dualization. The unitary representation U of G^X is called the energy representation of G^X. It has been studied in Refs. 74, 23, and 70 (see also Refs. 26 and 75 for some additional connections). $U(\psi)$ is a noncommutative realization of the corresponding commutative representation by multiplication, discussed above. In fact, the representation space $L^2(\mu)$ is obtained by the Gaussian measure μ associated with the Hilbert space H which corresponds to the space $H_1(\mathbb{R}^d; \mathbb{R})$ of the commutative case, hence to the free Markov field measure. Moreover, the positive definite function $\psi \to (1,U(\psi)1)$ characterizing the representation is given by [1 being the function identically one in $L^2(\mu)$]

$$(1,U(\psi)1) = \exp\left(-\frac{1}{2}\langle\beta(\psi),\beta(\psi)\rangle\right)$$
$$= \exp\left(-\frac{1}{2}\int \mathrm{Tr}(d\psi(x)\, d\psi(x)^*)\rho(x)\, dx\right.$$

which corresponds to the one $e^{-(1/2)(\psi,\psi)_D} = e^{-(1/2)(\psi,\psi)_1}$ of the commutative case (see Sec. 4). Inasmuch as $U(\psi)$ in the commutative case "yields the fields" we can look upon the present $U(\psi)$ as a realization of the concept of random field with values in a group. We shall see below that this is indeed an extension to multidimensional time and to a manifold of the Brownian motion on a Lie group.

First we want to point out a kind of minimality of the representation [a noncommutative property corresponding to the view of distributions as given by characters of irreducible representation of Abelian groups of mappings, again an analogy between a property of the random field $\langle\xi,\varphi\rangle$, $\varphi \in C_0^\infty(\mathbb{R}^d)$ and $U(\varphi)$]. Let $d = \dim X$. Ismagilov proved in Ref. 74a the irreducibility of the energy representation for $d \geq 5$. This was extended by Vershik, Gelfand, and Graev [74b] to $d \geq 4$ and by D. Testard and ourselves to $d \geq 3$ [70a]. Moreover, in the latter work it was shown that one has irreducibility for $d = 2$ if the lengths $|\alpha|$ of all roots satisfy $|\alpha| > \sup_x 4\sqrt{2\pi\rho(x)}$.

The method of proof of these results for $d \geq 2$ uses algebraic techniques of Ref. 70, together with the basic result on singularity of translates of Gaussian measures that we discussed in connection with quantum fields with exponential interaction in Sec. 4.3 (Lemma 4.1). The point is to use a maximal torus T in G and to look at the representation of T^X. The spectral measure for this representation is $\nu_0 * \nu$, where ν_0 is the Gaussian measure on the dual of t^X (t being the Cartan subalgebra of g corresponding to T) with Fourier transform $\exp[-(1/2)(dA,dA)]$, $A \in t^X$ (thus a Brownian field measure) and ν is the canonical Poisson measure with support on the space of all $x \in (t^X)'$ of the form $\langle x,A\rangle = \Sigma\, \alpha_j A(x_j)$, α_j being roots of g and x_j points in X.

Results on the support of convolutions of this form are given in Refs. 63, 70a, and 73. In particular it was shown that for $d \geq 3$, or $d = 2$ and α_j sufficiently large, $\nu_0 * \nu$ is singular with respect to ν_0. This then implies that the commutant of $U(\exp t^X)$ is in the set of decomposable operators of the integral representation obtained from the above decomposition, which then has as a consequence the irreducibility of the representation U in the cyclic component of 1 ("the vacuum"). Moreover, in the decomposition $U(\exp A) = W(dA) \times \exp(V(\exp A))$ with W, V acting in the Hilbert spaces for the energy representation of T^X [respectively, $(T^\perp)^X$], one has that W(dA) is in the von Neumann algebra generated by $U(G^X)$. This implies the cyclicity of the vacuum and thus, by the already proven irreducibility in the cyclic component of the vacuum, the irreducibility of the representation U.

This proof of the irreducibility of the energy representation is only one case of the interesting relations between commutative Markov fields, like the Brownian field given by ν_0, and the noncommutative ones.

The case where $\dim X = 1$, $X = \mathbb{R}$ (S^1, $\mathbb{R}$, $[0,1]$) has also been studied using a combination of algebraic, functional analytic, and probabilistic techniques (diffusion processes on manifolds) [70]. In this case to the representation U can also be given a realization in terms of the Brownian motion on a Lie groups, exhibiting in this way the "Markovian nature" of the representation. Let $\eta(t)$ be the standard Brownian motion on the semisimple compact Lie group G [defined as the nonanticipating solution of the stochastic equation $d\eta(t)\eta^{-1}(t) = dw(t)$, $\eta(0)$ being the unit in G and w the standard Brownian motion on the Lie algebra g]. Let $d\mu$ be the standard Wiener measure on C(X;G), i.e., the image of the standard Wiener measure on C(X,g) by the mapping $w \to \eta$. The following result has recently been proven.

THEOREM 6.1 [70a,b] For $X = \mathbb{R}, S^1, \mathbb{R}_+, [0,1]$, there are unitary representations U^R (respectively, U^L) in $L^2(\mu)$ of G^X by right (respectively, left) translations given by

$$(U^R(\psi)f)(\eta) \equiv \frac{d\mu(\eta\psi)}{d\mu(\eta)}^{1/2} f(\eta\psi)$$

$$(U^L(\psi)f)(\eta) \equiv \frac{d\mu(\eta^{-1}\psi)}{d\mu(\eta)}^{1/2} f(\psi^{-1}\eta)$$

These representations are unitarily equivalent to the cyclic component of the vacuum of the energy representation, the unitary equivalence being induced by $\psi \in G^X \to d\psi(t)\psi^{-1}(t)$ [respectively, $\psi^{-1}(t)\, d\psi(t)$]. U^L and U^R commute; in particular, the energy representation is reducible. For any torus $T \subset G$ one has the equivalence of U^L to a representation in $L^2(C(X,G/T),\mu_1) \otimes L^2(C(X,T),\mu_T)$, with $\mu = \mu_1 \times \mu_T$, μ_1 the Brownian measure on C(X;G/T). For G = SU(2) one has then a direct decomposition of U^L into irreducible components, μ_T a.s., U^L, U^R are factor representations of type III, at least for $X = \mathbb{R}$.

REMARK 1. The energy representation, in the present one-dimensional case, is associated naturally to the Brownian motion on the Lie group in much the same way as the representation of the free Markov field in one dimension is expressed in terms of the Brownian motion on $\mathbb{R}$ (μ_0 being in this case the measure of Brownian motion killed with constant exponential rate, given by the mass). In this sense, then, the energy representation, as a general concept for arbitrary Riemannian manifolds X and compact semisimple Lie groups, extends the concept of (global) Markov fields to fields with values on a group.

REMARK 2. The energy representation of the group of mappings G^X has, at least in the case dim X = 1, important connections with the study of graded Lie algebras. Does the extension to dim X > 1 bring corresponding results? A study of these questions is now being planned.

Another important connection of the energy representation is the one with representations of "current groups." In fact, it can be looked upon as a representation of the gauge group of mappings G^X associated with a gauge field. The study of classical and quantum gauge field has received great attention in recent years (see, e.g., Ref. 25). In particular, B. Gaveau and P. Trauber and Asorey and Mitter have studied space-regularized versions (i.e., regularized in d - 1 directions) of gauge fields on $\mathbb{R}^d$ as diffusion processes with values in the Lie algebra of a group. Much remains to be done to put all these objects and studies in an organic text. But certainly it can be said that here too the attempt to understand and construct complicated quantum mechanical and quantum field theoretical models has opened up entire new areas for the theory of Markov processes and fields. And vice versa, the latter theories have greatly influenced the shaping of the new theoretical concepts for the description of the quantum world.

ACKNOWLEDGMENTS

The first author is very grateful to Professor Mark Pinsky for his kind invitation to contribute to this book and for his patience with the practical difficulties we encountered in getting our manuscript ready for publication. It is a pleasure to thank Prof. P. Blanchard, Dr. A. Boukricha, J. Brasche, Dr. W. Kirsch, Dr. F. Martinelli, Prof. M. Pinsky, M. Röckner, K. Rullkötter, U. Spönemann, and N. Wielens for pointing out a number of misprints and corrections in the first draft of this article. Part of this work is an outgrowth of lectures held by the first author at the Cours de III Cycle de la Suisse Romande, Mathématiques/Physique, in Lausanne 1977, and the first author would like to thank Professors R. Cairoli, S. Chatterji, and Ph. Choquard for the kind invitation to hold that course. We are grateful to Professors M. Mebkhout and L. Streit for the great hospitality at the Faculté des Sciences, Université d'Aix-Marseille II, Luminy, and the Physics Department and the Zentrum für interdisziplinäre Forschung of the University of Bielefeld. Our work was also greatly facilitated by common stays at the Mathematics Departments of the Universities of Bielefeld, Bochum, and Oslo, at the Academy of Sciences in Moscow, and at the Centre de Physique Théorique, CNRS, Marseille. Financial support of the Deutsche Forschungsgemeinschaft, of research projects at Bielefeld and Bochum, and of the Norwegian Research Council for Science and the Humanities (NAVF, Program "Mathematical Seminar Oslo") is also gratefully acknowledged. We thank Mrs. L. Mischke and Mrs. B. Richter for their understanding and skill with a difficult manuscript.

NOTES

1. To keep the references within a reasonable length it is understood that "see X" means "see X and references

therein." Often, when we have the choice and we estimate that historical points are not essential in the given context, we prefer to give a newer reference containing older work of a given author rather than to list a series of references by the same author. We apologize for any omissions that may have occurred.

2. In applications to quantum mechanics the reality condition is usually tacitly dropped; however, it should be clear from the context which L^2-space is understood; in any case, the complex extension of the results needed is immediate.

3. We write "a.e." whenever there can be some doubt as to interpretation; otherwise, we drop writing "a.e.".

4. This conforms to terminology in the theory of Dirichlet forms. Actually, T-Markov is here what sometimes is understood as sub-Markov, in the sense that one does not necessarily have $T1 = 1$ (if T is Markov in our sense and satisfies $T1 = 1$, we say that T is Markov and conservative).

5. Let $\|\cdot\|$ be the norm in the Hilbert space. E is closed iff D(E) is complete in the norm $\|f\|_E \equiv [E(f) + \|f\|^2]^{1/2}$.

6. Or, in fact, any other function N mapping $\mathbb{R}$ into $[0,1]$ with $N(x) = x$ for $x \in [0,1]$.

7. That is, the closure of $E \upharpoonright FC^\infty$ is E, where $\upharpoonright$ means restriction.

8. $\text{ad } A(B) \equiv AB - BA$, $\text{ad}^n A(B) \equiv \text{ad ad}^{n-1} A(B)$.

9. $\nabla f \equiv \left\{\frac{\partial f}{\partial x_i}; i = 1,\ldots,d\right\}$ $\nabla f \cdot \nabla g \equiv \Sigma \frac{\partial f}{\partial x_i} \frac{\partial g}{\partial x_i}$

10. This has recently been improved by Brasche [38] to cover the situation where $\mathbb{R}^d$ is replaced by an arbitrary open subset Ω and $\rho > 0$ locally uniformly only outside some open zero measure set. That ρ, $\rho^{-1} \in L^1_{loc}(\mathbb{R}^d)$ is also sufficient for closability has been proven by Wielens [37a]. For other new results, covering also the case of arbitrary "speed measures," see [37b].

11. $S'(\mathbb{R}^d)$ is Schwartz' space of tempered distributions.

12. For diffusion processes on manifolds, see, e.g., [77] and references therein.

REFERENCES

1. D. Williams, Diffusions, Markov Processes, and Martingales, Wiley, New York, 1980.
2. N. Ikeda and S. Watanabe, Stochastic Differential Equations and Diffusion Processes, North-Holland/Kodansha, Amsterdam, 1981.
3. D. W. Stroock and S. R. S. Varadhan, Diffusion Processes in Several Dimensions, Springer-Verlag, Berlin, 1979.
4. (a) M. Fukushima, Dirichlet Forms and Markov Processes, North-Holland/Kodansha, Amsterdam, 1980.

   (b) M. Fukushima, On a representation of local martingale additive functionals of symmetric diffusions, pp. 110-118 in Stochastic Integrals, Proc. LMS Durham Symp., 1980, (D. Williams, ed.), Lect. Notes Math. 851, Springer-Verlag, Berlin, 1981.

   (c) H. Föllmer, Dirichlet processes, pp. 476-478 in Ref. 4b.
5. (a) M. L. Silverstein, Symmetric Markov Processes, Lect. Notes Math. 426, Springer-Verlag, Berlin, 1974.

   (b) M. L. Silverstein, Boundary Theory for Symmetric Markov Processes, Lect. Notes Math. 516, Springer-Verlag, Berlin, 1976.
6. (a) M. Fukushima, On a stochastic calculus related to Dirichlet forms and distorted Brownian motion, pp. 255-262 in New Stochastic Methods in Physics (C. De Witt-Morette and K. D. Elworthy, eds.), Phys. Rep. 77 (1981).

   (b) M. Fukushima, Markov processes and functional analysis, in Proc. Int. Math. Conf., Singapore, 1981, North-Holland, Amsterdam, 1983.

   (c) M. Fukushima, ed., Functional Analysis in Markov Processes, Proc. Katata Workshop and Kyoto Conf. 1981, Lect. Notes Math. 923, Springer-Verlag, Berlin, 1983.

   (d) M. L. Silverstein, On the closability of Dirichlet

forms, Z. Wahrscheinlichkeitsth. Verwend. Geb. 51:185-200 (1980).

(e) Y. Lejan, Quasi-continuous functions and Hunt processes, J. Math. Soc. Jap. 35:37-42 (1983).

(f) Y. Oshima, On stochastic differential equations characterizing some singular diffusion processes, Proc. Jap. Acad. 57:A151-154 (1981).

(g) Y. Oshima, Some singular diffusion processes and their associated stochastic differential equations, Z. Wahrscheinlichkeitsth. Verwend. Geb. 59:249-276 (1982); and contribution to Ref. 6c.

(h) M. Fukushima and D. W. Stroock, Reversibility of solutions to martingale problems, University of Osaka. University of Colorado, Boulder Preprint (1981).

(i) M. Tomisaki, A construction of diffusion processes with singular product measures, Z. Wahrscheinlichkeitsth. Verwend. Geb. 53:51-70 (1980).

(j) S. Albeverio, M. Fukushima, W. Karwowski, and L. Streit, Capacity and quantum mechanical tunneling, Commun. Math. Phys. 81:501-513 (1981).

(k) M. Tomisaki, Superpositions of diffusion processes, J. Math. Soc. Jap. 32:671-696 (1980).

(l) M. Tomisaki, Dirichlet forms and diffusion processes associated with product measures, Rep. Fac. Sci. Eng., Saga Univ., No. 9, 1-30 (1981); and contribution to Ref. 6c.

(m) S. Albeverio, R. Høegh-Krohn, and L. Streit, Regularization of Hamiltonians and processes, J. Math. Phys. 21:1636-1642 (1980).

(n) L. Streit, Energy forms: Schrödinger theory, processes, pp. 363-375 in Ref. 6a.

(o) S. Albeverio and R. Høegh-Krohn, Stochastic methods in quantum field theory and hydrodynamics, pp. 193-215 in Ref. 6a.

(p) S. Albeverio and R. Høegh-Krohn, Some Markov processes and Markov fields in quantum theory, group theory, hydrodynamics and C^*-algebras, pp. 497-540 in Ref. 4b.

(q) S. Albeverio and R. Høegh-Krohn, The structure of diffusion processes, CPT-CNRS Preprint, Marseille, 1977 (to appear).

(r) S. Albeverio and R. Høegh-Krohn, The Method of Dirichlet Forms, Proc. Volta Mem. Conf. Stochasticity Dyn. Syst., Como, 1977 (G. Casati and J. Ford, eds.), Lect. Notes Phys. 93, Springer-Verlag, Berlin, 1978.

(s) E. B. Dynkin. Green's and Dirichlet spaces associated with fine Markov processes, J. Funct. Anal. 47, 381-418 (1982).

(t) M. Fukushima, A note on irreducibility and ergodicity, pp. 200-207 in Stochastic Processes in Quantum Theory and Statistical Mechanics, Proc. Marseille 1981 (S. Albeverio, Ph. Combe, and M. Sirugue-Collin, eds.), Lect. Notes in Phys. 173, Springer-Verlag, Berlin, 1982.

(u) E. B. Dynkin, Green's and Dirichlet spaces for a symmetric transition function, in Lect. Notes London Math. Soc., Cambridge Univ. Press 1982.

7. (a) S. Albeverio and R. Høegh-Krohn, Quasi invariant measures, symmetric diffusion processes and quantum fields, pp. 11-59 in Proc. Int. Colloq. Math. Methods Quantum Field Theory, CNRS, 1976.

(b) S. Albeverio and R. Høegh-Krohn, Dirichlet forms and diffusion processes on rigged Hilbert spaces, Z. Wahrscheinlichkeitsth. Verwend. Geb. 40:1-57 (1977a).

(c) S. Albeverio and R. Høegh-Krohn, Topics in infinite dimensional analysis, pp. 279-303 in Mathematical Problems in Theoretical Physics, (G. Dell'Antonio, S. Doplicher, and G. Jona-Lasinio, eds.), Lect. Notes Phys.

80, Springer-Verlag, Berlin, 1978.
(d) S. Albeverio and R. Høegh-Krohn, Hunt processes and analytic potential theory on rigged Hilbert spaces, Z. Wahrscheinlichkeitsth. Verwend. Geb. 40:1-57 (1977).
(e) S. Albeverio and R. Høegh-Krohn, Canonical relativistic local quantum fields, Inst. H. Poincaré A (1985, in press).

8. (a) S. Kusuoka, Dirichlet forms and diffusion processes on Banach spaces, J. Fac. Sci. Univ. Tokyo Sect. IA, 29:79-95 (1982).
(b) S. Kusuoka, Analytic functionals of Wiener processes and absolute continuity, pp. 1-47 in Ref. 6c.
(c) S. Kusuoka, The support of Gaussian white noise and its applications, J. Fac. Sci. Univ. Tokyo IA, 29:387-400 (1982).

9. (a) Ph. Paclet, Espaces de Dirichlet et capacités fonctionelles sur triplets de Hilbert-Schmidt, Sem. Krée Exp. No. 5, 4e année 1977/78 (Université Pierre et Marie Curie, Paris).
(b) Ph. Paclet, Espaces de Dirichlet en dimension infinie, C. R. Ac. Sci., Paris, Ser. A:288, 981-983 (1979).

10. (a) W. Heisenberg and W. Pauli, Zur Quantendynamik der Wellenfelder, Z. Phys. 56:1-61 (1929); 59:168-190 (1930).
(b) F. Coester and R. Haag, Representation of states in a field theory with canonical variables, Phys. Rev. 117:1137-1145 (1960).
(c) H. Araki, Hamiltonian formalism and the canonical commutation relations in quantum field theory, J. Math. Phys. 1:492-504 (1960).
(d) L. Gross, Logarithmic Sobolev inequalities, Am. J. Math. 97:1061-1083 (1975).
(e) J. P. Eckmann, Hypercontractivity for anharmonic

oscillators, Helv. Phys. Acta 45:1074-1088 (1974).

(f) T. Hida, Brownian Motion, Springer-Verlag, Berlin. 1980. See also the references in Ref. 6o.

11. (a) E. Nelson, Dynamical Theories of Brownian Motion, Princeton University Press, Princeton, N. J., 1967.

(b) S. Albeverio and R. Høegh-Krohn, A remark on the connection between stochastic mechanics and the heat equation, J. Math. Phys. 15:1745-1747 (1974).

(c) F. Guerra, Structural aspects of stochastic mechanics and stochastic field theory, pp. 263-312 in Ref. 6a.

12. (a) B. Simon, Functional Integration and Quantum Physics, Academic Press, New York, 1979.

(b) J. Glimm and A. Jaffe, Quantum Physics: A functional Integral Point of View, Springer-Verlag, Berlin, 1982.

13. (a) D. Surgailis, On trajectories of Gaussian Markov random fields in Probability Theory, Banach Center Publ. 5:231-247 (1976).

(b) R. Marcus, Stochastic diffusion on an unbounded domain, Pac. J. Math. 84:143-153 (1979).

(c) W. G. Faris, G. Jona-Lasinio, Large fluctuations for a nonlinear heat equation with noise, J. Phys. A15, 3025-3055 (1982).

14. R. J. Adler, The Geometry of Random Fields, Wiley, Chichester, England, 1982.

15. S. K. Mitter, Filtering theory and quantum fields, Astérisque, 75-76:199-205 (1982).

16. B. Simon, The $P(\varphi)_2$ Euclidean (Quantum) Field Theory, Princeton University Press, Princeton, N. J., 1974.

17. S. Albeverio and R. Høegh-Krohn, Homogeneous random fields and statistical mechanics, J. Funct. Anal. 19:242-272 (1975).

18. I. M. Gelfand and N. Ya. Vilenkin, Generalized Functions,

Vol. 4, Academic Press, New York, 1964.

19. (a) Ch. Preston, Random Fields, Lect. Notes. Math. 534, Springer-Verlag, Berlin, 1976.
(b) H. O. Georgii, Canonical Gibbs Measure, Lect. Notes Math. 760, Springer-Verlag, Berlin, 1979.
(c) J. Fritz, J. L. Lebowitz, and D. Szasz, eds., Random Fields—rigorous results in statistical mechanics and quantum field theory, Proc. Esztergom Coll. Math. Soc. J. Bolyai 27, 1979.
(d) Yu. A. Rozanov, Markov Random Fields, Springer-Verlag, Berlin, 1982.

20. V. Isham, An introduction to spatial point processes and Markov random fields, Int. Stat. Rev. 49:21-43 (1981).

21. (a) D. Elworthy, Stochastic Differential Equations on Manifolds, London Math. Soc., Lect. Notes Series 70, Cambridge Univ. Press, 1981.
(b) M. A. Pinsky, Homogenization and stochastic parallel displacement, pp. 271-284 in Ref. 4b.
(c) M. A. Pinsky, Stochastic Taylor formulas and Riemannian geometry, Northwestern University Preprint.
(d) M. A. Pinsky, Stochastic Riemannian geometry, pp. 199-234 in Probabilistic Analysis and Related Topics, Vol. 1 (A. T. Bharucha-Reid, ed.), Academic Press, New York, 1978.

22. (a) R. Figari, R. Høegh-Krohn, and C. Nappi, Interacting relativistic bosonfields in the DeSitter universe with two space-time dimensions, Commun. Math. Phys. 44:265-278 (1975).
(b) J. Dimock, Algebras of local observables on a manifold, Commun. Math. Phys. 77:219-228 (1980).

23. S. Albeverio and R. Høegh-Krohn, The energy representation of Sobolev-Lie groups, Comp. Math. 36:37-52 (1978).

24. A. M. Vershik, I. M. Gelfand, and M. I. Graev, Repre-

sentation of the groups SL(2,R), where R is a ring of functions, Russ. Math. Surv. 28(5):83-128 (1973).

25. (a) R. F. Streater, Euclidean quantum mechanics and stochastic integrals, pp. 371-393 in Ref. 4b.
(b) M. Asorey and P. K. Mitter, Regularized continuous Yang-Mills process and Feynman-Kac functional, Commun. Math. Phys. 80:43-58 (1981).
(c) B. Gaveau and Ph. Trauber, Une approche rigoureuse à la quantification locale du champ de Yang-Mills avec cut-off, C. R. Acad. Sci. Paris A 291:673-676 (1980).

26. (a) I. B. Frenkel, Orbital theory of affine Lie algebras, Yale preprint, 1980; and references in Ref. 70b.
(b) V. G. Kac, Infinite Dimensional Lie Algebras, Birkhäuser, Boston (1983).

27. M. Brelot, Les étapes et les aspects multiples de la théorie du potentiel, L'Enseignement Math. 2nd ser. 18:1-36 (1972).

28. J. Deny, Methodes hilbertiennes en théorie du potentiel, pp. 121-203 in Potential Theory (M. Brelot, ed.), Stresa 1962, CIME, Edizione Cremonese, Rome, 1970.

29. (a) A. Beurling and J. Deny, Espaces de Dirichlet, Acta Math. 99:203-224 (1958).
(b) A. Beurling and J. Deny, Dirichlet spaces, Proc. Natl. Acad. Sci. USA 45:208-215 (1959).

30. (a) S. Albeverio, R. Høegh-Krohn, and L. Streit, Energy forms, Hamiltonians, and distorted Brownian paths, J. Math. Phys. 18:907-917 (1977).
(b) M. Reed and B. Simon, Methods of Modern Mathematical Physics, Vol. IV; Analysis of Operators, Chap. XIII, Sec. 12, Academic Press, New York, 1978.
(c) P. A. Deift, Applications of a commutation formula, Duke Math. J. 45:267-310 (1978).
(d) R. Carmona, Regularity properties of Schrödinger and Dirichlet semigroups, J. Funct. Anal. 33:259-296 (1979).

(e) I. G. Hooton, Dirichlet forms associated with hypercontractive semigroups, Trans. Am. Math. Soc. 253:237-256 (1979).

(f) G. Jona-Lasinio, F. Martinelli, and E. Scoppola, The semiclassical limit of quantum mechanics: a qualitative theory via stochastic mechanics, pp. 313-327 in Ref. 6a.

(g) P. Courrège and P. Renouard, Oscillateur anharmonique, mesures quasi-invariantes sur C(ℝ,ℝ) et théorie quantique des champs en dimension d = 1, Astérisque 22-23:1-245 (1976).

(h) S. Albeverio, J. E. Fenstad, R. Høegh-Krohn, and T. Lindstrøm, Nonstandard Methods in Stochastic Analysis and Mathematical Physics, Academic Press, New York, 1985 (in press).

(i) F. Gesztesy and L. Pittner, Two-body scattering for Schrödinger operators involving zero-range interactions, Rep. Math. Phys. 19:143-154 (1984).

(j) S. Albeverio and R. Høegh-Krohn, Point interactions as limits of short-range interactions, J. Operator Theory 61:313-339 (1981).

(k) S. Albeverio, F. Gesztesy, and R. Høegh-Krohn, The low energy expansion in nonrelativistic scattering theory, Ann. Inst. H. Poincaré A37:1-28 (1982).

(l) S. Albeverio, F. Gesztesy, R. Høegh-Krohn, and L. Streit, Charged particles with short range interactions, Ann. Inst. H. Poincare 38:263-293 (1983).

(m) W. Kirsch and F. Martinelli, On the spectrum of Schrödinger operators with a random potential, Commun. Math. Phys. 85:329-350 (1982).

(n) W. Kirsch and F. Martinelli, On the ergodic properties of the spectrum of general random operators, J. Reine Angew. Math. 334:141-156 (1982).

(o) S. Albeverio, R. Høegh-Krohn, W. Kirsch, and

F. Martinelli, The spectrum of the three-dimensional Kronig-Penney model with random point defects, Adv. Appl. Math. 3:435-440 (1982).

(p) S. Albeverio, R. Høegh-Krohn, and T. T. Wu, A class of exactly solvable three-body quantum mechanical problems and the universal low-energy behavior, Phys. Lett. 83A:105-109 (1981).

(q) A. Grossman, R. Høegh-Krohn, and M. Mebkhout, The one particle theory of periodic point interactions, Commun. Math. Phys. 77:87-110 (1980).

(r) W. Kirsch and F. Martinelli, pp. 223-244 in Ref. 6t.

(s) S. Albeverio, D. Bollé, F. Gesztesy, R. Høegh-Krohn, and L. Streit, Low energy parameters in non-relativistic scattering theory, Ann. of Phys. 148: 308-326 (1983).

(t) J. Zorbas, Perturbation of self-adjoint operators by Dirac distributions, J. Math. Phys. 21:840-847 (1980).

(u) S. Albeverio, L. S. Ferreira, F. Gesztesy, R. Høegh-Krohn, and L. Streit, Model dependence of Coulomb corrected scattering lengths, Phys. Rev. C29, 680-683 (1984).

31. T. Kato, Perturbation Theory for Linear Operators, Springer-Verlag, Berlin, 1966.

32. N. I. Portenko, Diffusion processes with unbounded drift coefficient, Theor. Prob. Appl. 20:27-37 (1976).

33. H. Ezawa, J. R. Klauder, and L. A. Shepp, A path space Picture for Feynman-Kac averages, Ann. Phys. 88:588-620 (1974).

34. (a) A. N. Kolmogorov, Über die analytischen Methoden in der Wahrschienlichkeitsrechnung, Math. Ann. 104: 415-458 (1931).

(b) M. Nagasawa, Segregation of a population in an environment, J. Math. Biol. 9:213-235 (1980).

(c) J. Kent, Time reversible diffusions, Adv. Appl. Prob. 10:819-835 (1978); 11:888 (1979).
(d) S. Albeverio, Ph. Blanchard, and R. Høegh-Krohn, Diffusions sur une variété riemanienne: barrières infranchissables et applications, Coll. Schwartz, Astérisque, 1984.

35. J. M. Hamza, Détermination des formes de Dirichlet sur $\mathbb{R}^n$, Thèse, Orsay, 1975.

36. K. Rullkötter and U. Spönemann, Dirichlet formen und Diffusions prozesse, Diplomarbeit, Bielefeld, 1982.

37. (a) N. Wielens, Eindeutigkeit von Dirichletformen und wesentliche Selbstadjungiertheit von Schrödinger operatoren mit stark singulären Potentialen, Diplomarbeit, Bielefeld, 1982 (to be published in J. Func. Anal. 1984).
(b) M. Röckner, and N. Wielens, Dirichlet forms closability and change of speed measure, Bielefeld Preprint, 1983 in Proc. Conf. Infinite Dimensional Analysis and Stochastic Processes, Bielefeld, 1983.

38. J. F. Brasche, Störungen von Schrödinger operatoren durch Masse, Staatsexamensarbeit, Bielefeld, 1982; and ZIF-Preprint, Bielefeld, 1984.

39. S. Albeverio, J. E. Fenstad, and R. Høegh-Krohn, Singular perturbations and nonstandard analysis, Trans. Am. Math. Soc. 252:275-295 (1979).

40. S. Albeverio, F. Gesztesy, R. Høegh-Krohn, and W. Kirsch, Point interactions in one dimension, Bochum preprint, 1982 to appear in J. Operat. Th. 1984.

41. K. Symanzik, Euclidean quantum field theory, pp. 152-219 in Local Quantum Theory (R. Jost, ed.), Academic Press, New York, 1972.

42. J. Westwater, On Edwards model for long polymer chains I, II, Commun. Math. Phys. 72:131-174 (1980).

43. (a) M. Aizenmann, Geometric analysis of φ^4 fields and

Ising models, pp. 37-46 in Mathematical Problems in Theoretical Physics, Proc. AMP Conf. Berlin, 1981 (R. Schrader, R. Seiler, and D. A. Uhlenbrock, eds.), Lect. Notes in Phys. 153, Springer-Verlag, Berlin, 1982.
(b) J. Fröhlich, Results and problems near the interface between statistical mechanics and quantum field theory, pp. 64-74 in Ref. 43c.

44. R. Cairoli and J. B. Walsh, Stochastic integrals in the plane, Acta Math. 134:111-183 (1975).
45. D. Williams, Some basic theorems on harnesses, pp. 349-363 in Stochastic Analysis (D. G. Kendall and E. F. Harding, eds.), Wiley, New York, 1973.
46. S. Albeverio and R. Høegh-Krohn, Diffusion, quantum fields and groups of mappings, in Ref. 6b.
47. S. Kusuoka, Markov fields and local operators, J. Fac. Sci., Univ. Tokyo IA, 26:199-212 (1979).
48. R. L. Dobrushin and R. A. Minlos, Polynomials in linear random functions, Russ. Math. Surv. 32:71-127 (1977).
49. R. Jost, The General Theory of Quantized Fields, American Mathematical Society, Providence, R. I., 1965.
50. (a) M. Röckner, Verallgemeinerte Markoffsche Felder und Potentialtheorie, Diplomarbeit, Bielefeld, 1982.
(b) M. Röckner, Self-adjoint harmonic spaces and Dirichlet forms, Bielefeld Preprint, 1983 (to appear in Hiroshima Math. J.).
(c) M. Röckner, Generalized Markov fields and axiomatic potential theory, Math. Ann. 264:153-177 (1983).
(d) M. Röckner, Generalized random fields and Dirichlet forms, Bielefeld Preprint, 1983 (to appear in Acta Appl. Math.)
51. H. Künsch, Gaussian Markov random fields, J. Fac. Sci., Univ. Tokyo IA 26:53-73 (1979).
52. (a) D. Preiss and R. Kotecky, Markoff property of

generalized random fields, Seventh Winter School, Czechoslovakia, 1979.

(b) R. Kotecky and D. Preiss, The use of projective limits in classical statistical mechanics and euclidean quantum field theory, Cz. J. Phys. B30:23-32 (1980).

53. (a) E. Nelson, Probability theory and euclidean field theory, pp. 94-104 in Constructive Quantum Field Theory (G. Velo and A. S. Wightman, eds.), Lect. Notes Phys. 25, Springer-Verlag, Berlin, 1973.

(b) Ch. Newman, The construction of stationary two-dimensional Markoff fields with an application to quantum field theory, J. Funct. Anal. 14:44-61 (1973).

54. J. Fröhlich, Schwinger functions and their generating functionals, Helv. Phys. Acta 94:265-306 (1976); Adv. Math. 23:119-180 (1977).

55. (a) Y. Okabe, On the general fields of stationary Gaussian processes with Markovian properties, J. Math. Soc. Jap. 28:86-95 (1976).

(b) S. Kotani, On a Markov property for stationary Gaussian processes with a multidimensional parameter, pp. 239-250 in Proc. 2nd Jap.-USSR Symp., Lect. Notes in Math. 330, Springer-Verlag, 1977.

56. (a) S. Albeverio and R. Høegh-Krohn, The exponential interaction, Marseille preprint 77/962, Nov. 1977.

(b) S. Albeverio and R. Høegh-Krohn, Local and global Markoff fields, Lect. Winter School Theor. Phys., Karpacz, Feb. 1978, Rep. Math. Phys. 19:225-248 (1984).

(c) S. Albeverio and R. Høegh-Krohn, Uniqueness and the global Markov property for euclidean fields. The case of trigonometric interactions, Commun. Math. Phys. 68:95-128 (1979).

(d) S. Albeverio and R. Høegh-Krohn, Uniqueness of Gibbs states and global Markov property for euclidean fields, Colloq. Math. Soc. J. Bolyai 27:Random Fields, Esztergom, Hungary, 1979.

(e) S. Albeverio and R. Høegh-Krohn, Uniqueness and global Markov property for euclidean fields and lattice systems, pp. 303-329 in Ref. 62.

(f) S. Albeverio, R. Høegh-Krohn, and G. Olsen, The global Markov property for lattice systems, J. Multiv. Anal. 11:599-607 (1981).

(g) S. Albeverio and R. Høegh-Krohn, Uniqueness of Gibbs states and global Markov property for euclidean fields, pp. 37-64 in Ref. 43a.

(h) S. Albeverio and R. Høegh-Krohn, Uniqueness and global Markov property for euclidean fields and lattice systems, pp. 303-320 in Quantum fields—Algebras, Processes (L. Streit, ed.), Springer-Verlag, Berlin, 1980.

(i) J. Bellissard and R. Høegh-Krohn, Compactness and the maximal Gibbs state for random Gibbs fields on a lattice, Commun. Math. Phys. 84:297-327 (1982).

57. (a) E. B. Dynkin, Markov processes, random fields and Dirichlet spaces, pp. 239-247 in Ref. 6a.

(b) E. B. Dynkin, Additive functionals of several time-reversible Markov processes, J. Funct. Anal. 42:64-101 (1981).

(c) E. B. Dynkin, Markov processes and random fields, Bull. Am. Math. Soc. 3:975-999 (1980).

58. (a) L. Accardi, Local perturbations of conditional expectations, J. Math. Anal. Appl. 72:34-69 (1979).

(b) A. Klein, Stochastic processes associated with quantum systems, pp. 329-337 in Ref. 6a.

(c) K. Osterwalder and R. Schrader, Axioms for euclidean Green's functions II, Commun. Math. Phys. 42:281-305 (1975).

(d) T. Hida and L. Streit, On quantum theory in terms

of white noise, Nagoya Math. J. 68:21-34 (1977).
(e) G. C. Hegerfeldt, From euclidean to relativistic fields and the notion of Markoff fields, Commun. Math. Phys. 35:155-171 (1974).

59. (a) F. Constantinesen and W. Thalheimer, Remark on generalized Markov processes, J. Funct. Anal. 23:33-38 (1976).
(b) S. P. Gudder, Gaussian random fields, Found. Phys. 8:295-302 (1978).

60. R. Gielerak, Verification of the global Markov property in some class of strongly coupled exponential interactions, J. Math. Phys. 24:347-355 (1983).

61. S. Albeverio and R. Høegh-Krohn, The Wightman axioms and the mass gap for strong interactions of exponential type in two-dimensional space-time, J. Funct. Anal. 16:39-82 (1974).

62. S. Albeverio and R. Høegh-Krohn, Martingale convergence and the exponential interaction on $\mathbb{R}^n$, pp. 331-353 in Quantum Fields — Algebras, Processes (L. Streit, ed.), Springer-Verlag, Berlin, 1980.

63. S. Albeverio, G. Gallavotti, and R. Høegh-Krohn, Some results for the exponential interaction in two or more dimensions, Commun. Math. Phys. 70:187-192 (1979).

64. E. P. Osipov, On the triviality of the quantum field theory in a finite volume, Novosibirsk preprint, 1979.

65. (a) G. Benfatto, G. Gallavotti, and F. Nicolò, On the massive sine-Gordon equation in the first few regions of collapse, Commun. Math. Phys. 83:387-410 (1982).
(b) E. P. Osipov, New approach to the construction of quantum field theory with exponential interaction in four dimensional space-time (Russ.), pp. 377-382 in Generalized functions and their applications in mathematical physics, Ak. Nauk SSR, Vych. TS. Moscow (1981).

66. G. C. Hegerfeldt, Representation of the canonical commutative relations of quantum field theory, pp. 140-165 in Functional Analysis, Surveys and Recent Results II (K. D. Bierstedt and B. Fuchsteiner, eds.), North Holland, Amsterdam, 1980.

67. K. R. Parthasarathy and K. Schmidt, A new method for constructing factorizable representations for current groups and current algebras, Commun. Math. Phys. 50: 167-175 (1976).

68. I. Marion, Etude du groupe G^X, représentations unitaires non localisées d'ordre > 0, Luminy preprint 1978.

69. K. R. Parthasarathy and K. Schmidt, Positive Definite Kernels, Continuous Tensor Products, and Central Limit Theorems of Probability Theory, Lect. Notes Math. 272, Springer-Verlag, Berlin, 1972.

70. (a) S. Albeverio, R. Høegh-Krohn, and D. Testard, Irreducibility and reducibility for the energy representation of the group of mappings of a Riemann manifold into a compact semisimple Lie group, J. Funct. Anal. 41:378-396 (1981).

(b) S. Albeverio, R. Høegh-Krohn, D. Testard, and A. Vershik, Factorial representation of path groups, Bochum preprint, 1981.

(c) S. Albeverio, Ph. Blanchard, and R. Høegh-Krohn, Some applications of functional integration, in Ref. 43a.

(d) J. Marion, Generalized energy representation for current groups, J. Funct. Anal. 59:1-17 (1983).

(e) S. Albeverio, R. Høegh-Krohn, and D. Testard, Factoriality of representations of the groups of paths on SU(n), J. Funct. Anal. 57:49-55 (1984).

(f) J. Marion, A survey on the unitary representations of gauge groups, and some remaining open questions, ZIF Preprint, 1984.

71. (a) S. Albeverio and R. Høegh-Krohn, Uniqueness of the physical vacuum and the Wightman functions in the infinite volume limit for some nonpolynomial interactions, Commun. Math. Phys. 30:171-200 (1973).

(b) J. Fröhlich and E. Seiler, The massive Thirring-Schwinger model $(QED)_2$: convergence of perturbation theory and particle structure, Helv. Phys. Acta 49:889-924 (1976).

(c) Y. M. Park, Massless quantum sine-Gordon equation in two space-time dimensions: correlation inequalities and infinite volume limit, J. Math. Phys. 18:2423-2426 (1977).

(d) R. Gielerak, On the grand canonical Gibbs ensemble in euclidean field theory, Fortschr. Phys. 29:19-34 (1981).

(e) S. Albeverio, Ph. Blanchard, Ph. Combe, R. Høegh-Krohn, and M. Sirugue, Local relativistic invariant flows for quantum fields, Commun. Math. Phys. 90:329-351 (1983).

72. (a) H. Englisch, The $(\varphi^{2n})_2$-quantum field as a limit of sine-Gordon fields, Karl Marx Universität preprint, 1979.

(b) H. Englisch, Remarks on McBryan's convergence proof for $\alpha^{-4}(\cos \alpha\varphi) - 1 + \alpha^2\varphi^2/2)_2$-quantum fields, Rep. Math. Phys. 18:387-397 (1980).

73. W. D. Wick, On the absolute continuity of a convolution with an infinite dimensional measure, University of Washington, Seattle preprint, 1979.

74. (a) R. S. Ismagilov, On unitary representation of the groups $C_0^\infty(X,G)$, $G = SU_2$, Mat. USSR Sb. 29:105-117 (1976).

(b) A. M. Vershik, I. M. Gelfand, and M. I. Graev, Representation of the group of functions taking values in a compact Lie group, Comp. Math. 42:217-243 (1981).

75. B. Gaveau and Ph. Trauber, Mesures et représentations non locales pour les groupes de Lie d'applications, C. R. Acad. Sci. Paris A291:575-578 (1980).

76. (a) T. Hida and L. Streit, Generalized Brownian functionals and the Feynman integral, Stoch. Proc. Appl. 16:55-69 (1983).
(b) T. Hida and L. Streit, Generalized Brownian functionals, pp. 285-287 in Ref. 43a.

77. (a) N. Ikeda and S. Watanabe, Stochastic differential equations and diffusion processes, North-Holland/ Kodansha, Amsterdam, Tokyo, 1981.
(b) D. Elworthy and A. Truman, Classical mechanics, the diffusion (heat) equation and the Schrödinger equation on a Riemannian manifold, J. Math. Phys. 22:2144-2166 (1981).

78. F. Y. Maeda, Dirichlet integrals on harmonic spaces, Lect. Notes in Math. 803, Springer-Verlag, Berlin 1980.

79. S. Albeverio, S. Kusuoka, L. Streit, in preparation.

80. B. Zegarlinski, Uniqueness and the global Markov property for euclidean fields. The case of general exponential interaction, to appear in Commun. Math. Phys.

81. W. Loges, Estimation of parameters for Hilbert-valued partially observable stochastic processes, Bochum Preprint 1982.

82. S. Albeverio, F. Gesztesy, R. Høegh-Krohn, and H. Holden, Solvable models in quantum mechanics, book in preparation.

83. W. Wirsch, Über Spektren stochastischer Schrödinger operatoren, Ph. D. Thesis, Bochum, 1981.

84. M. Takeda, On the uniqueness of Markovian extensions of diffusion operators on infinite dimensional spaces, ZIF-Bielefeld Preprint 1984.

2

Optimal Stochastic Control of Diffusion Processes with Jumps Stopped at the Exit of a Domain

ALAIN BENSOUSSAN/University of Paris, Dauphine, and INRIA, Paris, France

JOSÉ LUIS MENALDI*/University of Paris-Dauphine, Paris, France

1. INTRODUCTION

In a preceding article [1], one of the authors has considered the Hamilton-Jacobi-Bellman equation corresponding to the optimal control of diffusion processes with jumps in a bounded domain. This equation is a nonlinear integrodifferential equation. It was shown that the solution of the Bellman equation could be interpreted as the infimum of a functional depending on a diffusion process with jumps stopped at the exit of the domain, controlled by a Markov feedback. When one considers the same problem in the whole space, one can prove a better result, namely that the class of controls on which the infimum is taken is the class of nonanticipative controls. The situation is simpler in the whole space because the solution is more regular, hence Itô's calculus (at least in integrated form) can be applied.

In this chapter we show that one can also consider the class of nonanticipative controls in the case of a bounded domain. We rely on a penalty argument to be able to use the results which are true in the whole space.

*Current affiliation: Wayne State University, Detroit, Michigan

2. REVIEW OF RESULTS IN THE WHOLE SPACE

2.1 Assumptions-Notation

Let $a_{ij}(x)$, $a_i(x)$ be functions such that

$$a_{ij} = a_{ji} \in W^{1,\infty}(R^n) \qquad a_i \in L^\infty(R^n)$$
$$\Sigma\, a_{ij}\xi_i\xi_j \geq \alpha|\xi|^2 \qquad \text{for } \xi \in R^n,\ \alpha > 0 \tag{2.1}$$

We define the differential operator

$$A = -\sum_{i,j} \frac{\partial}{\partial x_i} a_{ij} \frac{\partial}{\partial x_j} + \sum_i a_i \frac{\partial}{\partial x_i} \tag{2.2}$$

Next let $m(dz)$ be a positive measure on R^n such that

$$\int_{\{|z|<1\}} |z|^2 m(dz) + \int_{\{|z|>1\}} |z| m(dz) < \infty \tag{2.3}$$

We consider a function $b(x,z)$, which is defined and measurable on $R^n \times R^n$ such that

$$0 \leq b(x,z) \leq 1 \tag{2.4}$$

We define the integrodifferential operator

$$B\varphi(x) = \int_{R^n} [\varphi(x+z) - \varphi(x) - z\cdot\nabla\varphi(x)\chi_{\{|z|<1\}}] b(x,z) m(dz) \tag{2.5}$$

which makes sense, for instance, for $\varphi \in \mathcal{D}(R^n)$.

Let $\mathcal{V}$ be a measurable space, and functions

$$\ell, a_1 : R^n \times \mathcal{V} \to R, \text{ measurable and bounded}$$
$$|\ell(x,v)| \leq \bar{\ell}(x) \in L^2(R^n) \cap L^\infty(R^n)$$
$$h: R^n \times \mathcal{V} \to R^n, \text{ measurable and bounded} \tag{2.6}$$
$$c: R^n \times \mathcal{V} \times R^n \to R, \text{ measurable and bounded}$$
$$c(x,v,z) \geq -1,\ |c(x,v,z)| \leq C|z|$$

We define again for $\varphi \in \mathcal{D}(R^n)$

$$H(\varphi)(x) = \inf_v \{\ell(x,v) + \nabla\varphi(x)\cdot h(x,v) - \varphi(x)a_1(x,v)$$
$$+ \int_{R^n} [\varphi(x+z) - \varphi(x) - z\cdot\nabla\varphi(x)\chi_{|z|<1}] \tag{2.7}$$
$$\times\, b(x,z)c(x,v,z)m(dz)\}$$

Also let $a_0(x)$ be such that

$$a_0 \text{ measurable and bounded, } a_0(x) \geq \gamma_0 \ \sup a_1^- = \gamma_1 \tag{2.8}$$

2.2 A Nonlinear Integrodifferential Equation

For $\mu \geq 0$ let

$$\beta_\mu(x) = \exp[-\mu(|x|^2 + 1)^{1/2}] \tag{2.9}$$

and $L^{p,\mu} = \{\varphi \mid \varphi\beta_\mu \in L^p(R^n)\}$, $1 < p < \infty$.

We define similarly

$$W^{2,p,\mu} = \left\{\varphi \in L^{p,\mu} \;\middle|\; \frac{\partial\varphi}{\partial x_i}, \frac{\partial^2\varphi}{\partial x_i \partial x_j} \in L^{p,\mu}\right\}$$

For $\mu = 0$, $W^{2,p,\mu}(R^n) = W^{2,p}(R^n)$.

Let $f \in L^p(R^n)$, $2 \leq p < \infty$; we consider the equation

$$Au - Bu - H(u) + a_0 u = f \tag{2.10}$$

Then we have:

THEOREM 2.1 There exists one and only one solution of (2.10) in the space $W^{2,p,\mu}(R^n)$, for all $\mu > 0$, under assumptions (2.1), (2.3), (2.4), (2.6), and (2.8).

This theorem is proved in Bensoussan and Lions [2]. For the sake of completeness, we outline the proof. Note that one expects the solution to be in $W^{2,p}$ since the data belong to L^p. We do not know how to prove this result, which seems likely, however.

We first need to state some properties of the operators B and H. We have

$$\begin{gathered} H: W^{1,p}(R^n) \to L^p(R^n) \\ |H(\varphi_1) - H(\varphi_2)|_{L^p(R^n)} \leq C\|\varphi_1 - \varphi_2\|_{W^{1,p}(R^n)} \end{gathered} \tag{2.11}$$

and more generally, if $\psi \in W^{1,p} \cap W^{1,q}$ and $\varphi \in W^{1,p}$, then $H(\varphi + \psi) - H(\varphi) \in W^{1,q} \cap W^{1,p}$ and

$$|H(\varphi + \psi) - H(\varphi)|_{L^q} \leq C|\psi|_{W^{1,q}} \tag{2.12}$$

where the constant C does not depend on φ or ψ or p, q.

Next,

$$\begin{gathered} B \in \mathcal{L}(W^{2,p}(R^n); L^p(R^n)) \quad \text{and} \\ |B\varphi|_{L^p} \leq C[\sigma(r)\|\varphi\|_{W^{2,p}} + \tau(r)\|\varphi\|_{W^{1,p}}] \end{gathered} \tag{2.13}$$

where $\sigma(r) \to 0$ as $r \to 0$, $\tau(r)$ may tend to $+\infty$ as $r \to 0$.

We also have

$$H: W^{1,p,\mu}(R^n) \cap L^\infty \to L^{p,\mu} \qquad 2 \le p < \infty,\ \mu > 0$$

and for

$$\psi \in W^{1,p,\mu} \cap W^{1,q,\mu} \cap L^\infty \qquad \varphi \in W^{1,q,\mu} \cap L^\infty \tag{2.14}$$

$$|H(\varphi + \psi) - H(\varphi)|_{L^{p,\mu}} \le C[\|\psi\|_{W^{1,p,\mu}} + \|\psi\|_{L^\infty}]$$

$$B \in (W^{2,p,\mu} \cap L^\infty;\ L^{p,\mu}) \qquad p \ge 2,\ \mu > 0$$

and $$|B\varphi|_{L^{p,\mu}} \le C[\sigma(r)\|\varphi\|_{W^{2,p,\mu}} + \tau(r)\|\varphi\|_{L^\infty}] \tag{2.15}$$

where $\sigma(r) \to 0$ as $r \to 0$. Property (2.13) [and analogously (2.15)] follows from considering (for $r < 1$)

$$B_r^1\varphi(x) = \int_{R^n}[\varphi(x + z) - \varphi(x) - z\cdot\nabla\varphi(x)]b(x,z)\chi_{\{|z|<r\}}m(dz)$$

$$= \sum_{i,j} \int_0^1 d\theta \int_0^\theta d\theta' \int_{R^n} \frac{\partial^2\varphi}{\partial x_i \partial x_j}(x + \theta' z)(z_i z_j b(x,z)\chi_{\{|z|<r\}}m(dz)$$

and

$$|B_r^1\varphi|_{L^p(R^n)} \le C\|\varphi\|_{W^{2,p}(R^n)} \int_{R^n} |z|^2\chi_{\{|z|<r\}}m(dz)$$

The steps to prove Theorem 2.1 are the following:

LEMMA 2.1 Consider for λ large enough

$$Au_\lambda - Bu_\lambda - H(u_\lambda) + (a_0 + \lambda)u_\lambda = f \tag{2.16}$$

Then there exists one and only one solution $u_\lambda \in W^{2,p}(R^n)$.

Proof. Considering

$$A\psi + (a_0 + \lambda)\psi = \varphi$$

we have the estimates

$$\|\psi_\lambda\|_{W^{2,p}} \le C|\varphi|_{L^p}$$

$$(\lambda - \lambda_0)|\psi_\lambda|_{L^p} \le C|\varphi|_{L^p}$$

where C does not depend on λ, for $\lambda > \lambda_0$.

For $z \in W^{2,p}$ we solve

$$Aw + (a_0 + \lambda)w = f + Bz + H(z), \qquad w \in W^{2,p}(R^n)$$

We thus have defined a mapping from $W^{2,p}$ into itself denoted by T_λ. Therefore, using (2.11) and (2.13) we obtain, considering z_1, z_2 and $w_1 = T_\lambda(z_1)$, $w_2 = T_\lambda(z_2)$,

$$\|w_1 - w_2\|_{W^{2,p}} \le C[\sigma(r)\|z_1 - z_2\|_{W^{2,p}} + \tau(r)\|z_1 - z_2\|_{W^{1,p}}]$$
$$\le C \quad \sigma(r) + \frac{\tau(r)\varepsilon}{2}\|z_1 - z_2\|_{W^{2,p}}$$
$$+ \frac{\tau(r)}{2\varepsilon}|z_1 - z_2|_{L^p}$$

and $(\lambda - \lambda_0)|w_1 - w_2|_{L^p}$ satisfies the same estimate, with C independent of λ. Consider on $W^{2,p}$ a new equivalent norm

$$|||\cdot|||_{W^{2,p}} = \|\cdot\|_{W^{2,p}} + (\lambda - \lambda_0)|\cdot|_{L^p}$$

Then choosing r and ε sufficiently small so that

$$2C\ \sigma(r) + \frac{\tau(r)\varepsilon}{2} < k$$

and λ large enough so that $2C[\tau(r)/2\varepsilon] < (\lambda - \lambda_0)k$, with k fixed, $0 < k < 1$, we obtain

$$|||T_\lambda z_1 - T_\lambda z_2|||_{W^{2,p}} < k|||z_1 - z_2|||_{W^{2,p}}$$

which proves that T_λ is a contraction, hence the result.

There is a useful variant of Lemma 2.1:

LEMMA 2.2 Let $\varphi \in L^p \cap L^q$, $2 \le p \le q < \infty$, and $\psi \in L^p$; then there exists one and only one solution of

$$A\zeta - B\zeta - H(\psi + \zeta) + H(\psi) + (\lambda + a_0)\zeta = \varphi$$
$$\zeta \in W^{2,p} \cap W^{2,q} \tag{2.17}$$

Proof. Use the fixed-point argument of Lemma 2.1, with property (2.12) in $W^{2,p} \cap W^{2,q}$.

LEMMA 2.3 Assume that in (2.17), $\varphi \in L^\infty$; then ζ belongs to L^∞ and

$$\|\zeta\|_{L^\infty} \le \frac{\|\varphi\|_{L^\infty}}{\lambda + \gamma_0 - \gamma_1} \tag{2.18}$$

Proof. Result (2.18) expresses the validity of the maximum principle for (2.16). One uses operators which approximate

B and H, denoted by B_ε and H_ε, and defined by the same formulas as B and H, with m(dz) changed to

$$m_\varepsilon(dz) = m(dz)\chi_{\{|z|>\varepsilon\}} \tag{2.19}$$

The measure m_ε is bounded. Now one considers

$$A\zeta_\varepsilon - B_\varepsilon\zeta_\varepsilon - H_\varepsilon(\psi + \zeta_\varepsilon) + H_\varepsilon(\psi) + (\lambda + a_0)\zeta_\varepsilon = \varphi$$
$$\zeta_\varepsilon \in W^{2,p} \tag{2.20}$$

The operators B_ε, H_ε satisfy the same properties as B, H with constants independent of ε. Then one checks that (2.18) is valid for ζ_ε. Let γ_ε be a constant to be chosen; set

$$\tilde\zeta = \tilde\zeta_\varepsilon = \zeta_\varepsilon - \gamma_\varepsilon$$

From (2.20) it follows that

$$\begin{aligned}
A\tilde\zeta + (\lambda + a_0)\tilde\zeta + \gamma_\varepsilon(a_0 + \lambda) &= \varphi + (B_\varepsilon + H_\varepsilon)(\psi + \tilde\zeta + \gamma_\varepsilon) \\
&\quad - (B_\varepsilon + H_\varepsilon)(\psi) \\
&= +\inf\{1 + \nabla\psi\cdot h - a_1\psi \\
&\quad + \int_{R^n}[\psi(x + z) - \psi(x) \\
&\quad - z\cdot\nabla\psi\chi_{|z|<1}]b(1 + c)m_\varepsilon(dz) \\
&\quad + \nabla\tilde\zeta\cdot h - \tilde\zeta a_1 + \int_{R^n}[\tilde\zeta(x + z) \\
&\quad - \tilde\zeta(x) - z\cdot\nabla\tilde\zeta\chi_{|z|<1}] \\
&\qquad b(1 + c)m_\varepsilon(dz) - \gamma_\varepsilon a_1\} \\
&\quad - \inf\{1 + \nabla\psi\cdot h - a_1\psi \\
&\quad + \int_{R^n}[\psi(x + z) - \psi(x) \\
&\quad - z\cdot\nabla\psi\chi_{|z|<1}]b(1 + c)m_\varepsilon(dz)\} \\
&\le \|\varphi\| + \gamma_\varepsilon\gamma_1 + C[|\nabla\tilde\zeta(x)| \\
&\quad + |\tilde\zeta(x)|][1 + m_\varepsilon(R^n)] \\
&\quad + C_1\int_{R^n}\tilde\zeta^+(x + z)m_\varepsilon(dz)
\end{aligned} \tag{2.21}$$

We take

$$\gamma_\varepsilon = \frac{\|\varphi\| + \gamma'_\varepsilon\|\zeta_\varepsilon\|}{\lambda + \gamma_0 - \gamma_1 + \gamma'_\varepsilon}$$

where γ_ε' is to be chosen. It follows from (2.21) that

$$A\tilde{\zeta} + (\lambda + a_0) + \gamma_\varepsilon')\tilde{\zeta} \le C[|\nabla\tilde{\zeta}(x)| + |\tilde{\zeta}(x)|](1 + m_\varepsilon(R^n)) + C_1 \int_{R^n} \tilde{\zeta}(x + z)m_\varepsilon(dz)$$

We multiply by $\tilde{\zeta}^+$ and integrate. We note that

$$(A\tilde{\zeta},\tilde{\zeta}^+) \ge \delta\|\tilde{\zeta}^+\|^2 - \lambda_0|\tilde{\zeta}^+|^2$$

hence

$$\delta\|\tilde{\zeta}^+\|^2 - \lambda_0|\tilde{\zeta}^+|^2 + (\lambda + \gamma_0 + \gamma_\varepsilon')|\tilde{\zeta}^+|^2 \le C_\varepsilon[\|\tilde{\zeta}^+\||\tilde{\zeta}^+| + |\tilde{\zeta}^+|^2]$$

and we can choose γ_ε' large enough to deduce that $\tilde{\zeta}^+ = 0$. From this we deduce that (2.18) holds for ζ_ε.

Using (2.14) and (2.15) we then check that ζ_ε remains in a bounded subset of $W^{2,p,\mu}$, $\mu > 0$. If one extracts a subsequence which converges to ζ in $W^{2,p,\mu}$ weakly and L^∞ weak star, it follows from (2.14) and (2.15) and the definition of B_ε, H_ε that

$$B_\varepsilon\zeta_\varepsilon \to B\zeta \qquad \text{in } L^{p,\mu} \text{ weakly}$$

$$H_\varepsilon(\psi + \zeta_\varepsilon) \to H(\psi + \zeta) \qquad \text{in } L^{p,\mu} \text{ weakly}$$

Passing to the limit we obtain that ζ is the solution (unique) of (2.17). We thus also have (2.18).

Proof of Theorem 2.1. One considers an iterative sequence as follows: $u_0 = 0$; then

$$Au_{k+1} - Bu_{k+1} - H(u_{k+1}) + (a_0 + \lambda)u_{k+1} = f + \lambda u_k \qquad (2.22)$$

which defines a sequence in $W^{2,p}$.

Consider

$$A(u_{k+1} - u_k) - B(u_{k+1} - u_k) - H(u_k + u_{k+1} - u_k) + H(u_k) + (a_0 + \lambda)(u_{k+1} - u_k) = \lambda(u_k - u_{k-1})$$

Applying Lemma 2.2, we note that $u_{k+1} - u_k$ is more regular than $u_k - u_{k-1}$. After a finite number of steps (exactly for $k \ge j + 1$, with $j = |n/2p|$), one obtains $u_k - u_{k-1} \in L^\infty$. Using Lemma 2.3, we obtain

$$\|u_{k+1} - u_k\|_{L^\infty} \le \rho\|u_k - u_{k-1}\| \qquad \text{for } k \ge j + 1$$

with $\rho < 1$. Therefore, for $k \geq j + 1$, $u_k - u_j$ is a Cauchy sequence in L^∞. Now from (2.22), using the fact that

$$A(u_{k+1} - u_{j+1}) - B(u_{k+1} - u_{j+1}) - H(u_{j+1} + u_{k+1} - u_{j+1})$$
$$+ H(u_{j+1}) + (a_0 + \lambda)(u_{k+1} - u_{j+1}) = \lambda(u_k - u_j)$$

one verifies that $u_k - u_{j+1}$ remains in a bounded subset of $W^{2,p,\mu}$. From that we deduce that $u_k - u_{j+1}$ converges in L^∞ and $W^{2,p,\mu}$ weakly, to $u - u_{j+1}$, and u is clearly a desired solution.

To prove uniqueness, we notice that if u, $\tilde{u}$ are two solutions,

$$A(u - \tilde{u}) - B(u - \tilde{u}) - H(\tilde{u} + u - \tilde{u}) + H(\tilde{u})$$
$$+ (a_0 + \lambda)(u - \tilde{u}) = \lambda(u - \tilde{u})$$

and by the reasoning of regularity improvement one checks that $u - \tilde{u} = L^\infty$. Then estimate (2.18) implies that

$$\|u - \tilde{u}\| \leq \frac{\lambda\|u - \tilde{u}\|}{\lambda + \gamma_0 - \gamma_1}$$

and since $\gamma_0 > \gamma_1$, $\|u - \tilde{u}\| = 0$.

REMARK 1. The method of regularity improvement used in the proof of Theorem 2.1 is due to P. L. Lions [4].

2.4 A Stochastic Control Problem for Diffusions with Jumps in the Whole Space

We consider equation (2.10) with $f = 0$ (to simplify the notation), and $a_i = \partial a_{ji}/\partial x_i$, which we can do without loss of generality, provided that we modify $\ell(x,v)$ in a suitable manner. We will also set

$$\bar{a}(x,v) = a_0(x) + a_1(x,v)$$

and incorporate a_0 into the operator H, by changing a_1 into $\bar{a}$ in the definition of H. We thus write (2.10) as

$$Au - Bu - H(u) = 0 \qquad u \in W^{2,p,\mu} \cap L^\infty, \text{ for all}$$
$$p \geq 2,\ p < \infty,\ \mu > 0 \qquad (2.23)$$

We will also assume

$$\mathcal{V} \text{ is a compact subset of a metric space} \tag{2.24}$$

$$\ell(x,v),\ h(x,v),\ \bar{a}(x,v),\ c(x,v,z) \text{ are lower semi-continuous in all variables} \tag{2.25}$$

Therefore, there exists a map $\hat{v}(x)$ which is Borel with values in $\mathcal{V}$ such that

$$H(u)(x) = L_u(x,\hat{v}(x)) \qquad \text{for all } x \tag{2.26}$$

where

$$L_u(x,v) = \ell(x,v) + \nabla u(x)\cdot h(x,v) - u(x)\bar{a}(x,v) + \int_{R^n}[u(x+z) - u(x) - z\cdot\nabla u(x)\chi_{|z|\le 1}] \times b(x,z)c(x,v,z)m(dz)$$

Notice that since in (2.23) p is arbitrary large, the function u is C^1.

Now let $\Omega = D([0,\infty);\ R^n)$ be the space of right-continuous functions having left limits, endowed with the Skorokhod metric. We note that $x(t;\omega) = \omega(t)$, the canonical process, $\mathcal{U}^t = \sigma(x(s),\ 0 \le s \le t)$, $\mathcal{U} = \mathcal{U}^\infty$, which coincides with the Borel σ-algebra on Ω_0. From D. Stroock [6] and Lepeltier and Marchal [5] it follows that there exists one and only one measure P^x on the $(\Omega,\mathcal{U})$ solution of the martingale problem for the operator A - B:

$$P^x[x(0) = x] = 1$$

$$\text{for all } \varphi \in \mathcal{D}(R^n),\ \varphi(x(t)) + \int_0^t (A - B)\varphi(x(s))\,ds \text{ is a } P^x,\ \mathcal{U}^t \text{ martingale} \tag{2.28}$$

Let $\mathcal{B}$ be the Borel σ-algebra on R^n, $\mathcal{B}_0$ the class of Borel sets whose closure does not contain 0. We define for $t > 0$, $A \in \mathcal{B}_0$,

$$\eta(t;A) = \text{number of jumps of the process } x(t) \text{ on } (0,t] \text{ whose value is in } A \tag{2.29}$$

Then $\eta(t;A)$ is an integer random measure, and

$$\zeta(t;A) = \eta(t;A) - \int_0^t \int_A b(x(s),z)m(dz)\,ds \tag{2.30}$$

is a measure martingale.

Moreover, there exists a standard n-dimensional Wiener process w(t) such that

$$x(t) = x + \int_0^t \sigma(x(s))\, dw(s) + \int_{\{|z|<1\}} z\zeta(t;dz) \qquad (2.31)$$
$$+ \int_{\{|z|>1\}} z\eta(t;dz) \qquad \text{for all } t, \text{ a.s. } P^x$$

where

$$\frac{\sigma^2}{2} = \text{matrix } a_{ij} \qquad (2.32)$$

We call an <u>admissible control</u> any stochastic process with values in V which is adapted to $\mathcal{U}^t$. Let v(t) be an admissible control; we will write

$$\begin{aligned} \ell_v(t) &= \ell(x(t),v(t)) \\ h_v(t) &= h(x(t),v(t)) \\ \bar{a}_v(t) &= \bar{a}(x(t),v(t)) \\ c_v(t,z) &= c(x(t),v(t),z) \end{aligned} \qquad (2.33)$$

We next define the process

$$\alpha_v(t) = \sigma^{-1}(x(t))[g_v(t) - \int_{\{|z|<1\}} zb(x(t),z)c_v(t,z)m(dz)]$$

and (2.34)

$$\chi_v(t) = \exp \int_0^t \alpha_v(s)\cdot dw(s) - \frac{1}{2}\int_0^t |\alpha_v(s)|^2\, ds$$
$$+ \int_0^t \int_{R^n} c_v(s,z)\zeta(ds,dz) \qquad (2.35)$$
$$- \int_0^t \int_{R^n} [c_v(s,z) - \ell n(1 + c_v(s,z))]\eta(ds,dz)$$

and one has

$$E^x \chi_v(t) = 1 \qquad \text{for all } t \qquad (2.36)$$

We then define a probability P_v^x on $\Omega, \mathcal{U}$ by setting

$$\left.\frac{dP_v^x}{dP^x}\right|_{\mathcal{U}_t} = \chi_v(t) \qquad (2.37)$$

Let us consider the functional

$$J^x(v(\cdot)) = E_v^x \int_0^\infty \ell(x(t),v(t)) \exp -\int_0^t \bar{a}(x(s),v(x))\, ds\, dt \qquad (2.38)$$

We define a particular admissible control $\hat{v}(t)$ by setting

$$\hat{v}(t) = \hat{v}(x(t)) \qquad (2.39)$$

Then we have:

THEOREM 2.1 We make the assumptions of Theorem 2.1 and (2.24), (2.25). Then the solution u of (2.23) is given explicitly by

$$u(x) = \inf_{v(\cdot)} j^x(v(\cdot)) \tag{2.40}$$

Moreover, $\hat{v}$ defined by (2.39) is an optimal control.

Details of the proof of Theorem 2.2 are given in Bensoussan and Lions [2]. An important step is to obtain an L^p estimate for P_v^x, independent of the control v, namely

$$|E_v^x \int_0^T \psi(x(t),t)\, dt| \leq C_T \, |\psi|_{L^p} \tag{2.41}$$

for any $\psi \in L^p(R^n \times (0,T))$, with $p > n + 2$, where the constant depends on T but not on ψ or on the control. This estimate is obtained from the Radon-Nikodym derivative (2.37) and a similar estimate which holds for P^x. Now L^p estimates allow us to derive a generalized Itô's formula in integrated form, which is valid for functions which are in $L^p(0,T;W^{2,p}_{loc}(R^n))$, with time derivatives in $L^p(0,T;L^p_{loc}(R^n))$. Then we can apply the verification method of dynamic programming, using (2.23), and obtain the desired result.

3. THE DIRICHLET PROBLEM FOR THE OPERATOR A - B - H

3.1 Setting of the Problem

Let $\mathcal{O}$ be an open domain of R^n such that

$$\mathcal{O} \text{ is bounded and the boundary } \Gamma = \partial\mathcal{O} \text{ is } C^2 \tag{3.1}$$

We will consider the operators A, B, H defined in Sec. 2.1, with the following additional assumption:

$$\frac{\partial b_i}{\partial x_k} \in L^\infty \qquad \text{for all } i, k \tag{3.2}$$

Now we are going to consider the operator A - B - H on functions defined on $\mathcal{O}$. However, since we are dealing with a nonlocal operator, we need to specify the value of the function in the whole space. This means that we have to define an extension of a function given on $\mathcal{O}$. Since we want to consider the Dirichlet problem relative to the operator $A - B - H$, it is natural to work with functions $u \in W_0^{1,p}(\mathcal{O})$, $2 \leq p < \infty$. The natural extension is then 0 outside $\mathcal{O}$. Such an extension is an isometry from $W_0^{1,p}(\mathcal{O})$ into $W^{1,p}(R^n)$. However, it has the following serious drawback. If $u \in W^{2,p}(\mathcal{O}) \cap W_0^{1,p}(\mathcal{O})$, then its extension does not belong to $W^{2,p}(R^n)$.

In the sequel when we consider functions in $W_0^{1,p}(\mathcal{O})$, we tacitly assume that they are extended by 0 outside $\mathcal{O}$, and belong to $W^{1,p}(R^n)$.

The operators B and H have some important properties:

LEMMA 3.1 We assume (2.3), (2.4), (2.6) and (3.2); then we have

$$B \in \mathcal{L}(W_0^{1,p};W^{-1,p}(\mathcal{O})) \qquad \text{for all } p \geq 2,\ p < \infty \tag{3.3}$$

and for all $r > 0$, B has the following decomposition property:

$$B = B_r^1 + B_r^2$$

where

$$B_r^1 \in \mathcal{L}(W_0^{1,p};W^{-1,p}) \qquad \|B_r^1\| \leq C\sigma(r)$$

with constants C and $\sigma(r)$ independent of $\mathcal{O}$ and p, and $\sigma(r) \to 0$ as $r \to 0$

$$B_r^2 \in \mathcal{L}(W_0^{1,p}(\mathcal{O});L^p(\mathcal{O}))$$

$$H: W_0^{1,p}(\mathcal{O}) \to L^p(\mathcal{O}) \qquad \text{for all } p \geq 2,\ p < \infty \tag{3.4}$$

$H(0) \in L^\infty$, and if $\varphi \in W_0^{1,p}$, $\psi \in W_0^{1,q}$, $q \geq p$, then $H(\varphi + \psi) - H(\varphi) \in L^q$ and

$$|H(\varphi + \psi) - H(\varphi)|_{L^q} \leq C\|\psi\|_{W_0^{1,q}}$$

where C does not depend on φ or ψ.

For all $\varphi, \psi \in H_0^1$ and $k \geq 0$, the following property holds:

$$\langle B\psi + H(\varphi + \psi) - H(\varphi), (\psi - k)^+\rangle$$
$$\leq C|(\psi - k)^+| \, \|(\psi - k)^+\| + k\gamma_1 \int_{\mathcal{O}} (\psi - k)^+ \, dx \quad (3.5)$$

Details of proof are in Bensoussan and Lions [2] or in Bensoussan [1]. Let us mention some key steps. We write B as

$$B = B_r^1 + B_r^2$$

where

$$B_r^1\varphi(x) = \int_{R^n} [\varphi(x + z) - \varphi(x) - z\cdot\nabla\varphi(x)]\chi_{|z|\leq r} \, bm(dz)$$

$$B_r^2\varphi(x) = -\nabla\varphi(x) \int_{R^n} z\chi_{r<|z|<1} bm(dz)$$
$$+ \int_{R^n} (\varphi(x + z) - \varphi(x))\chi_{r<|z|} bm(dz)$$

and for $\varphi, \psi \in \mathcal{D}(R^n)$, we can write by virtue of (3.2),

$$\int_{R^n} B_r^1\varphi(x)\psi(x) \, dx = \int_0^1 d\theta \int_0^\theta d\theta' \int_{R^n} dx \int_{R^n} m(dz)$$
$$\times [-\Sigma z_j z_k \frac{\partial\varphi}{\partial x_k} (x + \theta' z)$$
$$+ \psi(x) \frac{\partial b}{\partial x_j}(x,z) + \frac{\partial\psi}{\partial x_j}(x)b(x,z)]\chi_{|z|<r}$$

from which one deduces (3.3).

Property (3.5) is a maximum principle type of result.

We use, as in the proof of Lemma 2.3, the approximate operators B_ε, H_ε, and first prove (3.5) for these operators. An important step is the following property, interesting in itself:

$$\langle B_\varepsilon\varphi,\varphi\rangle \leq C|\varphi| \, \|\varphi\| \qquad \text{for all } \varphi \in H_0^1(\mathcal{O}) \quad (3.6)$$

3.2 Solution of the Dirichlet Problem

The main result is the following:

THEOREM 3.1 We assume (2.1), (2.3), (2.4), (2.6), (2.8) and (3.2). Let $f \in L^p(\mathcal{O})$, $2 < p < \infty$. Then there exists one and only one solution of

$$Au - Bu - H(u) + a_0u = f \qquad u \in W_0^{1,p}(\mathcal{O}) \quad (3.7)$$

This result is proved in Bensoussan and Lions [1]. The method is an analog to the one used for Theorem 2.1, with some important differences. Let us outline it for sake of completeness. First consider the equation

$$Au_\lambda - Bu_\lambda - H(u_\lambda) + (a_0 + \lambda)u_\lambda = f \qquad u_\lambda \in W_0^{1,p}(\mathcal{O}) \tag{3.8}$$

for λ sufficiently large, but fixed. Indeed, using (3.3) one finds $\lambda_1 > 0$ and $\delta > 0$ such that for all $\varphi_1, \varphi_2 \in H_0^1$, then

$$\langle (A - B - H)(\varphi_1) - (A - B - H)(\varphi_2),$$

$$\varphi_1 - \varphi_2\rangle \geq \|\varphi_1 - \varphi_2\|^2 - \lambda_1|\varphi_1 - \varphi_2|^2$$

Therefore, for $\lambda \geq \lambda_1$, $A - B - H + a_0 + \lambda$ is a monotone operator on H_0^1. This proves the existence and uniqueness of the solution of (3.8) in H_0^1.

The next step is to prove that the solution of (3.7) belongs to $W_0^{1,p}$. This is done as follows. Consider

$$Az - B_r^1 z = y \qquad g \in W^{-1,p}(\mathcal{O}) \tag{3.9}$$

Then at least for r small enough, using (3.3) it is easy to verify that B_r^1 is a small perturbation for the operator $A \in \mathcal{L}(W_0^{1,p}; W^{-1,p})$, which is invertible. Then rewriting (3.8) as

$$Au_\lambda - B_r^1 u_\lambda = f + H(u_\lambda) + B_r^2 u_\lambda - (a_0 + \lambda)u_\lambda$$

and using (3.9), properties of B_r^2 and H, it follows from a standard bootstrap regularity argument that the solution of (3.8) belongs to $W_0^{1,p}$.

In a similar way, one can solve

$$A\zeta - B\zeta - H(\zeta + \psi) + H(\psi) + (\lambda + a_0)\zeta = \varphi$$

$$\zeta \in W_0^{1,p}(\mathcal{O}) \tag{3.10}$$

where

$$\varphi \in L^p \qquad \psi \in H_0^1 \tag{3.11}$$

Moreover, when $\varphi \in L^\infty$, then using property (3.5) one can prove the following important estimate:

$$\|\zeta\|_{L^\infty} \le \frac{\|\varphi\|_{L^\infty}}{\lambda + \gamma_0 - \gamma_1} \tag{3.12}$$

Then one considers the iterative sequence

$$Au_{k+1} - Bu_{k+1} - H(u_{k+1}) + (\lambda + a_0)u_{k+1} = f + \lambda u_k$$
$$u_k \in W_0^{1,p}(\mathcal{O}) \tag{3.13}$$

The argument of regularity improvement for the sequence $u_{k+1} - u_k$, already used in Theorem 2.1, shows that for $k \ge j$, $u_k - u_{k-1}$ is bounded. From (3.12) one then deduces

$$\|u_{k+1} - u_k\|_{L^\infty} \le \rho\|u_k - u_{k-1}\|_{L^\infty} \qquad \rho < 1 \tag{3.14}$$

We complete the proof as in Theorem 2.1.

3.3 Penalization of the Domain

The solution of (3.7) does not belong to $W^{2,p}(R^n)$. Its restriction to $\mathcal{O}$ does not even belong to $W^{2,p}(\mathcal{O})$. We now show that we can approximate it by a function in $W^{2,p}(R^n)$, solutions of a problem (2.10), where we add in the equation a penalty term which vanishes only on $\mathcal{O}$. We define

$$q(x) = 1 - \chi_{\bar{\mathcal{O}}}(x) = \begin{cases} 0 & \text{if } x \in \bar{\mathcal{O}} \\ 1 & \text{if } x \notin \bar{\mathcal{O}} \end{cases} \tag{3.15}$$

We consider the following problem:

$$Au_\varepsilon - Bu_\varepsilon - Hu_\varepsilon + (a_0 + \frac{1}{\varepsilon}q)u_\varepsilon = f \qquad u_\varepsilon \in W^{2,p,\mu}(R^n) \tag{3.16}$$

where f is extended by 0 outside $\mathcal{O}$.

We are clearly in the situation of Theorem 2.1. Therefore, there exists one and only one solution of (3.16).

THEOREM 3.2 We make the assumptions of Theorem 3.1 and $f \in L^\infty$. Then we have [†]

$$u_\varepsilon \to u \qquad \text{in } H^{1,\mu}(R^n) \tag{3.17}$$

where u is the solution of (3.7), extended by 0 outside $\mathcal{O}$.

Proof. When $f \in L^\infty$, we have the following estimate:

---

† $H^{1,\mu} = H^{1,2,\mu}$.

$$\|u_\varepsilon\|_{L^\infty} \leq \frac{\|f\|_{L^\infty} + \|H(0)\|_{L^\infty}}{\gamma_0 - \gamma_1} \tag{3.18}$$

This follows from (2.22), which we write as

$$Au_{k+1} - Bu_{k+1} - H(u_{k+1}) + H(0) + (a_0 + \lambda)u_{k+1} = f + \lambda u_k + H(0)$$

and (2.18) implies that

$$\|u_{k+1}\|_{L^\infty} \leq \frac{\lambda\|u_k\|_{L^\infty} + \|f\| + \|H(0)\|}{\lambda + \gamma_0 - \gamma_1}$$

Letting k tend to $+\infty$, we deduce (3.18).

But then we can write (3.16) as

$$Au_\varepsilon - Bu_\varepsilon - H(u_\varepsilon) + \; a_0 + \lambda + \frac{1}{\varepsilon} q \; u_\varepsilon = f + \lambda u_\varepsilon \tag{3.19}$$

Multiplying by $\beta_\mu^2 u_\varepsilon$ and integrating over x, using property (3.3), one obtains, if λ is large enough,

$$\|u_\varepsilon\|_{H^{1,\mu}} \leq C \tag{3.20}$$

where we have used the fact that $|u_\varepsilon|_{L^{2,\mu}} \leq C$.

Moreover,

$$\frac{1}{\varepsilon}\int_{R^n - \mathcal{O}} |u_\varepsilon|^2 \beta_\mu^2 \, dx \leq C \tag{3.21}$$

We extract a subsequence such that

$$u_\varepsilon \to w \qquad \text{in } H^{1,\mu} \text{ weakly}$$

$$H(u_\varepsilon) \to \xi \qquad \text{in } L^{2,\mu} \text{ weakly}$$

From (3.21) it now follows that $w = 0$ on $R^n - \mathcal{O}$; hence its restriction to $\mathcal{O}$ belongs to $H_0^1(\mathcal{O})$.

Now

$$\int_{R^n} |u_\varepsilon - w|^2 \beta_\mu^2 \, dx \leq C\varepsilon + \int_{\mathcal{O}} |u_\varepsilon - w|^2 \beta_\mu^2 \, dx$$

and the right-hand side tends to 0, by a compactness argument. Hence

$$u_\varepsilon \to w \qquad \text{in } L^{2,\mu} \text{ strongly} \tag{3.22}$$

Multiplying (3.19) by $\beta_\mu^2 u_\varepsilon$ we then obtain

$$\langle Au_\varepsilon - Bu_\varepsilon, u_\varepsilon \rangle_\mu + \frac{1}{\varepsilon}\int_{R^n - \mathcal{O}} u_\varepsilon^2 \beta_\mu^2 \, dx \to (f,w)_\mu + (\xi,w)_\mu - (a_0 w, w)_\mu$$

Also,

$$\langle Au_\varepsilon - Bu_\varepsilon, w\rangle_\mu \to (f,w)_\mu + (\xi,w)_\mu - (a_0 w,w)_\mu$$

and

$$\langle Aw - Bw, w\rangle_\mu = (f,w)_\mu + (\xi,w)_\mu - (a_0 w,w)_\mu$$

From these relations, it follows that

$$\langle (A - B)(u_\varepsilon - w),\ u_\varepsilon - w\rangle_\mu \to 0$$

which with (3.22) allows us to conclude that $u_\varepsilon \to w$ in $H^{1,\mu}$ strongly. This and (3.18) imply, as can easily be checked, that $H(u_\varepsilon) \to H(w)$ in $L^{2,\mu}$ strongly. Hence $\xi = H(w)$. From this we conclude that $w = u$, and thus (3.17) holds.

REMARK 1. The assumption $f \in L^\infty$ is probably not necessary. For instance, if a_0 is sufficiently large, we can take $f \in L^p(\mathcal{O})$.

4. PROBABILISTIC INTERPRETATION OF THE SOLUTION OF THE DIRICHLET PROBLEM

4.1 Stochastic Control Problem

We use the notation of Sec. 2.3, take $f = 0$, introduce $\bar{a}(x,v)$, and write (3.7) as

$$Au - Bu - H(u) = 0 \qquad u \in W_0^{1,p}(\mathcal{O}) \qquad \text{for all } 2 \leq p < \infty \tag{4.1}$$

In particular, $u \in C^0(\mathcal{O})$. We need to introduce a map $\hat{v}(x)$ as in (2.26). However, since u is not C^1, we cannot proceed in the same way. We modify (2.25) as follows:

$\ell(x,v)$, $h(x,v)$, $\bar{a}(x,v)$, $c(x,v,z)$ are Lebesgue measurable in all variables and continuous with respect to v (4.2)

In that case, $L_u(x,v)$ becomes a Caratheodory function (see Ekeland and Temam [3], chap. VIII); hence there exists a (Lebesgue)-measurable $\hat{v}(x)$ such that (2.26) still holds.

Consider now the probabilistic setup described in Sec. 2.3. We will denote

$$\tau = \inf\{t \geq 0 \mid x(t) \text{ or } x(t-) \notin \mathcal{O}\} \tag{4.3}$$

Since $\mathcal{O}$ is open bounded and $x(t)$ is right continuous, τ is a $\mathcal{M}^t$ stopping time (of Dynkin [7]). Considering the probability P_v^x defined by (2.37), we define the functional

$$J^x(v(\cdot)) = E_v^x \int_0^\tau \ell(x(t)) \exp -\int_0^t \overline{a}(x(s),v(s))\, ds\, dt \tag{4.4}$$

We also define $\hat{v}(t)$ by (2.39). Then we have:

THEOREM 4.1 We make the assumptions of Theorem 3.1, (2.24), and (4.2). Then the solution of (4.1) is given explicitly by:

$$u(x) = \inf_{v(\cdot)} J^x(v(\cdot)) \qquad \text{defined by (4.4)} \tag{4.5}$$

Moreover, $\hat{v}$ is an optimal control.

4.2 Some Preliminary Results

Let us first notice the following property. If $\varphi \in L^\infty(R^n)$, and $\varphi = 0$ a.e., then

$$E^x\varphi(x(t)) = 0 \qquad \text{for all } t > 0 \tag{4.6}$$

This follows from the fact that the function

$$\psi(x,t) = E^x\varphi(x(t))$$

is the solution of the Cauchy problem

$$\frac{\partial\psi}{\partial t} + (A - B)\psi = 0$$
$$\psi(x,o) = \varphi(x) \tag{4.7}$$

We have $\psi \in L^2(0,T;H^1_\mu)$, ψ bounded, and $\psi \in W^{2,1,p,\mu}(\varepsilon,T)$ for all $\varepsilon > 0$. In particular, $\psi \in C^0(R^n \times [\varepsilon,T])$; hence (4.6) if $\varphi = 0$ a.e. A similar property holds for the probability P_v^x associated to any control, since

$$|E_v^x\varphi(x(t))| \leq C_t|E^x\varphi(x(t))|$$

Now consider

$$\hat{\tau} = \inf\{t \geq 0 \mid x(t) \notin \overline{\mathcal{O}}\} \tag{4.8}$$

Then $\hat{\tau}$ is a $\mathcal{M}^{t+0}$ stopping time (see Dynkin [7]).

We have:

LEMMA 4.1 $\tau = \hat{\tau}$, P^x a.s.

Proof. Let us define for λ fixed $\lambda > 0$:

$$\hat{\varphi}(x) = \frac{1}{\lambda} E^x(1 - e^{-\lambda\hat{\tau}}) \tag{4.9}$$

Next consider φ_ε to be the solution of

$$A\varphi_\varepsilon - B\varphi_\varepsilon + \lambda + \frac{1}{\varepsilon} q \varphi_\varepsilon = 1 \qquad \varphi_\varepsilon \in W^{2,p,\mu}(R^n)$$

$$2 \leq p < \infty \tag{4.10}$$

We have, as a particular case of Theorem 2.2,

$$\varphi_\varepsilon(x) = E^x\int_0^\infty e^{-\lambda t} \mid \exp - \frac{1}{\varepsilon}\int_0^t q(x(s))\, ds \mid dt \tag{4.11}$$

Now let us check that

$$\varphi_\varepsilon(x) \to \tilde{\varphi}(x) \qquad \text{for all } x, \text{ as } \varepsilon \to 0 \tag{4.12}$$

This follows from the convergence

$$\exp - \frac{1}{\varepsilon}\int_0^t q(x(s))\, ds \to \begin{cases} 0 & \text{if } t > \hat{\tau} \\ 1 & \text{if } t \leq \hat{\tau} \end{cases} \tag{4.13}$$

Let us prove (4.13). The process being right continuous, then for $t > \hat{\tau}$, for some t_0, $\hat{\tau} \leq t_0 < t$ such that $x(t_0) \notin \bar{\mathcal{O}}$; hence for some δ (random) such that $x(s) \notin \bar{\mathcal{O}}$ for $s \in [t_0, t_0 + \delta]$ and $t_0 + \delta \leq t$. But then

$$\exp - \frac{1}{\varepsilon}\int_0^t q(x(s))\, ds \leq \exp - \frac{\delta}{\varepsilon} \to 0$$

On the other hand, let $t \leq \hat{\tau}$; then

$$\exp - \frac{1}{\varepsilon}\int_0^t q(x(s))\, ds = 1$$

Now, from Theorem 3.2, in a particular case we have

$$\varphi_\varepsilon \to \varphi \qquad \text{in } H^{1,\mu} \tag{4.14}$$

where φ is the solution of

$$A\varphi - B\varphi + \lambda\varphi = 1 \qquad \varphi \in W_0^{1,p}(\mathcal{O}) \text{ for all } 2 \leq p < \infty \tag{4.15}$$

From (4.14) and (4.12) we deduce that

$$\varphi = \tilde{\varphi} \text{ a.e.} \tag{4.16}$$

Let us show that

$$\varphi(x) = \tilde{\varphi}(x) \qquad \text{for all } x \tag{4.17}$$

Indeed, using integrated Itô's formula applied to φ_ε we have

$$\varphi_\varepsilon(x) = E^x \int_0^t e^{-\lambda s} \exp - \frac{1}{\varepsilon} \int_0^s q(x(\lambda))\, d\lambda\, ds$$
$$+ \varphi_\varepsilon(x(t))e^{-\lambda t} \exp - \frac{1}{\varepsilon} \int_0^t q(x(\lambda))\, d\lambda \qquad \text{for all } t > 0$$

Passing to the limit as $\varepsilon \to 0$, using (4.12) and (4.13), we obtain

$$\tilde{\varphi}(x) = E^x \int_0^{t\wedge\hat{\tau}} e^{-\lambda s}\, ds + \tilde{\varphi}(x(t))e^{-\lambda t}\chi_{t<\hat{\tau}} \tag{4.18}$$

But

$$|E^x(\varphi - \tilde{\varphi})(x(t))\chi_{t\leq\hat{\tau}}| \leq E^x|\varphi - \tilde{\varphi}|(x(t)) = 0$$

by (4.6) and (4.17). Therefore, (4.18) implies that

$$\tilde{\varphi}(x) = E^x \int_0^{t\wedge\hat{\tau}} e^{-\lambda s}\, ds + e^{-\lambda t}\varphi(x(t))\chi_{t\leq\hat{\tau}} \qquad \text{for all } t > 0$$

Using (4.9) we deduce that

$$E^x\varphi(x(t))\chi_{t\leq\hat{\tau}}e^{-\lambda t} = E^x\int_{t\wedge\hat{\tau}}^{\hat{\tau}} e^{-\lambda s}\, ds \tag{4.19}$$

Now let $t \downarrow 0$; since φ is continuous and $s(t)$ right continuous we have

$$\varphi(x(t)) \to \varphi(x) \qquad P^x \text{ a.s.}$$

On the other hand, since $t > 0$,

$$\chi_{t\leq\hat{\tau}} \to \chi_{0<\hat{\tau}} \qquad \text{as } t \downarrow 0$$

if $x \in \bar{\mathcal{O}}$. Therefore, letting $t \downarrow 0$ in (4.19) we obtain

$$\varphi(x) = E^x \int_0^{\hat{\tau}} e^{-\lambda s}\, ds$$

which coincides with the right-hand side of (4.9). We thus obtained (4.17) for $x \in \bar{\mathcal{O}}$. For $x \in \bar{\mathcal{O}}$, we clearly have $\varphi = \tilde{\varphi} = 0$. Now let θ be an $\mathcal{M}^{t+0}$ stopping time. We can write

$$\varphi_\varepsilon(x) = E^x \int_0^\theta e^{-\lambda s} \exp - \frac{1}{\varepsilon} \int_0^s q(x(\lambda))\, d\lambda\, ds$$
$$+ \varphi_\varepsilon(x(\theta))e^{-\lambda\theta} \exp - \frac{1}{\varepsilon} \int_0^\theta q(x(\lambda))\, d\lambda$$

Letting $\varepsilon \to 0$, we deduce from (4.12) and (4.17) that

$$\varphi(x) = E^x \int_0^{\theta\wedge\hat{\tau}} e^{-\lambda s}\, ds + \varphi(x(\theta))e^{-\lambda\theta}\chi_{\theta<\hat{\tau}} \tag{4.20}$$

Consider a sequence $\mathcal{O}_k$ of open sets such that

$$\mathcal{O}_k \subset \mathcal{O}_{k+1} \qquad \bigcup_k \mathcal{O}_k = \mathcal{O} \qquad \bar{\mathcal{O}}_k \subset \mathcal{O}$$

Let τ^k be the first exit time of the process from $\bar{\mathcal{O}}_k$. We have $\tau^k \leq \tau \leq \hat{\tau}$. We apply (4.20) with $\theta = \tau^k$. We get

$$\varphi(x) = E^x \int_0^{\tau^k} e^{-\lambda s}\, ds + \varphi(x(\tau^k))e^{-\lambda\tau^k}$$

But $x(\tau^k) \notin \mathcal{O}_k$. Now since φ is uniformly continuous on R^n and $\varphi = 0$ on $R^n - \mathcal{O}$, we have

$$\sigma_k = \sup_{x \notin \mathcal{O}_k} |\varphi(x)| \to 0$$

Hence

$$\left|\varphi(x) - E^x \int_0^{\tau^k} e^{-\lambda s}\, ds\right| \leq \sigma_k$$

But

$$\varphi(x) = E^x \int_0^{\hat{\tau}} e^{-\lambda s}\, ds$$

Hence

$$E^x \int_{\tau^k}^{\hat{\tau}} e^{-\lambda s}\, ds \leq \sigma_k$$

Hence also, since $\tau^k \leq \tau \leq \hat{\tau}$,

$$E^x \int_{\tau}^{\hat{\tau}} e^{-\lambda s}\, ds \leq \sigma_k$$

which implies the desired result.

4.3 Proof of Theorem 4.1

We first notice that from Lemma 4.1 and the absolute continuity of P_v^x with respect to P^x on $\mathcal{M}^t$, we have

$$\tau \wedge t = \hat{\tau} \wedge t \qquad P_v^x \text{ a.s., for all } t$$

Hence also

$$\tau = \hat{\tau} \qquad P_v^x \text{ a.s.} \tag{4.21}$$

Therefore

$$J^x(v(\cdot)) = E_v^x \int_0^{\hat{\tau}} \ell(x(t),v(t)) \exp - \int_0^t \bar{a}(x(s),v(s))\, ds\; dt \tag{4.22}$$

Now consider the penalized problem associated with (4.1), i.e.,

$$Au_\varepsilon - Bu_\varepsilon - H(u_\varepsilon) + \frac{1}{\varepsilon} qu_\varepsilon = 0 \qquad u_\varepsilon \in W^{2,p,\mu}(R^n) \tag{4.23}$$

According to Theorem 2.2, we can interpret $u_\varepsilon(x)$ as

$$u_\varepsilon(x) = \inf_{v(\cdot)} E_v^x \int_0^\infty \ell_v(t) \exp\left[-\int_0^t \bar{a}_v(s)\, ds\right] \exp\left[-\frac{1}{\varepsilon}P\int_0^t q(x(s))\, ds\right] dt$$

$$= \inf_{v(\cdot)} J_\varepsilon^x(v(\cdot))$$

Now

$$J^x(v(\cdot)) - J_\varepsilon^x(v(\cdot)) = E_v^x \int_{\hat{\tau}}^{+\infty} \ell_v(t) \exp -\int_0^t \bar{a}_0\, ds$$
$$\times \exp -\frac{1}{\varepsilon}\int_0^t q\, ds \quad dt$$

Hence

$$|J^x(v(\cdot)) - J_\varepsilon^x(v(\cdot))| \le C\, E_v^x \int_{\hat{\tau}}^\infty e^{-(\gamma_0-\gamma_1)t} \exp -\frac{1}{\varepsilon}\int_0^t q\, ds\, dt$$
$$\le CE_v^x \int_{\hat{\tau}\wedge T}^\infty + C\, E_v^x \int_{\hat{\tau}\wedge T}^T e^{(\gamma_0-\gamma_1)T}$$

But

$$E_v^x \int_{\hat{\tau}\wedge T}^T = E^x \chi_v(T) \int_{\hat{\tau}\wedge T}^T$$
$$\le C_T E^x \int_{\hat{\tau}\wedge T}^T \exp -\frac{1}{\varepsilon}\int_0^t q\, ds \quad dt$$
$$= \mathcal{O}_T(\varepsilon) \to 0 \qquad \text{as } \varepsilon \to 0 \text{ for fixed } T$$

Therefore, gathering results, we obtain

$$|J_\varepsilon^x(v(\cdot)) - J^x(v(\cdot))| \le \mathcal{O}_T(\varepsilon) + \frac{C}{T}$$

the right-hand side being independent of v. Hence

$$u_\varepsilon(x) \to \tilde{u}(x) = \inf_{v(\cdot)} J^x(v(\cdot)) \tag{4.25}$$

From Theorem 3.2 we then have

$$u(x) = \tilde{u}(x) \qquad \text{a.e.} \tag{4.26}$$

We next prove that (4.26) holds everywhere. We use a reasoning similar to that of Lemma 4.1. We have indeed

$$E_v^x u_\varepsilon(x(t)) \exp -\int_0^t \bar{a}_v(s)\, ds \quad \exp -\int_0^t \frac{1}{\varepsilon} q(x(s))\, ds$$
$$= E_v^x \int_t^\infty \ell_v(s) \exp -\int_0^s \bar{a}_v(\lambda)\, d\lambda \quad \exp -\int_0^s \frac{q(x(\lambda))}{\varepsilon} \quad ds$$
$$+ E_v^x \int_t^\infty |H(u_\varepsilon)(x(s)) - L_{u_\varepsilon}(x(s),v(s)| \exp -\int_0^s \bar{a}_v(\lambda)\, d\lambda$$
$$\times \exp -\int_0^s \frac{q(x(\lambda))}{\varepsilon}\, d\lambda \quad ds$$

From the definition of H, it follows that

$$E_v^x u_\varepsilon(x(t)) \exp -\int_0^t \bar{a}_v(s)\, ds \quad \exp -\int_0^t \frac{1}{\varepsilon} q(x(s))\, ds \qquad (4.27)$$
$$\geq E_v^x \int_t^\infty \ell_v(s) \exp -\int_0^s \bar{a}_v(\lambda)\, d\lambda \quad \exp -\int_0^s \frac{q(x(\lambda))}{\varepsilon}\, d\lambda \quad ds$$

We let $\varepsilon \to 0$. From (4.25) and (4.13) we obtain

$$E_v^x \tilde{u}(x(t)) \exp -\int_0^t \bar{a}_v(s)\, ds\ \chi_{t\leq\hat{\tau}} \geq E_v^x \int_t^\infty \ell_v(s)$$
$$\times \exp -\int_0^s \bar{a}_v(\lambda)\, d\lambda\ \chi_{s\leq\hat{\tau}}\ ds$$

From (4.26) and the comments made before Lemma 4.1, we can assert that

$$E_v^x u(x(t)) \exp -\int_0^t \bar{a}_v(s)\, ds\ \chi_{t\leq\hat{\tau}} \geq E_v^x \int_t^\infty \ell_v(s)$$
$$\times \exp -\int_0^s \bar{a}_v(\lambda)\, d\lambda\ \chi_{s\leq\hat{\tau}}\ ds$$

Letting $t \downarrow 0$, and using the continuity of u, we obtain

$$u(x) \geq J_x(v(\cdot)) \qquad \text{for all } v(\cdot) \qquad (4.28)$$

Now we can write by definition of $\hat{v}$,

$$\hat{A}u - \hat{B}u + \bar{a}(x,\hat{v})u = \ell(x,\hat{v}) \qquad (4.29)$$

where $\hat{A}$ and $\hat{B}$ are the following modifications of A and B:

$\hat{A} = A - \nabla \cdot h(x,\hat{v})$

$\hat{B}$ = same definition as B, with b changed into $b(1 + c(x,\hat{v},z))$

From the linear theory [similar to what we have done in order to interpret (4.15)] we can write

$$u(x) = \hat{E}^x \int_0^\infty \ell(x(t),\hat{v}(t)) \exp -\int_0^t \bar{a}(x(s),\hat{v}(s))\, ds \quad dt$$

where $\hat{P}^x$ is the solution of the martingale problem for the operator $\hat{A} - \hat{B}$. But this is nothing other than

$$u(x) = J^x(\hat{v}(\cdot)) \qquad (4.30)$$

With (4.28) and (4.30) we conclude.

REFERENCES

1. A. Bensoussan, On the Hamilton Jacobi approach for the optimal control of diffusion processes with jumps, pp. 25-55, in Stochastic Analysis (A. Friedman and M. Pinsky, eds.), Academic Press, New York, 1978.
2. A. Bensoussan and J. L. Lions, Contrôle impulsionnel et inéquations quasi variationnelles, Dunod, Paris, 1982.
3. I. Ekeland and R. Temam, Convex Analysis and Variational Problems, North-Holland, Amsterdam, 1976.
4. P. L. Lions, Sur quelques classes d'équations aux dérivées partielles non linéaires et leur résolution numérique, Thèse d'État, Université de Paris VI, 1979.
5. J. P. Lepeltier and B. Marchal, Problèmes de martingales et équations différentielles stochastiques associées à un opérateur intégro-différentiel, Ann. Inst. H. Poincaré, B12: 43-103 (1976).
6. D. Stroock, Diffusion Processes associated with Levy generators, Z. Wahrscheinlichkeitsth. Verwend. Geb. 32: 209-244 (1975).
7. E. Dynkin, Theory of Markov Processes, Pergamon Press, Oxford, England, 1960.

3

Necessary Conditions on Optimal Markov Controls for Stochastic Processes

JURGEN FRANKE*/University of Heidelberg, Heidelberg, Federal Republic of Germany

1. INTRODUCTION

Most discussions of the control of stochastic processes are based more or less on dynamic programming principles. Bismut [2,3] presented a different method for deriving conditions on optimal stochastic controls for convex problems based on the convex optimization theory of Rockafellar. We take this functional analytic approach a step further. We demonstrate that locally convex optimization techniques, which have been developed in the last 20 years to a powerful tool for treating deterministic extremum problems, can also be applied to stochastic control problems.

This kind of approach lacks the intuitive component of the dynamic programming principle. It leads, however, to another type of result, sometimes being more general or holding under different conditions than those derived by dynamic programming (see, e.g., Bismut [2] and Franke [5,6]).

Let us consider the stochastic differential equation

$$d\xi_t = a(\xi_t, u(t,\xi_t))\, dt + b(\xi_t, u(t,\xi_t))\, dW(t) \qquad 0 \leq t \leq T$$

$$\xi_0 = x_0$$

*Current affiliation: University of Frankfurt, Frankfurt am Main, Federal Republic of Germany where a, b are nonrandom and $W(t)$ is an m-dimensional Brownian motion. We want to minimize a loss functional of Bolza type:

$$L(\underline{u}) = E[R(\xi_T) + \int_0^T S(\xi_t, u(t,\xi_t))\, dt]$$

where again R, S are nonrandom functions. The controls $\underline{u}$ are elements of some Banach space of functions of t, x.

We prove in the following that, under suitable conditions, the functional L is directional differentiable and its derivatives are bounded linear functionals; i.e., L behaves locally like a convex functional. We give explicit expressions for those derivatives. Then from results of locally convex optimization theory we get immediately necessary conditions on optimal controls submitted to various kinds of constraints. In particular, we derive a local stochastic maximum principle for a constraint

$$u(t,x) \in U(t) \qquad \text{for all } x, \text{ almost all } t$$

where $U(t)$, $0 \leq t \leq T$, is a sequence of convex subsets of R^q.

It is not the purpose of this chapter to prove a most general result. Principally, we want to demonstrate simply that the method works for a stochastic control problem without complicating the proofs with technical details. Therefore, we choose rather stringent smoothness conditions on a, b, R, S and the space of control functions. These conditions restrict the class of problems to which the results of this chapter may be applied. Arguing more carefully in the proofs it is possible to derive analogous results under weaker conditions.

Recently, Arkin and Saksonov [1] have discussed a related approach to problems of control of stochastic processes. They discuss a stochastic differential equation where the noise term does not depend explicitly on the control. The loss functional is of the form $R(E\xi_T)$. The controls are not of the Markov type we are considering here but are stochastic processes adapted to a given increasing sequence of σ-algebras. Using the Pontryagin maximum principle, which is related to locally convex optimization (e.g., Girsanov [9], chaps. 12 and 13), they derive necessary conditions on optimal controls for this kind of control problem.

The method that we present in the following is not applicable only to Markov control problems. It may also be used to get necessary conditions on optimal controls of the type considered by Arkin and Saksonov, even when the loss functional is of a more complicated form and the noise term also depends on the control. This will be the topic of a future paper.

1.1 Notation

Let $x \in R^n$, c be a real $n \times m$-matrix. $|x|$ ($|c|$) denotes the euclidean norm of x(c). x^T (c^T) denotes the transpose of the column vector x (of c). $c_{\cdot i}$ denotes the ith column of c. u denotes a vector in the control space R^q. $\underline{u}$ denotes a control function of time t and state x which assumes values in R^q. sup denotes the essential supremum if applied to a noncontinuous function.

Let a(x,u) [S(x,u)] be an R^n-valued (real-valued) differentiable function on $R^n \times R^q$. $\partial_x a(x,u)$ [$\nabla_x S(x,u)$] denotes the functional matrix (the gradient) of a (S) with respect to x; i.e., for j, k = 1,...,n,

$$[\partial_x a(x,u)]_{jk} = \frac{\partial}{\partial x_k} a_j(x,u), \qquad [\nabla_x S(x,u)]_k = \frac{\partial}{\partial x_k} S(x,u)$$

$\partial_u a(x,u)$ [$\nabla_u S(x,u)$] denotes the functional matrix (the gradient) of a (S) with respect to u.

Let H(t,x,u,dt), $0 \le t \le T$, $x \in R^n$, $u \in R^q$, be a stochastic differential given by

$$H(t,x,u,dt) = f(t,x,u)\,dt + g(t,x,u)\,dW(t)$$

where f(g) is R^n-valued ($n \times m$-matrix-valued) and a.s. differentiable with respect to x and u, and W(t), $0 \le t \le T$, is an m-dimensional Brownian motion. The n-dimensional stochastic differential $\langle \partial_x H(t,x,u,dt) | y \rangle$, $0 \le t \le T$, $x,y \in R^n$,

$u \in R^q$, is defined by

$$\langle \partial_x H(t,x,u,dt) | y \rangle = \partial_x f(t,x,u) y \, dt + \sum_{i=1}^{m} \partial_x g_{\cdot i}(t,x,u) y \, dW_i(t)$$

Analogously, the q-dimensional stochastic differential $\langle \partial_u H(t,x,u,dt) | v \rangle$, $0 \le t \le T$, $x \in R^n$, $u,v \in R^q$ is defined.

2. ANALYTIC PROPERTIES OF THE LOSS FUNCTIONAL

In the following we adopt the framework given by Bismut [3]. Let (Ω, F, P) be a complete probability space, $T > 0$ a constant, $\{F_t, 0 \le t \le T\}$ be an increasing, right-continuous sequence of complete sub-σ-algebras of F which has no time of discontinuity. $\mathcal{T}$ denotes the σ-algebra on $\Omega \times [0,T]$ of the well-measurable sets, $\mathcal{T}^*$ its completion for the measure $dP \otimes dt$. Let $W(t)$, $0 \le t \le T$, be an m-dimensional Brownian motion on (Ω, F, P), nonanticipating relative to $\{F_t, 0 \le t \le T\}$.

Let $G(x,u,dt)$, $x \in R^n$, $u \in R^q$, be a stochastic differential given by

$$G(x,u,dt) = a(x,u) \, dt + b(x,u) \, dW(t)$$

where $a(x,u)$ [$b(x,u)$] is a n-vector-valued ($n \times m$-matrix-valued) Borel-measurable function on $R^n \times R^q$.

For $x_0 \in R^n$ we consider the controlled stochastic differential equation without after effect:

$$d\xi_t = G(\xi_t, u(t,\xi_t), dt) \qquad 0 \le t \le T$$
$$\xi_0 = x_0 \tag{2.1}$$

We assume that the coefficients of (2.1) satisfy the following smoothness conditions:

a, b are continuously differentiable with respect to x, u and all their partial derivatives are uniformly bounded on $R^n \times R^q$ (2.2)

The control functions $\underline{u}$ are elements of a Banach space E of Borel-measurable functions on $[0,T] \times R^n$. The following results do not depend on a specific choice of E. For special problems of control different Banach spaces may provide an appropriate functional analytic framework. For example, if one is interested in homogeneous controls only, one may choose E as a space of functions which do not depend on time t.

However, the necessary conditions on optimal controls which we shall derive in this chapter are of a local nature; i.e., they involve spatial derivatives of the control functions. Therefore, the functions in E must at least be absolutely continuous with respect to x. In the following, $\|\cdot\|$ denotes the norm of E, and we assume that E satisfies the following condition:

i. E consists of Borel-measurable functions $\underline{u}$ on $[0,T] \times R^n$ with values in R^q.

ii. For $k = 1,\ldots,q$, $j = 1,\ldots,n$ and for all fixed x_i, $i \neq j$, and all t, $u_k(t,x)$ is an absolutely continuous function of x_j with density

$$\frac{\partial}{\partial x_j} u_k(t,x) \tag{2.3}$$

The functional matrix $\partial_x u(t,x)$ is a $q \times n$-matrix-valued, Borel-measurable, essentially bounded function on $[0,T] \times R^n$.

iii. For all $\rho > 0$ there exists $C_\rho > 0$ such that for all $\underline{u} \in E$ with $\|\underline{u}\| \leq \rho$,

$$\sup_t |u(t,0)| + \sup_{t,x} |\partial_x u(t,x)| \leq C_\rho \|\underline{u}\|$$

To have a specific situation in mind one may, e.g., think of E as the space of functions satisfying (2.3), endowed with the norm

$$\|\underline{u}\| = \sup_t |u(t,0)| + \sup_{t,x} |\partial_x u(t,x)|$$

As an immediate consequence of (2.3) the functions $\underline{u} \in E$ are linearly bounded and satisfy a uniform Lipschitz condition with respect to x. More precisely, for all $\rho > 0$ there exists C_ρ such that for all $\underline{u} \in E$ with $\|\underline{u}\| \leq \rho$

$$\sup_t |u(t,x)| \leq C_\rho \|\underline{u}\| (1 + n|x|)$$
$$\sup_t |u(t,x) - u(t,y)| \leq C_\rho \|\underline{u}\| n |x - y| \quad \text{for all } x,y \in R^n \tag{2 4}$$

Finally, let $S(x,u)$ [$R(x)$] be a continuously differentiable function on $R^n \times R^q$ (R^n) whose partial derivatives are uniformly bounded with respect to all arguments. (2.5)

By Lemma A2 of the Appendix there exists a solution ξ_t, $0 \leq t \leq T$, of (2.1) for every $\underline{u} \in E$. We consider the following loss functional on E:

$$L(\underline{u}) = E R(\xi_T) + E \int_0^T S(\xi_t, u(t,\xi_t))\, dt \tag{2.6}$$

An essential prerequisite for the applicability of general locally convex optimization techniques to stochastic control problems where loss functionals of the form (2.6) are involved is the following: L must behave locally like a convex functional. In particular, it has to be differentiable in an appropriate sense in the point of interest, and the derivative has to be a convex functional (Ref. 9, chap. 7). It is the main result of this chapter that L satisfies smoothness conditions of this kind. Theorem 2.1 presents the precise formulation. Due to this theorem the well-developed techniques of locally convex optimization can be applied to stochastic control problems of the type we are considering here. We illustrate this with some corollaries which provide necessary conditions on optimal controls under different types of constraints.

THEOREM 2.1 (i) L satisfies a local Lipschitz condition; i.e., for all $\rho > 0$ there exists $C_\rho > 0$ such that for all $\underline{u},\underline{v} \in E$ with $\|\underline{u}\|$, $\|\underline{v}\| \leq \rho$,

$$|L(\underline{u}) - L(\underline{v})| \leq C_\rho \|\underline{u} - \underline{v}\|$$

(ii) Let $\underline{u}^0 \in E$ such that $\partial_x u^0(t,x)$ is continuous in x.

Then for all $\underline{u} \in E$ there exists the directional derivative

$$L_0'(\underline{u}) = \lim_{\varepsilon \to 0} \frac{1}{\varepsilon} [L(\underline{u}^0 + \varepsilon\underline{u}) - L(\underline{u}^0)]$$

of L at $\underline{u}^0$ in direction $\underline{u}$. L_0' is a continuous linear functional on E.

(iii) Let the stochastic differential Γ^0 be defined by

$$\Gamma^0(t,x,dt) = G(x,u^0(t,x),dt)$$

Let ξ_t^0, $0 \leq t \leq T$, be the solution of (2.1) for control u^0. The linear homogeneous matrix equation

$$d\Phi(t) = \langle \partial_x \Gamma^0(t,\xi_t^0,dt) | \Phi(t) \rangle \qquad 0 \leq t \leq T$$

$$\Phi(0) = I_n \tag{2.7}$$

where $\Phi(t)$, $0 \leq t \leq T$, is a $n \times n$-matrix-valued process and I_n is the $n \times n$-unit matrix, has a unique solution. Furthermore, for all t there exists the inverse $\Phi^{-1}(t)$.

(iv) Let the functions β^0, σ^0 be defined by

$$\beta^0(t,x) = b(x,u^0(t,x)) \qquad \sigma^0(t,x) = S(x,u^0(t,x))$$

Let the stochastic differential g^0 be given by

$$g^0(t,x,u,dt) = G(x,u,dt) - \sum_{i=1}^{m} \{\partial_x \beta^0_{\cdot i}(t,x)\} b_{\cdot i}(x,u)\, dt$$

Let η_t, $0 \leq t \leq T$, be defined by

$$\eta_t^T = \nabla_x^T R(\xi_T^0)\Phi(T) + \nabla_x^T \sigma^0)t,\xi_t^0)\Phi(t)$$

Then for all $\underline{u} \in E$,

$$L_0'(\underline{u}) = \int_0^T \eta_t^T \int_0^t \Phi^{-1}(s) \langle \partial_u g^0(s,\xi_s^0,u^0(s,\xi_s^0),ds) | u(s,\xi_s^0) \rangle\, dt$$

$$+ E \int_0^T \nabla_u^T S(\xi_t^0, u^0(t,\xi_t^0)) u(t,\xi_t^0)\, dt \tag{2.8}$$

Proof. (a) (i) has been proven already as Lemma A5. From Lemma A6 we know that L is directionally differentiable at $\underline{u}^0$ in any direction $\underline{u} \in E$. Additionally, we have an explicit expression for $L_0'(\underline{u})$ depending on the stochastic process ξ_t', $0 \leq t \leq T$, defined in Lemma A3. This process, however, depends on $\underline{u}$. Now we derive an expression for $L_0'(\underline{u})$ from which the linearity of this functional can be seen immediately.

ξ_t', $0 \leq t \leq T$, is the solution of the linear, inhomogeneous stochastic differential equation on $[0,T]$:

$$d\xi_t' = \langle \partial_x \Gamma^0(t,\xi_t^0,dt) | \xi_t' \rangle + \langle \partial_u G(\xi_t^0,u^0(t,\xi_t^0),dt) | u(t,\xi_t^0) \rangle$$
$$\xi_0' = 0 \qquad (2.9)$$

(b) By part (d) of the proof of Lemma A4 the random coefficients of the linear stochastic differential $\langle \partial_x \Gamma^0(t,\xi_t^0,dt) | z \rangle$ are bounded $dP \otimes dt$ a.s. By Lemma 1 of Arkin and Saksonov [1], which in turn is a consequence of results of Bismut [2,3] about linear stochastic differential equations, the stochastic matrix equation (2.7) has a unique solution $\Phi(t)$, $0 \leq t \leq T$, which is continuous in t with probability 1. For all t there exists the inverse $\Phi^{-1}(t)$. Additionally,

$$E \sup_t |\Phi(t)|^k < \infty \qquad E \sup_t |\Phi^{-1}(t)|^k < \infty$$
$$\text{for all } k \geq 1 \qquad (2.10)$$

(c) By (2.2) the matrix coefficients of the linear stochastic differential $\langle \partial_u G(\xi_t^0,u^0(t,\xi_t^0),dt) | v \rangle$ are bounded $dP \otimes dt$ a.s. Furthermore, for $\underline{u} \in E$ there exists by (2.4) a constant C such that

$$|u(t,\xi_t^0)| \leq C(1 + |\xi_t^0|) \qquad dP \otimes dt \text{ a.s.}$$

From Lemma A2 and the Cauchy-Schwartz inequality we conclude that

$$E \int_0^T |u(t,\xi_t^0)|^k \, dt < \infty \qquad k = 1, 2$$

Therefore, we may apply Lemma 4 of Arkin and Saksonov [1], which provides a representation of the solution ξ_t', $0 \leq t \leq T$, of the linear, inhomogeneous equation (2.9) in terms of the fundamental matrix $\Phi(t)$ of the corresponding homogeneous equation:

$$
\begin{aligned}
\xi'_t &= \Phi(t) \int_0^t \Phi^{-1}(s)\partial_u a(\xi_s^0, u^0(s,\xi_s^0))u(s,\xi_s^0)\ ds \\
&\quad -\Phi(t) \int_0^t \Phi^{-1}(s) \sum_{i=1}^m \partial_x \beta^0_{\cdot i}(s,\xi_s^0)\partial_u b_{\cdot i}(\xi_s^0, u^0(s,\xi_s^0))u(s,\xi_s^0)\ ds \\
&\quad +\Phi(t) \int_0^t \Phi^{-1}(s) \sum_{i=1}^m \partial_u b_{\cdot i}(\xi_s^0, u^0(s,\xi_s^0))u(s,\xi_s^0)\ dW_i(s) \\
&= \Phi(t) \int_0^t \Phi^{-1}(s)\langle \partial_u g^0(s,\xi_s^0,u^0(s,\xi_s^0),ds) \,|\, u(s,\xi_s^0)\rangle \qquad (2.11)
\end{aligned}
$$

Combining (2.11) and Lemma A6 we get (2.8).

(d) To complete the proof we have to show that L_0' is a bounded, linear functional on E.

From (2.4) we know that there exists a constant C_1 such that

$$E\ \sup_t |u(t,\xi_t^0)|^k \le C_1^k \|\underline{u}\|^k E\ \sup_t (1 + n|\xi_t^0|)^k \qquad \text{for all } \underline{u} \in E \qquad (2.12)$$

By Lemma 3.8 of Gihman and Skorohod [8] the right-hand side is finite for all $k \ge 1$.

From (2.5) and (2.12) we conclude that there exists a constant C_2 such that

$$\left|E \int_0^T \nabla_u^T S(\xi_t^0, u^0(t,\xi_t^0))u(t,\xi_t^0)\ dt\right| \le C_2 \|\underline{u}\| \qquad \text{for all } \underline{u} \in E \qquad (2.13)$$

By (2.10) and the assumptions (2.5) on R, S,

$$E\ \sup_t |\eta_t|^k < \infty \qquad \text{for all } k \ge 1 \qquad (2.14)$$

By assumptions (2.2) on a, b the coefficients of v in the linear stochastic differential $\langle \partial_u g^0(t,x,u,dt)|v\rangle$ are uniformly bounded. Therefore, there exist C_3, C_4 such that for all $0 \le t \le T$,

$$
\begin{aligned}
&E\left| \int_0^t \Phi^{-1}(s)\langle \partial_u g^0(s,\xi_s^0,u^0(s,\xi_s^0),ds)\,|\,u(s,\xi_s^0)\rangle\right|^2 \\
&\le C_3\ E \int_0^T |\Phi^{-1}(s)|^2 |u(s,\xi_s^0)|^2\ ds \\
&\le C_4 \|\underline{u}\|^2
\end{aligned}
$$

where the last inequality follows from (2.10), (2.12), and the Cauchy-Schwartz inequality.

By (2.14), (2.15), and the Cauchy-Schwartz inequality

there exists C_5 such that

$$|L_0'(\underline{u}) - E\int_0^T \nabla_u^T S(\xi_t^0, u^0(t,\xi_t^0))u(t,\xi_t^0)\,dt| \leq C_5\|\underline{u}\| \quad \text{for all } \underline{u} \in E$$

and combining this and (2.13) we have

$$|L_0'(\underline{u})| \leq (C_2 + C_5)\|\underline{u}\| \qquad \text{for all } \underline{u} \in E \qquad \square$$

As a first example we consider the control problem
Minimize $L(\underline{u})$ under the constraint $\underline{u} \in D$!, where D is a given closed, convex subset of E with nonempty interior. (2.16)

By Theorem 2.1 we may apply standard results of locally convex optimization theory to get necessary conditions on an optimal control $\underline{u}^0$. Using Theorems 7.3, 10.2, and 10.5 of Girsanov [9] and the fundamental theorem of Dubovitskij and Milyutin (Ref. 9, theorem 6.1), we get immediately:

COROLLARY 2.1 Let $\underline{u}^0 \in D$ be a locally (with respect to the topology of E) optimal control for the problem (2.16). Then

$$L_0'(\underline{u} - \underline{u}^0) \geq 0 \qquad \text{for all } \underline{u} \in D \tag{2.17}$$

REMARK 1. Theorem 2.1 can be generalized to the situation where the stochastic differential $G(x,u,dt)$, which appears in (2.1), contains an additional jump term:

$$G(x,u,dt) = a(x,u)\,dt + b(x,u)\,dW(t) + \int c(x,u,y)M(dt,dy)$$

where $M(dt,dy)$ is a centered Poisson measure adapted to $\{\mathcal{F}_t,\ 0 \leq t \leq T\}$ and independent of the Brownian motion. $c(x,u,y)$ has to satisfy appropriate assumptions, of course.

It is not necessary for the validity of Theorem 2.1 that $u^0(t,x)$ is continuously differentiable in x. We needed this condition in the proof of Lemma A3 to apply Theorem II.8.5 of Gihman and Skorohod [7] directly. A careful check of the proof of this theorem shows that in our situation, in particular under the assumptions (2.2) and (2.3), the following weaker condition on u^0 suffices:

If ζ_t^ε, $0 \leq t \leq T$, is a sequence of F^*-measurable stochastic processes with

$\zeta_t^\varepsilon \to \xi_t^0$ $dP \otimes dt$ stochastically for $\varepsilon \to 0+$

then for $1 \leq k \leq q$, $1 \leq j \leq n$, (2.18)

$$\frac{u_k^0(t,\zeta_t^\varepsilon) - u_k^0(t,\xi_t^0)}{(\zeta_t^\varepsilon)_j - (\xi_t^0)_j} \to \frac{\partial}{\partial x_j} u_k^0(t,\xi_t^0)$$

$dP \otimes dt$ stochastically for $\varepsilon \to 0+$.

In the proof of Lemma A6 we may also replace continuous differentiability of $\underline{u}^0$ with respect to x by (2.18). Condition 2.18 holds, for instance, if $\underline{u}^0$ and the corresponding solution ξ_t^0, $0 \leq t \leq T$, of (2.1) satisfy the condition: For almost all t there exists an open subset $\mathcal{O}_t$ of R^n such that $\partial_x u^0(t,x)$ is continuous in x on each open, connected subset of $\mathcal{O}_t$ and $P\{\xi_t^0 \in \mathcal{O}_t\} = 1$.

3. ANOTHER REPRESENTATION OF THE DERIVATIVE OF THE LOSS FUNCTIONAL AND A LOCAL STOCHASTIC MAXIMUM PRINCIPLE

The process ξ_t', $0 \leq t \leq T$, which appears in the expression for $L_0'(\underline{u})$ given in Lemma A6 is the solution of a linear stochastic differential equation depending on $\underline{u}$. Therefore, we can apply the techniques which Bismut [3] used for treating the linear quadratic stochastic control problem. First, we define a dual process p_t, $0 \leq t \leq T$, which is the unique solution of a backward stochastic differential equation and which does not depend on $\underline{u}$. Then we express $L_0'(\underline{u})$ by means of this dual process.

Following Bismut we introduce some notation. Let $\mathcal{M}$ denote the space of square-integrable R^n-valued martingales stopped at T, vanishing at time 0. Let $\mathcal{M}_W$ denote the subspace of $\mathcal{M}$ generated by the stochastic integrals relative to W(t) of $n \times m$-matrix-valued $\mathcal{T}^*$-measurable functions H(t) with

$E \int_0^T |H(t)|^2 \, dt < \infty$. Let $M_W^{\perp}$ denote the orthogonal of M_W in M in the sense of Ref. 10, pp. 80ff. In particular, $M_W^{\perp} = \{0\}$ if $\{F_t, 0 \leq t \leq T\}$ is the family of σ-algebras generated by $W(t)$, $0 \leq t \leq T$, by a result of Itô (see Ref. 10, p. 135).

Under the conditions of Theorem 2.1 and using the same notation we have the following locally convex analog of Theorem 3.2 of Bismut [3]:

THEOREM 3.1 Let M_t, $0 \leq t \leq T$, be a martingale in $M_W^{\perp}$ and $H(t)$, $0 \leq t \leq T$, be a $n \times m$-matrix-valued, F^*-measurable process with

$$E \int_0^T |H(t)|^2 \, dt < \infty$$

Let p_t, $0 \leq t \leq T$, be the unique solution of the linear stochastic differential equation

$$dp_t = \{\nabla_x \sigma^0(t,\xi_t^0) - \partial_x^T \alpha^0(t,\xi_t^0) p_t - \sum_{i=1}^{m} \partial_x^T \beta^0_{\cdot i}(t,\xi_t^0) H_{\cdot i}(t)\} \, dt$$
$$+ H(t) \, dW(t) + dM_t$$
$$p_t = -\nabla_x R(\xi_T^0)$$

(i) For all $\underline{u} \in E$,

$$L_0'(\underline{u}) = E\int_0^T [\nabla_u^T S(\xi_t^0, u^0(t,\xi_t^0)) - p_t^T \partial_u a(\xi_t^0, u^0(t,\xi_t^0))] u(t,\xi_t^0) \, dt$$
$$- E\int_0^T \sum_{i=1}^{m} H_{\cdot i}^T(t) \partial_u b_{\cdot i}(\xi_t^0, u^0(t,\xi_t^0)) u(t,\xi_t^0) \, dt$$

(ii) If $\underline{u}^0$ is a locally optimal control for the control problem (2.16) without constraints, i.e., $D = E$, then $dP \otimes dt$ a.s.

$$\nabla_u^T S(\xi_t^0, u^0(t,\xi_t^0)) = p_t^T \partial_u a(\xi_t^0, u^0(t,\xi_t^0))$$
$$+ \sum_{i=1}^{m} H_{\cdot i}^T(t) \partial_u b_{\cdot i}(\xi_t^0, u^0(t,\xi_t^0))$$

Proof. Using (2.2) and (2.5) the existence and uniqueness of p_t, $0 \leq t \leq T$, follows as in the proof of Theorem 3.2 of Bismut [3]. From the same assumptions we may use proposition 1 of Bismut [3] and get

$$p_T^T \xi_T' = \int_0^T [\nabla_x \sigma^0(t,\xi_t^0) - \partial_x^T \alpha^0(t,\xi_t^0) p_t - \sum_{i=1}^m \partial_x^T \beta_{\cdot i}^0(t,\xi_t^0) H_{\cdot i}(t)]^T \xi_t' \, dt$$
$$+ \int_0^T p_t^T [\partial_x \alpha^0(t,\xi_t^0)\xi_t' + \partial_u a(\xi_t^0, u^0(t,\xi_t^0)) u(t,\xi_t^0)] \, dt$$
$$+ \int_0^T \sum_{i=1}^m H_{\cdot i}^T(t) [\partial_x \beta_{\cdot i}^0(t,\xi_t^0)\xi_t' + \partial_u b_{\cdot i}(\xi_t^0, u^0(t,\xi_t^0)) u(t,\xi_t^0)] \, dt$$

Equivalently,

$$E\, \nabla_x^T R(\xi_T^0)\xi_T' + E \int_0^T \nabla_x^T \sigma^0(t,\xi_t^0)\xi_t' \, dt$$
$$= -E \int_0^T p_t^T \partial_u a(\xi_t^0, u^0(t,\xi_t^0)) u(t,\xi_t^0) \, dt$$
$$- E \int_0^T \sum_{i=1}^m H_{\cdot i}^T(t) \partial_u b_{\cdot i}(\xi_t^0, u^0(t,\xi_t^0)) u(t,\xi_t^0) \, dt$$

From this and Lemma A6 follows (i).

If $\underline{u}^0$ is optimal for (2.16) without constraints then, by Corollary 2.1,

$$L_0'(\underline{u}) = 0 \qquad \text{for all } \underline{u} \in E \qquad \square$$

This and (i) imply (ii).

Let $U(t)$, $0 \leq t \leq T$, be a sequence of convex, closed subsets of R^q. We assume that $1_{U(t)}(u)$ is a Borel-measurable function of (t,u), and that there exists an essentially bounded, R^q-valued function $u(t)$, $0 \leq t \leq T$, such that for almost all t $u(t) \in \operatorname{int}\{U(t)\}$. If we consider the constraint

$$u(t,x) \in U(t) \qquad \text{for all } x, \text{ almost all } t \tag{3.1}$$

we get from Theorem 3.1 and Corollary 2.1 a local stochastic maximum principle for an optimal control. This result can be compared with the maximum principle derived by Arkin and Saksonov [1] for their control problem, which we discussed in the introduction.

COROLLARY 3.1 (Local Stochastic Maximum Principle) Let $\underline{u}^0 \in E$ be a local optimal control for the loss functional L given by (2.6) under the constraint (3.1). We assume that the conditions of Theorem 3.1 are satisfied and use the same notation. Let

$$\pi^T(t,x) = E\{p_t^T | \xi_t^0 = x\} \partial_u a(x,u^0(t,x)) - \nabla_u^T S(x,u^0(t,x)) + \sum_{i=1}^m E\{H_{\cdot i}^T(t) | \xi_t^0 = x\} \partial_u b_{\cdot i}(x,u^0(t,x))$$

If for all $\delta > 0$ there exists $\underline{u} \in E$ satisfying (3.1) such that

$$E \int_0^T \left[\sup_{u \in U(t)} \pi^T(t,\xi_t^0)u - \pi^T(t,\xi_t^0)u(t,\xi_t^0) \right] dt \leq \delta$$

Then

$$\pi^T(t,\xi_t^0)u^0(t,\xi_t^0) = \sup_{u \in U(t)} \pi^T(t,\xi_t^0)u \qquad dP \otimes dt \text{ a.s.} \tag{3.2}$$

Proof. From well-known properties of the conditional expectation (e.g., Breiman [4], chap. 4), Theorem 3.1, and Corollary 2.1 we conclude from the local optimality of $\underline{u}^0$,

$$L_0'(\underline{u} - \underline{u}^0) = \int_0^T E\, \pi^T(t,\xi_t^0)[u^0(t,\xi_t^0) - u(t,\xi_t^0)]\, dt \geq 0$$

for all $\underline{u}$ satisfying (3.1). Then, by assumption, we have for all $\delta > 0$

$$-\delta \leq \int_0^T E \left[\pi^T(t,\xi_t^0)u^0(t,\xi_t^0) - \sup_{u \in U(t)} \pi^T(t,\xi_t^0)u \right] dt$$

As the integrand is nonpositive we conclude (3.2). □

APPENDIX

We subdivide the proof of Theorem 2.1 into several parts. First, we formulate some lemmas about the existence and properties of solutions of (2.1). In the Appendix we always tacitly assume that dW(t) is an m-dimensional Wiener measure as described at the beginning of Sec. 1 and that the coefficients a, b of (2.1) satisfy (2.2). Furthermore, the functions R, S determining the loss functional L satisfy (2.5).

As the following relations, which are an immediate consequence of (2.2), (2.5), and the mean-value theorem, are used quite often in the following, we formulate them in an original lemma:

LEMMA A1 There exists $C > 0$ such that for all u, v, x, z,

(i) $|a_k(x,u) - a_k(z,v)|^2 \leq C\{|x - z|^2 + |u - v|^2\}$, $k = 1,\ldots,n$

(ii) $|b_{k\ell}(x,u) - b_{k\ell}(z,v)|^2 \leq C\{|x - z|^2 + |u - v|^2\}$, $k = 1,\ldots,n;\ \ell = 1,\ldots,m$

(iii) $|R(x) - R(z)|^2 \leq C \cdot |x - z|^2$

(iv) $|S(x,u) - S(z,v)|^2 \leq C\{|x - z|^2 + |u - v|^2\}$

LEMMA A2 Let $\underline{u} \in E$. Then there exists a unique solution $\{\xi_t, 0 \leq t \leq T\}$ of (2.1) and $E(\sup_t \xi_t)^2 < \infty$.

Proof. As is well known (e.g., Ref. 8) there exists a unique solution of the stochastic differential equation (2.1) if a(x,u(t,x)) and b(x,u(t,x)) as functions of t, x are linearly bounded and satisfy a uniform Lipschitz condition. Additionally, the supremum over $t \in [0,T]$ of this solution is square integrable with respect to P.

(a) By (2.4) and Lemma A1(i) there exists $C > 0$ depending on $\|\underline{u}\|$ only such that for all x and almost all t,

$$|a_k(x,u(t,x)) - a_k(0,0)|^2 \leq C[|x|^2 + |u(t,x)|^2] \leq C[|x|^2 + \|\underline{u}\|^2(1 + n|x|)^2], \quad k = 1,\ldots,n$$

Therefore, a(x,u(t,x)) is linearly bounded as function of x and t; i.e., there exists $C' > 0$, which depends on $\|\underline{u}\|$ only, such that

$$|a(x,u(t,x))| \leq C'(1 + |x|) \qquad \text{for all } x, \text{ almost all } t$$

(b) Likewise, by (2.4) and Lemma A1(i) there exists $C > 0$ depending on $\|\underline{u}\|$ only such that for all x, y and almost all t, $k = 1,\ldots,m$,

$$|a_k(x,u(t,x)) - a_k(y,u(t,y))|^2 \leq C[|x - y|^2 + |u(t,x) - u(t,y)|^2] \leq C[|x - y|^2 + (\|\underline{u}\|n|x - y|)^2]$$

Therefore, a(x,u(t,x)) satisfies a uniform Lipschitz condition with a Lipschitz constant depending on $\|\underline{u}\|$ only.

(c) Analogously to (a), (b) it follows from (2.4) and Lemma A1(ii) that b(x,u(t,x)) as a function of x, t is linearly bounded and satisfies a uniform Lipschitz condition with constants depending on $\|\underline{u}\|$ only. □

LEMMA A3 Let $\underline{u}^0$, $u \in E$, $\underline{u}^\varepsilon = \underline{u}^0 + \varepsilon u$ for $\varepsilon \geq 0$. Let $\{\xi_t^\varepsilon, 0 \leq t \leq T\}$ be the solution of (2.1) for control $\underline{u}^\varepsilon$, $0 \leq \varepsilon \leq 1$.

(i) $E \sup_t |\xi_t^\varepsilon - \xi_t^0|^2 \to 0$ for $\varepsilon \to 0+$.

(ii) If $\partial_x u^0(t,x)$ is continuous in x, then ξ_t^ε is mean-square differentiable with respect to ε in $\varepsilon = 0$ uniformly in t; i.e., there exist ξ_t', $0 \leq t \leq T$, such that

$$\sup_t E \left|\frac{1}{\varepsilon}(\xi_t^\varepsilon - \xi_t^0) - \xi_t'\right|^2 \to 0 \qquad \text{for } \varepsilon \to 0+$$

(iii) Under the assumption of (ii), $\{\xi_t', 0 \leq t \leq T\}$ solves the linear inhomogeneous stochastic differential equation on $[0,T]$:

$$d\xi_t' = \langle \partial_x \Gamma^0(t,\xi_t^0,dt) | \xi_t' \rangle + \langle \partial_u G(\xi_t^0, u^0(t,\xi_t^0),dt) | u(t,\xi_t^0) \rangle$$

$$\xi_0' = 0$$

where $\Gamma^\varepsilon(t,x,dt) = G(x,u^\varepsilon(t,x),dt)$ for $0 \leq \varepsilon \leq 1$.

Proof. The lemma follows by applying theorems of Gihman and Skorohod about continuity and mean-square differentiability of solutions of stochastic differential equations depending on a parameter.

(a) We set for $0 \leq \varepsilon \leq 1$ and all t, x,

$$\alpha^\varepsilon(t,x) = a(x,u^\varepsilon(t,x)) \qquad \beta^\varepsilon(t,x) = b(x,u^\varepsilon(t,x))$$

We know from the proof of Lemma A2 that the coefficients α^ε, β^ε of the stochastic differential equation

$$d\xi_t^\varepsilon = \Gamma^\varepsilon(t,\xi_t^\varepsilon,dt) \qquad \xi_0^\varepsilon = x_0$$

are linearly bounded and satisfy a uniform Lipschitz condition where the constants can be chosen uniformly in $0 \leq \varepsilon \leq 1$ as $\|\underline{u}^\varepsilon\| \leq \|\underline{u}^0\| + \|\underline{u}\|$ for all $0 \leq \varepsilon \leq 1$.

(b) As $a_k(x,u)$, $k = 1,\ldots,n$, is continuously differentiable we have by the mean-value theorem

$a_k(x,u^\varepsilon(t,x)) - a_k(x,u^0(t,x)) = \varepsilon\nabla_u^T a_k(x,u^0(t,x))u(t,x) + A_k^\varepsilon(t,x)$ where $(1/\varepsilon)A_k^\varepsilon(t,x)$ converges to 0 for $\varepsilon \to 0+$. By assumption (2.2), $\nabla_u a_k(x,u)$ is uniformly bounded, and by Lemma A1(i) there exists $C' > 0$ such that

$$|a_k(x,u^\varepsilon(t,x)) - a_k(x,u^0(t,x))| \leq C'\varepsilon|u(t,x)| \quad \text{for all } t,\ x,\ \varepsilon$$

We conclude that there exists $C > 0$ such that for all t, x, ε,

$$\alpha^\varepsilon(t,x) - \alpha^0(t,x) = \varepsilon\partial_u a(x,u^0(t,x))u(t,x) + A^\varepsilon(t,x) \quad \text{(A2)}$$

where

$$\frac{1}{\varepsilon} A^\varepsilon(t,x) \to 0 \text{ for } \varepsilon \to 0+$$

$$\left|\frac{1}{\varepsilon} A^\varepsilon(t,x)\right| \leq C\ |u(t,x)|$$

Analogously, we have for suitably chosen $C > 0$,

$$\beta_{k\ell}^\varepsilon(t,x) - \beta_{k\ell}^0(t,x) = \varepsilon \sum_{j=1}^{q} \frac{\partial}{\partial u_j} b_{k\ell}(x,u^0(t,x))u_j(t,x) + B_{k\ell}^\varepsilon(t,x) \quad \text{(A3)}$$

where

$$\frac{1}{\varepsilon} B_{k\ell}^\varepsilon(t,x) \to 0 \text{ for } \varepsilon \to 0+$$

$$\left|\frac{1}{\varepsilon} B_{k\ell}^\varepsilon(t,x)\right| \leq C\ |u(t,x)|$$

for all t, x, ε; $k = 1,\ldots,n$; $\ell = 1,\ldots,m$. As an immediate consequence of (A2), (2.2), and (2.4) there exists $C' > 0$ such that

$$|\alpha^\varepsilon(t,x) - \alpha^0(t,x)| \leq \varepsilon C'|u(t,x)| \leq \varepsilon C'\|\underline{u}\|(1 + n|x|)$$

From this relation we conclude that for any $\rho > 0$, $\alpha^\varepsilon(t,x) - \alpha^0(t,x)$ converges to 0 uniformly in t and $|x| \leq \rho$, and by an analogous argument the same is true for $\beta^\varepsilon(t,x) - \beta^0(t,x)$. Due to this fact we may apply theorem II.8.3 of Gihman and Skorohod [7], from which follows (i).

(c) From (A2) and (A3) we know that $\alpha^\varepsilon(t,x)$, $\beta^\varepsilon(t,x)$ are differentiable with respect to ε in $\varepsilon = 0$, and we also know their derivatives. By (A2) and (2.4),

$$\left|\frac{1}{\varepsilon}(\alpha^{\varepsilon}(t,\xi_t^0) - \alpha^0(t,\xi_t^0)) - \partial_u a(\xi_t^0,u^0(t,\xi_t^0))u(t,\xi_t^0)\right|^2$$
$$\leq C^2|u(t,\xi_t^0)|^2$$
$$\leq C^2\|\underline{u}\|^2(1 + n|\xi_t^0|)^2 \qquad dP \otimes dt \text{ a.s.}$$

By (A2) the left-hand side converges to 0 for $\varepsilon \to 0+$ $dP \otimes dt$ a.s. and, as by Lemma A2, $\xi_t^0 \in L^2(dP \otimes dt)$, it also converges in $L^2(dP \otimes dt)$. From (A3) we conclude an analogous result for the convergence of $(1/\varepsilon)(\beta^{\varepsilon}(t,\xi_t^0) - \beta^0(t,\xi_t^0))$.

(d) By assumption on $\underline{u}^0$, $\alpha^0(t,x)$ is continuously differentiable with respect to $x_1,\ldots,x_n$ and

$$\partial_x\alpha^0(t,x) = \partial_x a(x,u^0(t,x)) + \partial_u a(x,u^0(t,x))\partial_x u^0(t,x)$$

By (2.2) and (2.4), $\partial_x\alpha^0(t,x)$ is uniformly bounded in t, x. $\beta^0(t,x)$ possesses analogous properties.

(e) By parts (c) and (d) we may apply theorem II.8.5 of Gihman and Skorohod [7], from which follows (ii). The uniformity with respect to t of the convergence is not stated explicitly in the theorem mentioned but it follows from the last step of its proof, where theorem II.8.3 of Ref. 7 is applied. Again from theorem II.8.5 follows

$$\xi_0' = 0$$
$$d\xi_t' = [\partial_x\alpha^0(t,\xi_t^0)\xi_t' + \frac{\partial}{\partial\varepsilon}\alpha^0(t,\xi_t^0)]\,dt$$
$$+\left[\sum_{k=1}^{n}\frac{\partial}{\partial x_k}\beta^0(t,\xi_t^0)(\xi_t')_k + \frac{\partial}{\partial\varepsilon}\beta^0(t,\xi_t^0)\right]dW(t)$$
$$= \langle\partial_x\Gamma^0(t,\xi_t^0,dt)|\xi_t'\rangle + \langle\partial_u G(\xi_t^0,u^0(t,\xi_t^0),dt)|u(t,\xi_t^0)\rangle$$

by (A2) and (A3). □

LEMMA A4 Let $\underline{u},\underline{v} \in E$. Let $\{\xi_t,\ 0 \leq t \leq T\}$, $\{\eta_t,\ 0 \leq t \leq T\}$ denote the solution of (2.1) for control $\underline{u}$, $\underline{v}$. For all $\rho > 0$ there exists C_ρ such that $\|\underline{u}\|$, $\|\underline{v}\| \leq \rho$ implies that

$$\sup_t E\,|\xi_t - \eta_t|^2 \leq C_\rho\|\underline{u} - \underline{v}\|^2$$

Proof. (a) By (2.4) we have for some constant C_0

depending on $\|\underline{u}\|$ only,

$$|u(t,\xi_t) - v(t,\eta_t)| \le |u(t,\xi_t) - u(t,\eta_t)| + |u(t,\eta_t) - v(t,\eta_t)|$$
$$\le C_0\|\underline{u}\| n|\xi_t - \eta_t| + C_0\|\underline{u} - \underline{v}\|(1 + n|\xi_t|)$$
$$dP \otimes dt \text{ a.s.} \quad \text{(A4)}$$

By Lemma A1(ii) there exists C such that $dP \otimes dt$ a.s.

$$|b(\xi_t,u(t,\xi_t)) - b(\eta_t,v(t,\eta_t))|^2$$
$$\le C[|\xi_t - \eta_t|^2 + |u(t,\xi_t) - v(t,\eta_t)|^2]$$
$$\le C_1[|\xi_t - \eta_t|^2 + \|\underline{u} - \underline{v}\|^2(1 + n|\eta_t|)^2]$$

where by (A4) C_1 is a suitable constant depending on ρ only.

(b) As $b(x,u(t,x))$ is linearly bounded as function of t, x and as $\xi_t \in L^2(dP \otimes dt)$ by Lemma A2 we conclude that $b(\xi_t,u(t,\xi_t)) \in L^2(dP \otimes dt)$. The same argument applies to $\underline{v}$, η_t. Therefore, we have (e.g., Ref. 8, chap. III.1):

$$E|\int_0^t[b(\xi_s,u(s,\xi_s)) - b(\eta_s,v(s,\eta_s))]\, dW(s)|^2$$
$$= E\int_0^t |b(\xi_s,u(s,\xi_s)) - b(\eta_s,v(s,\eta_s))|^2\, ds$$
$$\le C_1\{E\int_0^t |\xi_s - \eta_s|^2\, ds + \|\underline{u} - \underline{v}\|^2 E\int_0^t(1 + n|\eta_s|)^2\, ds\}$$
$$\le C_1\{E\int_0^t |\xi_s - \eta_s|^2\, ds + \|\underline{u} - \underline{v}\|^2 T \sup_t E(1 + n|\eta_t|)^2\}$$
$$\le C_2\{E\int_0^t |\xi_s - \eta_s|^2\, ds + \|\underline{u} - \underline{v}\|^2\} \quad \text{(A5)}$$

where C_2 depends on ρ only. For the last inequality we have used the fact that by the proof of Lemma A2 the coefficients of (2.1) are linearly bounded with a constant depending on the norm of the control function only; from this follows by Lemma 3.6 of Gihman and Skorohod [8] that $E|\eta_t|^2$ is bounded uniformly in t by a constant depending monotonically on $\|\underline{v}\|$, and by the Cauchy-Schwartz inequality this property is shared by $E|\eta_t|$.

(c) By the same arguments we can choose C_3 depending on ρ only such that for all $0 \le t \le T$,

$$E\left| \int_0^t [a(\xi_s,u(s,\xi_s)) - a(\eta_s,v(s,\eta_s))]\, ds \right|^2$$
$$\leq C_3[E \int_0^t |\xi_s - \eta_s|^2\, ds + \|\underline{u} - \underline{v}\|^2] \qquad (A6)$$

(d) $\xi_t = x_0 + \int_0^t a(\xi_s,u(s,\xi_s))\, ds + \int_0^t b(\xi_s,u(s,\xi_s))\, dW(s)$

From this representation and its analog for η_t, $\underline{v}$ we get by (A5) and (A6) that there exists C_4 depending on ρ only such that

$$E|\xi_t - \eta_t|^2 \leq C_4[\int_0^t E|\xi_s - \eta_s|^2\, ds + \|\underline{u} - \underline{v}\|^2]$$
$$0 \leq t \leq T$$

This relation implies Gronwall's inequality (e.g., Lemma 3.5 of Ref. 8):

$$E|\xi_t - \eta_t|^2 \leq C_4\|\underline{u} - \underline{v}\|^2 + C_\rho \int_0^t \exp[C_4(t - s)]\|\underline{u} - \underline{v}\|^2\, ds$$
$$\leq C_\rho\|\underline{u} - \underline{v}\|^2 \qquad 0 \leq t \leq T$$

where C_ρ depends on ρ only. □

LEMMA A5 L satisfies a local Lipschitz condition on E; i.e., for all $\rho > 0$ there exists C'_ρ such that

$$|L(u) - L(v)| \leq C'_\rho \|\underline{u} - \underline{v}\| \qquad \text{for all } \underline{u}, \underline{v} \text{ with } \|\underline{u}\|, \|\underline{v}\| \leq \rho$$

Proof. Let ξ_t, η_t, $0 \leq t \leq T$, be as in Lemma A4, $\rho > 0$. By Lemma A1(iii) and (iv) there exists C such that

$$|L(\underline{u}) - L(\underline{v})|^2 \leq 2E|R(\xi_T) - R(\eta_T)|^2 + 2E \int_0^T |S(\xi_t,u(t,\xi_t)) - S(\eta_t,v(t,\eta_t))|^2\, dt$$
$$\leq 2E|\xi_T - \eta_T|^2 + CE \int_0^T [|\xi_t - \eta_t|^2 + |u(t,\xi_t) - v(t,\eta_t)|^2]\, dt \qquad (A7)$$

By (A4) and (A5) there exists C_1 depending on ρ only such that

$$E|u(t,\xi_t) - v(t,\eta_t)|^2 \leq C_1[E|\xi_t - \eta_t|^2 + \|\underline{u} - \underline{v}\|^2]$$
$$0 \leq t \leq T$$

Combining this relation, (A7) and Lemma A4 the local Lipschitz property of L follows. □

LEMMA A6 Let $\underline{u}^0$, $\underline{u} \in E$ and let $\partial_x u^0(t,x)$ be continuous in x. L is differentiable in $\underline{u}^0$ in direction $\underline{u}$ with derivative

$$L_0'(\underline{u}) = \lim_{\varepsilon \to 0+} \frac{1}{\varepsilon} [L(\underline{u}^0 + \varepsilon\underline{u}) - L(\underline{u}^0)]$$

$$= E[\nabla_x^T R(\xi_T^0)\xi_T'] + E \int_0^T \nabla_x^T \sigma^0(t,\xi_t^0)\xi_t' \, dt$$

$$+ E \int_0^T \nabla_u^T S(\xi_t^0, u^0(t,\xi_t^0))u(t,\xi_t^0) \, dt$$

where ξ_t', $0 \le t \le T$, is the same process as in Lemma A3 and $\sigma^0(t,x) = S(x,u^0(t,x))$.

Proof. Let $\underline{u}^\varepsilon$, ξ_t^ε, $0 \le t \le T$, $0 \le \varepsilon \le 1$, be as in Lemma A3.

(a) As R is continuously differentiable there exists by the mean-value theorem a random variable $\theta^\varepsilon \in [0,1]$ such that

$$R(\xi_T^\varepsilon) - R(\xi_T^0) = \nabla_x^T R((1 - \theta^\varepsilon)\xi_T^0 + \theta^\varepsilon \xi_T^\varepsilon)(\xi_T^\varepsilon - \xi_T^0) \quad \text{a.s.} \tag{A8}$$

[Checking the proof of the mean-value theorem it can be seen that $\theta^\varepsilon = f(\xi_T^\varepsilon, \xi_T^0)$, where f(z,x) is a Borel-measurable function; therefore, θ^ε is measurable indeed.]

As θ^ε is bounded we have by Lemma A3(i),

$$(1 - \theta^\varepsilon)\xi_T^0 + \theta^\varepsilon \xi_T^\varepsilon \to \xi_T^0 \quad \text{for } \varepsilon \to 0+ \text{ in mean square and therefore in probability}$$

As $\nabla_x R(x)$ is bounded and continuous we conclude by Lebesgue's theorem of dominated convergence that

$$\nabla_x R((1 - \theta^\varepsilon)\xi_T^0 + \theta^\varepsilon \xi_T^\varepsilon) \to \nabla_x R(\xi_T^0) \quad \text{for } \varepsilon \to 0+ \text{ in mean square} \tag{A9}$$

By Lemma A3(ii), $(1/\varepsilon)(\xi_T^\varepsilon - \xi_T^0)$ converges in mean square to ξ_T'. From this, (A8), and (A9) we conclude that

$$\frac{1}{\varepsilon} E[R(\xi_T^\varepsilon) - R(\xi_T^0)] \to E \nabla_x^T R(\xi_T^0)\xi_T' \quad \text{for } \varepsilon \to 0+$$

(b) As S is continuously differentiable there exists by the mean-value theorem $\theta_t^\varepsilon \in [0,1]$, $0 \le t \le T$, such that for $0 \le t \le T$,

$$S(\xi_t^\varepsilon, u^\varepsilon(t,\xi_t^\varepsilon)) - S(\xi_t^0, u^0(t,\xi_t^0)) = \nabla_x^T S(\zeta_t^\varepsilon, \nu_t^\varepsilon)(\xi_t^\varepsilon - \xi_t^0) + \nabla_u^T S(\zeta_t^\varepsilon, \nu_t^\varepsilon)(u^\varepsilon(t,\xi_t^\varepsilon) - u^0(t,\xi_t^0)) \quad \text{(A10)}$$

with

$$\zeta_t^\varepsilon = (1 - \theta_t^\varepsilon)\xi_t^0 + \theta_t^\varepsilon \xi_t^\varepsilon$$

$$\nu_t^\varepsilon = (1 - \theta_t^\varepsilon)u^0(t,\xi_t^0) + \theta_t^\varepsilon u^\varepsilon(t,\xi_t^\varepsilon)$$

As θ_t^ε is $dP \otimes dt$ essentially bounded, we have from Lemma A3(i)

$|\zeta_t^\varepsilon - \xi_t^0| \to 0$ for $\varepsilon \to 0+$ in $L^2(dp \otimes dt)$ and therefore $dP \otimes dt$ stochastically

By (A4) and (A5) there exists a constant C depending on $\|\underline{u}\| + \|\underline{u}^0\|$ only such that

$$E \int_0^T |u^\varepsilon(t,\xi_t^\varepsilon) - u^0(t,\xi_t^0)|^2 \, dt \le C[E \int_0^T |\xi_t^\varepsilon - \xi_t^0|^2 \, dt + \|\underline{u}^\varepsilon - \underline{u}^0\|^2]$$

and from this follows by Lemma A3(i),

$|u^\varepsilon(t,\xi_t^\varepsilon) - u^0(t,\xi_t^0)| \to 0 \qquad |\nu_t^\varepsilon - \nu_t^0| \to 0$

for $\varepsilon \to 0+$ in $L^2(dP \otimes dt)$ and therefore $dP \otimes dt$ stochastically

As $\nabla_x S(x,u)$, $\nabla_u S(x,u)$ are continuous and bounded, we conclude from Lebesgue's theorem of dominated convergence that

$$|\nabla_x S(\zeta_t, \nu_t^\varepsilon) - \nabla_x S(\xi_t^0, u^0(t,\xi_t^0))| \to 0$$

$$|\nabla_u S(\zeta_t, \nu_t^\varepsilon) - \nabla_u S(\xi_t^0, u^0(t,\xi_t^0))| \to 0$$

$$\text{for } \varepsilon \to 0+ \text{ in } L^2(dP \otimes dt) \quad \text{(A11)}$$

By Lemma A3(ii) we have

$$|\tfrac{1}{\varepsilon}(\xi_t^\varepsilon - \xi_t^0) - \xi_t'| \to 0 \qquad \text{for } \varepsilon \to 0+ \text{ in } L^2(dP \otimes dt) \quad \text{(A12)}$$

As $\underline{u}^0$ is continuously differentiable with respect to $x_1,\ldots,x_n$ there exist $\delta_t \in [0,1]$, $0 \le \varepsilon \le 1$, for almost all t such that

$$\frac{1}{\varepsilon}[u^0(t,\xi_t^\varepsilon) - u^0(t,\xi_t^0)] = \frac{1}{\varepsilon}\, \partial_x u^0(t,(1 - \delta_t^\varepsilon)\xi_t^0 + \delta_t^\varepsilon \xi_t^\varepsilon)(\xi_t^\varepsilon - \xi_t^0)$$

As $\partial_x u^0(t,x)$ is bounded, we conclude from (A12) that

$$\left|\frac{1}{\varepsilon}\{u^0(t,\xi_t^\varepsilon) - u^0(t,\xi_t^0)\} - \partial_x u^0(t,\xi_t^0)\xi_t'\right| \to 0$$

$$\varepsilon \to 0+ \text{ in } L^2(dP \otimes dt) \tag{A13}$$

Finally, as $|\xi_t^\varepsilon - \xi_t^0| \to 0$ in $L^2(dP \otimes dt)$ and by (2.4) we have

$$|u^\varepsilon(t,\xi_t^\varepsilon) - u^\varepsilon(t,\xi_t^0)| \to 0 \qquad \text{for } \varepsilon \to 0+ \text{ in } L^2(dP \otimes dt)$$

From this and (A10) to (A13) we conclude that

$$\begin{aligned}
&\lim_{\varepsilon\to 0+} \frac{1}{\varepsilon} E \int_0^T [S(\xi_t^\varepsilon,u^\varepsilon(t,\xi_t^\varepsilon)) - S(\xi_t^0,u^0(t,\xi_t^0))] \, dt \\
&= E \int_0^T \nabla_x^T S(\xi_t^0,u^0(t,\xi_t^0))\xi_t' \, dt \\
&\quad + E \int_0^T \nabla_u^T S(\xi_t^0,u^0(t,\xi_t^0))[u(t,\xi_t^0) + \partial_x u^0(t,\xi_t^0)\xi_t'] \, dt \\
&= E \int_0^T \nabla_x^T \sigma^0(t,\xi_t^0)\xi_t' + E \int_0^T \nabla_u^T S(\xi_t^0,u^0(t,\xi_t^0))u(t,\xi_t^0) \, dt
\end{aligned}$$

ACKNOWLEDGMENT

This work was supported by the Deutsche Forschungsgemeinschaft in connection with SFB 123, "Stochastische Mathematische Modelle."

REFERENCES

1. B. I. Arkin and M. T. Saksonov, To the maximum principle theory for problems of control of stochastic differential equations, pp. 255-263, in Stochastic Differential Systems, Filtering and Control, Lect. Notes Control Inf. Sci. 25, Springer-Verlag, Berlin, 1980.
2. J. M. Bismut, Conjugate convex functions in optimal stochastic control, J. Math. Anal. Appl. 44:384-404 (1973).
3. J. M. Bismut, Linear quadratic optimal stochastic control with random coefficients, SIAM J. Control 14:419-444 (1976).
4. L. Breiman, Probability, Addison-Wesley, Reading, Mass., 1968.
5. J. Franke, Optimal navigation with random terminal time in the presence of phase constraints, Z. Wahrscheinlich-

keitstheorie verw. Gebiete 60:453-484 (1982).

6. J. Franke, The intuitive dynamic programming approach to optimal stochastic navigation, Z. Wahrscheinlichkeitstheorie verw. Gebiete 60:485-495 (1982).

7. I. I. Gihman and A. V. Skorohod, Stochastic Differential Equations, Springer-Verlag, Berlin, 1972.

8. I. I. Gihman and A. V. Skorohod, Controlled Stochastic Processes, Springer-Verlag, Berlin, 1979.

9. I. V. Girsanov, Lectures on Mathematical Theory of Extremum Problems, Lect. Notes Econ. Math. Syst. 67, Springer-Verlag, Berlin, 1972.

10. P. A. Meyer, Intégrales stochastiques, pp. 72-142 in Séminaire de probabilités I, Lect. Notes Math. 39, Springer-Verlag, Berlin, 1967.

4

On Wavefront Propagation in Periodic Media

MARK I. FREIDLIN/Moscow, Union of Soviet Socialist Republics

1. INTRODUCTION

Mathematical models reducing to differential equations of the form

$$\frac{\partial u}{\partial t} = D \frac{\partial^2 u}{\partial x^2} + f(u) \tag{1.1}$$

are used for the description of a number of processes in chemical kinetics, population genetics, and in some other fields. If the function u(t,x) is interpreted as a concentration of some particles at a point x at time t, then the evolution process u(t,x) described by equation (1.1) is determined by two factors: diffusion of particles with the diffusion coefficient D and "multiplication" of the particles with rate f(u). Generally, the multiplication rate f(u) may be looked upon as possessing the following properties: $f(0) = f(1) = 0$, $f(u) < 0$ for $u \notin [0,1]$. Various hypotheses may be made on the behavior of f(u) inside the interval (0,1). For instance, it is sometimes natural to consider f(u) a positive smooth function in this interval and $f'(0) = \max_{0 \leq u \leq 1} f'(u)$. Such a case is treated in Ref. 1. If the function

$$u(0,x) = g(x) = \begin{cases} 1 & x < 0 \\ 0 & x > 0 \end{cases}$$

is chosen as an initial condition, then, as shown in Ref. 1, for large t the function $u(t,x)$ is close to the solution $V(x - \alpha t)$ of equation (1.1), which may be imagined as a wave expanding from the left to the right with velocity α. The shape of the wave $V(\xi)$ is defined by the equation

$$DV''(\xi) + \alpha V'(\xi) + f(V(\xi)) = 0 \qquad -\infty < \xi < \infty$$
$$V(-\infty) = 1 \qquad V(\infty) = 0 \tag{1.2}$$

The velocity $\alpha = 2\sqrt{Df'(0)}$.

This result, specifically, implies that for any $h > 0$,

$$\lim_{t\to\infty} u(t, (\alpha + h)t) = 0 \qquad \lim_{t\to\infty} u(t, (\alpha - h)t) = 1$$

Therefore, the domain of large (close to 1) values of the function $u(t,x)$ expands with velocity α.

This remark enables us to separate the simpler problem of propagation velocity from the more delicate one of the wave shape. Furthermore, the problem of propagation velocity of the domain of large values of the function $u(t,x)$ may be considered in a more general situation, for example, in the case of both the diffusion coefficients and multiplication rate depending on space variables. Certainly, in the case of arbitrary diffusion coefficients and the nonlinear term $f(x,u)$, there is no hope that, for large t, a constant propagation velocity of domain of large values of the function $u(t,x)$ will be reached. For this constant velocity to be reached, it is necessary to make hypotheses on the homogeneity (in a sense) of the diffusion coefficients and multiplication rate. The following two types of such hypotheses are conceived as being natural. First, one can consider the periodic media case, i.e., look upon the diffusion coefficients and the multiplication rate as functions periodic in space variables. Second, one can consider random homogeneous media, i.e., think of these functions as random homogeneous fields. Some results concerning both cases without detailed proofs are stated in Ref. 2. In this chapter we consider in detail the case of many-dimensional diffusion with periodic coefficients and periodic multiplication rate. We do not touch on a number of interesting questions, such as that of the nature of the transition area between the domain of large values and the domain where $u(t,x)$ is close to zero [in the case of problems (1.1) and (1.2) this question is reduced to that of the wave shape]. We also leave aside the case of function $f(u)$ having zeros inside the interval $(0,1)$.

Just as in Ref. 3, our research is based on applying the Kac formula and limit theorems for large deviation probabilities. Wave propagation velocity is defined both by logarithmic asymptotics of probability that, at large time t, the diffusing particle will travel away at a distance of αt, $\alpha > 0$, and by the exponent of the exponential function characterizing the rate of growth of particle number. In doing so, large deviations connected both with abnormally fast motion of the Brownian particle and with deviation of this motion from typical behavior on one space period should be taken into account. The asymptotic invariance with respect to translations enables us to restrict ourselves to large deviations for families of finite dimensional random variables. The propagation law of large value domain is described in terms of spectral problems for the corresponding differential operators.

2. EXPONENTIAL BOUNDS FOR FAMILIES OF RANDOM VECTORS

In Ref. 3 large deviation theorems in the space of continuous functions are used substantially when studying the wave propagation for quasi-linear equations with small diffusion.

Calculation of the asymptotic wave propagation velocity for the equations to be considered in the present chapter is also connected with large deviation theorems. However, unlike Ref. 3, we apply large deviation theorems for some families of (finite dimensional) random vectors, rather than for diffusion processes with small parameter. Let us formulate the corresponding results to be used in the sequel.

Let $(\Omega_\theta, B_\theta^t, P_\theta^t)$ be a family of probability spaces, where t runs over the positive half-line $(0,\infty)$ and the parameter θ varies over an arbitrary nonempty set Θ. Consider a family of d-dimensional random vectors η_θ^t defined on the corresponding measurable spaces $(\Omega_\theta^t, B_\theta^t)$, $t \in (0,\infty)$, $\theta \in \Theta$. Suppose that, for some positive function $\varepsilon(t)$ such that $\varepsilon(t) \to 0$ as $t \to \infty$ and for all $z \in R^d$, a limit independent of $\theta \in \Theta$,

$$G(z) = \lim_{t\to\infty} \varepsilon(t)\, \ell n\, E_\theta^t \exp \frac{1}{\varepsilon(t)}(z, \eta_\theta^t) \qquad (2.1)$$

exists uniformly in the parameter θ. Here E_θ^t designates the expectation with respect to the probability measure P_θ^t and $(\cdot,\cdot)$ denotes scalar product in the space R^d. We also admit that the expectation in (2.1) as well as the limit G(z) can take the value $+\infty$.

Let us introduce the action function $S\colon R^d \to [0,\infty]$ by the equality

$$S(y) = [\sup_{z\in R^d} (y,z) - G(z)] \qquad y \in R^d$$

The function S is convex and lower continuous. Let D(G) be the set of all points of the form $\nabla G(z)$. Here z runs over all points of the space R^d where the function G is finite and differentiable. Denote by $\overline{D}(G)$ the set of all points $y \in R^d$ for which there exists a sequence $\{y_n\}$, $y_n \in D(G)$, such that simultaneously $y_n \to y$ and $S(y_n) \to S(y)$ as $n \to \infty$. If the function G is differentiable at all points of the space R^d, then $\bar{D}(G)$ coincides with the domain of finiteness of the action function S (see Ref. 11). For any $s \geq 0$, we set

$$\Phi(s) = \{y \in R^d: S(y) \leq s\}$$

The sets $\Phi(s)$ are closed and convex. If 0 is an interior point of the domain of finiteness of the function G, then the sets $\Phi(s)$, $s \geq 0$, are bounded.

Let ρ be the euclidean metric in the space R^d.

THEOREM 2.1 Suppose that for some $s \geq 0$, the set $\Phi(s)$ is nonempty and bounded. Then, for any $\delta > 0$, $h > 0$, we can choose $t_0 > 0$ such that for $t > t_0$ and for all $\theta \in \Theta$ the bound

$$P_\theta^t \; \rho(\eta_\theta^t, \Phi(s)) > \delta \;\; \lesssim \exp\left[-\frac{s-h}{\epsilon(t)}\right]$$

holds.

THEOREM 2.2 For any $\delta > 0$, $h > 0$ and for all $y \in \bar{D}(G)$ there exists $t_0 > 0$ such that for $t > t_0$ and for all $\theta \in \Theta$

$$P_\theta^t \; \rho(\eta_\theta^t, y) < \delta \;\; \geq \exp\left[-\frac{S(y)+h}{\epsilon(t)}\right]$$

The proofs of Theorems 2.1 and 2.2 are mainly carried out in a similar way to those in Ref. 2 for the particular case when the random vectors η_θ^t and the probability spaces $(\Omega_\theta^t, B_\theta^t, P_\theta^t)$ do not depend on the parameter θ and besides the function G is finite and differentiable. The details of the proofs are omitted.

3. CALCULATION OF THE ACTION FUNCTION

Consider the Cauchy problem for the quasi-linear equation

$$\frac{\partial u}{\partial t}(t,x) = Lu(t,x) + f(x,u(t,x)) \qquad (t,x) \in (0,\infty) \times R^d \quad (3.1)$$

with the initial condition

$$u(0,x) = \varphi(x) \quad (3.2)$$

Here L denotes a linear elliptic second-order differential

operator

$$L = \frac{1}{2} \sum_{i,j=1}^{d} a^{ij}(x) \frac{\partial^2}{\partial x^i \partial x^j} + \sum_{i=1}^{d} b^i(x) \frac{\partial}{\partial x^i}$$

We shall assume the matrix $a(x) = \{a^{ij}(x)\}$ to be nonsingular, the coefficients $a^{ij}(x)$, $b^i(x)$, and the function $f(x,u)$, $x = (x^1,\ldots,x^d) \in R^d$, $u \in R^1$, satisfy the Lipschitz condition with respect to x and u and depend on the variables $x^1,\ldots,x^d$ periodically with period 1. From now on the function φ is supposed nonnegative, continuous, not identically equal to zero, and as having a compact support. Under these assumptions, problem (3.1) with condition (3.2) is known (see Ref. 5) to have a unique classic solution $u(t,x)$.

We are going to study the asymptotic behavior of the function $u(t,x)$ as $t \to \infty$. Reference 6 establishes that if L is the Laplace operator and the function $f(x,u)$ does not depend on x, vanishes for $u = 1$ and $u = 0$, and has some additional properties, then there is a positive constant c^* such that the following assertion holds: For any $c > c^*$, the function $u(t,x)$ vanishes as $t \to \infty$ in the region $|x| > ct$, and for any $c \in (0,c^*)$ it tends to 1 in the region $|x| < ct$. Thus c^* can be interpreted as an asymptotic wavefront propagation velocity.

If one wants the function u to behave similarly in the case when the coefficients of equation (3.1) depend periodically on the space variable x, it is necessary to impose some additional assumptions on the function f. Namely, we suppose that for any $x \in R^d$,

$$f(x,0) = f(x,1) = 0 \tag{3.3}$$

$$f(x,u) > 0 \quad \text{for } 0 < u < 1 \qquad f(x,u) \leq 0 \quad \text{for } u > 1$$

What is more, we suppose that for every $h \in (0,1)$ there is a point $x_h \in R^d$ such that

$$f(x_h,u) > 0 \qquad \text{for } u \in (0,h) \tag{3.4}$$

Furthermore, let the function

$$c(x,u) = \frac{f(x,u)}{u} \qquad (x,u) \in R^d \times (0,\infty)$$

be extendable to the hyperplane u = 0, and in this hyperplane c(x,u) as function of u attains its absolute maximum c(x):

$$c(x) = c(x,0) = \sup_{u>0} c(x,u)$$

From (3.3) and the properties of the initial function φ, it follows, in particular, that the solution of Cauchy's problem (3.1) with condition (3.2) is nonnegative, bounded from above by the quantity $1 \vee \max_{x \in R^d} \varphi(x)$, and satisfies the relation

$$\overline{\lim_{t\to\infty}} \sup_{x \in R^d} u(t,x) \leq 1 \tag{3.6}$$

Henceforth we shall utilize the following integral equation for the function u(t,x). Consider a d-dimensional process X_t, $t \geq 0$, obeying the Itô stochastic differential equation

$$dX_t = b(X_t)\,dt + \sigma(X_t)\,dW_t$$

where $b(x) = \{b^i(x)\}$, $\sigma(x) = \{a^{ij}(x)\}^{1/2}$, and W_t is the d-dimensional Wiener process defined on a probability space (Ω,B,P). Then by the Kac formula the function u(t,x) is the unique bounded solution of the equation

$$u(t,x) = E_x\varphi(X_t) \exp \left[\int_0^t c(X_s,\, u(t-s,\, X_s))\, ds\right] \qquad (t,x) \in [0,\infty) \times R^d \tag{3.7}$$

(see Ref. 7).

For further considerations it will be helpful to note that due to periodicity of the drift vector b(x) and the diffusion matrix a(x), for any $x \in R^d$ and any d-dimensional vector z with integer components, the distribution in path space of the process X_t with respect to the probability measure P_{x+z} coincides with the distribution of the process $X_t + z$ with respect to P_x.

The equality

$$P_x^t(A) = \frac{E_x \chi_a \exp\left[\int_0^t c(X_s)\,ds\right]}{E_x \exp\left[\int_0^t c(X_s)\,ds\right]} \qquad A \in B \tag{3.8}$$

defines a family of probability measures P_x^t, $t > 0$, $x \in R^d$, in the measurable space (Ω, B). We will make use of large deviation theorems for the family of random vectors

$$\eta_x^t = \frac{1}{t}(x - X_t) \tag{3.9}$$

defined on probability spaces (Ω, B, P_x^t), $t > 0$, $x \in R^d$, respectively (see Sec. 2). Let $C_\pi(R^d)$ be the Banach space of all periodic of period 1 continuous functions in R^d with the uniform norm. To apply Theorems 2.1 and 2.2 to our case, we shall need the following:

LEMMA 3.1 For any $z = (z_1, \ldots, z_d) \in R^d$,

$$\lim_{t\to\infty} \frac{1}{t} \ln E_x^t \exp [t(z, \eta_x^t)] = \lambda(z) - \lambda(0) \tag{3.10}$$

exists uniformly in $x \in R^d$, with $\lambda(z)$ being the eigenvalue of the differential operator

$$L^z = L - \sum_{i,j=1}^d a^{ij}(x) z_i \frac{\partial}{\partial x^j} + c(x) + \frac{1}{2} \sum_{i,j=1}^d a^{ij}(x) z_i z_j \tag{3.11}$$

in the space $C_\pi(R^d)$ corresponding to a positive eigenfunction. The function $\lambda(z)$, $z \in R^d$, is differentiable.

Proof. For any $z \in R^d$, we define a d-dimensional diffusion process X_t^z, $t \geq 0$, by the following stochastic differential equation:

$$dX_t^z = [b(X_t^z) - a(X_t^z)z]\,dt + \sigma(X_t^z)\,dW_t$$

The equality

$$Q_t^z \psi(x) = E_x \exp \{\int_0^t [c(X_s^z) - b(X_s^z), z) + \frac{1}{2}(a(X_s^z)z, z)]\,ds\} \Psi(X_t^z)$$
$$\Psi \in C_\pi(R^d) \tag{3.12}$$

defines a continuous semigroup Q_t^z of linear bounded operators in the space $C_\pi(R^d)$ dependent on the parameter z. Note that the operator L^z is the restriction of the infinitesimal operator of the semigroup Q_t^z to the set of all twice-continuously differentiable periodic functions of period 1. The space $C_\pi(R^d)$ can be identified in a natural way with the Banach space $C(T^d)$ of all continuous functions on the d-dimensional unit torus T^d, considered as a factor group R^d by the integer lattice. Denote by $\bar{a}^{ij}(\bar{x})$, $\bar{b}^i(\bar{x})$, and $\bar{c}(\bar{x})$ the restrictions of the periodic functions $a^{ij}(x)$, $b^i(x)$, and $c(x)$, respectively, to the unit torus T^d. Let $\bar{X}_t^z$, $t \geq 0$, be a diffusion process on the torus T^d governed by the differential operator

$$\bar{L}^z = \frac{1}{2}\sum_{i,j=1}^{d} \bar{a}^{ij}(\bar{x}) \frac{\partial^2}{\partial \bar{x}^i \, \partial \bar{x}^j} + \sum_{i=1}^{d} [b^j(\bar{x}) - \bar{a}^{ij}(\bar{x})z_i] \frac{\partial}{\partial \bar{x}^j}$$

Then the semigroup $\bar{Q}_t^z$ of linear bounded operators corresponding to Q_t^z in the space $C(T^d)$ is acting according to the formula

$$\bar{Q}_t^z\bar{\Psi}(\bar{x}) = E_x\bar{\Psi}(\bar{X}_t^z) \exp \{\int_0^t [\bar{c}(\bar{X}_s^z) - (b(\bar{X}_s^z),z) + \frac{1}{2}(\bar{a}(\bar{X}_s^z)z,z)] \, ds\}$$

$$\psi \in C(T^d)$$

It is straightforward to check that the quasi-transition probabilities $\bar{Q}(t,\bar{x},\Gamma) = \bar{Q}_t^z\chi_\Gamma(\bar{x})$, with $t > 0$, $\bar{x} \in T^d$, and Γ being a Borel subset of T^d, are strictly positive and satisfy the Doeblin condition. So there is a number $\lambda(z)$ such that, for all $t > 0$, $e^{t\lambda(z)}$ is a simple eigenvalue of the operator $\bar{Q}_t^z$ with a (independent of t) strictly positive eigenfunction $\bar{u}^z$ (see Ref. 8). Hence $e^{t\lambda(z)}$ is an eigenvalue of the operator Q_t^z and

$$Q_t^z u^z = e^{t\lambda(z)} u^z \tag{3.13}$$

where $u^z \in C_\pi(R^d)$ corresponds to $\bar{u}^z \in C(T^d)$. By differentiating equation (3.13) at $t = 0$, we conclude that $\lambda(z)$ is an eigenvalue of the operator L^z in $C_\pi(R^d)$ corresponding to a positive eigenfunction. The process X_t^z differs from the process X_t only by the drift. Thus the measures μ^z and μ, induced by this processes in path space in the interval $[0,t]$, are absolutely continuous and the density of one measure with respect to another has the form

$$\frac{d\mu^z}{d\mu}(X.) = \exp\left[-\int_0^t (z,\sigma(X_s))dW_s - \frac{1}{2}\int_0^t (a(X_s)z,z)\, ds\right]$$

(see, e.g., Ref. 9). Using this and the definition of the random vectors η_x^t, (3.12) can be rewritten in the form

$$Q_t^z\Psi(x) = E_x \exp\left[\int_0^t c(X_s)\, ds + t(z,\eta_x^t)\right]\Psi(X_t) \qquad (3.14)$$

Since the function u^z is strictly positive and periodic, one can deduce from (3.14) via equation (3.13) that

$$\lim_{t\to\infty} \frac{1}{t} \ln E_x \exp\left[\int_0^t c(X_s)\, ds + t(z,\eta_x^t)\right]$$

$$= \lim_{t\to\infty} \frac{1}{t} \ln (Q_t^z 1)(x) = \lambda(z) \qquad (3.15)$$

the convergence being uniform in $x \in R^d$. From this, recalling the definition of the probability measures P_x^t, we deduce the relation (3.10). It remains to establish the differentiability of the function $\lambda(z)$. As follows from perturbation theory of linear operators [10], for this purpose it is sufficient to check differentiability of the operator function $z \to Q_1^z$ in uniform operator topology. The latter is done with representation (3.14). Thus Lemma 3.1 is proved.

Lemma 3.1 implies that, for probability measures (3.8) and random vectors (3.9), exponential bounds of Theorems 2.1 and 2.2 are valid for all $x \geq 0$ and all $y \in R^d$ with $\varepsilon(t) = 1/t$. The sets $\Phi(s)$ are compact and the action function has the form

$$S(y) = H(y) + \lambda(0)$$

where

$$H(y) = \sup_{z\in R^d} [(y,z) - \lambda(z)] \qquad y \in R^d$$

From (3.12) and (3.15), we derive the estimate

$$\lambda(z) \geq \min_{x \in R^d} [c(x) - (b(x),z) + \frac{1}{2}(a(x)z,z)]$$

This yields that $\lambda(z)/|z| \to \infty$ as $|z| \to \infty$. Hence [10] the function $H(y)$ is finite for all $y \in R$ and, moreover, being convex, it is continuous. Furthermore, observe that the set $\{y \in R^d : H(y) \leq 0\}$ is nonempty, whereas $\lambda(0) > 0$, and that the action function $S(y)$ vanishes for $y = \nabla\lambda(0)$. The inequality $\lambda(0) > 0$ follows from relation (3.15) with $z = 0$ if one takes into account that by virtue of conditions (3.4) and (3.5), $C(x) \geq 0$ and $C(x) \not\equiv 0$.

4. ASYMPTOTIC WAVEFRONT PROPAGATION VELOCITY IN PERIODIC MEDIUM

We now proceed to study asymptotic behavior of the solution $u(t,x)$ of problem (3.1) with condition (3.2).

LEMMA 4.1 For any closed set $F \subset \{y \in R^d : H(y) > 0\}$,

$$\overline{\lim_{t\to\infty}} \frac{1}{t} \ln \sup_{y \in F} u(t,ty) \leq -\min_{y \in F} H(y)$$

Proof. Let a number s be chosen in such a way that $0 < s < \min_{y \in F} H(y)$. As the closed set F does not intersect the compact set $\Psi(s) = \{y \in R^d : H(y) \leq s\}$, the distance 2δ between them is positive. For t sufficiently large, the support of the initial function φ is contained in the δt-neighborhood of the point 0. Therefore, on account of condition (3.5), for such t, equation (3.7) implies the estimate

$$u(t,ty) \leq \|\varphi\| E_{ty} \exp [\int_0^t c(X_s)\, ds] \chi_{X_t \in U_{\delta t}(0)} \tag{4.1}$$

with $\|\varphi\| = \sup_{x \in R^d} |\varphi(x)|$. From this, using definitions (3.8) and (3.9) of probability measures P_x^t and random vectors η_x^t, we obtain

$$\sup_{y\in F} u(t,ty) \le \|\varphi\| \sup_{y\in F} P^t_{ty}[\rho(\eta^t_{ty},y) < \delta] E_{ty} \exp [\int_0^t c(X_s)\, ds]$$

$$\le \|\varphi\| \sup_{x\in R^d} P^t_x[\rho(\eta^t_x,\psi(s)) > \delta] \sup_{x\in R^d} E_x \exp [\int_0^t c(X_s)\, ds]$$

The expressions on the right-hand side of this inequality can be estimated from below via relation (3.15) with z = 0 and the exponential bound contained in Theorem 2.1. As a result we arrive at the following inequality:

$$\overline{\lim_{t\to\infty}} \frac{1}{t} \ln \sup_{y\in F} u(t,ty) \le -s$$

Since s can be chosen arbitrarily close to $\min_{y\in F} H(y)$, this fact leads to the assertion of Lemma 4.2.

Let $Q = [0,1]^d$ be a unit cube and $U_\delta(y)$ be the δ-neighborhood of a point y in the space R^d. Now we are going to cite an asymptotic bound of the function u, which is, in a sense, opposite to (4.1).

LEMMA 4.2 For all $y \in R^d$ for which $H(y) > 0$:

$$\overline{\lim_{t\to\infty}} \frac{1}{t} \ln \inf_{\tilde{y}\in U_\delta(y)} u(t,t\tilde{y})$$

$$\ge \underline{\lim_{t\to\infty}} \frac{1}{t} \ln \inf_{x\in Q,\tilde{y}\in U_{2\delta}(y)} E_x(\exp [\int_0^t c(X_s)\, ds] \chi_{X_t} \in U_{\delta y}(x - t\tilde{y})$$

provided that $\delta > 0$ is sufficiently small.

Proof. We put

$$\ell = \lim_{t\to\infty} \frac{1}{t} \ln \inf_{\tilde{y}\in U_\delta(y)} u(t,t\tilde{y})$$

We first verify that $\ell > -\infty$. By the Markov property of the process X_t, we obtain, for any $\tilde{y} \in R^d$, the following bound:

$$E_{t\tilde{y}}\varphi(X_t) \ge \prod_{k=1}^{[t]-1} \inf_{x\in U_\delta((t-k+1)\tilde{y})} P_x[X_t \in U_\delta((t-k)\tilde{y})]$$

$$\times \inf_{x\in U_\delta((t-[t]+1)\tilde{y})} E_x\varphi(X_{t-[t]+1})$$

where [t] is the integer part of t. Taking into account the periodicity of the process X_t, we conclude that

$$\inf_{\tilde{y}\in U_\delta(y)} E_{t\tilde{y}}\varphi(X_t) \geq \inf_{x\in Q,\tilde{y}\in U_{2\delta}(y)} P_x[X_1 \in U_\delta(x - \tilde{y})]$$

$$\times \inf_{\substack{x\in U_{2|y|+3\delta}(0) \\ 1 \leq u \leq 2}} E_x\varphi(X_u)$$

With all the factors on the right-hand side being positive, we have

$$\overline{\lim_{t\to\infty}} \frac{1}{t} \ln \inf_{\tilde{y}\in U_\delta(y)} E_{t\tilde{y}}\varphi(X_t) > -\infty$$

From this, noting equation (3.7), we derive that $\ell > -\infty$. For all $x \in R^d$ and any positive numbers η and t, let us introduce the Markov times:

$$\sigma_\eta(t) = \inf\{s \geq 0 : |X_s - (t - s)y| > \eta t\}$$
$$\tau_{x,\eta}(t) = \inf\{s \geq 0 : |X_s - x + sy| > \eta t\}$$

Choose $\varepsilon > 0$ so that the 2ε-neighborhood of the point y is contained entirely in the region in which the function H is positive, and fix $\delta \in (0,\varepsilon/3)$ and $h \in (0,1)$ in an arbitrary manner. If $\sigma_\varepsilon(t) > t$, then by Lemma 4.1, $u(t - s,X_s) \leq h$, from some t on. Hence $c(X_s,u(t - s,X_s) > C_h(X_s)$ for $s \in [0,t/2]$, where

$$C_h(x) = \inf_{u\in[0,h]} c(x,u)$$

This and equation (3.7) together result, via the Markov property, in the following bound:

$$\inf_{\tilde{y}\in U_\delta(y)} u(t,t\tilde{y})$$

$$\geq \inf_{\tilde{y}\in U_\delta(y)} E_{t\tilde{y}}\exp[\int_0^{\varkappa t} c_\varkappa(X_s)ds]\chi_{\sigma_\varepsilon(t)>\varkappa t,X_{\varkappa t}\in U_{(1-\varkappa)\delta t}((1-\varkappa)ty)}$$

$$\times \inf_{\tilde{y}\in U_\delta(u)} u((1 - \varkappa)t,(1 - \varkappa)t\tilde{y})$$

$$\geq \inf_{x\in Q,\hat{y}\in U_{2\delta}(y)} E_x\exp[\int_0^{\varkappa t} c_\varkappa(X_s)ds]\chi_{\tau_{x,\delta/\varkappa}(\varkappa t)>t,X_{\varkappa t}\in U_{\varkappa\delta t}(x-\varkappa t\hat{y})}$$

$$\times \inf_{\tilde{y}\in U_\delta(y)} u((1-\varkappa)t,(1-\varkappa)t\tilde{y}) \tag{4.2}$$

for any $\varkappa \in (0,1/2)$.

The second part of this bound is obtained by translating the trajectories of the process X_s by the integer part of the vector $t\tilde{y}$, if one makes use of the "periodicity" of the process and the following implications:

$$U_t(x - t\tilde{y} + (t-s)y) \supset U_{\delta t}(x - sy)$$

$$U_{(1-\varkappa)\delta t}(x - t\tilde{y} + (1-\varkappa)ty) \supset U_{\varkappa\delta t}(x - \varkappa t\hat{y})$$

where x is the fractional part of $t\tilde{y}$ and $\hat{y} = y + 2(\tilde{y} - y)$. Since $\ell > -\infty$, (4.2) implies the inequality

$$\ell \geq \overline{\lim_{t\to\infty}} \frac{P_{h,\delta/\varkappa}(t)}{t} \tag{4.3}$$

where

$$P_{h,\eta}(t) = \ell n \inf_{x\in Q,\tilde{y}\in U_{2\delta}(y)} E_x \exp[\int_0^t c_h(X_s)\, ds]\chi_{\tau_{\varkappa,\eta}>t,X_t\in U_{\delta t}(x-t\tilde{y})} \tag{4.4}$$

Again by the Markov property and the periodicity of the process X_t, we establish semiadditivity of the function $P_{h,\eta}(t)$:

$$P_{h,\eta}(s+t) \geq P_{h,\eta}(s) + P_{h,\eta}(t)$$

Moreover, it is plain that $P_{h,\eta}(t) \leq t \max_x c(x)$, $t > 0$. These properties of the function $P_{h,\eta}(t)$ are known [12] to yield the equality

$$\lim_{t\to\infty} \frac{P_{h,\eta}(t)}{t} = \sup_{t>0} \frac{P_{h,\eta}(t)}{t}$$

Thus (4.3) leads to the bound

$$\ell \geq \sup_{t>0} \frac{P_{h,\delta/\varkappa}(t)}{t} \tag{4.5}$$

We now introduce another function,

$$p(t) = \ell n \inf_{x\in Q,\tilde{y}\in U_{2\delta}(y)} E_x \exp[\int_0^t c(X_s)\, ds]\chi_{X_t\in U_{\delta t}(x-t\tilde{y})} \tag{4.6}$$

In accordance with conditions (3.4) and (3.5) we have $c_h(x) \uparrow c(x)$ as $h \downarrow 0$. Consequently, by the Fatou lemma the expectation on the right-hand side of (4.4) converges monotonically to the expectation on the right-hand side of (4.6) as $h \downarrow 0$, $\eta \uparrow \infty$. With this expectation depending continuously on x and $\tilde{y}$, we have

$$p_{h,\eta}(t) \uparrow p(t) \qquad \text{as } h \downarrow 0,\ \eta \uparrow \infty$$

Therefore, passing to limit in (4.5) as $h \downarrow 0$, $\varkappa \downarrow 0$, we get

$$\ell \geq \sup_{t>0} \frac{p(t)}{t}$$

From this the proof is immediate.

LEMMA 4.3 For any compact set $K \subset \{y \in R^d : H(y) > 0\}$

$$\overline{\lim_{t\to\infty}} \frac{1}{t} \ell n \inf_{y\in K} u(t,ty) \geq -\max_{y\in K} H(y)$$

Proof. Due to the compactness of the set K, it is sufficient to show that, for any $y \in R^d$, with $H(y) > 0$, and for any $\varepsilon > 0$ there exists $\delta > 0$ such that

$$\overline{\lim_{t\to\infty}} \frac{1}{t} \ell n \inf_{\tilde{y}\in U_\delta(y)} u(t,ty) \geq H(y) - \varepsilon$$

By virtue of Lemma 4.2 and continuity of the function H, for the above inequality to be valid, it is sufficient, in its turn, that for small $\delta > 0$,

$$\overline{\lim_{t\to\infty}} \frac{1}{t} \ell n\, I(t) \geq - \sup_{\tilde{y}\in U_{2\delta}(y)} H(\tilde{y}) \tag{4.7}$$

where

$$I(t) = \inf_{x\in Q, \tilde{y}\in U_{2\delta}(y)} E_x \exp\left[\int_0^t c(X_s)\, ds\right] \chi_{X_t \in U_{\delta t}}(x - t\tilde{y})$$

Taking into account definitions (3.8) and (3.9) of measures P_x^t and random vectors η_x^t, we arrive at

$$I(t) \geq \inf_{\tilde{y}\in U_{2\delta}(y)} \inf_{x\in Q} P_x^t\{\eta_x^t \in U_0(\tilde{y})\} \inf_{x\in Q} E_x \exp\left[\int_0^t c(X_s)\, ds\right]$$

The first factor on the right-hand side of this inequality

may be estimated by Theorem 2.2, and the second by limit relation (3.15) with $z = 0$. As a result we get the required bound (4.7), completing the proof.

We now proceed to formulate the main result of this chapter.

THEOREM 4.1 (i) For any closed set $F \subset \{y \in R^d : H(y) > 0\}$,

$$\lim_{t\to\infty} u(t,ty) = 0$$

uniformly in $y \in F$.

(ii) For any compact set $K \subset \{y \in R^d : H(y) < 0\}$,

$$\lim_{t\to\infty} u(t,ty) = 1$$

uniformly in $y \in K$.

Proof. Assertion (i) follows directly from Lemma 4.1. Let us prove assertion (ii). We set $\psi(s) = \{y \in R^d : H(y) = s\}$, $\Psi(s) = \{y \in R^d : H(y) \leq s\}$, $\overline{C} = \int_{T^d} \overline{C}(\overline{x})\overline{\mu}(d\overline{x})$, where $\overline{\mu}$ is the normalized invariant measure of the process $\overline{X}_t$ on the torus T^d. Next we introduce for any $\delta > 0$ and all $T > 1$

$$\Gamma_T = [\{1\} \times \Psi(\delta)] \cup [\bigcup_{1\leq t\leq T} \{t\} \times (t\Psi(\delta))]$$

Thus the set Γ_T consists of the lateral surface of the truncated cone and one of its bases, that is, the set $\Psi(\delta)$. It follows readily from equation (3.7) that $u(1,x) > 0$ for all $x \in R^d$. Therefore, relying on Lemma 3.4 we have that for sufficiently large t

$$u(s,x) \geq e^{-2\delta t} \qquad \text{for all } (s,x) \in \Gamma_t \tag{4.8}$$

Next, for any positive t, h, η, we introduce the Markov times:

$$\sigma_\Gamma(t) = \min \{s \geq 0 : (t - s, X_s) \in \Gamma_t\}$$

$$\sigma_h(t) = \min \{s \geq 0 : u(t - s, X_s) \geq h\}$$

$$\tau_\eta(t) = \min \{s \geq 0 : |X_s - X_0 - s\overline{C}| \geq \eta t\}$$

Given $u(t - s, X_s) < h$ for all $s \in [0,t]$, we set $\sigma_h(t) = \infty$. Using Itô's equation for the process X_t one can easily prove that for any $\eta > 0$,

$$\sup_{x \in R^d} P_x[\tau_\eta(t) \leq t] \to 0 \tag{4.9}$$

as $t \to \infty$. Choose η so that the distance between the η-neighborhood of the set K and the set $\{H \geq 0\}$ is positive. Then there is a number $\varkappa \in (0,1)$ such that for all $y \in K$

$$t - 1 > \sigma_\Gamma(t) > \varkappa^t \qquad \text{if } \tau_\eta(t) > t \tag{4.10}$$

By the strong Markov property of the process X_t, we obtain from equation (3.7) that for any Markov time τ

$$u(t,x) = E_x \exp[\int_0^{\tau\wedge t} c(X_s, u(t - s, X_s))\, ds] u(t - (t \wedge t), X_{\tau \wedge t}) \tag{4.11}$$

Relying on condition (3.3) and the definition of the Markov time $\sigma_h(t)$ the relation (4.11), with $\tau = \sigma_h(t)$, yields the bound

$$u(t,x) \geq h \cdot P_x[\sigma_h(t) \leq t]$$

Because of relations (3.6) and (4.9) we remark that in order for assertion (ii) to be proven, it is sufficient to verify that for any $h \in (0,1)$,

$$P_{ty}[\sigma_h(t) > t,\ \tau_\eta(t) > t] \to 0 \tag{4.12}$$

uniformly in $y \in K$ as $t \to \infty$. Using inequalities (4.8) and (4.10) we get such a bound:

$$\begin{aligned} P_{ty}[\sigma_h(t) > t,\ \tau_\eta(t) > t] &\leq P_{ty}[\varkappa t < \sigma_\Gamma(t) \leq \sigma_h(t)] \\ &\leq e^{\delta t} E_{ty} \exp[\tfrac{1}{2}\int_0^{\sigma_\Gamma(t)} c(X_s, u(t-s, X_s))ds] u(t-\sigma_\Gamma(t), X_{\sigma_\Gamma(t)})^{1/2} \\ &\times \exp[-\tfrac{1}{2}\int_0^{\sigma_\Gamma(t)} c(X_s, u(t-s,X_s))ds] \chi_{\varkappa t < \sigma_\Gamma(t) < \sigma_h(t)} \\ &\leq e^{\delta t} E_{ty} \exp[\tfrac{1}{2}\int_0^{\sigma_\Gamma(t)} c(X_s, u(t-s,X_s))ds][u(t-\sigma_\Gamma(t), X_{\sigma_\Gamma(t)})]^{1/2} \\ &\times \exp[-\tfrac{1}{2}\int_0^{\varkappa t} c_h(X_s)\, ds] \end{aligned}$$

where as before $c_h(x) = \inf_{u \in [0,h]} c(x,u)$. By the Hölder inequality and equality (4.11) with $\tau = \sigma_\Gamma(t)$,

$$P_{ty}[\sigma_h(t) > t,\ \tau_\eta(t) > t]$$
$$\leq e^{\delta t}[u(t,ty)]^{1/2}\{E_{ty} \exp [-\int_0^{\varkappa t} c_h(X_s)\, ds]\}^{1/2} \qquad (4.13)$$

The function u is bounded. Since [by condition (3.4)] the function $c_h(x)$ is nonnegative and not identically equal to zero,

$$\lambda_h = \lim_{t\to\infty} \frac{1}{t} \ln \sup_{x\in R^d} E_x \exp [-\int_0^t c_h(X_s)\, ds] < 0 \qquad (4.14)$$

Namely, as is seen from relation (3.15) of Lemma 3.1, λ_h is the eigenvalue of the operator $L - c_h(x)$ in $C_\pi(R^d)$ such that the corresponding eigenfunction is positive. Clearly, this eigenvalue cannot be equal to zero. If we now select $\delta < (\varkappa/2)|\lambda_h|$, then (4.13) and (4.14) together imply (4.12), the convergence being uniform in $y \in K$. This completes the proof.

COROLLARY 4.1

$$\lim_{t\to\infty} u(t,ty) = \begin{cases} 0 & \text{for } H(y) > 0 \\ 1 & \text{for } H(y) < 0 \end{cases}$$

The set $t\{y \in R^d : H(y) = 0\}$ may thereby be interpreted as the front of the wave which becomes stabilized for large t.

If

$$\min_{z\in R^d} \lambda(z) > 0 \qquad (4.15)$$

then $H(0) < 0$. In this case, for all unit vectors $e \in R^d$, the equation $H(ve) = 0$ has a unique positive solution $v = v^*(e)$. The definition of the function H yields

$$v^*(e) = \inf \frac{\lambda(z)}{(e,z)}$$

where the infimum is taken over all $z \in R^d$ for which $(e,z) > 0$. Here $v^*(e)$ can be thought of as the asymptotic wave propagation velocity in the direction of the vector e.

It is easily seen that condition (4.15), in particular, holds if the differential operator has the self-conjugate form

$$L = \frac{1}{2} \sum_{i,j=1}^{d} \frac{\partial}{\partial x^i}\left[a^{ij}(x) \frac{\partial}{\partial x^j}\right]$$

REFERENCES

1. A. Kolmogorov, I. Petrovsky, and N. Piscounov, Étude de l'équation de la diffusion avec croissance de la quantité de la matière et son application à un problème biologique, Mosc. Univ. Bull. Math. 1:1-25 (1937).
2. Yu. Gertner and M. I. Freidlin, On the propagation of concentration waves in periodic random media, Sov. Math. Dokl. 20(6):1282-1286 (1979).
3. M. I. Freidlin, Propagation of a concentration wave in the presence of random motion associated with the growth of a substance, Sov. Math. Dokl. 20(3):503-507 (1979).
4. A. D. Wentzell and M. I. Freidlin, Fluctuations in Dynamical Systems Caused by Small Random Perturbations, Nauka, Moscow, 1979 (in Russian; English trans. to appear).
5. O. A. Ladyženskaja, V. A. Solonnikov, and N. N. Ural'ceva, Linear and Quasi-linear Equations of Parabolic Type, Transl. Math. Monogr., Vol. 23, American Mathematical Society, Providence, R. I., 1968.
6. D. G. Aronson and H. F. Weinberger, Multidimensional nonlinear diffusion arising in population genetics, Adv. Math. 30(1):33-76 (1978).
7. M. I. Freidlin, Quasi-linear parabolic equations and measures on a function space, Funkc. Anal. Prilozh. 1(3):74-82 (1967) (in Russian).
8. L. Parch, On an ergodic theorem for the Markov processes with finite lifetime, Teor. Veroyatn. Primen. 16(4):

711-714 (1971) (in Russian).

9. A. Friedman, Stochastic Differential Equations and Applications, Vol. 1, Academic Press, New York, 1975.

10. T. Kato, Perturbation Theory for Linear Operators, Springer-Verlag, Berlin, 1966.

11. R. T. Rockafellar, Convex Analysis, Princeton University Press, Princeton, N. J., 1970.

12. E. Hille and R. S. Phillips, Functional Analysis and Semi-groups, Am. Math. Soc. Colloq. Publ. 31, American Mathematical Society, Providence, R. I., 1957.

5

Wiener-like Integrals for Gaussian Processes and the Linear Estimation Problem

MASUYUKI HITSUDA*/Faculty of Integrated Arts and Science, Hiroshima University, Hiroshima, Japan

1. INTRODUCTION

A Gaussian process $X = \{X(t);\ 0 \leq t \leq 1\}$ has common properties with Brownian motion $B = \{B(t);\ 0 \leq t \leq 1\}$, when the covariance Γ_X of X is equivalent to the covariance Γ_B of B in the following sense: both $\Gamma_X - c_1\Gamma_B$ and $c_2\Gamma_B - \Gamma_X$ are positive definite for some constants $c_1, c_2 > 0$. Such a relation between Γ_B and Γ_X is written $c_1\Gamma_B < \Gamma_X < c_2\Gamma_B$, or simply $\Gamma_X \sim \Gamma_B$. One of the remarkable consequences of $\Gamma_X \sim \Gamma_B$ is the coincidence of their reproducing kernel Hilbert spaces $H(X)$ and $H(B)$ generated by Γ_X and Γ_B, respectively. Using the property $H(X) = H(B)$, we can define a Wiener-like integral $\int_0^1 f(u)\, dX(u)$, which is analogous to the Wiener integral $\int_0^1 f(u)\, dB(u)$. This is illustrated in Sec. 2.

In the main part of the present chapter, Sec. 3, we treat the problem of signal-noise type

$$X(t) = B(t) + Y(t) \qquad 0 \leq t \leq 1 \tag{1.1}$$

The problem is motivated by the following observation. When the signal $Y = \{Y(t);\ 0 \leq t \leq 1\}$ is a Gaussian process independent of the noise B, we wish to estimate the exact message

*Current affiliation: Faculty of Science, Kumamoto University, Kumamoto, Japan

from the receipt X. In the case where the reproducing kernel Hilbert space $\mathcal{H}(Y)$ satisfies the relation $\mathcal{H}(Y) \cap \mathcal{H}(B) = \{0\}$, the noise and the signal are completely separated, so we can get the complete information on Y. The case has been discussed in the author's previous paper [5]. If $\mathcal{H}(Y) \cap \mathcal{H}(B)$ is nontrivial, there are some difficult problems since the information obtained by X is much less than the joint information of Y and B. Therefore, we need to estimate the real signal Y by the use of the receipt X. In particular, when Y is expressed in the form

$$\begin{aligned} &Y(t) = \int_0^t y(s)\, ds \qquad 0 \leq t \leq 1 \\ &E \int_0^1 y(s)^2\, ds < \infty \end{aligned} \tag{1.2}$$

the complete results have been obtained by many authors in connection with the Kalman-Bucy filtration and Wiener-Hopf type equation (see Hitsuda [4], Kailath [6], and Liptzer and Shiryaev [7]). Our concerns in Sec. 3 are to construct the conditional expectation $Y(t) = E[Y(t)|X(s);\ 0 \leq s \leq t]$ in terms of the Wiener-like integral defined in Sec. 2, under the condition that the reproducing kernel Hilbert space $\mathcal{H}(Y)$ for Y is included in $\mathcal{H}(B)$. Since the condition $\mathcal{H}(Y) \subset \mathcal{H}(B)$ is satisfied if Y has an expression (1.2), the results are a generalization of the case (1.2). As a natural consequence of Theorem 3.1, we determine the general form of Gaussian martingale with respect to $\mathcal{B}_t(X) = \sigma\{X(s);\ s \leq t\}$. Although the author has expected that the Gaussian process X has a single innovation with a spectral measure equivalent to Lebesgue measure, the desired conjecture has, regretfully, not been proved in the present chapter. Some results on the spectral multiplicity are presented in Sec. 4, but they are not yet the concrete results. Theorem 4.1 only clarifies the necessary and sufficient conditions for the existence of the single innovation. This problem is very closely related to the Volterra-type factorization of the positive definite operator in the Hilbert space $L^2[0,1]$ (see Gohberg and Krein [2]), so the conditions in Theorem 4.1 are in some degree analogous to the one of Sakhnovich's papers [8]. The author believes that the innovation problems on the Gaussian process X are quite important. The problems have been originated by Hida [3] connected with the canonical representation of X.

2. REPRODUCING KERNEL HILBERT SPACES AND WIENER-LIKE INTEGRALS

Let $X = \{X(t);\ 0 \leq t \leq 1\}$ be a Gaussian process satisfying the covariance condition $\Gamma_X \sim \Gamma_B$, where $B = \{B(t);\ 0 \leq t \leq 1\}$ is a Brownian motion. We cite some results on the reproducing kernels from Aronszajn's original paper [1].

I. $\Gamma_X \sim \Gamma_B$ if and only if $\mathcal{H}(X) = \mathcal{H}(B)$, where $\mathcal{H}(X)$ and $\mathcal{H}(B)$ are the reproducing kernel Hilbert spaces for Γ_X and Γ_B, respectively.

II. The mapping Λ: $\Gamma_B(t,\cdot) \to \Gamma_X(t,\cdot)$ can be extended to the linear operator from $\mathcal{H}(B)$ to $\mathcal{H}(X)$, and the extended Λ gives a linear isomorphism between $\mathcal{H}(B)$ and $\mathcal{H}(X)$.

These results are naturally translated to the linear spaces spanned by the processes B and X.

II'. The mapping Λ: $B(t) \to X(t)$ is extended to the linear operator from H(B) to H(X), and the extended Λ is a linear isomorphism, where H(B) and H(X) are the linear spans by $\{X(t);\ 0 \leq t \leq 1\}$ and $\{B(t);\ 0 \leq t \leq 1\}$, respectively.

As is well known, each $\xi \in H(B)$ is expressed in the form of a Wiener integral

$$\xi = \int_0^1 f(s)\, dB(s) \qquad f \in L^2[0,1]$$

Applying the isomorphism Λ, we can define the Wiener-like integrals with respect to X by

$$\int_0^1 f(s)\, dX(s) = \Lambda\left(\int_0^1 f(s)\, dB(s)\right) \tag{2.1}$$

It is obvious that the integral (2.1) is the same as the integral defined by the constructive method starting from the construction for a step function, since

$$\int_s^t dX(u) = X(t) - X(s) = \Lambda(B(t) - B(s)) = \Lambda(\int_s^t dB(u))$$

We note that

$$H(X) = \{\int_0^1 f(s)\, dX(s);\ f \in L^2[0,1]\} \tag{2.2}$$

REMARK 1. For any $f \in L^2[0,1]$, the Wiener-like integral $\int_0^1 f(s)\, dX(s)$ is defined under a mild condition $\Gamma_X < c\Gamma_B$ for a constant $c > 0$ (or Γ_X is subordinate to Γ_B). In fact, the extension of Λ: $\Gamma_B(t,\cdot) \to \Gamma_X(t,\cdot)$ becomes a bounded operator from $\mathcal{H}(B)$ to $\mathcal{H}(X)$. In this case, we cannot state the fact (2.2), in general, because H(X) may not be covered with Wiener-like integrals. But it is easy to prove that the set of Wiener-like integrals

$$H(X) = \{\int_0^1 f(s)\, dX(s);\ f \in L^2[0,1]\}$$

is dense in H(X).

DEFINITION. Let us define the inner product $(f,g)_X$ and the norm $\|f\|_X$ for $f,g \in L^2[0,1]$ by

$$(f,g)_X = E[\int_0^1 f(s)\, dX(s) \int_0^1 g(s)\, dX(s)] \quad \text{and}$$
$$\|f\|_X^2 = (f,f)_X \tag{2.3}$$

PROPOSITION 2.1. The norm $\|\cdot\|_X$ defined by (2.3) is equivalent to the usual L^2-norm $\|\cdot\|$ (or $c_1\|f\| < \|f\|_X < c_2\|f\|$ for some constants $c_1, c_2 > 0$), under the condition $\Gamma_X \sim \Gamma_B$. The proof is easy, so it is omitted.

COROLLARY 2.1. Under the same condition, there exists an isomorphism G_X: $L^2[0,1] \to L^2[0,1]$ such that

$$(f, G_X g) = (f,g)_X \tag{2.4}$$

Furthermore, G_X is symmetric and positive definite.

Let us define the subspace $H_t(X)$, $0 \le t \le 1$, of H(X) by

$$H_t(X) = \{\int_0^t f(s)\, dX(s);\ f \in L^2[0,1]\}$$
$$= \{\int_0^1 P_t f(s)\, dX(s);\ f \in L^2[0,1]\}$$

where P_t is the usual projection defined by $P_t f(s) = f(s)$, $s \leq t$, and $P_t f(s) = 0$, $s > t$. The subspace $H_t(B)$ of $H(B)$ is defined analogously. The subspaces $H_t(X)$ and $H_t(B)$ are linear spans by $\{X(s);\ s \leq t\}$ and $\{B(s);\ s \leq t\}$, respectively, under the condition $\Gamma_X \sim \Gamma_B$. The following is also easily proved.

PROPOSITION 2.2. The operator Λ gives an isomorphism between subspaces $H_t(B)$ and $H_t(X)$ for each $t \in [0,1]$.

Let us introduce the orthogonal projection $Q_t = Q_t^X$ to the subspace $H_t(X)$ from $H(X)$. This will play an important role in the following section.

3. GAUSSIAN MARTINGALES CORRESPONDING TO X AND ESTIMATION PROBLEMS

In this section we treat a communication system

$$X(t) = B(t) + Y(t), \qquad 0 \leq t \leq 1 \tag{3.1}$$

where a Brownian motion B and a Gaussian process Y are mutually independent, and the covariance Γ_Y of Y is subordinate to Γ_B ($\Gamma_Y < \Gamma_B$). Then the equality $\Gamma_X = \Gamma_B + \Gamma_Y$ holds, since B and Y are independent. This means that $\Gamma_X \sim \Gamma_B$. It should be noted that the relation (3.1) is useful not only for the communication system but also for the analysis of the process X treated in Sec. 2, because there exists a Gaussian process Y with covariance $\Gamma_Y = \Gamma_X - c_1\Gamma_B$ and X is expressed by $X(t) = \sqrt{c_1}B(t) + Y(t)$ using independent B and Y, if $\Gamma_X \sim \Gamma_B$.

Let us introduce some notations:

$$H(B,Y) = H(B) \oplus H(Y)$$

$$H_t(X)^{\perp} = H(B,Y) \ominus H_t(X)$$

$$H_t(Y) = \text{the linear span by } \{Y(s);\ s \leq t\}$$

$$H_t(B,Y) = H_t(B) \oplus H_t(Y)$$

As we have seen in Sec. 2, there exists an operator G_Y satisfying

$$(f, G_Y g) = (f,g)_Y = E(\int_0^1 f\, dY \int_0^1 g\, dY)$$

Then it is easy to see that $G_X = I + G_Y$.

LEMMA 3.1. The random variable $\xi = \int_0^1 g_1\, dB + \int_0^1 g_2\, dY$ belongs to $H_t(X)^\perp$ if and only if $P_t(g_1 + G_Y g_2) = 0$ or $g_1(s) = -(G_Y g_2)(s)$ for almost all $s \leq t$.

Proof. Since $\eta \in H_t(X)$ is expressed in the form $\eta = \int_0^t f\, dX$, for an $f \in L^2[0,1]$, the equality $E(\eta\xi) = 0$ yields that

$$E[(\int_0^t f\, dX)(\int_0^1 g_1\, dB + \int_0^1 g_2\, dY)] = (P_t f, g_1) + (P_t f, g_2)_Y$$
$$= (P_t f,\, g_1 + G_Y g_2) = 0$$

COROLLARY 3.1. $\int_0^t g_1\, dB + \int_0^t g_2\, dY \in H_t(B,Y) \ominus H_t(X)$ if and only if $P_t g_1 = -G_Y^t g_2$, where $G_Y^t = P_t G_Y P_t$.

REMARK 1. The operator $P_t(I + G_Y)P_t = P_t G_X P_t$ gives an isomorphism between $L^2[0,t] = P_t L^2[0,1]$ and itself, because $\|P_t f\|^2 \leq \|P_t(I + G_Y)P_t f\|^2 \leq c^2\|P_t f\|^2$, where c is the operator norm of $G_X = I + G_Y$. By this fact we can define the inverse operator $(P_t + G_Y^t)^{-1} = (G_X^t)^{-1}$ on $L^2[0,t]$.

Now we will try to decompose the random variable of the form

$$\xi = \int_0^t f_1\, dB + \int_0^t f_2\, dY \in H_t(B,Y) \qquad f_1, f_2 \in L^2[0,1] \tag{3.2}$$

into the sum of elements of $H_t(X)$ and $H_t(B,Y) \ominus H_t(X)$.

THEOREM 3.1. ξ of (3.2) is decomposed in the form

$$\xi = [\int_0^t (G_X^t)^{-1}(G_Y^t f_2 + P_t f_1)\, dX]$$
$$+ [-\int_0^t G_Y^t(G_X^t)^{-1} P_t(f_2 - f_1)\, dB + \int_0^t (G_X^t)^{-1} P_t(f_2 - f_1)\, dY] \tag{3.3}$$

where the first term belongs to $H_t(X)$ and the second belongs to $H_t(B,Y) \ominus H_t(X)$.

Proof. Taking account of Corollary 3.1, we put

$$\xi = \int_0^t f_1\, dB + \int_0^t f_2\, dY = \int_0^t f\, dX + [\int_0^t (-G_Y^t g)\, dB + \int_0^t g\, dY] = \xi_1 + \xi_2$$

From this, if there exists a solution of the equation

$$\begin{aligned} P_t f_1 &= P_t f - G_Y^t g \\ P_t f_2 &= P_t f + P_t g \end{aligned} \tag{3.4}$$

the desired decomposition (3.3) is derived. Since

$$P_t(f_1 - f_2) = -(P_t + G_Y^t)g = -(G_X^t)g$$

we get

$$P_t g = (G_X^t)^{-1} P_t(f_2 - f_1)$$

and therefore

$$P_t f = (G_X^t)^{-1}(G_Y^t f_2 + P_t f_1)$$

is derived.

LEMMA 3.2. Let $M_f(t) = Q_t[\int_0^t f(s)\, dB(s)]$ for $f \in L^2[0,1]$. Then $M_f = \{M_f(t);\ 0 \le t \le 1\}$ is a Gaussian martingale adapted to $(F_t)_{t\in[0,1]}$, where F_t is the σ-algebra generated by $\{X(s);\ s \le t\}$ and null sets.

Proof. Denote by $Q_t^{(B,Y)}$ the projection to the subspace $H_t(B,Y)$. As it holds that

$$M_f(t) = Q_t[Q_t^{(B,Y)} \int_0^1 f(s)\, dB(s)] = Q_t[\int_0^1 f(s)\, dB(s)]$$

we get

$$Q_s M_f(t) = Q_s Q_t[\int_0^1 f(s)\, dB(s)] = Q_s[\int_0^1 f(s)\, dB(s)] = M_f(s)$$

for $s \le t$.

COROLLARY 3.2. M_f is expressed in the form

$$M_f(t) = \int_0^t (G_X^t)^{-1} P_t f\, dX \tag{3.5}$$

Proof. Put $f_1 = f$ and $f_2 = 0$ in (3.3), and apply the projection Q_t. Then the result follows from Theorem 3.1.

We note that the expression (3.5) depends only on X. The following corollary is easily derived.

COROLLARY 3.3. $Q_t(\int_0^1 f\,dX) = \int_0^t (G_X^t)^{-1}P_tG_Xf\,dX$, in particular,

$$Q_tX(s) = \int_0^t (G_X^t)^{-1}P_tG_XP_s1\,dX$$

From this corollary, we get:

THEOREM 3.2. Let $M = \{M(t);\ 0 \leq t \leq 1\}$ be a Gaussian martingale adapted to $(\mathcal{F}_t)_{t\in[0,1]}$ and suppose that each $M(t)$ belongs to $H(X)$. Then there exists an $f \in L^2[0,1]$ such that $M(t) = M_f(t)$.

Proof. $\xi = M(1)$ is in $H(X)$, so $M(1)$ is expressed uniquely in the form $M(1) = \int_0^1 g\,dX$ for a $g \in L^2[0,1]$. By putting $f = (G_X)^{-1}g$, the desired result follows from Corollary 3.3. The estimation of the functional Y is given as follows by the use of Theorem 3.1.

COROLLARY 3.4. For the functional of the form $\xi = \int_0^1 f\,dY$, the estimation $Q_t\xi$ conditioned by the receipt $\{X(s);\ s \leq t\}$ is expressed in the form

$$Q_t\xi = \int_0^t (G_X^t)^{-1}G_Yf(u)\,dX(u) \tag{3.6}$$

Under the assumption $\Gamma_Y < \Gamma_B$, $H(Y)$ is not covered with the Wiener-like integrals $\int_0^1 f\,dY$, so (3.6) cannot be directly applied. But any $\xi \in H(Y)$ is approximated by a sequence $\{\xi_n\}$, $\xi_n = \int_0^1 f_n\,dX$ $(n = 1,2,\ldots)$ (see Remark 1 of Sec. 2), so that $\{G_Y^tf_n\}$ converges to a function $g \in L^2[0,t] = P_tL^2[0,1]$. Therefore, $Q_t\xi$ is given by the form $\int_0^t (G_X^t)^{-1}g\,dX$.

4. SPECTRAL MULTIPLICITY IN CANONICAL REPRESENTATION

The first formulation and the general scheme for the canonical representation have been given by Hida [3]. We discuss this topic in this section. In the preceding section we have determined the possible type of Gaussian martingales belonging to $H(X)$. Using these results, we can derive a necessary and sufficient condition for the existence of the single Brownian innovation. In other words, we get a condition for X to be canonically represented by a single Brownian motion.

THEOREM 4.1. Let X be a Gaussian process satisfying the condition $\Gamma_X \sim \Gamma_B$. Then X has a single Brownian innovation if and only if the following conditions are satisfied:

(i) There exists an element f_0 in $L^2[0,1]$ such that $((G_X^s)^{-1}P_s f_0, P_s g) = 0$, $s \leq t$, yields $g(s) = 0$, $s \leq t$, for each $t \in [0,1]$.

(ii) For the f_0 in (i), the measure $dA(t)$ defined by the increasing function

$$A(t) = ((G_X^t)^{-1}P_t f_0, P_t f_0) = \|(G_X^t)^{-1}P_t f_0\|_X^2 \tag{4.1}$$

is equivalent to the Lebesgue measure.

REMARK 1. The $A(t)$ of (4.1) is an increasing function, because we see that $A(t) = \langle M_{f_0}\rangle_t = E(M_{f_0}(t)^2)$ by Corollary 3.2.

Proof. If X has a single innovation $B = \{B(t);\ 0 \leq t \leq 1\}$, the Brownian motion B is expressed in the form

$$B(t) = M_{f_0}(t) = \int_0^t (G_X^t)^{-1}P_t f_0 \, dX(u)$$

for some $f_0 \in L^2[0,1]$, by virtue of Theorem 3.2. Suppose that

$$\begin{aligned}((G_X^s)^{-1}P_s f_0, P_s g) &= ((G_X^s)^{-1}P_s f_0, (G_X^s)^{-1}P_s g)_X \\ &= E[M_{f_0}(s)M_g(s)] = 0 \qquad s \leq t\end{aligned}$$

for a fixed $t \in [0,1]$. Then $M_g(s)$ is orthogonal to the subspace $H_t(M_{f_0})$ spanned by $\{M_f(s);\ s \leq t\}$. Therefore, $M_g(s) = 0$ for $s \leq t$. In particular, $E[M_g(t)^2] = \|(G_X^t)^{-1}P_t g\|_X^2 = 0$. G_X^t is an isomorphism on $L^2[0,t]$, so $g(s) = 0$, $s \leq t$. The condition (ii) is easily proved, since $A(t) = t$. The converse is easy: (i) means that $H_t(M_{f_0}) = H_t(X)$, and from (ii) we get a Brownian motion

$$B = \{B(t) = \int_0^t \alpha(s)\, dM_{f_0}(s);\ 0 \le t \le 1\}$$

where the function $\alpha(s)$ is the square root of the density $\alpha(s)^2$ such that $t = \int_0^t \alpha(s)^2\, dA(s)$. For the Brownian motion B, we can see that the linear span $H_t(B)$ by $\{B(s);\ s \le t\}$ is equal to $H_t(M_{f_0})$, since $\alpha(t)^2 > 0$ for almost all $t \in [0,1]$ with respect to the measure $dA(t)$. Thus the proof is complete.

REMARK 2. The condition (i) states that the function

$$G(s,u) = (G_X^s)^{-1} P_s f_0(u) \qquad 0 \le u \le s \le 1$$

is the canonical kernel for the Gaussian process in Hida's sense [3]. By the method of operator theory, the analogous conditions have been given by Sakhnovich [8]. The conditions (i) and (ii) are simpler than the ones in Ref. 8.

REFERENCES

1. N. Aronszajn, Theory of reproducing kernels, Trans. Am. Math. Soc. 68:337-404 (1950).
2. I. C. Gohberg and M. G. Krein, Theory and Applications of Volterra Operators in Hilbert Spaces, American Mathematical Society, Providence, R. I., 1970 (original Russian edition, 1967).
3. T. Hida, Canonical representation of Gaussian processes and their applications, Mem. Colloq. Sci. Univ. Kyoto 33:109-155 (1960).
4. M. Hitsuda, Representation of Gaussian processes equivalent to Wiener process, Osaka J. Math. 5:299-312 (1968).
5. M. Hitsuda, Multiplicity of some classes of Gaussian processes, Nagoya Math. J. 52:39-46 (1973).
6. T. Kailath, Likelihood ratios for Gaussian processes IEEE Trans. Inf. Theory IT-16:276-288 (1970).

7. R. Sh. Liptzer and A. N. Shiryaev, Statistics of Stochastic Processes, Springer-Verlag, Berlin, 1977 (original Russian edition, 1974).

8. L. A. Sakhnovich, Factorization of operators in $L^2(a,b)$, Func. Anal. Appl. 13:187-192 (1980) (original Russian edition, 1979).

6

Stochastic Flows of Diffeomorphisms

NOBUYUKI IKEDA/Faculty of Science, Osaka University, Toyonaka, Osaka, Japan

SHINZO WATANABE/Faculty of Science, Kyoto University, Kyoto, Japan

1. INTRODUCTION

Let M be a σ-compact, orientable, and connected C^∞-manifold and $A_0, A_1, \ldots, A_r$ be a given system of C^∞-vector fields. Let $(W_0^r, F, (F_t), P^W)$ be the standard r-dimensional Wiener space: W_0^r is the space of all continuous paths $w = (w(t)): [0,\infty) \to R^r$ such that $w(0) = 0$, P^W is the standard Wiener measure, F is the completion of the Borel σ-field on W_0^r, and (F_t) is the natural filtration defined by $F_t = \sigma[w(s); s \leq t] \vee \underline{N}$, where $\underline{N}$ is the collection of all P^W-null sets. Then $w(t) = (w^\alpha(t))_{\alpha=1}^r$ is a system of continuous (F_t)-martingales with the property $dw^\alpha(t)\, dw^\beta(t) = \delta^{\alpha\beta}\, dt(\delta^{\alpha\beta}$: Kronecker's δ). We consider the following stochastic differential equation (SDE) on M:

$$dX(t) = \sum_{\alpha=1}^r A_\alpha(X(t))\circ dw^\alpha(t) + A_0(X(t))\, dt \qquad (1.1)$$
$$X(0) = x \in M$$

where $\circ$ is the stochastic differential in the Stratonovich sense. Then, for each $x \in M$, the solution $X = (X(t,x,w))$ of (1.1) exists uniquely. To be precise, $X = (X(t,x,w))$ exists and satisfies the following conditions: For almost all $w(P^W)$, $X(0,x,w) = x$, $0 < e(w) \leq \infty$ exists such that $[0,e(w)) \ni t \to X(t,x,w) \in M$ is continuous, $\lim_{t\uparrow e(w)} X(t,x,w) = \Delta$ (: = the point at infinity of M) if $e(w) < \infty$ and, for every C^∞-function f on M with a compact support [by setting $f(\Delta) = 0$ and $X(t,x,w) = \Delta$ for $t \geq e(w)$] $t \to f(X(t,x,w))$ is a continuous (F_t)-semimartingale such that

$$f(X(t,x,w)) - f(x) = \sum_{\alpha=1}^{r} \int_0^t (A_\alpha f)(X(s,x,w))\circ dw^\alpha(s) + \int_0^t (A_0 f)(X(s,x,w))\, ds \tag{1.2}$$

Furthermore, if $X' = (X'(t,x,w))$ also satisfies these conditions, then with probability 1, $X(t,x,w) = X'(t,x,w)$ for all $t \geq 0$ (see Ref. 7 for details).

As for the dependence of solutions $X(t,x,w)$ in x, we have the following: With probability 1, we can choose the above solutions $X(t,x,w)$ such that $[0,\infty) \times M \ni (t,x) \to X(t,x,w) \in M \cup \{\Delta\}$ is continuous; furthermore, for fixed t and almost all $w(P^W)$ such that $\{t < e(w)\}$, $x \to X(t,x,w)$ is a diffeomorphism between a neighborhood of x and a neighborhood of $X(t,x,w)$. To avoid undesirable complications, we shall impose the following assumption on the system of vector fields in the future, although almost all assertions below also hold with some necessary modifications in the general case of possible finite explosions of solutions.

ASSUMPTION 1. By embedding M into a higher-dimensional euclidean space R^m, vector fields A_α, $\alpha = 0, 1, \ldots, r$, are restrictions on M of smooth vector fields $\tilde{A}_\alpha$ on R^m whose coefficients in the euclidean coordinates are C^∞ with bounded derivatives of all orders α with $|\alpha| \geq 1$.

This assumption holds for any system of vector fields if,

in particular, M is compact. Let Diff(M) be the collection of all orientation preserving diffeomorphisms of M onto M. We endow the topology of Diff(M) by the systems of seminorms

$$\rho_{D,\ell}(\varphi_1,\varphi_2) = \sum_{|\alpha|\leq \ell}^{m} \sum_{j=1}^{m} \sup_{x\in D} |D^{\alpha}\varphi_1^j(x) - D^{\alpha}\varphi_2^j(x)|$$

where $\varphi_1, \varphi_2 \in$ Diff(M), D is a relatively compact domain in M, (φ^j) is the euclidean coordinate of φ induced by the embedding, and

$$D^{\alpha} = \frac{\partial^{|\alpha|}}{(\partial x^1)^{\alpha_1} \cdots (\partial x^m)^{\alpha_m}} \qquad \alpha = (\alpha_1, \alpha_2, \ldots, \alpha_m)$$

$$|\alpha| = \alpha_1 + \alpha_2 + \cdots + \alpha_m$$

Then Diff(M) is a topological group. Let $W_{\mathrm{Diff}(M)} = C([0,\infty) \to \mathrm{Diff}(M))$ with the topology of uniform convergence on compact intervals. Since Diff(M) is metrizable, $W_{\mathrm{Diff}(M)}$ is also metrizable.

THEOREM 1.1. Under Assumption 1 we can choose the solution $X(t,x,w)$ of (1.1) such that, for almost all $w(P^W)$, $e(w) = \infty$ for every x and $[0,\infty) \ni t \to [x \to X(t,x,w)] \in \mathrm{Diff}(M)$ is continuous, i.e., $t \to X_t := [x \to X(t,x,w)] \in W_{\mathrm{Diff}(M)}$.

For the proof of this theorem, see Refs. 7 and 8.

DEFINITION This $W_{\mathrm{Diff}(M)}$-valued random variable $X = (X_t)$ defined on the Wiener space is called the flow of diffeomorphisms defined by the SDE (1.1).

Clearly, the orbit of $x \in M$ defines a path of the L-diffusion process on M starting at x where $L = (1/2)\Sigma_{\alpha=1}^{r} A_{\alpha}^2 + A_0$. But the flow of diffeomorphisms has much finer structure than the L-diffusion process; in fact, it happens often that two different flows of diffeomorphisms induce the same L-diffusion process; i.e., the laws of the orbits on $W_M := C([0,\infty) \to M)$ coincide for every x, although the laws on $W_{Diff(M)}$ are different. The purpose of the present exposition is to give a survey of the results for SDE from a viewpoint of flows of diffeomorphisms.

2. ITÔ'S FORMULA

First, we introduce some notions and notations which will be necessary in the future. Each $\varphi \in \mathrm{Diff}(M)$ induces a transformation on tensor fields, denoted by φ^*, which is defined as follows: If

$$T(x) = \left(T^{i_1,\ldots,i_p}_{j_1,\ldots,j_q}(x) \right)$$

in a local coordinate,

$$\varphi^*(T)(x) = \left(\sum_{k_1,\ldots,k_q=1}^{m} \sum_{\ell_1,\ldots,\ell_p=1}^{m} T^{\ell_1,\ldots,\ell_p}_{k_1,\ldots,k_q}(\varphi(x)) \frac{\partial\varphi^{k_1}}{\partial x^{j_1}} \cdots \frac{\partial\varphi^{k_q}}{\partial\varphi^{j_q}} \left(\frac{\partial\varphi}{\partial x}\right)^{-1\, i_1}_{\ell_1} \cdots \left(\frac{\partial\varphi}{\partial x}\right)^{-1\, i_p}_{\ell_p} \right)$$

In particular, if T is a vector field A, $\varphi^*(A)(x) = (\varphi^{-1})_*(A(\varphi(x)))$, where $(\varphi^{-1})_*$ is the differential of the mapping φ^{-1} and if T is a differential form ω, $\varphi^*(\omega)$ is the pullback of ω. If X is a complete vector field and φ_t is a one-parameter subgroup of Diff(M) generated by X, $L_X(T) = \lim_{t\to 0}(\varphi_t^*(T) - T)/t$ is called the <u>Lie derivative</u> of a tensor field T. Note that if T = f is a scalar field, $L_X(f) = X(f)$, and if T = A is a vector field, $L_X(A) = [X,A]$, the Poisson bracket. If $T = \omega$ is a differential form, $L_X(\omega)$ is given by Cartan's formula $L_X(\omega) = i(X)d\omega + d[i(X)\omega]$, where i is the interior product and d is the exterior derivative.

Let $A_0, A_1, \ldots, A_r$ be a system of vector fields

satisfying Assumption 1 and $X = (X_t)$ be the flow of diffeomorphisms defined by the SDE (1.1). We denote $[x \to X(t,x,w)] \in \mathrm{Diff}(M)$ by X_t and thus $X_t(x) = X(t,x,w)$. A fundamental formula is the following Itô's formula: For every C^∞-function f on M and $t > s \geq 0$,

$$f(X_t \circ X_s^{-1}(x)) - f(x)$$

$$= \sum_{\alpha=1}^{r} \int_s^t (A_\alpha f)(X_u \circ X_s^{-1}(x)) \circ dw^\alpha(u) + \int_s^t (A_0 f)(X_u \circ X_s^{-1}(x))\, du$$

$$= \sum_{\alpha=1}^{r} \int_s^t (A_\alpha f)(X_u \circ X_s^{-1}(x))\, dw^\alpha(u) + \int_s^t (Lf)(X_u \circ X_s^{-1}(x))\, du \tag{2.1}$$

where $L = \Sigma_{\alpha=1}^r A_\alpha^2/2 + A_0$. Also, the following backward Itô's formula holds (see Refs. 15 and 9): For every C^∞-function f and $t > s \geq 0$,

$$f(X_t \circ X_s^{-1}(x)) - f(x)$$

$$= \sum_{\alpha=1}^{r} \int_s^t A_\alpha[f(X_t \circ X_u^{-1}(x))] \circ \widehat{dw^\alpha(u)} + \int_s^t A_0[f(X_t \circ X_u^{-1}(x))]\, du$$

$$= \sum_{\alpha=1}^{r} \int_s^t A_\alpha[f(X_t \circ X_u^{-1}(x))]\, \widehat{dw(u)} + \int_s^t L[f(X_t \circ X_u^{-1}(x))]\, du \tag{2.2}$$

Here A_α and L operate on the composite function $F(x) = f(X_t \circ X_u^{-1}(x))$ and stochastic integrals $\circ\widehat{dw(u)}$ and $\widehat{dw(u)}$ are defined in a usual way with the reversed direction of time: Since $X_t \circ X_u^{-1}(x)$ is $\sigma[w(v) - w(u);\ t \geq v \geq u]$-measurable these integrals can be defined by the usual Itô calculus. Recently, Kunita [9,10] studied this backward Itô's formula in detail. In particular, he obtained the backward SDE determining $X_t(x)$ as follows:

$$X_t \circ X_s^{-1}(x) = x + \sum_{\alpha=1}^{r} \int_s^t [((X_t \circ X_u^{-1})^{-1})^* A_\alpha](X_t \circ X_u^{-1}(x)) \circ \widehat{dw(u)}$$

$$+ \int_s^t [((X_t \circ X_u^{-1})^{-1})^* A_0](X_t \circ X_u^{-1}(x))\, du \tag{2.3}$$

Itô's formulas (2.1) and (2.2) can be easily extended to the case when f is a tensor field: In this case, $f(X_t \circ X_s^{-1}(x))$

is replaced by $(X_t \circ X_s^{-1})^*(f)$ and A_α and L are replaced by L_{A_α} and $L = \Sigma_{\alpha=1}^r L_{A_\alpha}^2/2 + L_{A_0}$, respectively.

EXAMPLE 1. If f = A is a vector field, then

$$X_t^*(A) - A = \sum_{\alpha=1}^r \int_0^t X_u^*([A_\alpha, A]) \circ dw^\alpha(u) + \int_0^t X_u^*([A_0, A])\, du$$

EXAMPLE 2. If f = ω is a p-form, then

$$X_t^*(\omega) - \omega = \sum_{\alpha=1}^r \int_0^t X_u^*(L_{A_\alpha}\omega) \circ dw^\alpha(u) + \int_0^t X_u^*(L_{A_0}\omega)\, du$$

If c is a p-chain, it is clear that

$$\int_c X_t^*(\omega) = \int_{X_t(c)} \omega$$

and by a Fubini-type theorem (Ref. 7, p. 116),

$$\int_c [\int_0^t X_u^*(L_{A_\alpha}\omega) \circ dw^\alpha(u)] = \int_0^t (\int_{X_u(c)} L_{A_\alpha}\omega) \circ dw^\alpha(u)$$

Thus,

$$\int_{X_t(c)} \omega - \int_c \omega$$

$$= \sum_{\alpha=1}^r \int_0^t (\int_{X_u(c)} L_{A_\alpha}\omega) \circ dw^\alpha(u) + \int_0^t (\int_{X_u(c)} L_{A_0}\omega)\, du \tag{2.4}$$

Consider another system of vector fields $\tilde{A}_\alpha$, $\alpha = 0, 1, \ldots, r$ which also satisfies Assumption 1. Then $X_t^*(\tilde{A}_\alpha)$ is a vector field on M which depends on (t,w) and we can consider the following SDE:

$$dY(t) = \sum_{\alpha=1}^r (X_t^*(\tilde{A}_\alpha))(Y(t)) \circ dw^\alpha(t) + X_t^*(\tilde{A}_0)(Y(t))\, dt$$

$$Y(0) = x \tag{2.5}$$

By the same arguments as in the case of vector fields which are independent of (t,w), we can show the existence and uniqueness of solutions Y = (Y(t,x,w)) of (2.5) and also, $Y_t = [x \to Y(t,x,w)] \in \mathrm{Diff}(M)$. The following <u>composition</u>

theorem is due to Kunita [8] and Bismut [2].

THEOREM 2.1. Let $Z_t(x) = (Y_t \circ X_t)(x)$. Then $\{Z_t\}$ coincides with the flow of diffeomorphisms defined by the following SDE:

$$dZ(t) = \sum_{\alpha=1}^{r} (A_\alpha + \tilde{A}_\alpha)(Z(t)) \circ dw^\alpha(t) + (A_0 + \tilde{A}_0)(Z(t))\, dt$$
$$Z(0) = x \tag{2.6}$$

3. APPROXIMATION AND SUPPORT THEOREMS

Let S be a metric space with the metric d_S and $B(S)$ be the Borel σ-fields on S. A mapping $F: W_0^r \to S$ which is $F/B(S)$-measurable is called an S-valued Wiener functional. Two S-valued Wiener functionals F_1 and F_2 are identified (and written as $F_1 = F_2$) if they coincide almost everywhere. An S-valued Wiener functional F is said to have a continuous modification $\tilde{F}$ if there exists a continuous mapping $\tilde{F}: W_0^r \to S$ such that $F(w) = \tilde{F}(w)$ almost all $w(P^W)$. Note that the topology of W_0^r is given by the metric

$$\rho(w_1,w_2) = \sum_{n=0}^{\infty} \frac{1}{2^n} \min\left(1, \max_{0\le t\le n} |w_1(t) - w_2(t)|\right)$$
$$w_1, w_2 \in W_0^r$$

Clearly, a continuous modification is unique if it exists. Many interesting Wiener functionals as obtained by Itô's calculus, e.g., stochastic integrals or solutions of SDE, flows of diffeomorphisms in particular, do not usually have continuous modifications and therefore we cannot apply to these functionals the standard Fréchet differential calculus.

Let H be a subspace of W_0^r formed of all $\varphi = (\varphi^\alpha(t)) \in W_0^r$ such that each $\varphi^\alpha(t)$ is absolutely continuous and $\int_0^t |\dot{\varphi}^\alpha(s)|^2\, ds < \infty$ for each $t > 0$. We define the topology of H by the following metric ρ_H on H:

$$\rho_H(\varphi_1,\varphi_2) = \sum_{n=1}^{\infty} \frac{1}{2^n} \min\left(1, \sqrt{\sum_{\alpha=1}^{r} \int_0^n |\dot{\varphi}_1^\alpha(s) - \dot{\varphi}_2^\alpha(s)|^2 \, ds}\right)$$

Then $H \subsetneq W_0^r$; i.e., H is a linear subspace of W_0^r and the injection i: $H \to W_0^r$ is continuous with a dense range. But since $P^W(H) = 0$, the restriction of a Wiener functional on H has no meaning unless it has the continuous modification.

For such Wiener functionals as obtained by Itô's calculus, however, there are smooth functionals on H naturally associated to them. To see this, we have to introduce the following notion. A sequence of H-valued Wiener functionals $\varphi_n(w)$ is called an H-approximation of the Wiener process if for every $\delta > 0$, $P^W[\rho(\varphi_n(w),w) > \delta] \to 0$ as $n \to \infty$. The following are typical examples.

EXAMPLE 1. (Polygonal Approximation). Let $\Delta : 0 = t_0 < t_1 < \cdots < t_n < \cdots \to \infty$ and $\varphi_\Delta(w) \in H$, $w \in W_0^r$, be defined by

$$\varphi_\Delta(w)(t) = \begin{cases} w(t_i) & t = t_i \\ \text{linear} & t_i \le t \le t_{i+1} \end{cases}$$

If a sequence Δ_n of partitions satisfies

$$|\Delta_n|_T := \max_{t_i < T} |t_i - t_{i-1}| \to 0 \qquad \text{as } n \to \infty \text{ for every } T > 0$$

then $\varphi_n = \varphi_{\Delta_n}$ is an H-approximation of the Wiener process. Such an approximation is called a polygonal approximation.

EXAMPLE 2. (Mollifier Approximation). Let $\eta \in C^\infty((0,1))$ such that $\eta \ge 0$ and $\int_0^1 \eta(s)\, ds = 1$. Set $\varphi_{\eta,\varepsilon}(w) \in H$, $w \in W_0^r$, $\varepsilon > 0$ by

$$\varphi_{\eta,\varepsilon}(w)(t) = \frac{1}{\varepsilon} \int_0^\varepsilon w(s+t)\eta\left(\frac{s}{\varepsilon}\right) ds$$

Then for a sequence ε_n such that $\varepsilon_n \downarrow 0$, $\varphi_n = \varphi_{\eta,\varepsilon_n}$ is an H-approximation which is called a mollifier approximation.

It seems natural to say that an S-valued Wiener functional $F(w)$ and a continuous S-valued function $\hat{F}(\varphi)$ on H are <u>associated</u> if the following two conditions are satisfied:

i. For a polygonal approximation φ_n or a mollifier approximation φ_n, we have, for every $\delta > 0$,
$$P^W[d_S(\hat{F}(\varphi_n(w)),F(w)) > \delta] \to 0 \qquad \text{as } n \to \infty \tag{3.1}$$

ii. There exists a dense subset H_0 of H such that for every $\varphi \in H_0$ and $\delta > 0$,
$$P^W[d_S(\hat{F}(\varphi),F(w)) > \delta \mid \rho(w,\varphi) < \varepsilon] \to 0 \qquad \text{as } \varepsilon \to 0 \tag{3.2}$$

If F and $\hat{F}$ are associated in this sense, the support in S of the law $F_*(P^W)$ of F coincides with the closure in S of $\{\hat{F}(\varphi);\ \varphi \in H\}$. It is also clear that if $F(w)$ has the continuous modification $\tilde{F}(w)$, then by setting $\hat{F} = \tilde{F}|H$, F and $\hat{F}$ are associated in the above sense; in fact, (3.1) holds for any H-approximation and (3.2) holds with $H_0 = H$. For those functionals obtained by Itô's calculus, the associated H-functionals are obtained by the ordinary differential calculus and this is what Malliavin [11, 12] called a <u>transfer principle</u>. Here we are interested in our flow of diffeomorphisms $X = (X_t)$ defined by SDE (1.1) and so we shall concentrate on it from now on. X is an S-valued Wiener functional where $S = W_{\text{Diff}(M)}$. As we remarked, $W_{\text{Diff}(M)}$ is metrizable and we fix a metric d_S. We define an S-valued smooth functional $\tilde{X}(\ )$ on H by the following ordinary differential equation (ODE): For $\varphi \in H$ and $x \in M$, let $\xi = (\xi(t,x,\varphi)) \in W_M$ be the solution of ODE:

$$\begin{aligned} \dot{\xi}(t) &= \sum_{\alpha=1}^{r} A_\alpha(\xi(t))\dot{\varphi}^\alpha(t) + A_0(\xi(t)) \\ \xi(0) &= x \end{aligned} \tag{3.3}$$

Then $\tilde{X}(\varphi) = (\tilde{X}_t(\varphi)) \in W_{\text{Diff}(M)}$ is defined by

$$[0,\infty) \ni t \to [x \to \xi(t,x,\varphi)] \in \text{Diff}(M)$$

THEOREM 3.1. X and $\tilde{X}$ are associated in the above sense. To be more precise, we can take any polygonal approximation or any mollifier approximation in (3.1) and (3.2) is satisfied if we take $H_0 = C^\infty([0,\infty) \to R^r)$.

For a proof of (3.1), see Ikeda and Watanabe [7] and Bismut [2] for polygonal approximations, and Malliavin [12] and Shu [14] for mollifier approximations. A proof of (3.2) was essentially given in Ref. 7, chap. VI, sec. 8: Apply the result obtained there to SDE for $X(t,x,w)$ and their derivatives $D^\alpha X(t,x,w)$ put together.

Using this approximation theorem, a proof of (2.2) and Theorem 2.1 is easily obtained. Another example of applications is the following time reversion formula of Malliavin [11] (see also Ref. 7, p. 250). Let $\hat{X}(t,x,w)$ be the flow of diffeomorphisms defined by SDE:

$$\begin{aligned} d\hat{X}(t) &= \sum_{\alpha=1}^{r} A_\alpha(\hat{X}(t)) \circ dw^\alpha(t) - A_0(\hat{X}(t))\, dt \\ \hat{X}(0) &= x \end{aligned} \tag{3.4}$$

Then for any $T > 0$, we have

$$X(T - t, x, w) = \hat{X}(t, X(T,x,w), \hat{w}) \tag{3.5}$$

for every $t \in [0,T]$ and $x \in M$, a.s. (P^W). Here $\hat{w}$ is another Wiener process defined by

$$\hat{w}(t) = w(T - t) - w(T) \qquad 0 \leq t \leq T$$

4. EXAMPLES

We shall now discuss several examples.

EXAMPLE 1. (Brownian Motion on Lie Groups). Let G be a Lie group. A stochastic process $\{g(t)\}$ on G is call a right-invariant Brownian motion if it satisfies the following condition: (i) with probability 1, $g(0) = e :=$ identity and

$t \to g(t)$ is continuous; (ii) for every $t \geq s$, $g(t)g(s)^{-1}$ and $\sigma[g(u), u \leq s]$ are independent; and (iii) for every $t \geq s$, $g(t)g(s)^{-1}$ and $g(t - s)$ are equally distributed. Let $A_0, A_1, \ldots, A_r$ be a system of right-invariant vector fields on G and consider the following SDE on G over the r-dimensional Wiener space W_0^r:

$$\begin{aligned} dg(t) &= \sum_{\alpha=1}^{r} A_\alpha(g(t)) \circ dw^\alpha(t) + A_0(g(t))\, dt \\ g(0) &= g \end{aligned} \tag{4.1}$$

Then, for every $g \in G$, a solution $g(t,g,w)$ of (4.1) exists uniquely and globally. We denote $g(t,e,w)$ by $g^0(t,w)$. Then $g^0(t,w)$ is a right-invariant Brownian motion on G and conversely, every right-invariant Brownian motion can be obtained in this way for some r. The flow of diffeomorphisms $\{g(t,g,w)\}$ defined by the SDE (4.1) is given by $g(t,g,w) = g^0(t,w)g$.

Generally, if M is a compact manifold, the system of diffeomorphisms g_t: $x \to X(t,x,w)$ defined by SDE (1.1) can be viewed as a right-invariant Brownian motion on the infinite dimensional Lie group Diff(M) (see Refs. 4 and 13). Even if M is not compact, a situation will be simple if the system of vector fields $A_0, A_1, \ldots, A_r$ in the SDE (1.1) generates a finite dimensional Lie algebra with respect to the Poisson bracket $[X,Y] = XY - YX$. In this case, a Lie group G exists which acts on M as a transformation group of M; i.e., there exists a homomorphism $G \ni g \to \tilde{g} \in \mathrm{Diff}(M)$, and also there exists a system of right-invariant vector fields $\bar{A}_0, \bar{A}_1, \ldots, \bar{A}_r$ on G such that if $g^0(t,w)$ is the solution of the SDE:

$$\begin{aligned} dg(t) &= \sum_{\alpha=1}^{r} \bar{A}_\alpha(g(t)) \circ dw^\alpha(t) + \bar{A}_0(g(t))\, dt \\ g(0) &= e \end{aligned} \tag{4.2}$$

then the flow of diffeomorphisms $X = (X_t)$ defined by SDE (1.1) is given simply by

$$X(t,x,w) = g^0(t,w)(x) \tag{4.3}$$

For a more detailed study of flows of diffeomorphisms in this case, see Gaveau [5], Yamato [16], and Kunita [8].

EXAMPLE 2. We shall give a typical example of a SDE for which the Lie algebra of vector fields generated by $A_0, A_1, \ldots, A_r$ is not finite dimensional.

Let $M = \underline{T}^2 := R^2/2\pi\underline{Z}^2$ be the two-dimensional torus. Consider the following system of vector fields A_1, A_2, A_3, A_4:

$$\begin{aligned} A_1(x) &= \sin x^1 \frac{\partial}{\partial x^2} & A_2(x) &= \text{cox } x^1 \frac{\partial}{\partial x^2} \\ A_3(x) &= \sin x^2 \frac{\partial}{\partial x^1} & A_4(x) &= \cos x^2 \frac{\partial}{\partial x^1} \\ & & x &= (x^1, x^2) \in \underline{T}^2 \end{aligned} \tag{4.4}$$

and the following SDE on $\underline{T}^2$ over the Wiener space (W_0^4, F, P^W):

$$\begin{aligned} dX(t) &= \sum_{\alpha=1}^{4} A_\alpha(X(t)) \circ dw^\alpha(t) \\ X(0) &= x \end{aligned} \tag{4.5}$$

More generally, let A be a vector field on $\underline{T}^2$ of the form

$$A(x) = a^1(x) \frac{\partial}{\partial x^1} + a^2(x) \frac{\partial}{\partial x^2}$$

where

$$a^1(x) = \frac{\partial \varphi}{\partial x^2}(x) \quad \text{and} \quad a^2(x) = -\frac{\partial \varphi}{\partial x^1}(x)$$

for some $\varphi \in C^\infty(\underline{T}^2)$. The collection of all such vector fields forms a Lie algebra $\mathfrak{g}_0$; indeed, by writing $A(x) = A_\varphi(x)$, $[A_{\varphi_1}, A_{\varphi_2}] = A_{\{\varphi_1,\varphi_2\}}$, where

$$\{\varphi_1, \varphi_2\} = \frac{\partial \varphi_1}{\partial x^1}\frac{\partial \varphi_2}{\partial x^2} - \frac{\partial \varphi_1}{\partial x^2}\frac{\partial \varphi_2}{\partial x^1}$$

This Lie algebra generates the following subgroup $S_0\text{Diff}(\underline{T}^2)$

of $\mathrm{Diff}(\underline{T}^2)$ (see Arnold [1]):

$$S_0\mathrm{Diff}(\underline{T}^2) = \{g \in \mathrm{Diff}(\underline{T}^2);\ g \text{ preserves the area } dx = dx^1 \wedge dx^2 \text{ and the center of gravity of } \underline{T}^2\}$$

That is, $g \in S_0\mathrm{Diff}(\underline{T}^2)$ if and only if $g^*(dx) = dx$ and

$$\frac{\int_{[0,2\pi]^2} x^i\, dx}{(2\pi)^2} = \frac{\int_{[0,2\pi]^2} \tilde{g}^i(x)\, dx}{(2\pi)^2} \qquad (\mathrm{mod}\ 2\pi) \quad i = 1, 2$$

where $\tilde{g}(x) = (\tilde{g}^1(x), \tilde{g}^2(x))$ is the unique (up to a translation by $a \in 2\pi\underline{Z}^2$) diffeomorphism of the universal covering surface R^2 of $\underline{T}^2$ such that $g \circ \pi = \pi \circ \tilde{g}$, where π: $R^2 \to \underline{T}^2$ is the projection

$$\begin{array}{ccc} R^2 & \xrightarrow{\tilde{g}} & R^2 \\ \pi\downarrow & \circlearrowright & \downarrow\pi \\ \underline{T}^2 & \xrightarrow[g]{} & \underline{T}^2 \end{array}$$

Indeed, if

$$A(x) = a^1(x)\frac{\partial}{\partial x^1} + a^2(x)\frac{\partial}{\partial x^2}$$

is a vector field on $\underline{T}^2$, then the one-parameter subgroup g_t of $\mathrm{Diff}(\underline{T}^2)$ generated by A is contained in $S_0\mathrm{Diff}(\underline{T}^2)$ if and only if $A \in \underline{g}_0$.

The system of vector fields A_i, $i = 1, 2, 3, 4$, clearly belongs to $\underline{g}_0$ and it can be shown that the Lie algebra generated by this system coincides with $\{A_\varphi;\ \varphi$ is a trigonometric polynomial$\}$ and hence it is dense in $\underline{g}_0$. Let $X = (X_t)$ be the flow of diffeomorphisms defined by the SDE (4.5). Then, by the support theorem, this flow of diffeomorphisms lies, with probability 1, in the subgroup $S_0\mathrm{Diff}(\underline{T}^2)$, i.e., $X \in W_{S_0\mathrm{Diff}(\underline{T}^2)}$ a.s., and also for every smooth curve $[t \to g_t] \in W_{S_0\mathrm{Diff}(\underline{T}^2)}$ such that g_0 = identity, there is a positive probability that $X \in U$ for every neighborhood U in $W_{S_0\mathrm{Diff}(\underline{T}^2)}$ of $[t \to g_t]$. Note that for every $g \in S_0\mathrm{Diff}(\underline{T}^2)$, there exists a smooth curve $[t \to g_t] \in W_{S_0\mathrm{Diff}(\underline{T}^2)}$ such that g_0 = identity and $g_1 = g$ (see Ref. 1, p. 419). That $X_t \in S_0\mathrm{Diff}(\underline{T}^2)$ for all t can be also proved directly by using Itô's formula. Indeed, if

$$A(x) = a^1(x)\frac{\partial}{\partial x^1} + a^2(x)\frac{\partial}{\partial x^2} \in \underline{g}_0$$

then

$$L_A(dx^1 \wedge dx^2) = \left(\frac{\partial a^1}{\partial x^1} + \frac{\partial a^2}{\partial x^2}\right) dx^1 \wedge dx^2 = 0$$

and hence

$$X_t^*(dx^1 \wedge dx^2) - X_0^*(dx^1 \wedge dx^2) = \sum_{\alpha=1}^{4} \int_0^t X_s^*(L_{A_\alpha}(dx^1 \wedge dx^2))\circ dw^\alpha(s)$$

$$= 0$$

Set $\omega^i = x^i\, dx^1 \wedge dx^2/(2\pi)^2$, $i = 1, 2$. By extending $a^1(x)$ and $a^2(x)$ onto R^2 as periodic functions and setting

$$\tilde{A}(x) = a^1(x)\frac{\partial}{\partial x^1} + a^2(x)\frac{\partial}{\partial x^2}$$

we have

$$L_{\tilde{A}}(\omega^1) = \left[a^1(x) + x^1\left(\frac{\partial a^1}{\partial x^1} + \frac{\partial a^2}{\partial x^2}\right)\right] dx^1 \wedge dx^2/(2\pi)^2$$

$$= a^1(x)\, dx^1 \wedge dx^2/(2\pi)^2$$

$$= \frac{\partial \varphi}{\partial x^2}(x)\, dx^1 \wedge dx^2/(2\pi)^2$$

where φ is also periodic. Therefore, $\int_c L_{\tilde{A}}(\omega^1) = 0$ on any fundamental domain c in R^2 of $\underline{T}^2$. By (2.4),

$$\int_{\tilde{X}_t([0,2\pi]^2)} \omega^1 - \int_{\tilde{X}_0([0,2\pi]^2)} \omega^1$$

$$= \sum_{\alpha=1}^{4} \int_0^t \left[\int_{\tilde{X}_s([0,2\pi]^2)} L_{\tilde{A}_\alpha}(\omega^1)\right]\circ dw^\alpha(s)$$

$$= 0$$

The proof of

$$\int_{X_t([0,2\pi]^2)} \omega^2 = \int_{\tilde{X}_0([0,2\pi]^2)} \omega^2$$

is similar. Since

$$\sum_{\alpha=1}^{4} A_\alpha^2 = \frac{\partial^2}{(\partial x^1)^2} + \frac{\partial^2}{(\partial x^2)^2}$$

we see that each orbit of this flow is a sample path of the Brownian motion on $\underline{T}^2$. Another example of a flow of diffeomorphisms on $\underline{T}^2$ which induces the Brownian motion on $\underline{T}^2$ is a flow of translations: $X(t,x,w) = x + w(t) \pmod{2\pi\underline{Z}^2}$, i.e., the case $r = 2$ and $A_1 = \partial/\partial x^1$, $A_2 = \partial/\partial x^2$. As diffeomorphisms, however, this flow is completely different from that discussed above.

EXAMPLE 3. (Stochastic Moving Frame). Let M be a Riemannian manifold of the dimension n with the metric $ds^2 = \Sigma_{i,j=1}^n g_{ij}(x)\, dx^i\, dx^j$ and GL(M) be the bundle of linear frames on M. Thus a point $r \in GL(M)$ is a pair $r = (x, \underline{e} = [e_1, e_2, \ldots, e_n])$, where $x \in M$ and $\underline{e}$ is a basis of the tangent space $T_x(M)$. A local coordinate of r is given as $r = (x^i, e_\alpha^i)$, where (x^i) is a local coordinate of x and

$$e_\alpha = \sum_{i=1}^{n} e_\alpha^i \frac{\partial}{\partial x^i} \qquad \alpha = 1, 2, \ldots, n$$

Let O(M) be the bundle of orthonormal frames; i.e., it is a submanifold of GL(M) given by

$$O(M) = \{r = (x^i, e_\alpha^i);\ \sum_{i,j=1}^{n} g_{ij}(x) e_\alpha^i e_\beta^j = \delta_{\alpha\beta},$$
$$\alpha, \beta = 1, 2, \ldots, n, \text{ or equivalently, } \Sigma_{\alpha=1}^n e_\alpha^i e_\alpha^j = g^{ij}(x)\}$$

where $(g^{ij}(x)$ is the inverse matrix of $(g_{ij}(x))$. Let L_1, $L_2, \ldots, L_n$ be the system of basic vector fields on O(M), i.e.,

$$(L_\alpha f)(r) = \frac{\lim_{t\downarrow 0}[f(\varphi^{(\alpha)}(t),\underline{e}^{(\alpha)}(t)) - f(r)]}{t}$$
$$\alpha = 1,2,\ldots,n$$

$$f \in C^\infty(O(M)),\ r = (x,\ \underline{e} = [e_1,e_2,\ldots,e_n])$$

where $\varphi^{(\alpha)}(t)$ is the geodesic such that $\varphi^{(\alpha)}(0) = x$ and $\dot{\varphi}^{(\alpha)}(0) = e_\alpha$ and $\underline{e}^{(\alpha)}(t)$ is the parallel translate of the frame $\underline{e}$ along the curve $\varphi^{(\alpha)}(t)$. Consider the following SDE on $O(M)$ over the Wiener space (W_0^n,F,P^W):

$$dr(t) = \sum_{\alpha=1}^{n} L_\alpha(r(t))\circ dw^\alpha(t) \tag{4.6}$$

The solution $r(t) = (X(t),\underline{e}(t))$ of (4.6) is called the <u>stochastic moving frame</u> over M. In a local coordinate, (4.6) is given by

$$\begin{aligned} dX^i(t) &= \sum_{\alpha=1}^{n} e_\alpha^i(t)\circ dw^\alpha(t) \\ de_\alpha^i(t) &= -\sum_{m,k=1}^{n} \Gamma^i_{m,k}(X(t))e_\alpha^k(t)\circ dX^m(t) \end{aligned} \tag{4.6'}$$

where $\{\Gamma^k_{ij}\}$ are the Christoffel symbols. An intuitive meaning of (4.6) or (4.6'), obviously, is that the infinitesimal motion of $X(t)$ on the tangent space, which is a euclidean space with respect to the frame $\underline{e}(t)$, is that of the n-dimensional standard Wiener process and the infinitesimal motion of the frame $\underline{e}(t)$ is the parallel translation along the curve $X(t)$. $X(t)$ defines a sample path of the Brownian motion on M, i.e., a $(1/2)\Delta_M$-diffusion process on M, where Δ_M is the Laplacian on M (see Ikeda and Watanabe [7] for details).

In the following, we consider the stochastic moving frame over a complex manifold, especially on a Kähler manifold. Let M be a complex manifold of the (complex) dimension n on which, we assume, is given a Riemannian metric such that it induces a Hermitian metric $\Sigma^n_{\alpha,\beta=1} g_{\alpha\bar{\beta}}\, dz^\alpha\, d\bar{z}^\beta$. Since M is a Riemannian manifold anyway, the stochastic moving frame was defined as above. Suppose further that M is Kählerian, i.e., a (1,1)-tensor field J defining the almost complex structure of M satisfies $\nabla J = 0$, where ∇ is the covariant derivative. In this case, the stochastic moving frame is described more conveniently by using the complex structure.

Let (W_0^{2n}, F, P^W) be the standard 2n-dimensional Wiener space and define a system of complex-valued martingales $z(t) = (z^\alpha(t))_{\alpha=1}^n$ by

$$z^\alpha(t) = \frac{1}{\sqrt{2}}(w^{2\alpha-1}(t) + \sqrt{-1}w^{2\alpha}(t)) \qquad \alpha = 1, 2, \dots, n$$

Then z(t) is a system of complex-valued continuous (F_t)-martingales such that $dz_t^\alpha\, dz_t^\beta = 0$ and $dz_t^\alpha\, d\overline{z}_t^\beta = \delta^{\alpha\beta}\, dt$, $\alpha, \beta = 1, 2, \dots, n$. In general, a system $\zeta(t) = (\zeta^i(t))_{i=1}^m$ of m continuous complex-valued (local) martingales defined on a probability space with a filtration (F_t) is said to be <u>an m-dimensional conformal martingale</u> if it satisfies $d\zeta_t^i\, d\zeta_t^j = 0$ for every i, j. Thus the above z(t) is a typical example of n-dimensional conformal martingale. The notion of conformal martingales is always meaningful if $\zeta(t)$ is a continuous (F_t)-adapted process on complex manifold M of dimension n; namely, it is called a <u>conformal martingale on</u> M if $\zeta(t) = (\zeta^i(t))_{i=1}^n$, in a local coordinate, is an n-dimensional conformal martingale. It is clear that this property is independent of a particular choice of local coordinates.

Let M be a Kähler manifold and U(M) be the bundle of frames $r = (z, \underline{e})$, where $z \in M$ and $\underline{e} = [e_1, e_2, \dots, e_n]$ is any system of complex tangent vectors $e_\alpha \in T_z^C(M)$ such that $g_z(e_\alpha, \overline{e}_\beta) = \delta_{\alpha\beta}$, $\alpha, \beta = 1, 2, \dots, n$ and each e_α is of holomorphic type; i.e., it has no component of $\partial/\partial\overline{z}^\beta$, $\beta = 1, 2, \dots, n$. Since M is Kählerian, U(M) is clearly invariant under the parallel translation and so the stochastic moving frame can be defined as a flow of diffeomorphisms on U(M) obtained by the following SDE:

$$dr(t) = \sum_{\alpha=1}^{n} L_\alpha(r(t))\circ dz^\alpha(t) \tag{4.7}$$

where $\{L_\alpha\}_{\alpha=1}^{n}$ is the system of basic vector fields defined as above. In a local coordinate, it is given as follows:

$$\begin{aligned} d\zeta^i(t) &= \sum_{\alpha=1}^{n} e_\alpha^i(t)\circ dz^\alpha(t) \\ de_\alpha^i(t) &= -\sum_{m,\beta=1}^{n} \Gamma_{m\beta}^i(\zeta(t))e_\alpha^\beta(t)\circ d\zeta^m(t) \\ &\quad -\sum_{m,\beta=1}^{n} \Gamma_{\bar{m}\beta}^i(\zeta(t))e_\alpha^\beta(t)\circ d\bar{\zeta}^m(t) \\ &= -\sum_{m,\beta=1}^{n} \Gamma_{m\beta}^i(\zeta(t))e_\alpha^\beta(t)\circ d\zeta^m(t) \end{aligned} \tag{4.8}$$

since $\Gamma_{\bar{m}\beta}^i = 0$ in a Kähler manifold. This, combined with $dz^\alpha\, dz^\beta = 0$, yields at once that $de_\alpha^i(t)\, dz^\alpha(t) = 0$ and hence (4.8) is equivalent to the following in which the Stratonovich differential is replaced by the Itô differential in the first line:

$$\begin{aligned} d\zeta^i(t) &= \sum_{\alpha=1}^{n} e_\alpha^i(t)\, dz^\alpha(t) \\ de_\alpha^i(t) &= -\sum_{m,\beta=1}^{n} \Gamma_{m\beta}^i(\zeta(t))e_\alpha^\beta(t)\circ d\zeta^m(t) \end{aligned} \tag{4.8'}$$

Again the process $\zeta(t)$ on M is a sample path of the Brownian motion on M. The immediate consequences of equation (4.8') are the following.

COROLLARY 4.1. $\zeta(t)$ is a conformal martingale on M. In particular, if $f: M \to C$ is holomorphic, $f(\zeta(t))$ is a one-dimensional conformal martingale.

COROLLARY 4.2. The Laplacian Δ_M on M is given by

$$\Delta_M = \sum_{\alpha,\beta=1}^{n} g^{\alpha\bar{\beta}} \frac{\partial^2}{\partial z^\alpha \partial \bar{z}^\beta}$$

It is natural to call $\zeta(t)$ the Kähler diffusion on M. Interesting applications of Kähler diffusions are given, e.g., in Debiard and Gaveau [3] and Gaveau [6].

REFERENCES

1. V. I. Arnold, Mathematical Methods of Classical Mechanics, Springer-Verlag, Berlin, 1978.
2. J. M. Bismut, Mécanique aléatoire, Lect. Notes Math. 866, Springer-Verlag, Berlin, 1981.
3. A. Debiard and B. Gaveau, Frontière de Silov de domains faiblement pseudoconvexes de C^n, Bull. Sci. Math., 2nd sér., 100:17-31 (1976).
4. K. D. Elworthy, Stochastic dynamical systems and their flows, pp. 79-95 in Stochastic Analysis, Academic Press, New York, 1978.
5. B. Gaveau, Principe de moindre action, propagation de la chaleur et estimées sous elliptiques sur certains groupes nilpotents, Acta Math. 139:95-153 (1977).
6. B. Gaveau, Méthodes de contrôle optimal en analyse complexe; 1. Résolution d'équations de Monge Ampère, J. Funct. Anal. 25:391-411 (1977).
7. N. Ikeda and S. Watanabe, Stochastic Differential Equations and Diffusion Processes, North-Holland/Kodansha, Amsterdam/Tokyo, 1981.
8. H. Kunita, On the decomposition of solution of stochastic differential equations, pp. 213-255 in Stochastic Integrals, Lect. Notes Math. 851, Springer-Verlag, Berlin, 1981.

9. H. Kunita, On backward stochastic differential equations, Stochastics 6:293-313 (1982).
10. H. Kunita, Stochastic partial differential equations connected with non-linear filtering, pp. 100-169 in Proc. CIME, Session on "Non-linear Filtering and Stochastic Control", Lect. Notes Math. 972, Springer-Verlag, Berlin, 1982.
11. P. Malliavin, Stochastic calculus of variation and hypoelliptic operators, pp. 195-263 in Proceedings of the International Symposium on Stochastic Differential Equations, Kyoto (K. Itô, ed.), Wiley, New York/ Kinokuniya, Tokyo, 1978.
12. P. Malliavin, Géométrie différentielle stochastique, Les Presses de l'Université de Montréal, Montréal, 1978.
13. J. Marsden, D. Ebin, and A. Fisher, Diffeomorphism groups, hydrodynamics and relativity, Proc. 13th Bienniel Semin., Can. Math. Congr. 1:135-279 (1972).
14. J. G. Shu, On the mollifier approximation for solutions of stochastic differential equations, J. Math. Kyoto Univ. 22:243-254 (1982).
15. S. Watanabe, Flows of diffeomorphisms defined by stochastic differential equations on manifolds and their differentials and variations, Sūriken-Kōkyūroku, RIMS Kyoto Univ., 391:1-23 (1980) (in Japanese).
16. Y. Yamato, Stochastic differential equations and nilpotent Lie algebras, Z. Wahrscheinlichkeitsth. Verwend. Geb. 47:213-229 (1979).

7

Differentiability in Measure for Stochastic Differential Equations

JEAN JACOD*/Mathematical Research Institute, University of Rennes, Rennes, France

1. INTRODUCTION

Consider a family of stochastic differential equations

$$X_t = K_t^u + \int_0^t g^u(X)_s \, dZ_s^u \qquad (1.1)$$

on a given filtered probability space $(\Omega,\underline{F},\underline{F}_t,P)$, indexed by the elements u of a subset U of $\mathbb{R}^q$. These are equations of "Doléans-Dade and Protter" type; each Z^u is an m-dimensional semimartingale, each K^u is a d-dimensional process which is right-continuous and admits left-hand limits, and each g^u is a d x m-matrix-valued process. We shall assume that each equation (1.1) admits a solution process (or "strong solution") X^u. Assuming that K^u and Z^u are differentiable in u, and that $g^u(X)$ is differentiable in (u,X), in a sense that will be made precise later, we want to prove the differentiability of X^u in u.

Of course, when $g^u = g$ and $Z^u = Z$ do not depend on u, and when g has the form $g(\omega,X)_s = h_s(\omega,X_{s-})$, this kind of result is not new: Since the early work of Gikhman [4], many papers have been devoted to studying differentiability

*Current affiliation: University of Paris, Paris, France.

in the L^p-sense for $p > 0$ (see the book [5] of Gikhman and Skorokhod, and also the recent paper [1] by Bichteler, in which g^u is general but $Z^u = Z$ has independent increments). Recently, almost-sure pathwise differentiability has been proved for the Itô equation (Malliavin [8], Bismut [2], Kunita [7]) and for a general semimartingale Z by Meyer [11].

Here, our results are not quite comparable with the ones noted above; on the one hand, they are weaker, since the differentiability obtained for X^u is "in measure" (but of course the differentiability assumptions on K^u, g^u, Z^u are also in measure); on the other hand, unlike for L^p-differentiability, which necessitates rather good estimates for $p > 0$, here the assumptions look "minimal" for the obtained results. As for pathwise differentiability, it seems very unlikely that it can be obtained unless $Z^u = Z$ does not depend on u.

In fact, the results of this chapter are indeed very simple corollaries of the stability theorem of Emery [3], a theorem that will be slightly ameliorated in Sec. 2 below in order to suit our needs. In particular, Emery's topology of semimartingales plays a central role.

Now, we must recognize the fact that if equation (1.1) is simple enough, it is not well suited for applications, especially for discontinuous Markov processes: A nicer equation to consider would be one involving Poisson random measures or more general integer-valued random measures, and continuous martingales (see Ref. 6, chap. XIV). But in order to study such an equation, it is necessary to introduce a topology on the set of random measures that would play the role of Emery's topology.

Let me end this paragraph by gratefully acknowledging K. Bichteler for letting me use his preprint [1]; although his results are of a different sort than mine, his preprint has been the main impetus for this work.

2. CONTINUITY RESULTS

We adopt the standard notations and terminology (see Ref. 6 or 10). Recall that a dot "$\circ$" denotes stochastic integration (i.e., $H\circ Y_t$ stands for $\int_0^t H_s\, dY_s$). For any (multidimensional) process X, we set $X_t^* = \sup_{s\leq t} |X_s|$. The filtered probability space $(\Omega,\underline{F},\underline{F}_t,P)$ is given, and we always identify two processes that are indistinguishable. We introduce the following topological vector spaces:

$\underline{S}^p$, the space of p-dimensional semimartingales, endowed with Emery's topology (see Ref. 3); this topology can be defined by the following quasinorm, where $\underline{H}^p$ denotes the set of all predictable $\mathbb{R}^p$-valued processes bounded by 1, and where $H\circ Y = \Sigma_{(i)} H^i\circ Y^i$ if $Y \in \underline{S}^p$, $H \in H^p$:

$$\|Y\|_{\underline{S}^p} = \sum_{n\geq 1} 2^{-n} \sup_{H\in\underline{H}^p} E\{1 \wedge (H\circ Y)_n^*\}$$

$\underline{D}^p$, the space of $(\underline{F}_t)$-adapted processes with values in $\mathbb{R}^p$ that are right continuous and admit left-hand limits; this space is endowed with the topology of uniform convergence on compact sets of $\mathbb{R}_+$, in measure; this topology is defined by the quasinorm

$$\|X\|_{D^p} = \sum_{n\geq 1} 2^{-n} E\{1 \wedge X_n^*\}$$

$\tilde{\underline{D}}^p$, the space of $(\underline{F}_t)$-predictable processes X with values in $\mathbb{R}^p$, which are <u>locally bounded</u>, i.e., such that $X_t^* < \infty$ for every $t > 0$; this space is endowed with the topology of uniform convergence on compact sets, in measure.

Let U be a subset of $\mathbb{R}^q$. For every $u \in U$ we are given

$Z^u \in \underline{S}^m$

$K^u \in \underline{D}^d$ (sometimes $K^u \in \underline{S}^d$)

g^u, a mapping $\underline{D}^d \to \tilde{\underline{D}}^{d\times m}$, such that for every finite stopping time T and all $X,Y \in \underline{D}^d$ with $X_t = Y_t$ for $t < T$, we have

$$g^u(X)_T = g^u(Y)_T \tag{2.1}$$

We consider equation (1.1) which can also be written componentwise as follows:

$$X^i = K^{u,i} + \sum_{j=1}^{m} g^{u,ij}(X)\circ Z^{u,j} \qquad i = 1, \ldots, d$$

We will be interested in the continuity or the differentiability at a particular point $v \in U$. When we write $u \to v$, we mean that u tends to v in $\mathbb{R}^q$, while staying in U. For this particular point $v \in U$ we make the following "local Lipschitz" assumptions:

(H1) There exists a sequence $(A^{v,n})_{n\geq 1}$ of finite-valued increasing processes such that $|g^v(X)_t - g^v(Y)_t| \leq A_t^{v,n}(X-Y)^*_{t-}$ for all $X,Y \in \underline{D}^d$ and all $t > 0$ such that $X^*_{t-} \leq n$, $Y^*_{t-} \leq n$.

Under (H1), a theorem of Emery [3] shows that equation (1.1) for $u = v$ admits a unique solution process, which in addition is nonexploding, and we denote this solution by X^v.

THEOREM 2.1. Assume the following hypotheses: (H1), and
(H2) $u \rightsquigarrow Z^u$ is continuous in $\underline{S}^m$ at point v.
(H3) $u \rightsquigarrow K^u$ is continuous in $\underline{D}^d$ (respectively, in $\underline{S}^d$, in which case we of course suppose that $K^u \in \underline{S}^d$) at point v.
(H4) For every $u \in U$, $u \neq v$, there exists a sequence $(A^{u,n})_{n\geq 1}$ of increasing processes with values in $[0,\infty]$, with the following properties:
(i) $|g^u(X)_t - g^u(Y)_t| \leq A_t^{u,n}(X-Y)^*_{t-}$ for all $X,Y \in \underline{D}^d$ and all $t > 0$ such that $X^*_{t-} \leq n$, $Y^*_{t-} \leq n$;
(ii) For each $t > 0$, there exists a sequence $(B_t^n)_{n\geq 1}$ of finite-valued random variables, such that for all $n \geq 1$, $\varepsilon > 0$ we have $\lim_{u\to v, u\neq v} P(A_t^{n,u} > B_t^n + \varepsilon) = 0$.
(H5) For every $u \in U$, equation (1.1) admits a nonexploding solution process X^u.

(H6) $g^u(X^v) \to g^v(X^v)$ in $\tilde{\underline{D}}^{d\times m}$ when $u \to v$.

Then, $u \rightsquigarrow X^u$ is continuous in $\underline{D}^d$ (respectively, in $\underline{S}^d$) at point v.

Note that, because of (H1), we have (H4i) also for u = v, and if B_t^n is replaced by $B_t^n \vee A_t^{v,n}$, (H4ii) is valid with $\lim_{u\to v} P(A_t^{n,u} > B_t^n + \varepsilon) = 0$. Note also that for $u \neq v$ we may have $A_t^{u,n} = +\infty$, so g^u is not locally Lipschitz: hence equation (1.1) may have several solution processes. However, if X'^u is another solution and if

$$T(u,n) = \inf\ (t : |X_t^u| \geq n \text{ or } |X_t'^u| \geq n)$$

then $X^u = X'^u$ on the stochastic interval $[\![0,T(u,n)]\!] \cap \{A^{u,n} < \infty\}$; hence (H4ii) implies a sort of "asymptotic uniqueness" when $u \to v$.

Let us recall that Theorem 2.1 is due to Emery [3] when (H4) is replaced by:

(H4') There exists a finite-valued increasing process B such that

$$|g^u(X)_t - g^u(Y)_t| \leq B_t(X - Y)^*_{t-} \qquad \text{for all } X,Y \in \underline{D}^d,\ t \geq 0,\ u \in U$$

[we have (H4') → (H1), (H4), (H5)]. In fact, Emery assumes that $g^u(X)$ is left continuous, with right-hand limits, but when $g^u(X)$ is locally bounded this assumption is not necessary for his proof.

The reader will obviously feel that weakening assumption (H4') into (H1) + (H4) + (H5) has no interest, and we share that opinion! However, for obtaining the forthcoming differentiability result, we will need the above theorem.

Proof of Theorem 2.1. (a) We shall give a unified proof for the two sorts of assumptions in (H3), and to this effect we introduce the following notations:

1. If $u \rightsquigarrow K^u$ is continuous in $\underline{D}^d$, we put $\underline{D}' = \underline{D}^d$ and $H = \{H(i); i = 1, \ldots, d\}$, where $H(i)$ is the $\mathbb{R}^d$-valued process whose jth component is $H(i)^j = 0$ (respectively, $= 1$) if $j \neq i$ (respectively $i = j$). If $Y \in \underline{D}'$, we put $H(i)\cdot Y_t = Y_t^i$ (a kind of trivial stochastic integral).
2. If $u \rightsquigarrow K^u$ is continuous in $\underline{S}^d$, we put $\underline{D}' = \underline{S}^d$ and $\underline{H}$ is the set of all predictable $\mathbb{R}^d$-valued processes bounded by 1.

With these notations, the theorem amounts to proving that $X^u \to X^v$ in $\underline{D}'$ when $u \to v$, for the following quasinorm:

$$Y \in \underline{D}' \rightsquigarrow \|Y\|_{\underline{D}'} = \sum_{n \geq 1} 2^{-n} \sup_{H \in \underline{H}} E\{1 \wedge (H\cdot Y)_n^*\}$$

In other words, it is sufficient to prove the following. Let $t > 0$ and $\eta > 0$, and let (u_n) be a sequence in U going to v; then one can find an infinite subset $\mathbb{N}' \subset \mathbb{N}$ such that

$$\lim_{n \in \mathbb{N}'} \sup_{H \in \underline{H}} P\{(H\cdot X^{u_n} - H\cdot X^v)_t^* > \eta\} = 0 \tag{2.2}$$

(b) In the sequel, $t > 0$, $\eta > 0$, and the sequence (u_n) are fixed. For every $p \in \mathbb{N}$ we define $h^p\colon \mathbb{R}_+ \to [0,1]$ by $h^p(x) = 1 \wedge (p + 1 - x)^+$, and we put

$$T_p = \inf (s : |X_s^v| \geq p) \tag{2.3}$$

$$T_p(u) = \inf (s : |X_s^u| \geq p) \tag{2.4}$$

$$g^{u,p}(X)_s = g^u(X)_s h^p(X_{s-}^*) 1_{\{s \leq t \wedge T_p\}} \quad \text{for } X \in \underline{D}^d \tag{2.5}$$

$$\alpha^{u,p} = (2p + 2)A_t^{u,p+1} + [g^u(X^v)]_t^* \tag{2.6}$$

We are going to prove that for all $X, Y \in \underline{D}^d$, $p \in \mathbb{N}$, $u \in U$, $s \leq t$,

$$|g^{u,p}(X)_s - g^{u,p}(Y)_s| \leq \alpha^{u,p}(X - Y)_{s-}^* \tag{2.7}$$

By (2.5), this inequality is trivial if $X_{s-}^* \geq p + 1$ and $Y_{s-}^* \geq p + 1$. It is then sufficient to prove (2.7) for $X_{s-}^* \leq Y_{s-}^*$ and $X_{s-}^* < p + 1$. We have for $s \leq t$,

$$|g^{u,p}(X)_s - g^{u,p}(Y)_s| = 1_{\{s \leq T_p\}} |g^u(X)_s h^p(X_{s-}^*) - g^u(Y)_s h^p(Y_{s-}^*)| \leq |g^u(X)_s - g^u(Y)_s| h^p(Y_{s-}^*)$$

$$+ |g^u(X)_s| [h^p(X^*_{s-}) - h^p(Y^*_{s-})] 1_{\{s \le T_p\}}$$

Using the facts that $X^*_{s-} < p + 1$, that $h^p(Y^*_{s-}) = 0$ if $Y^*_{s-} \ge p + 1$, and (H4i), and the inequalities $0 \le h^p(X^*_{s-}) - h^p(Y^*_{s-}) \le (X - Y)^*_{s-}$ and $h^p \le 1$, we obtain that

$$|g^{u,p}(X)_s - g^{u,p}(Y)_s|$$
$$\le A^{u,p+1}_s (X - Y)^*_{s-} + |g^u(X)_s - g^u(X^v)_s| 1_{\{s \le T_p\}} (X - Y)^*_{s-}$$
$$+ |g^u(X^v)_s| (X - Y)^*_{s-}$$
$$\le [A^{u,p+1}_s + A^{u,p+1}_s (X - X^v)^*_{s-} 1_{\{s \le T_p\}}$$
$$+ [g^u(X^v)]^*_s](X - Y)^*_{s-}$$

Now, since we assumed that $X^*_{s-} < p + 1$, and $(X^v)^*_{s-} \le p$ if $s \le T_p$, in the above expression we have $1_{\{s \le T_p\}}(X - X^v)^*_{s-} \le 2p + 1$. Then (2.7) follows from (2.6) from the previous inequality, and from $s \le t$.

(c) Using (H6) and (H4ii) one can find a subsequence $\mathbb{N}' \subset \mathbb{N}$ such that for each $p \in \mathbb{N}$,

$$\lim_{n \in \mathbb{N}'} [g^{u_n}(X^v) - g^v(X^v)]^*_t = 0 \qquad P \text{ a.s.}$$

$$\limsup_{n \in \mathbb{N}'} A^{u_n,p+1}_t \le B^{p+1}_t \qquad P \text{ a.s.}$$

hence, by (2.6) we obtain for each $p \in \mathbb{N}$:

$$\limsup_{n \in \mathbb{N}'} \alpha^{u_n,p} \le (2p + 2) B^{p+1}_y + [g^v(X^v)]^*_t \qquad P \text{ a.s.}$$

On the other hand, we know that $\alpha^{v,p} < \infty$ a.s., and $\lim_{(p)} \uparrow T_p = \infty$ a.s. Thus for every $\varepsilon > 0$ there exist $p,q \in \mathbb{N}$ such that

$$P(T_{p-1} \le t) \le \varepsilon$$
$$P(\Omega_q) \ge 1 - \varepsilon \qquad \text{with} \tag{2.8}$$
$$\Omega_q = \{\alpha^{v,p} \le q\} \cap [\cap_{n \in \mathbb{N}', n \ge q} \{\alpha^{u_n,p} \le q\}]$$

(d) Let $p,q \in \mathbb{N}$ as above, and let

$U' = \{v\} \cup \{u_n : n \in \mathbb{N}', n \geq q\}$. By (2.5), $g^{u,p}(X^u) = g^u(X^u)$ on the interval $[0, t \wedge T_p \wedge T_p(u)]$; hence

$$X^u = K^u + g^{u,p}(X^u) \circ Z^u \qquad \text{on } [0, t \wedge T_p \wedge T_p(u)] \qquad (2.9)$$

Now, set $P_q(\cdot) = P(\cdot \cap \Omega_q)/P(\Omega_q)$, and consider the following equations on the space $(\Omega, \underline{F}, \underline{F}_t, P_q)$:

$$X = K^u + g^{u,p}(X) \circ Z^u \qquad u \in U' \qquad (2.10)$$

Since $\alpha^{u,p} \leq q$, P_q a.s. by the definition (2.8) of Ω_q, for $u \in U'$, then (2.7) implies that $g^{u,p}$ is P_q a.s. Lipschitz with a constant equal to q. Hence (2.10) admits a unique solution $X^u(p,q)$ and by comparison with (2.9), which is valid P a.s. and thus P_q a.s., we obtain

$$X^u(p,q) = X^u \text{ on } [0, t \wedge T_p \wedge T_p(u)] \qquad P_q \text{ a.s.,} \quad \text{for } u \in U' \qquad (2.11)$$

On the space $(\Omega, \underline{F}, \underline{F}_t, P_q)$, the set of equations (2.10) obviously satisfies (H2) and (H3) (with U replaced by U'); we have seen that it satisfies (H4') with $B_t(\omega) = q$, hence it satisfies (H1) and (H5). Moreover, (2.11) implies that $X^v(p,q) = X^v$ P_q a.s. on $[0, t \wedge T_p]$; thus $g^{u,p}(X^v(p,q)) = g^u(X^v)$ (respectively, $= 0$) on $[0, t \wedge T_p]$ (respectively, on $]t \wedge T_p, \infty[$) P_q a.s. for $u \in U'$, and (H6) implies that the set of equations (2.10) also satisfies (H6). Therefore, we can apply Emery's theorem [3], which gives that $X^{u_n}(p,q) \to X^v(p,q)$ in $\underline{D}^d(P_q)$ [respectively, in $\underline{S}^d(P_q)$] when $n \to \infty$, $n \in \mathbb{N}'$. With the notations of part (a) of the proof, this implies that for every $\eta' > 0$,

$$\lim_{n \in \mathbb{N}'} \sup_{H \in \underline{H}} P_q[(H \circ X^{u_n}(p,q) - H \circ X^v(p,q))^*_t > \eta'] = 0 \qquad (2.12)$$

(e) Using (2.11), we obtain for $u \in U'$:

$$\begin{aligned} P_q(T_p(u) \wedge T_p \leq t) &\leq P_q(T_p(u) \leq t \text{ or } T_{p-1} \leq t) \\ &\leq P_q(T_{p-1} > t,\ T_p(u) \leq t) \\ &\quad + P_q(T_{p-1} \leq t) \end{aligned}$$

$$= P_q(T_{p-1} > t,\ (X^u)_t^* \geq p) + P_q(T_{p-1} \leq t)$$

$$\leq P_q[(X^v(p,q)_t^* < p - 1, (X^u(p,q))_t^* \geq p] + P_q(T_{p-1} \leq t)$$

$$\leq P_q[(X^v(p,q) - X^u(p,q))_t^* > 1/2] + P_q(T_{p-1} \leq t)$$

Using (2.8) and (2.12), it follows that

$$\limsup_{n \in \mathbb{N}'} P_q(T_p(u_n) \wedge T_p \leq t) \leq P_q(T_{p-1} \leq t) \leq \frac{\varepsilon}{1-\varepsilon} \tag{2.13}$$

If $H \in \underline{H}$, (2.11) yields $H \circ X^u = H \circ X^u(p,q)$, P_q a.s. on $[0, T_p \wedge T_p(u) \wedge t]$ for $u \in U'$. Thus (2.12) and (2.13) imply that for $\eta > 0$:

$$\limsup_{n \in \mathbb{N}'} \sup_{H \in \underline{H}} P_q[(H \circ X^{u_n} - H \circ X^v)_t^* > \eta] \leq \frac{\varepsilon}{1-\varepsilon}$$

Then (2.8) yields

$$\limsup_{n \in \mathbb{N}'} \sup_{H \in \underline{H}} P[(H \circ X^{u_n} - H \circ X^v)_t^* > \eta] \leq \frac{\varepsilon}{1-\varepsilon} P(\Omega_q) + P(\Omega \backslash \Omega_q) \leq \varepsilon$$

Since $\varepsilon > 0$ in (2.8) is arbitrary, we obtain (2.2), and the theorem is proved.

3. DIFFERENTIABILITY RESULTS

Our first task is to define what we mean by "differentiability in measure" for K^u, X^u, Z^u, g^u. For the mappings $u \rightsquigarrow K^u$, $u \rightsquigarrow X^u$, $u \rightsquigarrow Z^u$, it is exactly Fréchet differentiability, for mappings from U into the metrizable vector spaces $\underline{D}^d$ or $\underline{S}^d$ (and we call it differentiability "in measure" because the convergence in $\underline{D}^d$ or in $\underline{S}^d$ is in some sense "in measure"). More precisely, we set:

DEFINITION A mapping $u \rightsquigarrow Y^u$ from U into $\underline{D}^d$ (respectively, $\underline{S}^d$) is D <u>differentiable</u> (respectively, S <u>differentiable</u>) at point $v \in U$ if there exists a process $(DY)^v \in \underline{D}^{d \times q}$ (respectively, $\in \underline{S}^{d \times q}$), called the derivative of $Y^{\cdot}$ at point v, such that if we put

$$\alpha Y(v,u) = Y^u - Y^v - (DY)^v \cdot (u - v) \tag{3.1}$$

[or, componentwise:

$$\alpha Y(v,u)^i = Y^{u,i} = Y^{v,i} = \sum_{r=1}^{q} (DY)^{v,ir}(u^r - v^r) \qquad i = 1, \ldots, d$$

then $\frac{\alpha Y(v,u)}{|u - v|} \to 0$ in $\underline{D}^d$ (respectively, in $\underline{S}^d$) when $u \to v$.

Sometimes, the component $(DY)^{v,ir}$ will be denoted by $(\partial/\partial u^r)Y^{v,i}$. Note that if $Y^\cdot$ is D (respectively, S)differentiable at point v, it is also continuous in $\underline{D}^d$ (respectively, in $\underline{S}^d$) at point v; if $Y^\cdot$ is S differentiable, it is also D differentiable and the two kinds of derivatives coincide.

As for the coefficient, we will consider the differentiability of the mapping $(u,X) \rightsquigarrow g^u(X)$ from $U \times \underline{D}^d$ into $\underline{\tilde{D}}^{d\times m}$ in the Fréchet sense, except that since $\underline{D}^d$ is not a normed vector space for the topology that we consider here, we must give a sort of ad hoc definition:

DEFINITION The coefficient $g^u(X)$ is said to be C <u>differentiable</u> (C for coefficient) at point $(v,Y) \in U \times \underline{D}^d$ if there exists a linear mapping $Dg(v,y)$: $\mathbb{R}^q \times \underline{D}^d \to \underline{\tilde{D}}^{d\times m}$ with the following properties:

(i) It is <u>continuous</u>, in the sense that there exists a finite-valued increasing process G such that

$$[Dg(v,Y)\cdot(h,V)]_t^* \leq G_t(|h| + V_{t-}^*) \tag{3.2}$$

for all $h \in \mathbb{R}^q$, $V \in \underline{D}^d$, $t > 0$; in (3.2), $Dg(v,Y)\cdot(h,V)$ is the process in $\underline{\tilde{D}}^{d\times m}$ that is the image of the point $(h,V) \in \mathbb{R}^q \times \underline{D}^d$ by the mapping $Dg(v,Y)$.

(ii) It is the <u>derivative of</u> g <u>at point</u> (v,Y), in the sense that if

$$\beta g(v,y;u,X) = g^u(X) - g^v(Y) - Dg(v,Y)\cdot(u - v,\ X - Y) \tag{3.3}$$

we have $[\beta g(v,Y;u,X)]_t^*/[|u - v| + (X - Y)_{t-}^*] \to 0$ in measure for all $t > 0$, when $u \to v$ and $X \to Y$ in $\underline{D}^d$.

Obviously, the existence of $Dg(v,Y)$ amounts to the existence of:

1. A process $D_u g(v,Y) \in \underline{D}^{d\times m\times q}$, called the u-partial derivative of g, which can be considered as a linear mapping of $\mathbb{R}^q$ into $\underline{\tilde{D}}^{d\times m}$ acting as follows:

$$[D_u g(v,Y)\cdot h]^{ij} = \sum_{r=1}^{q} D^u g(v,Y)^{ijr}\, h^r \qquad i = 1, \ldots, d \quad j = 1, \ldots, m$$

[and we shall also write $\frac{\partial}{\partial u^r} g^{v,ij}(Y)$ for $D_u g(v,Y)^{ijr}$

2. A linear mapping $D_X g(v,Y)$ from $\underline{D}^d$ into $\underline{\tilde{D}}^{d\times m}$, called the X-partial derivative of g, such that those two terms satisfy:

$$Dg(v,Y)\cdot(h,V) = D_u g(v,Y)\cdot h + D_X g(v,Y)\cdot V \tag{3.4}$$

Since $D_u g(v,Y)$ is locally bounded by definition of $\underline{\tilde{D}}^{d\times m\times q}$, condition (3.2) amounts to the existence of a finite-valued increasing process G such that for all $t > 0$, $V \in \underline{D}^d$;

$$[D_X g(v,Y)\cdot V]_t^* \le G_t V_{t-}^* \tag{3.5}$$

Note also that the linearity of $D_X g(v,Y)$ means that for all $V, V' \in \underline{D}^d$, $a \in \mathbb{R}$, then $D_X g(v,Y)\cdot(aV + V') = a\, D_X g(v,Y)\cdot V + D_X g(v,Y)\cdot V'$, up to an evanescent set.

REMARK 1. Suppose that g has the form $g^u(X)_t = h_t(u, X_{t-})$, where h_t is a mapping: $U \times \mathbb{R}^d \to \mathbb{R}^{d\times m}$. Then if

(a) for every $t > 0$, h_t is differentiable on $U \times \mathbb{R}^d$;

(b) for every $u \in U$, the partial derivatives
$\frac{\partial}{\partial u^r} h_t(u,x)$ and $\frac{\partial}{\partial x^i} h_t(u,x)$

are bounded in (t,x) on every compact subset of $\mathbb{R}_+ \times \mathbb{R}^d$, then g is C differentiable at every point $(v,Y) \in U \times \underline{D}^d$, and

$$D_u g(v,Y)_t^{ijr} = \frac{\partial}{\partial u^r} h_t^{ij}(v, Y_{t-})$$

$$[D_X g(v,Y)\cdot V]_t^{ij} = \sum_{k=1}^{d} \frac{\partial}{\partial x^k} h_t^{ij}(v, Y_{t-}) V_{t-}^k$$

REMARK 2. Suppose that the linearity of $D_X g(v,Y)$ is pathwise, that is, without exceptional sets. Then there exist d Radon measures $\mu_t^k(\omega,ds)$ on $\mathbb{R}_+$, with values in $\mathbb{R}^{d\times m}$, such that

$$[D_X g(v,Y)\cdot V]_t(\omega) = \sum_{k=1}^{d} \int \mu_t^k(\omega,ds)V_s^k(\omega) \tag{3.6}$$

for every $V \in \underline{D}^d$ with <u>continuous</u> paths; moreover, since $D_X g(v,Y)\cdot V$ is predictable, $\mu_t^k(\omega,ds)$ does not change interval $[t,\infty[$; for instance, in the example of Remark 1, we have

$$\mu_t^k(\omega,ds) = \varepsilon_t(ds) \frac{\partial}{\partial x^k} h_t(v,Y_{t-}(\omega))$$

But, of course, equation (3.6) does not extend to processes V that are discontinuous at time t.

Now, we state the main theorem of this chapter.

THEOREM 3.1. Assume (H1), (H4), (H5), and
(K1) $u \rightsquigarrow Z^u$ is S differentiable at point v;
(K2) $u \rightsquigarrow K^u$ is D differentiable (respectively, S differentiable) at point v;
(K3) $g^u(X)$ is C differentiable at point (v,X^v).
Then $u \rightsquigarrow X^u$ is D differentiable (respectively S differentiable) at point v, and its derivative

$$(DX)^v = \left(\frac{\partial}{\partial u^r} X^{v,i}\right)_{i\leq d, r\leq q}$$

is the unique solution of the following system of linear equations, where $W = (W^{ir})$ is the unknown process

$$W^{ir} = \Big[\frac{\partial}{\partial u^r} K^{v,i} + \sum_{j=1}^{m} \frac{\partial}{\partial u^r} g^{v,ij}(X^v)\circ Z^{v,j} + \sum_{j=1}^{m} g^{v,ij}(X^v)\circ \frac{\partial}{\partial u^r} Z^{v,j}\Big] + \sum_{j=1}^{m} (D_X g(v,X^v)\circ W^{\cdot r})^{ij}\circ Z^{v,j}$$

$$1 \le i \le d;\ 1 \le r \le q \tag{3.7}$$

Note that (3.7) is the equation that is obtained if one differentiates (1.1) at point v in a "naive" way.

Of course, instead of (H1), (H4), (H5), one could assume (H4'), thus obtaining a less general but simpler statement. However, in the course of the following proof, one needs Theorem 2.1 in its full force, even in this simple case.

Proof of Theorem 3.1. (a) Throughout all the proof, we will use shorthand notations, which we hope will have a clear meaning from the context: For example, the process

$$[\sum_{j\le m} \frac{\partial}{\partial u^r} g^{v,ij}(X^v)\circ Z^{v,j}]_{i\le d, r\le q}$$

will be denoted by $D_u g(v,X^v)\circ Z^v$.

We define the mapping h: $\underline{D}^{d\times q} \to \tilde{\underline{D}}^{(d\times q)\times m}$ by $h(W)^{irj} = (D_X g(v,X^v)\cdot W^{\cdot r})^{ij}$. Hence a compact form for equation (3.7) reads as follows:

$$W = [(DK)^v + D_u g(v,X^v)\circ Z^v + g^v(X^v)\bullet(DZ)^v] + h(W)\circ Z^v \tag{3.8}$$

By (K3) we have (3.5) for some finite-valued increasing process G. Hence

$$[h(W) - h(W')]_t^* \le G_t\ (W - W')_{t-}^*$$

for all $W,W' \in \underline{D}^{d\times q}$, by the linearity of $D_X g(v,X^v)$. That is, h satisfies a Lipschitz condition (H1), and in particular if $W_t = W'_t$ for all $t < T$ for some stopping time T, then $h(W)_T = h(W')_T$. Therefore, equation (3.8) admits a unique solution, which is nonexploding, and which will be denoted by W. We always have $W \in \underline{D}^{d\times q}$, and if $K^{\cdot}$ is S differentiable we have $W \in \underline{S}^{d\times q}$.

(b) We define the following processes:

$$\Psi^u = X^u - X^v - W\cdot(u - v) \qquad \varphi^u = \frac{\Psi^u}{|u - v|}$$

with the convention $0/0 = 0$; thus $\varphi^v = 0$. For our theorem, all that we need to prove is that

$$\varphi^u \to 0 \text{ in } \underline{D}^d \text{ (respectively, } \underline{S}^d\text{) when } u \to v \tag{3.9}$$

and for this purpose we will show that the φ^v's are solutions of a family of stochastic differential equations to which we can apply Theorem 2.1.

With the notations (3.1) and (3.3), equations (1.1) and (3.8) yield

$$\begin{aligned} X^u - X^v &= K^u - K^v + g^v(X^v)\circ(Z^u - Z^v) \\ &\quad + [g^u(X^u) - g^v(X^v)]\circ Z^v + [g^u(X^u) - g^v(X^v)]\circ(Z^u - Z^v) \\ \Psi^u &= \alpha K(v,u) + g^v(X^v)\circ\alpha Z(v,u) + [D_X g(v,X^v)\cdot\Psi^u]\circ Z^v \\ &\quad + \beta g(v,X^v;u,X^u)\circ Z^v + [g^u(X^u) - g^v(X^v)]\circ(Z^u - Z^v) \end{aligned} \tag{3.10}$$

If $u \neq v$, we define $\gamma^u \in \tilde{\underline{D}}^{d\times m}$ by putting

$$\gamma^u = \frac{\beta g(v,X^v;u,X^u)}{|u - v| + (X^u - X^v)^*_-} \tag{3.11}$$

Hence $\beta g(v,X^v;u,X^u) = \gamma^u|u - v| + \gamma^u(X^u - X^v)^*_-$. Set $\gamma^v = 0$. Equation (3.10) yields, for $u \neq v$:

$$\begin{aligned} \varphi^u &= \frac{1}{|u-v|}\,[\alpha K(v,u) + g^v(X^v)\circ\alpha Z(v,u) \\ &\quad + (g^u(X^u) - g^v(X^v))\circ(Z^u - Z^v)] + \gamma^u\circ Z^v \\ &\quad + [D_X g(v,X^v)\cdot\varphi^u]\circ Z^v + \frac{1}{|u-v|}\,(X^u - X^v)^*_-\,\gamma^u\circ Z^v \end{aligned} \tag{3.12}$$

Now we define (with the convention $0/0 = 0$)

$$Z'^u = Z^v$$

$$K'^u = \frac{1}{|u-v|}\,[\alpha K(v,u) + g^v(X^v)\circ\alpha Z(v,u) + (g^u(X^u) - g^v(X^v))\circ(Z^u - Z^v)] + \gamma^u\circ Z^v$$

$$V \in \underline{D}^d \rightsquigarrow g'^u(V) = D_X g(v,X^v)\cdot V + [V + \frac{1}{|u-v|}\,W\cdot(u-v)]^*_-\gamma^u$$

Those terms $(Z'^u, K'^u, g'^u)_{u\in U}$ clearly satisfy (2.1), and $K'^v = 0$, and $g'^v(V) = D_X g(v,X^v)\cdot V$. Since by definition of φ^u we have

$$\varphi^u + \frac{1}{|u-v|}\,W\cdot(u - v) = \frac{(X^u - X^v)}{|u-v|} \qquad \text{if } u \neq v$$

by comparison with (3.12) we see that φ^u is a solution of the equation

$$V = K'^u + g'^u(V)\circ Z'^u \tag{3.13}$$

for $u \neq v$. Since $\varphi^v = 0$, $K'^v = 0$, and $g'^v(0) = 0$, we see that φ^v is also a solution of (3.13) for $u = v$.

(c) To prove (3.9) it is then sufficient to prove that $(Z'^u, K'^u, g'^u)_{u\in U}$ satisfies assumptions (H1) to (H6) of Theorem 2.1. This is trivial for (H2), for (H1) it follows from (3.5), and for (H5) it follows from the fact that φ^u is a nonexploding solution of (3.13).

We have the implications (K1) $\to$ (H2), (K2) $\to$ (H3), and (K3) $\to$ (H6), for $(Z^u, K^u, g^u)_{u\in U}$. Hence by Theorem 2.1 we have

$$X^u \to X^v \text{ in } \underline{D}^d \text{ (respectively, in } \underline{S}^d\text{) when } u \to v \tag{3.14}$$

Then (K3) and (3.11) yield

$$Y^u \to Y^v = 0 \qquad \text{in } \underline{\tilde{D}}^{d\times m} \tag{3.15}$$

We have $\varphi^v = 0$, and

$$g'^u(0) - g'^v(0) = \frac{1}{|u - v|}\,[W\cdot(u - v)]^*_- Y^u$$

and W is locally bounded; hence by (3.15), $g'^u(\varphi^v) \to g'^v(\varphi^v)$ in $\underline{\tilde{D}}^{d\times m}$ and $(Z'^u, K'^u, g'^u)_{u\in U}$ satisfies (H6).

Let $V, V' \in \underline{D}^d$. For every $u \in U$ we have, with $0/0 = 0$,

$$g'^u(V) - g'^u(V') = D_X g(v, X^v)\cdot(V - V')$$

$$+ \quad \left(V + \frac{W\cdot(u - v)}{|u - v|}\right)^*_- - \left(V' + \frac{W\cdot(u - v)}{|u - v|}\right)^*_- \; Y^u$$

and since $[(V + V'')^*_- - (V' + V'')^*_-] \le 2(V - V')^*_-$, it follows that

$$|g'^u(V)_t - g'^u(V')_t| \le [G_t + 2(Y^u)^*_t](V - V')^*_{t-}$$

We deduce that $(Z'^u, K'^u, g'^u)_{u\in U}$ satisfies (H4) with $A'^{u,n}_t = G_t + 2(Y^u)^*_t$ and $B'^n_t = G_t$ for every $n \in \mathbb{N}$ [use (3.15) again].

(d) It remains to prove that $K'^u \to K'^v = 0$ in $\underline{D}^d$ (respectively, in $\underline{S}^d$). By (K2) we have $\alpha K(v,u)/|u - v| \to 0$ in $\underline{D}^d$ (respectively, in $\underline{S}^d$) and we will prove that all three other terms showing in the definition of K'^u tend to 0 in $\underline{S}^d$. For this purpose, we first recall a result due to Mémin [9]:

If $\hat{Z}^u \to \hat{Z}^v$ in $\underline{S}^m$, and if $\hat{H}^u \to \hat{H}^v$ in $\tilde{\underline{D}}^{d\times m}$, then
$\hat{H}^u \circ \hat{Z}^u \to \hat{H}^v \circ \hat{Z}^v$ in $\underline{S}^d$ (3.16)

Applying (3.16) and (K1) [respectively, and (3.15)], we obtain that $(1/|u - v|)g^v(X^v)\circ \alpha Z(u,v)$ (respectively, $\gamma^u \circ Z^v$) tends to 0 in $\underline{S}^d$.

(K3) and (3.14) yield $g^u(X^u) \to g^v(X^v)$ in $\tilde{\underline{D}}^{d\times m}$. Hence (3.16) and (K1) imply that

$$\frac{1}{|u - v|}\,[g^u(X^u) - g^v(X^v)]\circ \alpha Z(v,u) \to 0 \text{ in } \underline{S}^d$$

It remains to prove that

$$\rho(u) := \frac{1}{|u - v|}\,[g^u(X^u) - g^v(X^v)]\circ[(DZ)^v\cdot(u - v)]$$

tends to 0 in $\underline{S}^d$. This property would again be a consequence of (3.16) if we knew that $(1/|u - v|)(DZ)^v\cdot(u - v)$ was going to some limit in $\underline{S}^d$, a property which amounts to the fact that $(u - v)/|u - v|$ has a limit in $\mathbb{R}^q$. Of course, this is not true; but from any sequence (u_n) going to v one can extract a subsequence $(u_{n'})$ such that $(u_{n'} - v)/|u_{n'} - v|$ converges, and it follows that $\rho(u_{n'}) \to 0$ in $\underline{S}^d$. Since $\underline{S}^d$ is metrizable, this property implies that $\rho(u) \to 0$ in $\underline{S}^d$ when $u \to v$, and the proof of the theorem is finished.

REFERENCES

1. K. Bichteler, Stochastic integrators with stationary independent increments, Z. Wahrscheinlichkeitsth. Verwand. Geb. 58:529-548 (1982).
2. M. Bismut, A generalized formula of Itô and some other properties of stochastic flows, Z. Wahrscheinlichkeitsth. Verwand. Geb. 55:331-350 (1981).

3. M. Emery, Equations différentielles stochastiques lipschitziennes, pp. 281-293 in Séminaire de probabilités XIII, Lect. Notes Math. 721, Springer-Verlag, Berlin, 1979.

4. I. I. Gikhman, On the theory of differential equations of random processes, Ukr. Math. J. 2(4):37-63 (1950).

5. I. I. Gikhman and A. V. Skorokhod, Stochastic Differential Equations, Springer-Verlag, Berlin, 1972.

6. J. Jacod, Calcul stochastique et problèmes de martingales, Lect. Notes Math. 714, Springer-Verlag, Berlin, 1979.

7. H. Kunita, On the decomposition of solutions of stochastic differential equations, pp. 213-255 in Stochastic Integrals, Proc. LMS Durham Symp., 1980 (D. Williams, ed.), Lect. Notes Math. 851, Springer-Verlag, Berlin, 1981.

8. P. Malliavin, Stochastic calculus of variations and hypoelliptic operators, pp. 195-263 in Proceedings of the International Symposium on Stochastic Differential Equations, Kyoto, 1976 (K. Itô, ed.), Wiley, New York, 1978.

9. J. Mémin, Espaces de semimartingales et changements de probabilité, Z. Wahrscheinlichkeitsth. Verwand. Geb. 52:9-40 (1980).

10. P. A. Meyer, Un cours sur les intégrales stochastiques, pp. 245-400 in Séminaire de probabilités, X, Lect. Notes Math. 511, Springer-Verlag, Berlin, 1976.

11. P. A. Meyer, Flot d'une équation différentielle stochastique, pp. 103-117 in Séminaire de probabilités XV, Lect. Notes Math. 850, Springer-Verlag, Berlin, 1981.

8

Generalized Feynman Integrals Using Analytic Continuation in Several Complex Variables

GOPINATH KALLIANPUR/University of North Carolina, Chapel Hill, North Carolina

C. BROMLEY/University of North Carolina, Chapel Hill, North Carolina

1. INTRODUCTION

The need to provide a mathematical foundation for the Feynman integral has led to a number of different approaches toward a rigorous definition. By now, the literature on the subject has grown to such an extent that it is not possible here to give even a brief sketch of the major directions of its development. In the recent monograph of Albeverio and Høegh-Krohn [1], the reader will find an excellent historical survey of the mathematical theory of Feynman integrals as well as an extensive bibliography. Indeed, as we shall presently describe, this work has been a major stimulus for the present chapter.

Our interest in the Feynman integral stems from its formal similarity to an integral on function space with respect to a nonexistent Wiener measure with an imaginary variance parameter, noted first by Feynman himself [1]. Probabilists have found this idea attractive because of the possibility of providing a stochastic process background, however tenuous, to Feynman's original definition. A precise formulation of this approach is the definition of the Feynman integral via the method of analytic continuation due to Cameron, Daletskii, Nelson, and many others [1]. Cameron and his co-workers, in particular, have devoted several papers to different aspects of this theory.

In the present work we shall be concerned specifically with a recent paper of Cameron and Storvick [3] and the work of Albeverio and Høegh-Krohn referred to above, which appeared before Ref. 3. Albeverio and Høegh-Krohn define a class of integrals which they call Fresnel integrals. This definition is not based on the technique of analytic continuation, but appears to be distribution theoretic in spirit—an analog of tempered distributions in the setting of a space of test functions given on a Hilbert space. In their theory the integrands are Fourier transforms of bounded complex Borel measures on an infinite dimensional, separable Hilbert space. In Ref. 1 the authors show the versatility of the Fresnel-Feynman integrals by displaying them as solutions of important problems of quantum mechanics and of quantum field theory.

Essentially three kinds of Fresnel integrals have been developed in Ref. 1; the first involves a positive definite (invertible) quadratic form. As we show in Sec. 2, it is equivalent to the analytic Feynman integral of Cameron and Storvick studied in Ref. 2. However, this integral is not general enough for applications to certain problems of quantum mechanics in which the action functional determines a self-adjoint operator which is invertible but not positive, and to problems of quantum field theory, where a further difficulty is encountered in that the operator involved may not be invertible. Albeverio and Høegh-Krohn have extended their Fresnel integrals to handle such problems. According to Ref. 1, the analytic continuation theory of Feynman integrals does not seem adequate for the treatment of a large class of problems in physics typified by the ones mentioned above.

It is our purpose here to reexamine the analytic continuation approach from this point of view. We propose to show that this approach, appropriately generalized, yields a theory of Feynman integrals which can be used to solve the physical problems discussed in Ref. 1. Two such problems whose solution is expressed in terms of (extended) Fresnel integrals in Ref. 1 are discussed afresh in Secs. 3 and 5, where they are treated using analytic or extended analytic Feynman integrals defined in Secs. 3 and 5. The generalization presented relies on analytic continuation of functions of several complex variables (instead of a single complex variable as in the existing theory). The principles on which we base our work are explained at the beginning of Sec. 2.

In the existing theory of Feynman integrals via analytic continuation, the Cameron-Storvick paper [3] most closely resembles and invites comparison with some of Albeverio and Høegh-Krohn's results. We begin our work by undertaking this comparison. However, since the Fresnel integrals of Ref. 1 are defined in terms of an arbitrary, separable Hilbert space, we need first to find a suitable probabilistic framework which facilitates comparison at the most general level with the work of Ref. 1. The most natural setup for this is abstract Wiener space. In Sec. 2 we generalize the main results of Ref. 3 (i.e., those pertaining to the Banach algebra S of Ref. 3) in terms of abstract Wiener space. It is shown in Propositions 2.2 and 2.3 that the Banach algebra $\mathcal{F}$ of functionals on the abstract Wiener space (B,P) is algebra isomorphic to the Banach algebra $\mathcal{F}(H)$ of Ref. 1 and that the analytic Feynman integral of any member of $\mathcal{F}$ exists and has the same value as the Fresnel integral of the corresponding member of $\mathcal{F}(H)$. It is also shown that for the particular choice of n-dimensional Wiener measure over $C_0^n[a,b]$ the definitions of the analytic Feynman integrals of Ref. 3 and Sec. 2 coincide.

The use of abstract Wiener space has the additional advantage that it lends flexibility and generality to the treatment, since the spaces considered in Ref. 3, the probability space of pinned Wiener space (or Brownian bridge), and the space of the (multiparameter) Yeh-Wiener process are all examples of abstract Wiener spaces.

In Sec. 3 we extend our definition of the Feynman integral to cover situations where its value is expressed in terms of an indefinite bilinear form on the Hilbert space. The definition involves analytic continuation in two complex variables. Here, too, we prove a corresponding equivalence theorem with the Fresnel integral approach of Ref. 1 (Proposition 3.3). It may be mentioned that this integral contains as a special case the integral of Sec. 2. As an application we work out the example (Ref. 1, p. 65) of the quantum mechanical anharmonic oscillator with n degrees of freedom.

The problem of describing the quasi-free states of the harmonic oscillator with an infinite number of degrees of freedom (Ref. 1, p. 90) has motivated much of the work of Sec. 5. The lack of invertibility of the operators occurring naturally in this problem has led us to define an extended Feynman integral. The natural region Ω of analyticity in this case turns out to be a region of $\mathbb{C}^4$ which is strictly larger than $\{z \in \mathbb{C}^4 : \operatorname{Re} z_m > 0,\ m = 1, 2, 3, 4\}$. The significance of this fact is that, unlike the previous analytic Feynman integrals, we obtain integrals which are limiting values at points on the boundary of Ω for which $\operatorname{Re} z_m < 0$ for some m. In other words, the original analogy that Feynman's integral is a heuristic integral involving Wiener process with a pure imaginary variance parameter breaks down. Sec. 5 concludes with a rather detailed (and we hope, self-contained) discussion of the problem mentioned above. The expression for the quasi-free states is given as a limit of extended Feynman integrals.

In Sec. 4 we define a Feynman integral with reference to an important class of indefinite metric spaces, Krein spaces. Technically, this section is closely related to Sec. 3. It is clearly possible to take the theory of Feynman integrals on Krein spaces further than we have done, but we have let Sec. 4 stand by itself to show the possibilities in this direction. It is not clear to us whether such efforts will lead to significantly new results. We would like to note, however, that in a seemingly different context in quantum mechanics, Dirac and Pauli suggested the use of indefinite metric spaces as long ago as 1943 [8]. Although these ideas have been examined by many physicists since that time, as far as we know, there have been no attempts to apply them to the definition of Feynman integrals (see the references in [2]).

2. ANALYTIC WIENER AND FEYNMAN INTEGRALS IN ABSTRACT WIENER SPACE

Throughout this section H will denote a real separable infinite dimensional Hilbert space. The inner product on H will be written (h_1,h_2) and the norm as $|h|$. The triple (ι,H,B) or, equivalently, (B,P) will denote an abstract Wiener space (see Kuo [6] chap. I, sec. 4). Briefly, there is a unique finitely additive probability measure $\tilde{P}$ on the Borel subsets of H such that for $h_1, \ldots, h_k \in H$, $k \geq 1$, the random vector $h \to ((h_1,h), \ldots ,(h_k,h))$ induces the k-dimensional Gaussian distribution with mean 0 and covariance $\Gamma = (\Gamma_{ij})$, $\Gamma_{ij} = E[(h_i,h)(h_j,h)] = (h_i,h_j)$. The Banach space B is the completion of H under a fixed measurable norm and $\iota : H \to B$

is the natural injection which induces a countably additive Gaussian probability measure P on the Borel subsets of B according to a well-known result of Gross [6]. The adjoint operator i^* maps the strong dual B^* continuously, one to one, onto a dense subspace of H^*. Each $y \in B^*$ defines a random variable $y(x)$, $x \in B$, which is Gaussian with mean 0 and variance $|i^*(y)|^2$ under P. There then exists a unique isometry of the Hilbert spaces $\psi: H^* \to L^2(B,P)$ such that for every $y \in B$, the random variable defined by y as a continuous linear functional on B belongs to the equivalence class $\psi(i^*y)$. In the sequel we shall identify H^* with H and write $B^* \subseteq H \subseteq B$.

For each $h \in H$, we may select a version of the random variable $\psi(h)$ defined a.s. P, to obtain a function of both $h \in H$ and $x \in B$: $[\psi(h)](x)$. We will require a jointly measurable version of this function.

LEMMA 2.1. The inner product (h_1,h_2) on $H \times H$ extends to a jointly measurable mapping $(h,x)^\sim$ of $H \times B$ into $\mathbb{R}$ such that

(i) The random variable $x \to (h,x)^\sim$ belongs to $\psi(h)$ for each $h \in H$.

(ii) $(h,\lambda x)^\sim = \lambda(h,x)^\sim$ for any real scalar λ and $(h,x) \in H \times B$.

In particular $x \to (h,x)^\sim$ is Gaussian with mean 0 and variance $|h|^2$.

Proof. Since B^* is dense in H, there exists a countable set in B^* which is dense in H. By Gram-Schmidt orthogonalization, there is a sequence $\{e_j\}$, $j \geq 1$, in B^* which forms a complete orthonormal system in H. For $(h,x) \in H \times B$, define

$$(h,x)^\sim = \begin{cases} \lim_{m\to\infty} \sum_{j=1}^{m} (h,e_j)e_j(x) & \text{if this limit exists} \\ 0 & \text{otherwise} \end{cases}$$

Joint measurability is ensured by this construction and we

also have the linearity condition $(h,\lambda x)^{\sim} = \lambda(h,x)^{\sim}$ for all real λ, $h \in H$, $x \in B$. If both h and x belong to H, then Parseval's identity gives $(h,x)^{\sim} = (h,x)$.

Since $\{e_j\}$ is complete and ψ is an isometry, $\psi(\Sigma_1^m(h,e_j)e_j) \to \psi(h)$ in $L^2(B,P)$. On the other hand, by Levy's theorem the sequence of random variables $\xi_m(x) = \Sigma_1^m(h,e_j)e_j(x) = [\psi(\Sigma_1^m(h,e_j)e_j](x)$ a.s. P converges for almost every x. Hence for any $h \in H$, $x \to (h,x)^{\sim}$ is a version of $\psi(h)$.

Before proceeding with specific definitions of analytic Wiener and analytic Feynman integrals in the context of abstract Wiener spaces, we shall consider the general pattern that such definitions will follow and their ingredients.

First, in each case we shall specify a domain Ω, an open connected subset of some complex k-dimensional space $\mathbb{C}^k$, k being a fixed positive integer. This is the region over which analytic continuation is to be carried out and we shall always select Ω so that it contains all real vectors $(\lambda_1, \ldots, \lambda_k)$ when $\lambda_m > 0$ for $1 \leq m \leq k$.

Second, we require a well-defined class of measurable complex-valued functionals $F(x_1, \ldots, x_k)$ for $(x_1, \ldots, x_k) \in B^k$ which have two properties:

i. $J_F(\lambda_1, \ldots, \lambda_k) = \int_{B^k} F(\lambda_1^{-1/2}x_1, \ldots, \lambda_k^{-1/2}x_k)\, dP^k(x)$ exists for all real $\lambda_m > 0$. (P^k is the product measure $P \times \cdots \times P$ on $B^k = B \times \cdots \times B$.)

ii. There is a (necessarily unique) analytic function of k complex variables $J_F^*(z)$, $z = (z_1, \ldots, z_k) \in \Omega$ such that $J_F^*(\lambda_1, \ldots, \lambda_k) = J_F(\lambda_1, \ldots, \lambda_k)$ when all λ_m are positive real numbers.

Here and throughout, "measurable" means measurable with respect to the product Borel σ-field on B^k. Condition (i) will be construed to mean that the integral in (i) is finite.

When these conditions are realized, the number $I^z(F) = J_F^*(z)$ deserves to be called the analytic Wiener integral of F over B^k for the parameter z, with respect to Ω. For certain functionals F and parameters $z \in \partial\Omega$ (the boundary of Ω) it may happen that $\lim_{z' \to z, z' \in \Omega} I^{z'}(F)$ exists. In this case it is reasonable to denote this limit also by $I^z(F)$. In the special case when all components of $z \in \partial\Omega$ are purely imaginary, i.e., $z = (-iq_1, \ldots, -iq_k)$ where $q \in \mathbb{R}^k$, we shall write $I^q(F)$ for $I^z(F)$ and call this the analytic Feynman integral of F over B^k for the parameter q.

This then will be the format in which Definitions 2.1, 3.1, 4.1, and 5.1 will be given, depending on the choice of the domain Ω considered in each section. This apparent redundancy of definition adds a new flexibility to the method of analytic continuation in two important ways. First, for certain applications such as Example 2 in Sec. 3, it seems necessary to exploit more fully the analytic continuation device by introducing more than one complex variable (see also Ref. 5). Second, there are physical problems where we must consider a domain Ω which permits some of the complex parameters to have negative real parts. A single definition where the domain contains only $z \in \mathbb{C}^k$ having $\text{Re}(z_m) > 0$, $1 \leq m \leq k$, would be inadequate (see Example 1 of Sec. 5).

Throughout the rest of this section we take k = 1 and $\Omega = \{z \in \mathbb{C} : \text{Re}(z) > 0\}$. We shall show how the above program can be carried through for a certain Banach algebra of integrands F(x). The result is a generalization to the abstract Wiener space framework of the classical Wiener space approach of Ref. 3 which at the same time yields an analytic Feynman integral equivalent to the formally defined Fresnel integral of Ref. 1, Sec. 2.

DEFINITION 2.1 Let F be a measurable complex-valued

functional on B such that these conditions hold:

(i) $J_F(\lambda) = \int_B F(\lambda^{-1/2}x)\, dP(x)$ exists for each real $\lambda > 0$.

(ii) There is an analytic function J_F^* on Ω such that $J_F^*(\lambda) = J_F(\lambda)$ for real $\lambda > 0$.

Then we will define $I^z(F) = J_F^*(z)$ and call $I^z(F)$ the <u>analytic Wiener integral</u> of F over B with parameter z.

If $\lim_{z\to -iq, z\in\Omega} I^z(F)$ exists for some q ($\neq 0$) real, we will denote the value of this limit by $I^q(F)$, called the <u>analytic Feynman integral</u> of F over B with parameter q.

Before proceeding further, we should draw attention to a peculiarity of the integrals $I^z(F)$. We have defined $J_F(\lambda)$ and $I^z(F)$ for a single functional F on B. In analogy with the usual practice in integration theory it is natural to expect that if G is a functional on B such that $F(x) = G(x)$ a.s. P, then $J_F(\lambda) = J_G(\lambda)$ for all $\lambda > 0$. However, this is not the case. For n-dimensional Wiener measure on $C^n[a,b]$ this point has been noted in the work of Cameron and Storvick [3] and has led them to modify the definition of equivalence classes of functionals for which J_F can be defined. A similar situation obtains for the more general setup of the abstract Wiener space considered here.

For $\alpha > 0$ let $\tilde{P}_\alpha$ be the canonical Gauss measure on the Borel sets of H with variance parameter α, i.e., the random vector $[(h_1,h), \ldots ,(h_k,h)]$ is Gaussian with mean zero and covariance matrix $\alpha\Gamma$, where $\Gamma = (\Gamma_{ij})$ and $\Gamma_{ij} = (h_i,h_j)$. For any k-dimensional Borel set E and $y_j \in B^*$, $j = 1, \ldots ,k$ arbitrary, set

$$P_\alpha\{x \in B: [y_1(x), \ldots ,y_k(x)] \in E\}$$
$$= \tilde{P}_\alpha\{h \in H: [(i^*y_1,h), \ldots ,(i^*y_k,h)] \in E\}$$

Then P_α extends uniquely to a probability measure (denoted by P_α) on the Borel sets of B. P_α is called the abstract Wiener measure with variance parameter of α. (Note that

$P_1 = P$.) From the theory of equivalence and singularity of Gaussian measures it is known that for $\alpha, \alpha' > 0$, P_α and $P_{\alpha'}$ are singular if $\alpha \neq \alpha'$ (see Kuo [6], theorem 5.2). Using this fact, it is easy to find functionals F and G defined on B such that $F(x) = G(x)$ a.s. P, but for which it is not true that $J_F(\lambda) = J_G(\lambda)$ for all $\lambda > 0$.

These considerations lead to the definition of s-equivalence for functionals F(x) on B. Given two complex-valued functionals F(x) and G(x) on B, we will say F and G coincide s almost surely with respect to P if for each $\alpha > 0$

$$P(\{x \in B: F(\alpha x) = G(\alpha x)\}) = 0$$

Two such functionals are then called s-equivalent on (B,P).

In the future we shall identify any two functionals which are s-equivalent. It is clear that $J_F(\lambda)$ and $J_G(\lambda)$ exist simultaneously and coincide if $F = G$, s-a.s. Hence the analytic Wiener and analytic Feynman integral defined above depends only on the s-equivalence class of F.

A similar consideration applies to the jointly measurable extension of the inner product $(h,x)^\sim$ constructed in Lemma 2.1. We will say that a jointly measurable map $(h,x)^\approx$ of $H \times B$ into $\mathbb{R}$ is an <u>equivalent extension</u> of the inner product if $(h,x)^\approx = (h,x)$ when both h and x are in H and $P(\{x \in B : (h,\alpha x)^\approx = (h,\alpha x)^\sim\}) = 1$ for all $h \in H$ and real scalars $\alpha > 0$.

REMARKS 1. A different choice of basis $\{e_j\}$ in the construction of Lemma 2.1 leads to an equivalent extension of the inner product in this sense.

2. For $h \in H$ and $\alpha > 0$, we have

$$(h,\alpha x)^\approx = (h,\alpha x)^\sim = \alpha(h,x)^\sim = \alpha(h,x)^\approx \qquad \text{a.s. } P$$

Hence $x \mapsto (h,\alpha x)^\approx$ is Gaussian with mean 0 and variance $\alpha^2|h|^2$ when $(h,x)^\approx$ is any equivalent extension.

3. In terms of the abstract Wiener measures with

variance parameter α introduced above as P_α, these equivalence notions can be expressed: $F = G$ a.s. P_α for all $\alpha > 0$ and $(h,\alpha x)^{\approx} = (h,\alpha x)^{\sim}$ a.s. P_α for all $\alpha > 0$.

4. We can now make clear the reason for choosing $\{e_j\}$ to belong to B^* in Lemma 2.1. If $\{h_j\}$ is any basis in H, a representative $y_j(x)$ of the class $\psi(h_j)$ can be fixed in $L^2(B,P)$ and the jointly measurable extension $(h,x)^{\approx} = \lim_{m\to\infty} \Sigma_1^m (h,h_j)y_j(x)$ defined as before, but we would have no guarantee that $y_j(\lambda x)^{\approx} = \lambda y_j(x)$ and so $(h,x)^{\approx}$ may not be equivalent to $(h,x)^{\sim}$.

DEFINITION 2.2 Denote by $\mathcal{F}$ the class of all functionals on B of the form

$$F(x) = \int_H e^{i(h,x)^{\sim}} d\mu(h) \tag{2.1}$$

for μ a finite complex measure on the Borel subsets of H. More precisely, since we shall identify functionals which coincide s-a.s. on B, $\mathcal{F}$ is regarded as the space of all s-equivalence classes of functionals of the form (2.1).

It is important to note that this set $\mathcal{F}$ does not depend on the choice of measurable extension $(h,x)^{\sim}$. In fact, if $(h,x)^{\approx}$ is an equivalent extension and $\alpha > 0$ is given, $(\mu \times P)(\{(h,x) \in H \times B : (h,\alpha x)^{\approx} = (h,\alpha x)^{\sim}\}) = 1$. Thus, $\int_H e^{i(h,\alpha x)^{\approx}} d\mu(h) = \int_H e^{i(h,\alpha x)^{\approx}} d\mu(h)$ a.s. P. Hence a different $(h,x)^{\approx}$ gives the same functional (s-a.s.) in (2.1).

Let $\mathcal{M}(H)$ denote the space of finite complex measures μ on H. $\mathcal{M}(H)$ is then a Banach algebra over the complex numbers under convolution, with the norm $\|\mu\|$ equal to the total variation of μ.

PROPOSITION 2.1. Under the operations induced on the s-equivalence classes of $\mathcal{F}$ by pointwise addition, pointwise multiplication, and complex scalar multiplication, $\mathcal{F}$ is an algebra over $\mathbb{C}$. Equation (2.1) establishes a one-to-one correspondence between $\mathcal{M}(H)$ and $\mathcal{F}$ which is an isomorphism of Banach algebras when we define the norm on $\mathcal{F}$ so that $\|F\| = \|\mu\|$. This isomorphism is unaffected if any other equivalent extension $(h,x)^{\approx}$ is used in (2.1).

Proof. For the purposes of this proof, we will write $[F]$ for the s-equivalence class of a functional F. It is easily verified that the operations of pointwise addition, pointwise multiplication, and scalar multiplication can be regarded as operations on the equivalence classes of $\mathcal{F}$. For example, given $F_1 = G_1$, s-a.s. and $F_2 = G_2$, s-a.s., it follows that for any $\alpha > 0$, $P(\{x : F_1(\alpha x)\cdot F_2(\alpha x) \neq G_1(\alpha x)\cdot G_2(\alpha x)\}) = 0$. Hence, $F_1\cdot F_2 = G_1\cdot G_2$, s-a.s. on B. Denoting by F_μ the functional defined by (2.1) for a given $\mu \in \mathcal{M}(H)$, clearly $F_{\mu_1} + F_{\mu_2} = F_{\mu_1+\mu_2}$, $F_{\mu_1}\cdot F_{\mu_2} = F_{\mu_1 * \mu_2}$, and $zF_{\mu_1} = F_{z\mu_1}$, for all μ_1, μ_2 in $\mathcal{M}(H)$ and $z \in \mathbb{C}$. Hence

$$\mu \to [F_\mu] \tag{2.2}$$

defines a map of $\mathcal{M}(H)$ onto $\mathcal{F}$ which is an algebra homomorphism. As we noted above, this map is unchanged if $(h,x)^{\sim}$ is replaced by an equivalent extension $(h,x)^{\approx}$.

It remains only to show that (2.2) is one to one and it suffices to show that $F_\mu = 0$ a.s. P implies that $\mu = 0$. Let $\{e_j\}$ be the basis of H in B^* used in the proof of Lemma 2.1 and let π_n be the projection of H onto the subspace spanned by $\{e_1, \ldots, e_m\}$: $\pi_m(h) = \Sigma_1^m (h,e_j)e_j$.

Fix any $h_0 \in H$. For $m \geq 1$ and $\varepsilon > 0$, define

$$S_{m,\varepsilon} = \{x \in B : |e_j(x) - (e_j,h_0)| \leq \varepsilon,\ 1 \leq j \leq m\}$$

Since $e_j(x)$ are independent Gaussian random variables, $N_{m,\varepsilon} = P(S_{m,\varepsilon}) \neq 0$. From $F_\mu = 0$ a.s. P and Fubini's theorem, it follows that

$$\int [N_{m,\varepsilon}^{-1} \int_{S_{m,\varepsilon}} e^{i(h,x)^{\sim}}\, dP(x)]\, d\mu(h) = 0 \tag{2.3}$$

For each $m \geq 1$ and h,

$$(h,x)^{\sim} = \sum_1^m (h,e_j)e_j(x) + \sum_{m+1}^{\infty} (h,e_j)e_j(x)$$

and the random variables $\xi^h_{m,\varepsilon}(x) = I_{S_{m,\varepsilon}}(x)e^{i\Sigma_1^m(h,e_j)e_j(x)}$ and $\eta^h_m(x) = e^{i\Sigma_{m+1}^{\infty}(h,e_j)e_j(x)}$ are independent. We can therefore rewrite (2.3) as

$$\int_H (N^{-1}_{m,\varepsilon}E\xi^h_{m,\varepsilon} \cdot (E\eta^h_m)\, d\mu(h) = 0 \tag{2.4}$$

Note that both factors are bounded by 1 and

$$\begin{aligned}|N^{-1}_{m,\varepsilon}E\xi^h_{m,\varepsilon} - e^{i(\pi_m h,h_0)}| &\leq N^{-1}_{m,\varepsilon}\int_{S_{m,\varepsilon}} |e^{i\Sigma_1^m(h,e_j)e_j(x)} - e^{i\Sigma_1^m(h,e_j)(e_j,h_0)}|\, dP(x) \\ &\leq \sup_{x\in S_{m,\varepsilon}} |e^{i\Sigma_1^m(h,e_j)[e_j(x)-(e_j,h_0)]} - 1|\end{aligned}$$

which converges to 0 as $\varepsilon \downarrow 0$ (m and h being held fixed). Furthermore, $\Sigma_{m+1}^{\infty}(h,e_j)e_j(x)$ is Gaussian with mean 0 and variance $\Sigma_{m+1}^{\infty}|(h,e_j)|^2 = |h|^2 - (\pi_m h,h)$. By the dominated convergence theorem, letting $\varepsilon \downarrow 0$ in (2.4) yields

$$\int_H e^{i(\pi_m h,h_0)-1/2[|h|^2-(\pi_m h,h)]}\, d\mu(h) = 0.$$

Using the dominated convergence theorem with $m \to \infty$, we have $\int_H e^{i(h,h_0)}\, d\mu(h) = 0$. Now h_0 is an arbitrary element of H, so the unicity property of the Fourier transform of finite complex measures on H implies that $\mu = 0$. (This property is easily derived from the uniqueness of the Fourier transform on finite complex measures on $\mathbb{R}$, or see Ref. 1, pp. 17, 18.)

REMARKS

5. A different proof of the one-to-oneness of (2.2) can be given by generalizing an argument of Cameron and Storvick (Ref. 3, theorem 2.1). We will, in effect, do this in Sec. 3,

where we present the analogous result for functionals of two complex variables.

6. The proof above gives something slightly stronger than what is stated in the proposition. It shows that if $F_{\mu_1} = F_{\mu_2}$ a.s. P for two finite complex measures μ_1 and μ_2, then $F_{\mu_1} = F_{\mu_2}$, s-a.s. P. This is pointed out in theorem 2.1 of Ref. 3.

We now evaluate the analytic Wiener and Feynman integrals of Definition 2.1 for functionals F in $\mathcal{F}$.

PROPOSITION 2.2. For each functional F in $\mathcal{F}$, the analytic Wiener integral $I^z(F)$ exists for any parameter $z \in \Omega$ and

$$I^z(F) = J_F^*(z) = \int_H e^{-(1/2z)|h|^2} \, d\mu(h) \tag{2.5}$$

The analytic Feynman integral of F over B for the parameter $q \in \mathbb{R}$ exists provided that $q \neq 0$ and

$$I^q(F) = \int_H e^{-(i/2q)|h|^2} \, d\mu(h) \tag{2.6}$$

Furthermore, the values of $I^z(F)$ and $I^q(F)$ depend only on the s-equivalence class defined by F in $\mathcal{F}$.

Proof. The fact that $I^z(F)$ and $I^q(F)$ depend on the s-equivalence class of F has already been noted when we defined the notion of s-equivalence. For (2.5) and (2.6) it suffices to show that for F of the form (2.1), the function

$$\bar{J}_F^*(z) = \int_H e^{-(1/2z)|h|^2} \, d\mu(h) \tag{2.7}$$

is analytic on $z \in \Omega$ and continuous on $z \in \bar{\Omega} - \{0\}$ and that $\bar{J}_F^*(\lambda) = J_F(\lambda)$ for all real $\lambda > 0$. Indeed, Fubini's theorem shows that

$$\begin{aligned} J_F(\lambda) &= \int_B \int_H e^{i(h,\lambda^{-1/2}x)^\sim} \, d\mu(h) \, dP(x) \\ &= \int_H \int_B e^{i(h,\lambda^{-1/2}x)^\sim} \, dP(x) \, d\mu(h) \\ &= \int_H \int_B e^{i\lambda^{-1/2}(h,x)^\sim} \, dP(x) \, d\mu(h) \\ &= \int_H e^{-(1/2\lambda)|h|^2} \, d\mu(h) \end{aligned}$$

The function defined by (2.7) clearly exists for all $z \in \overline{\Omega} - \{0\} = \{z \neq 0$ such that $\mathrm{Re}(z) \geq 0\}$ and coincides with J_F when $z = \lambda > 0$. By dominated convergence $\overline{J}_F^*$ is continuous on $\overline{\Omega} - \{0\}$. When $\mathrm{Re}(z) > 0$ we can also use dominated convergence to show that

$$\frac{d}{dz}\overline{J}_F^*(z) = \int_H \frac{|h|^2}{2z^2} e^{-(1/2z)|h|^2} d\mu(h)$$

The restriction of $\overline{J}_F^*$ to Ω is therefore an analytic extension of J_F. There is, of course, at most, one such extension. Equations (2.5) and (2.6) follow immediately from Definition 2.1.

REMARK 7. Let A be a nonnegative self-adjoint operator on the Hilbert space H and μ any finite complex measure. The functional

$$F(x) = \int_H e^{i(A^{1/2}h,x)^\sim} d\mu(h)$$

belongs to F because it can be rewritten as $\int_H e^{i(h,x)^\sim} d\nu(h)$ for $\nu = \mu \circ (A^{1/2})^{-1}$. For the parameter $q = 1$, the analytic Feynman integral of F over B with respect to Ω has the form

$$I^1(F) = \int_H e^{-(i/2)(Ah,h)} d\mu(h)$$

Expressions of this form when A is no longer required to be nonnegative or even self-adjoint will motivate the definitions given in Sec. 3, 4, and 5.

We will now show how the analytic Feynman integral given in Definition 2.1 in the context of abstract Wiener space provides the link between the Fresnel integral of Albeverio and Høegh-Krohn (Ref. 1, sec. 2) and the analytic continuation approach for classical Wiener space considered by Cameron and Storvick in Ref. 3.

First we briefly recapitulate the terminology of Ref. 1. Let $F(H)$ be the space of functionals f(h) on H of the form

$$f(h) = \int_H e^{i(h',h)}\, d\mu(h') \tag{2.8}$$

Then the map $\mu \mapsto f$ defines an isomorphism of Banach algebras between $\mathcal{M}(H)$ and $\mathcal{F}(H)$, where $\|f\|$ is defined to be $\|\mu\|$ and the operations on $\mathcal{F}$ are pointwise addition, pointwise multiplication, and scalar multiplication over $\mathbb{C}$. $\mathcal{F}(H)$ is called the class of Fresnel integrable functions and the Fresnel integral of $f \in \mathcal{F}(H)$ is defined to be

$$\mathcal{F}(f) = \int_H e^{-(i/z)|h|^2}\, d\mu(h)$$ when it is related to f through (2.8)

PROPOSITION 2.3. The correspondence

$$f \mapsto \mu \mapsto [F] \tag{2.9}$$

sets up an isomorphism of Banach algebras between $\mathcal{F}(H)$ and $\mathcal{F}$, where f and μ are related by (2.8) and F and μ are related by (2.1) and [F] is the s-equivalence class of F in $\mathcal{F}$. When f and F are related by (2.9), then

$$\mathcal{F}(f) = I^1(F) \tag{2.10}$$

Proof. That (2.9) establishes a Banach algebra isomorphism follows from Proposition 2.1 and proposition 2.2 of Ref. 1. The equality (2.10) follows from the definition of $\mathcal{F}(f)$ and (2.6) with $q = 1$.

Next, before considering extensions of the method of analytic continuation introduced in Secs. 3 and 5, let us compare our Definition 2.1 with that given by Cameron and Storvick in Ref. 3. To do this, replace (1) (B,P) by the usual n-dimensional Wiener space $C_0^n[a,b]$ with n-dimensional Wiener measure P_w and (2) H by the reproducing kernel Hilbert space of the n-dimensional Wiener process (X_t), $a \le t \le b$, as described below.

EXAMPLE 1. Let $n \ge 1$ be an integer and $a < b$ fixed real numbers. Let $L_2^n([a,b])$ be the separable real Hilbert space of square-integrable $\mathbb{R}^n$-valued functions on [a,b] and define

H to be the space of $\mathbb{R}^n$-valued functions $h(t) = (h_m(t))$, $1 \leq m \leq n$, $a \leq t \leq b$, such that h is absolutely continuous, $h(a) = 0$, and $h' = dh/dt \in L_2^n([a,b])$. H is a separable real Hilbert space under the norm

$$|h|^2 = \sum_1^n \int_a^b [h_m'(s)]^2 \, ds$$

There is an obvious isometric isomorphism of Hilbert spaces between $L_2^n([a,b])$ and H given by $h \overset{\theta}{\longmapsto} \int_a^{\cdot} h'(s)\,ds$. Then $\|h\| = \sup_{a\leq t\leq b, 1\leq m\leq n} |h_m(t)|$ is a measurable norm on H and the Banach space B which is the completion of H under $\|\cdot\|$ can be identified with $C_0^n([a,b])$, the space of continuous functions from $[a,b]$ to $\mathbb{R}^n$ which vanish at a. We can also identify B^* as $M_0^n([a,b])$, the n-dimensional product of the space of finite signed measures on $[a,b]$ having no mass at $\{a\}$. (A further discussion may be found in Ref. 6.) Taking $B = C_0^n([a,b])$ and $\imath$ as the natural injection of H into B, we have a realization of the abstract Wiener space $(\imath,H,B)$ where $\tilde{P}$ is the weak Gaussian distribution on H and $P = P_w$ the standard Wiener measure on the Borel sets of $C_0^n([a,b])$.

Let $\{\varphi_j\}$ be a sequence of functions of bounded variation on $[a,b]$ which form a complete orthonormal system of $L_2^n([a,b])$. We will now show that the Paley-Wiener-Zygmund integral $(\varphi) \int_a^b v(s)\,\tilde{d}x(s)$ defined in Refs. 3 and 7 for $v \in L_2^n([a,b])$ and almost all $x \in C_0^n([a,b])$ (with respect to P_w) corresponds under the isomorphism θ to the extension of the inner product $(h,x)^{\sim}$ constructed in Lemma 2.1.

Let φ be a function of bounded variation on $[a,b]$. For each $x \in B = C_0^n([a,b])$ define

$$\int_a^b \varphi(s)\,\tilde{d}x(s) = \varphi(b)x(b) - \int_a^b x(s)\,d\varphi(s) \tag{2.11}$$

(see Ref. 7, definition 1). The right-hand side is bounded by constant $\|x\|$ and since $B^* = M_0^n([a,b])$ there exists a unique $\nu \in M_0^n([a,b])$ such that $\int_a^b \varphi(s)\,\tilde{d}x(s) = \int_a^b x(s)\,d\nu(s)$ for all $x \in C_0^n([a,b])$. Taking a sequence in $C_0^n([a,b])$

converging pointwise to $I_{(s,b]}$, it is easily seen that

$$\nu(s,b] = \varphi(s+) \tag{2.12}$$

for all $a \le s \le b$. Conversely, any two members of $\mathcal{M}_0^n([a,b])$ with this property coincide on sets of the form $(c,d]$, $a \le c < d \le b$, and hence on all Borel sets in $[a,b]$ by the monotone class theorem. Moreover, φ is right-continuous a.e., so $\nu(s,b] = \varphi(s)$ for almost all $s \in [a,b]$.

Before proceeding, we shall try to clarify the relationship between φ and ν. It is customary to associate with φ the measure μ on $[a,b]$ defined by

$$\mu[a,s] = \varphi(s+) - \varphi(a)$$

for all $a \le s < b$ and $\mu(\{b\}) = \varphi(b) - \varphi(b-)$. If we normalize μ by subtracting the jump at $\{a\}$, we get a member μ_φ of $\mathcal{M}_0^n([a,b])$. In fact, μ_φ is just the measure which is defined by the Lebesgue-Stieltjes integral on the right-hand side of (2.11): $\int_a^b x(s)\, d\mu_\varphi(s) = \int_a^b x(s)\, d\varphi(s)$ for all $x \in C_0^n([a,b])$. The relation between μ_φ and ν as defined by (2.12) is simply

$$\mu_\varphi(s,b] = \varphi(b) - \nu(s,b], \qquad a \le s < b$$

Now to every φ_j in the basis of $L_2^n([a,b])$ used by Cameron and Storvick, the above construction associates $\nu_j \in \mathcal{M}_0^n([a,b])$. Defining $\Phi_j(t) = \int_a^t \varphi_j(s)\, ds$, $\{\Phi_j\}$ forms a complete orthonormal system of H and

$$\begin{aligned}(h, i^*\nu_j)_H &= \nu_j(ih) \\ &= \int_a^b h(u)\, d\nu_j(u) \\ &= \int_a^b \int_a^b I_{[a,u)}(s) h'(s)\, ds\, d\nu_j(u) \\ &= \int_a^b \int_a^b I_{(s,b]}(u) h'(s)\, d\nu_j(u)\, ds \\ &= \int_a^b h'(s)\nu_j(s,b]\, ds \\ &= \int_a^b h'(s)\varphi_j(s)\, ds \\ &= (h,\varphi_j)_H\end{aligned}$$

for all $h \in H$. This shows that $i^*\nu_j = \Phi_j$ for all $j \ge 1$.

Given $v \in L_2^n([a,b])$, the value of $(\varphi) \int_a^b v(s)\, \tilde{d}x(s)$ is defined in Ref. 3 as the limit of $\int_a^b [\Sigma_1^m (v,\varphi_j)_{L_2} \varphi_j](s)\, \tilde{d}x(s)$ whenever this limit exists. We shall take its value to be zero when the limit does not exist. (See also definition 2 of Ref. 7.) For any $(v,x) \in L_2^n([a,b]) \times C_0^n([a,b])$,

$$\int_a^b [\sum_1^m (v,\varphi_j)_{L_2}](s)\, \tilde{d}x(s) = \sum_1^m (v,\varphi_j)_{L_2} \int_a^b \varphi_j(s)\, \tilde{d}x(s)$$

$$= \sum_1^m (\theta v,\theta\varphi_j)_H \int_a^b x(s)\, d\nu_j(s)$$

$$= \sum_1^m (\theta v,\Phi_j)_H \nu_j(x)$$

This shows $(\varphi) \int_a^b v(s)\, \tilde{d}x(s)$ is precisely $(\theta v,x)^{\sim}$ as defined in Lemma 2.1 for the basis $e_j = \Phi_j = \iota^*(\nu_j)$. That is,

$$(\varphi) \int_a^b h'(s)\, \tilde{d}x(s) = (h,x)^{\sim}$$

for all $(h,x) \in H \times B$.

Turning now to the class S defined in Ref. 3 to contain s-equivalence classes of functionals

$$F(x) = \int_{L_2^n([a,b])} e^{i(\varphi)\int_a^b v(s)\tilde{d}x(s)}\, d\mu(v) \qquad (2.13)$$

where μ is a finite complex measure on $L_2^n([a,b])$, we see that

$$F(x) = \int_{L_2^n([a,b])} e^{i(\theta v,x)^{\sim}}\, d\mu(v)$$

$$= \int_H e^{i(h,x)^{\sim}}\, d(\mu \circ \theta^{-1})(h)$$

Since $\mathcal{M}(L_2^n([a,b])$ and $\mathcal{M}(H)$ are in one-to-one correspondence under θ, the spaces S and $\mathcal{F}$ are identical.

Finally, the definition of analytic Wiener and analytic Feynman integrals given in sec. 1 of Ref. 3 coincides with Definition 2.1 above in the case where $B = C_0^n[a,b]$. The definition given in this section in the context of abstract Wiener space therefore generalizes the treatment of the analytic Feynman integral on S presented in Ref. 3 to an arbitrary Gaussian process on a separable Banach space.

3. ANALYTIC WIENER AND FEYNMAN INTEGRALS INVOLVING INDEFINITE BILINEAR FORMS

The expression

$$I^1(F) = \int_H e^{-(i/2)(Ah,h)}\, d\mu(h) \tag{3.1}$$

appears as the analytic Feynman integral of the functional

$$F(x) = \int_H e^{i(A^{1/2}h,x)^{\sim}}\, d\mu(h)$$

according to Definition 2.1 provided that A is a bounded nonnegative self-adjoint operator on H. As we shall see below, however, a successful treatment of certain physical problems by means of a Feynman integral (e.g., the anharmonic oscillator of Ref. 1, sec. 5) requires (3.1) to be a Feynman integral when A is self-adjoint but not nonnegative. This would allow the case where A has the form

$$A = A^+ - A^- \tag{3.2}$$

and both A^+ and A^- are bounded nonnegative self-adjoint operators. In order to widen the scope of the analytic continuation technique to treat such cases, one of us [5] introduced a definition analogous to that of Sec. 2, but involving two complex variables. We present that definition here in a slightly modified form and apply it to the example mentioned above.

In line with the general scheme of our definitions outlined in Sec. 2, we first need to describe the domain of analyticity required for this case. Throughout this section we will let

$$\Omega = \{(z_1,z_2) \in \mathbb{C} \times \mathbb{C} : \mathrm{Re}(z_m) > 0,\ m = 1,\ 2\}$$

DEFINITION 3.1 Suppose that $F(x_1,x_2)$ is a measurable complex-valued functional on B × B such that:

(i) $J_F(\lambda_1,\lambda_2) = \int_{B\times B} F(\lambda_1^{-1/2}x_1,\lambda_2^{-1/2}x_2)\, d(P\times P)(x_1,x_2)$ is finite for all $\lambda_1 > 0$, $\lambda_2 > 0$.

(ii) There is an analytic function of two complex variables $J_F^*(z)$, $z = (z_1,z_2) \in \Omega$ such that

$$J^*_F(\lambda_1,\lambda_2) = J_F(\lambda_1,\lambda_2)$$

for all $\lambda_1 > 0$, $\lambda_2 > 0$.

Then we shall call $I^z_2(F) = J^*_F(z)$ the analytic Wiener integral of F over B × B with the parameter $z = (z_1,z_2)$ relative to Ω.

If $\lim_{z'\to z, z'\in\Omega} I^z_2(F)$ exists for some $z = (-iq_1,-iq_2)$, $q = (q_1,q_2) \in \mathbb{R}^2$, we will denote the limit $I^q_2(F)$, calling it the analytic Feynman integral of F over B × B with parameter $q = (q_1,q_2)$. We assume here that $q \neq (0,0)$.

REMARK 1. The analytic extension $J^*(z)$ is unique if it exists on Ω. (See the Appendix, Theorem A1.)

As in Sec. 2 we need a notion of s-almost sure equivalence, or s-equivalence, for integrand functionals $F(x_1,x_2)$. Two complex-valued functionals F and G on B × B are said to be s-a.s. equal if

$$F(\alpha_1 x_1,\alpha_2 x_2) = G(\alpha_1 x_1,\alpha_2 x_2) \qquad \text{a.s. } [P \times P]$$

for all $\alpha_1 > 0$ and $\alpha_2 > 0$. The definition of equivalent extensions of the inner product $(h,x)^\sim$ will be retained just as it was given in Sec. 2. Note then that for two functionals F and G which coincide on B × B s-almost surely, the integrals $I^z_2(F)$ and $I^z_2(G)$ [respectively, $I^q_2(F)$ and $I^q_2(G)$] either coincide or fail to exist simultaneously. Hence $I^z_2(F)$ and $I^q_2(F)$ depend only on the s-equivalence class of F on B × B.

We now define spaces of functionals $F(x_1,x_2)$ where I^z_2 and I^q_2, $q \neq (0,0)$, can be shown to exist.

DEFINITION 3.2 Let A_1 and A_2 be two bounded nonnegative self-adjoint operators on H. Let F_{A_1,A_2} be the space of all s-equivalence classes of functionals F which for some $\mu \in M(H)$ have the form

$$F(x_1,x_2) = \int_H e^{i[(A_1^{1/2}h,x_1)^{\sim}+(A_2^{1/2}h,x_2)^{\sim}]} d\mu(h) \qquad (3.3)$$

The s-equivalence class defined by F is unchanged if an equivalent extension of the inner product $(h,x)^{\approx}$ is used in (3.3). For convenience we shall often write $\mathcal{F}_A$ for $\mathcal{F}_{A_1,A_2}$. In this context A represents the pair of operators A_1 and A_2, rather than a single operator on H. For these classes of integrands $\mathcal{F}_A$ the analytic Wiener and analytic Feynman integrals can be explicitly evaluated:

PROPOSITION 3.1 For any pair of bounded nonnegative self-adjoint operators A_1, A_2 on H and any functional $F(x_1,x_2)$ belonging to $\mathcal{F}_A$, the analytic Wiener integral of F over B × B with parameter $z = (z_1,z_2) \in \Omega$ exists and

$$I_2^z(F) = \int_H e^{-(1/2)\Sigma_1^2 z_m^{-1}(A_m h,h)} d\mu(h) \qquad (3.4)$$

The analytic Feynman integral of F over B × B with parameter $q = (q_1,q_2)$ exists provided that $q_1 \neq 0$ and $q_2 \neq 0$:

$$I_2^q(F) = \int_H e^{-(i/2)\Sigma_1^2 q_m^{-1}(A_m h,h)} d\mu(h) \qquad (3.5)$$

In particular,

$$I_2^{(1,-1)}(F) = \int_H e^{-(i/2)((A_1-A_2)h,h)} d\mu(h) \qquad (3.6)$$

These integrals depend only on the s-equivalence class of F in $\mathcal{F}_A$.

Proof. The random variables $(A_1^{1/2}h,x_1)^{\sim}$ and $(A_2^{1/2}h,x_2)^{\sim}$ on B × B are independent Gaussian with mean 0, respective variances $|A_1^{1/2}h|^2$ and $|A_2^{1/2}h|^2$ under P × P. Using Fubini's theorem we may therefore write

$$J_F(\lambda_1,\lambda_2) = \int_{B\times B} \int_H \exp[i\Sigma_1^2 (A_m^{1/2}h,\lambda_m^{-1/2}x_m)^{\sim}\, d\mu(h)] \times dP \times P(x_1,x_2)$$

$$= \int_H \int_{B\times B} \exp[i\Sigma_1^2 \lambda_m^{-1/2}(A_m,h,x_m)^\sim]\, dP \times P(x_1,x_2)\, d\mu(h)$$

$$= \int_H e^{-(1/2\lambda_1^{-1}|A_1^{1/2}h|^2 - (1/2\lambda_2^{-1}|A_2^{1/2}h|^2}\, d\mu(h)$$

$$= \int_H e^{-(1/2)\Sigma_1^2 \lambda_m^{-1}(A_m h,h)}\, d\mu(h)$$

for all real $\lambda_1 > 0$ and $\lambda_2 > 0$. Define

$$J^*(z) = \int_H e^{-(1/2)\Sigma_1^2 z_m^{-1}(A_m h,h)}\, d\mu(h)$$

which exists for all $z \in \{(z_1,z_2) \in \mathbb{C} \times \mathbb{C} : z_1 \neq 0, z_2 \neq 0, \mathrm{Re}(z_1) \geq 0 \text{ and } \mathrm{Re}(z_2) \geq 0\}$ and is continuous on this set by the dominated convergence theorem. Since $J^*(\lambda_1,\lambda_2) = J(\lambda_1,\lambda_2)$ for all real $\lambda_1 > 0$ and $\lambda_2 > 0$, equations (3.4) to (3.6) will follow immediately from Definition 3.1 once we show the restriction of J^* to Ω is an analytic function. Just as in the proof of Proposition 2.2, dominated convergence can be used to show $J^*(z_1,z_2)$ is differentiable in each variable separately in Ω. From the theory of analytic functions of several complex variables, it is known that this implies analyticity of J^* on Ω. (See the Appendix, Theorem A1.)

REMARK 2. Suppose that A is a bounded self-adjoint operator on H. Then we may write A in the form (3.2) where A^+ and A^- are each bounded and nonnegative self-adjoint. Take $A_1 = A^+$ and $A_2 = A^-$ in the definitions above. For any $F(x_1,x_2)$ in $\mathcal{F}_A$, (3.6) becomes

$$I_2^{(1,-1)}(F) = \int_H e^{-(i/2)(Ah,h)}\, d\mu(h) \tag{3.7}$$

Also in this situation, if $A^- = 0$, then the right-hand side of (3.7) coincides with (3.1). In particular, when $A^- = 0$ and A^+ is the identity, then $\mathcal{F}_A$ is essentially $\mathcal{F}$ of Sec. 2 and $I_2^{(z_1,z_2)}(F) = I_2^{z_1}(F')$ and $I_2^{A(q_1,q_2)}(F) = I^{q_1}(F')$ where $F_1'(x_1) = F(x_1,x_2)$ for all $x_1 \in B$, $x_2 \in B$. In this sense the definition given in this section includes Definition 2.1 as a special case.

If the two operators A_1 and A_2 in (3.3) are sufficiently singular, it is obvious that the functional $F(x_1,x_2)$ cannot determine μ uniquely. We will present a necessary and sufficient condition for the uniqueness of μ with the aid of the following result.

LEMMA 3.1 For any two bounded nonnegative self-adjoint operators A_1 and A_2 on H,

$$\overline{R(A_1^{1/2}) + R(A_2^{1/2})} = \overline{R(A_1 + A_2)}$$

[Here R indicates the range of an operator, $\bar{R}$ its closure.]

Proof. We use the fact that for any self-adjoint operator A, $N(A)^\perp = \overline{R(A)}$, where $N(A)$ is the null space or kernel of A. Since $N(A^{1/2}) \subseteq N(A)$, we have $N(A^{1/2})^\perp \supseteq N(A)^\perp$. That is,

$$\overline{R(A^{1/2})} \supseteq \overline{R(A)}$$

It is easy to see that

$$\overline{R(A_1 + A_2)} \subseteq \overline{R(A_1^{1/2}) + R(A_2^{1/2})}$$

Conversely, assume that $h \in N(A_1 + A_2)$. Then we have

$$((A_1 + A_2)h,h) = 0$$

$$(A_1h,h) + (A_2h,h) = 0$$

$$|A_1^{1/2}h|^2 + |A_2^{1/2}h|^2 = 0$$

Hence

$$h \in N(A_1^{1/2}) \cap N(A_2^{1/2})$$

$$h \in \overline{R(A_1^{1/2})}^\perp \cap \overline{R(A_2^{1/2})}^\perp$$

that is,

$$h \in \overline{R(A_1^{1/2}) + R(A_2^{1/2})}^\perp$$

This shows that $\overline{R(A_1^{1/2}) + R(A_2^{1/2})} \subseteq N(A_1 + A_2)^{\perp} = \overline{R(A_1 + A_2)}$.

REMARK 3. The proof above is easily generalized to any finite number of operators A_m, $1 \leq m \leq k$, which are nonnegative and self-adjoint.

We now treat the analog of Proposition 2.1. For $\mu \in M(H)$ and F defined as in (3.3), we shall denote the s-equivalence class of F by [F]. Let the operations of addition, multiplication, and complex scalar multiplication on the s-equivalence classes of F_A be generated by the elementwise operations as in Sec. 2. From (3.3) and the fact that $M(H)$ is an algebra under convolution it follows that F_A is an algebra.

PROPOSITION 3.2 For A_1 and A_2 bounded nonnegative self-adjoint operators on H, the map

$$\mu \mapsto [F] \tag{3.8}$$

defined by (3.3) sets up an algebra homomorphism between $M(H)$ and F_A which does not depend on the equivalent extension $(h,x)^{\sim}$ used in (3.3). This is an isomorphism if and only if $R(A_1 + A_2)$ is dense in H. In this case, F_A becomes a Banach algebra under the norm $\|F\| = \|\mu\|$ and (3.8) is an isomorphism of Banach algebras.

Proof. All the assertions above will follow just as in the proof of Proposition 2.1 once we show (3.8) is one to one if and only if $R(A_1 + A_2)$ is dense in H. By Lemma 3.1 this condition is equivalent to $R(A_1^{1/2}) + R(A_2^{1/2})$ being dense in H. We give two proofs of the "if" part below, the first using the same type of reasoning as in the uniqueness part of Proposition 2.1, and the other adapted from the argument of Cameron and Storvick in their proof of Theorem 2.1 [3]. Assume that $R(A_1^{1/2}) + R(A_2^{1/2})$ is dense and suppose for some $\mu \in M(H)$ the functional F in (3.3) vanishes a.s. on

$B \times B$. It suffices to show that $\mu = 0$.

First method. For any two Borel sets S^1 and S^2 in B, $E[I_{S^1 S^2} \cdot F] = 0$. By Fubini's theorem, then

$$\int_H \left[\int_{S^1} e^{i(A_1^{1/2}h,x_1)^\sim} dP(x_1)\right] \left[\int_{S^2} e^{i(A_2^{1/2}h,x_2)^\sim} dP(x_2)\right] d\mu(h) = 0 \quad (3.9)$$

Fix any two vectors $h_k \in R(A_k^{1/2})$, $k = 1, 2$, and let $\{e_j\}$ be the basis of H in B^* used to construct $(h,x)^\sim$ in Lemma 2.1. For every integer $m \geq 1$ and $\varepsilon > 0$, define Borel sets

$$S^1_{m,\varepsilon} = \{x \in B : |e_j(x) - (e_j,h_1)| \leq \varepsilon,\ 1 \leq j \leq m\}$$
$$S^2_{m,\varepsilon} = \{x \in B : |e_j(x) - (e_j,h_2)| \leq \varepsilon,\ 1 \leq j \leq m\}$$

and normalizing constants $N^k_{m,\varepsilon} = P(S^k_{m,\varepsilon})$, $k = 1, 2$. From $(h,x)^\sim = \Sigma_1^m (h,e_j)e_j(x) + \Sigma_{m+1}^\infty (h,e_j)e_j(x)$ a.s. P and the independence of $I_{S^k_{m,\varepsilon}}(x) \cdot e^{i\Sigma_1^m(h,e_j)e_j(x)}$ and $e^{i\Sigma_{m+1}^\infty(h,e_j)e_j(x)}$ as random variables on (B,P) we can write (3.9) for $S^1 = S^1_{m,\varepsilon}$ and $S^2 = S^2_{m,\varepsilon}$ as

$$\int_H \prod_{k=1}^{2} \left[(N^k_{m,\varepsilon})^{-1} \int_{S^k_{m,\varepsilon}} e^{i\Sigma_1^m(h,e_j)e_j(x)} dP(x)\right]$$
$$\left[\times \int_B e^{i\Sigma_{m+1}^\infty(h,e_j)e_j(x)} dP(x) \quad d\mu(h)\right] = 0$$

For m and h held fixed,

$$(N^k_{m,\varepsilon})^{-1} \int_{S^k_{m,\varepsilon}} e^{i\Sigma_1^m(h,e_j)e_j(x)} dP(x)$$
$$\rightarrow e^{i\Sigma_1^m(h,e_j)(e_j,h_k)} = e^{i(\pi_m h,h_k)}$$

as $\varepsilon \downarrow 0$ for both $k = 1$ and $k = 2$. Also,

$$\int_B e^{i\Sigma_{m+1}^\infty(h,e_j)e_j(x)} dP(x) = e^{-(1/2)[|h|^2 - (\pi_m h,h)]}$$

Dominated convergence then gives

$$\int_H \prod_{k=1}^{2} e^{i(\pi_m h, h_k)-(1/2)[|h|^2-(\pi_m h,h)]} d\mu(h) = 0$$

A second application of dominated convergence as $m \to \infty$ yields

$$\int_H e^{i(h,(h_1+h_2))} d\mu(h) = 0$$

But $h_1 + h_2$ is any element of $R(A_1^{1/2}) + R(A_2^{1/2})$. Since this space is assumed to be dense and since

$$h_0 \mapsto \int_H e^{i(h,h_0)} d\mu(h)$$

is continuous on H, we conclude that

$$\int_H e^{i(h,h_0)} d\mu(h) = 0$$

for all $h_0 \in H$ so that $\mu = 0$ by the uniqueness property of the Fourier transform of finite complex measures on H.

Second method. Let (α_1,β_1) and (α_2,β_2) be open intervals with finite endpoints. For m = 1, 2 write

$$g_m(s) = I_{(\alpha_m,\beta_m)}(s)$$

and for $\delta > 0$ let $g_{m,\delta}$ be the function which coincides with g_m on (α_m,β_m), is linear on $(\alpha_m - \delta, \alpha_m)$ and $(\beta_m, \beta_m + \delta)$, and vanishes elsewhere. Set

$$g^*_{m,\delta}(t) = g_{m,\delta}(s)e^{-(1/2)s^2}$$

and

$$G^*_{m,\delta}(t) = \frac{1}{\sqrt{2\pi}} \int_{-\infty}^{\infty} e^{ist} g^*_{m,\delta}(s)\, ds \qquad m = 1, 2$$

Then $G^*_{m,\delta}$ is integrable and Fourier inversion gives

$$g^*_{m,\delta}(s) = \frac{1}{\sqrt{2\pi}} \int_{-\infty}^{\infty} e^{-ist} G^*_{m,\delta}(t)\, dt$$

Fix two unit vectors u_1 and u_2 in H. Letting E indicate integration on $(B \times B, P \times P)$, we have

$$E\{G^*_{1,\delta}[-(u_1,x_1)^\sim]G^*_{2,\delta}[-(u_2,x_2)^\sim]$$
$$\times\ e^{(1/2)[(u_1,x_1)^\sim]^2+(1/2)[(u_2,x_2)^\sim]^2}$$
$$\times \int_H e^{i\Sigma_1^2(A_m^{1/2}h,x_m)^\sim}\ d\mu(h)\} = 0 \tag{3.10}$$

(Here we have used the hypothesis F = 0 a.s. P × P.) The integrability of the left-hand side in (3.10) follows from the independence of the Gaussian random variables (u_m,x_m) and the fact that $\int_{-\infty}^{\infty} |G^*_{m,\delta}(t)|dt < \infty$ m = 1, 2. This also justifies using Fubini's theorem to write (3.10) as

$$\int_H EQ(h;x_1,x_2)\ d\mu(h) = 0 \tag{3.11}$$

where

$$Q(h;x_1,x_2) = G^*_{1,\delta}[-(u_1,x_1)^\sim]G^*_{2,\delta}[-(u_2,x_2)^\sim]$$
$$\times \exp \sum_{m=1}^{2} \{\tfrac{1}{2}[(u_m,x_m)^\sim]^2 + i[(h_m,x_m)^\sim$$
$$+ (A_m^{1/2}u_m,h)(u_m,x_m)^\sim]\}$$

and $h_m = A_m^{1/2}h - (u_m,A_m^{1/2}h)u_m$ for every $h \in H$, m = 1, 2. In view of the fact that $h_m \perp u_m$ for m = 1, 2, the random variables $\xi_1 = (u_1,x_1)^\sim$, $\xi_2 = (u_2,x_2)^\sim$, $\eta_1 = (h_1,x_1)^\sim$, and $\eta_2 = (h_2,x_2)^\sim$ are independent Gaussian random variables. Hence for each h held fixed,

$$EQ(h;x_1,x_2) = \prod_{m=1}^{2}\Big\{E\Big[G^*_{m,\delta}(\xi_1)e^{(1/2)\xi_1^2+i(A_m^{1/2}u_m,h)\xi_m}\Big]$$
$$\times E\Big[e^{i\eta_m}\Big]\Big\}$$
$$= \prod_{m=1}^{2}\Big[\frac{1}{\sqrt{2\pi}}\int_{-\infty}^{\infty} G^*_{m,\delta}(t)e^{i(A_m^{1/2}u_m,h)t}\ dt$$
$$\times\ e^{-(1/2)|h_m|^2}\Big]$$
$$= \prod_{m=1}^{2}\Big\{ g^*_{m,\delta}[(A_m^{1/2}u_m,h)]$$

$$\times\, e^{-(1/2)[|A_m^{1/2}h|^2-(A_m^{1/2}u_m,h)^2]}\Big\}$$

$$= \prod_{m=1}^{2} g_{m,\delta}[(A_m^{1/2}u_m,h)]e^{-(1/2)|A_m^{1/2}h|^2}$$

Hence (3.11) becomes

$$\int_H g_{1,\delta}[(A_1^{1/2}u_1,h)]g_{2,\delta}[(A_2^{1/2}u_2,h)]\; d\mu'(h) = 0$$

where the finite complex measure μ' is defined on H by

$$d\mu'(h) = e^{-(1/2)\Sigma_1^2(A_m^{1/2}h,h)}\; d\mu(h)$$

Now, by letting δ tend to zero, the dominated convergence theorem gives

$$\int_H g_1[(A_1^{1/2}u_1,h)]g_2[A_2^{1/2}u_2,h)]\; d\mu'(h) = 0$$

i.e.,

$$\mu'(\{h \in H: \alpha_1 \le (A_1^{1/2}u_1,h) \le \beta_1,$$
$$\alpha_2 \le (A_2^{1/2}u_2,h) \le \beta_2\}) = 0$$

Since α_1, β_1, α_2 and β_2 are arbitrary, it follows that $\mu' = 0$ on the smallest σ-field with respect to which the functions $(A_1^{1/2}u_1,\cdot)$ and $(A_2^{1/2}u_2,\cdot)$ are measurable on H. Then for any two scalars a, b, $\mu' = 0$ on the smallest σ-field with respect to which the function $(A_1^{1/2}au_1 + A_2^{1/2}bu_2,\cdot) = a(A_1^{1/2}u_1,\cdot) + b(A_2^{1/2}u_2,\cdot)$ is measurable. This shows that $\mu' = 0$ on the σ-field generated by $(h,\cdot)$ for any $h \in \mathcal{R}(A_1^{1/2}) + \mathcal{R}(A_2^{1/2})$. In particular, $\int_H e^{i(h',h)}\, d\mu'(h') = 0$ for each $h \in \mathcal{R}(A_1^{1/2}) + \mathcal{R}(A_2^{1/2})$. Just as with the proof via the first method used above, the continuity and uniqueness for the Fourier transform of finite complex measures on H shows $\mu' = 0$. Then since

$$\frac{d\mu'}{d\mu}(h) = e^{-(1/2)\Sigma_1^2(A_m^{1/2}h,h)}$$

never vanishes on H, we conclude that $\mu = 0$ too. This completes the proof of uniqueness of μ by the second method.

The following example shows that the homomorphism of $\mathcal{M}(H)$ and $\mathcal{F}_A$ defined by (3.8) cannot be one to one unless $\mathcal{R}(A_1 + A_2)$ is dense.

EXAMPLE 1. Fix $h_0 \perp \overline{\mathcal{R}(A_1 + A_2)}$, which is nonzero. Take $\mu = \delta_{h_0}$, unit mass concentrated at h_0. By Lemma 3.1, $h_0 \perp \overline{\mathcal{R}(A_1^{1/2}) + \mathcal{R}(A_2^{1/2})}$. Then $h_0 \perp \overline{\mathcal{R}(A_1^{1/2})}$ and $h_0 \perp \overline{\mathcal{R}(A_2^{1/2})}$, so that $h_0 \in \mathcal{N}(A_1^{1/2}) \cap \mathcal{N}(A_2^{1/2})$. For the functional F associated to $\mu = \delta_{h_0}$ by (3.3),

$$\begin{aligned} F(x_1,x_2) &= \int_H e^{i\Sigma_1^2 (A_m^{1/2}h,x_m)^\sim} \, d\mu(h) \\ &= e^{i\Sigma_1^2 (A_m^{1/2}h_0,x_m)^\sim} \\ &= 1 \text{ s-a.s. } P \times P \end{aligned}$$

But the measure $\mu \equiv 0$ also gives this functional $F = 1$ in (3.3).

From Remark 2 it is clear that $I_2^{(1,-1)}(F)$ extends the original definition of the analytic Feynman integral $I^1(F)$ given in Sec. 2. In sec. 4 of Ref. 1, Albeverio and Høegh-Krohn present for their Fresnel integral $\mathcal{F}(f)$, $f \in \mathcal{F}(H)$ (discussed above in connection with Proposition 2.3) a generalization called the Fresnel integral normalized relative to a symmetric bilinear form Δ on H (see Ref. 1, pp. 50, 51, 59 and 60). The Fresnel integral relative to Δ is defined for all functionals f in the Banach algebra $\mathcal{F}(H)$ [see equation (2.8)]. The value of this integral is denoted $\mathcal{F}_\Delta(f)$ and when Δ is a bounded symmetric nondegenerate bilinear form, $\mathcal{F}_\Delta(f)$ is defined to be

$$\int_H e^{-(i/2)\Delta(h,h)} \, d\mu(h)$$

for f of the form

$$f(h) = \int_H e^{i(h',h)}\, d\mu(h') \tag{3.12}$$

We can now compare $I_2^{(1,-1)}$ and F_Δ in this case. Let A be a bounded invertible self-adjoint operator on H. Let A^+ and A^- be any two bounded nonnegative operators satisfying $A = A^+ - A^-$ [condition (3.2)]. Then by (3.7), $I_2^{(1,-1)}(F) = \int_H e^{-(i/2)(Ah,h)}\, d\mu(h)$ for F of the form

$$F(x_1,x_2) = \int_H e^{i\Sigma_1^2 (A_m^{1/2}h,x_m)^\sim}\, d\mu(h)$$

in F_{A^+,A^-}. We summarize this as follows:

PROPOSITION 3.3 For a bounded invertible self-adjoint operator A on H take $A_1 = A^+$ and $A_2 = A^-$ in (3.2) and set $\Delta(h_1,h_2) = (Ah_1,h_2)$. Then $f \mapsto \mu \mapsto [F]$ establishes an isomorphism of Banach algebras $F(H)$ and F_A, where f and μ correspond according to (3.12) and F and μ are related by (3.3). When f and F correspond under this isomorphism the values of the Fresnel integral of f relative to Δ and the analytic Feynman integral of F relative to Ω coincide, i.e.,

$$F_\Delta(f) = I_2^{(1,-1)}(F) = \int_H e^{-(i/2)(Ah,h)}\, d\mu(h) \tag{3.13}$$

Proof. The only detail to be checked is that $\mu \mapsto [F]$ is indeed one to one. But since A is invertible, $N(A^+ + A^-) \subseteq N(A^+ - A^-) = \{0\}$ and therefore $\overline{R(A^+ + A^-)} = N(A^+ + A^-)^\perp = H$. Now apply Proposition 3.2.

We turn now to a physical example drawn from Ref. 1, sec. 5, where the Feynman integrals (3.13) find application.

EXAMPLE 2 (The Anharmonic Oscillator). For the purposes of this example, take H to be the real separable Hilbert space of paths $\gamma: [0,t] \to \mathbb{R}^n$ such that $\gamma(t) = 0$ and $\gamma(s) = -\int_s^t \gamma'(u)\, du$, $0 \le s \le t$, for some $\gamma' \in L^2([0,t])$. For the inner product on H define

$$(\gamma_1,\gamma_2) = \int_0^t \gamma_1'(u)\cdot\gamma_2'(u)\, du$$

The number t will be fixed throughout our discussion. (The dot $\cdot$ stands for inner product in $\mathbb{R}^n$.) Let $V(x)$ and $\varphi(x)$, $x \in \mathbb{R}^n$, be functions which are the Fourier transforms of two complex measures on $\mathbb{R}^n$ (V may be assumed real for physical applications). Furthermore, we shall assume that $\varphi \in L^2(\mathbb{R}^n)$.

Next, let $a = (a_{ij})$ be an $n \times n$ real matrix which is symmetric and positive definite and such that

$$\det [\cos(a^{1/2}t)] \neq 0 \tag{3.14}$$

(This condition is equivalent to the assumption that no eigenvalue of a has the form $[k + \frac{1}{2})\frac{\pi}{t}]^2$ for any integer k. See the treatment in Ref. 1, which is closely followed here.)

Introduce on H the indefinite bilinear form

$$\langle \gamma_1, \gamma_2 \rangle = \int_0^t [\gamma_1(u)\cdot\gamma_2(u) - a\gamma_1(u)\cdot\gamma_2(u)]\, d\mu$$

This form determines uniquely a self-adjoint operator A such that $(\tilde{A}\gamma_1, \gamma_2) = \langle \gamma_1', \gamma_2' \rangle$ and (3.14) above guarantees that $\tilde{A}$ is invertible. Let $A = \tilde{A}^{-1}$.

Now let $\psi(x,t)$ be the solution of the Schrödinger equation for the anharmonic oscillator

$$i \frac{\partial}{\partial t} \psi(x,t) = - \frac{1}{2}\Delta\psi(x,t) + \frac{1}{2}(ax\cdot x)\psi(x,t) + V(x)\psi(x,t)$$

with initial condition $\psi(x,0) = \varphi(x)$. Our purpose here is to represent $\psi(x,t)$ as an analytic Feynman integral.

Let β: $[0,t] \to \mathbb{R}^n$ be any absolutely continuous function such that $\int_0^t |\beta'(u)|^2\, du < \infty$ and $\beta(t) = x$. Let

$$f(\gamma) = \exp[\tfrac{i}{2}\langle \beta,\beta \rangle + i\langle \gamma,\beta \rangle - i \int_0^t V(\gamma(u) + \beta(u))\, du]$$
$$\times\ \varphi(\gamma(0) + \beta(0))$$

Then it is easy to see from the conditions on V and φ that f is the Fourier transform of some $\mu \in M(H)$,

$$f(\gamma) = \int_H e^{i(\gamma,\tilde{\gamma})}\, d\mu(\tilde{\gamma})$$

Hence $f \in F(H)$, the space of Fresnel integrable functionals.

Take $A_1 = A^+$ and $A_2 = A^-$ where $A = A^+ - A^-$ and let Δ be the bilinear form $\Delta(\gamma_1,\gamma_2) = (A\gamma_1,\gamma_2)$. By theorem 5.1 of Ref. 1, $\psi(x,t)$ has the following representation in terms of the Fresnel integral of f normalized with respect to Δ:

$$\psi(x,t) = (\cos|a^{1/2}t|)^{-1/2}F_\Delta(f)$$

Now define $F \in F_A$ from μ, as in (3.3), i.e.,

$$F(x_1,x_2) = \int_H e^{i(A^+h_1x_1)^\sim+(A^-h_1x_2)^\sim} d\mu(h)$$

Then $F \in F_A$ and Proposition 3.3 yields

$$\psi(x,t) = (\cos|a^{1/2}t|)^{-1/2}I_2^{(1,-1)}(F)$$

[Note: As in theorem 5.1 of Ref. 1, the factor $(\cos|a^{1/2}t|)^{-1/2}$ serves to indicate a suitably chosen fourth root of $[\cos(\det(a^{1/2}t))]^{-2}$. For t near 0, for example, this factor is just $|\cos(\det(a^{1/2}t))|^{-1/2}$.]

4. FEYNMAN INTEGRALS ON KREIN SPACES

The definition of the analytic Wiener and analytic Feynman integral given in the preceding section is based on a pair of A_1, A_2 of nonnegative, self-adjoint operators on the Hilbert space H. In fact, the dependence of the class of integrands in Proposition 3.1 on the operators A_1, A_2 is made explicit in our notation. According to this point of view—and it is the one adopted in almost all investigations on the Feynman integral known to us—the Hilbert space H is regarded as primary and the requirements of the particular physical problem give rise to the operators A_1, A_2.

An alternative point of view is set forth in this section, in which an indefinite metric space or, more precisely, a special type of such space known as a Krein space is taken to be the natural mathematical model of the physical problem of interest. The use of indefinite metric spaces in physics is not new. In connection with the study of quantum field theory, it first appeared in a paper of Dirac and was subsequently discussed by Pauli (see Ref. 8 and also the bibliography in Ref. 2).

Our aim in this section is limited to pointing out that the principal results of Sec. 3 can be formulated in the context of a Krein space. To do this, we recapitulate briefly some definitions and properties of indefinite inner product spaces taken from Ref. 2.

A linear vector space E over $\mathbb{R}$ is an inner product space if E is equipped with an inner product $(\cdot,\cdot)$ which satisfies the usual axioms except that for $h \in E$, (h,h) may be positive, negative, or zero. If E contains elements h, h' for which $(h,h) > 0$ and $(h',h') < 0$, E is called an indefinite inner product (or indefinite metric) space and $(\cdot,\cdot)$ is said to be an indefinite inner product.

The notion of orthogonality in E is introduced in the usual manner: For $h_1, h_2 \in E$, $h_1 \perp h_2$ if $(h_1,h_2) = 0$. Let $E_+ = \{h \in E: (h,h) > 0 \text{ or } h = 0\}$, $E_- = \{h \in E: (h,h) < 0 \text{ or } h = 0\}$, and $E_0 \subset \{h \in E: (h,h) = 0\}$. An indefinite inner product space E has a fundamental decomposition if it can be written in the form $E = E_0 \oplus E_+ \oplus E_-$, where E_0, E_+, E_-, defined as above, are mutually orthogonal subspaces. A subspace L of E is definite if $(h,h) = 0$ implies that $h = 0$. Note that E_+ and E_- are definite subspaces. A definite subspace L of E is <u>intrinsically complete</u> if it is complete with respect to the norm $|h|_L = |(h,h)|^{1/2}$ $(h \in L)$.

We are now ready to introduce the particular indefinite inner product space which is of interest to us in this section, a Krein space.

DEFINITION 4.1 An indefinite inner product space E is a <u>Krein space</u> if it admits a fundamental decomposition of the form

$$E = E_+ \oplus E_- \tag{4.1}$$

where the subspaces are intrinsically complete.

Krein spaces and operators defined on them have been studied extensively. We shall now state some properties of Krein spaces which are essential for our purpose (for proofs, see Ref. 2).

A fundamental decomposition of the form (4.1) gives rise to the following orthogonal projection operators Π_+, Π_- defined by the relations $\Pi_+ h = h_+$, $\Pi_- h = h_-$ $(h \in E)$, where $h = h_+ + h_-$, $h_+ \in E_+$, and $h_- \in E_-$ is the decomposition of the element h corresponding to (4.1). The operators Π_+, Π_- are called the fundamental projectors belonging to the decomposition (4.1). The operator $K = \Pi_+ - \Pi_-$ is called the fundamental symmetry corresponding to the decomposition (4.1). Then K is symmetric, completely invertible, isometric, and $K^{-1} = K$. Furthermore, the K-inner product $(h,h')_K = (Kh,h')$ for h, $h' \in E$ converts E into a Hilbert space. We shall assume that this Hilbert space is separable and infinite dimensional. Then norm $|\cdot|_K$ is called a natural norm on E.

A Krein space E can have more than one fundamental decomposition. However, the three operators Π_+, Π_-, and K are uniquely associated with each fundamental decomposition of E. It is known that the natural norms $|\cdot|_K$ corresponding to all fundamental symmetries K are equivalent and the topology they define is called the strong topology of E.

We shall denote by $\mathcal{M}(E)$ the family of all finite complex measures μ on the Borel sets of the strong topology of E. Let us now fix a fundamental decomposition of E and let K be the associated fundamental symmetry. Then the relations $K = \Pi_+ - \Pi_-$ and $\Pi_+ + \Pi_- = I = K^2$ show that K uniquely determines Π_+ and Π_-. Let $\|\cdot\|$ be a measurable norm on E regarded as a Hilbert space under the norm $|\cdot|_K$. Let (B,P) be the abstract Wiener space corresponding to $\|\cdot\|$. The analytic

Feynman integral of a functional $F(x_1,x_2)$ on $B \times B$ is given in Definition 3.1. We shall illustrate the role of the indefinite metric space in the definition of the Feynman integral by considering the problem for the class $\mathcal{F}_K$ of functionals to be defined below. This is done in very much the same manner as in Sec. 3, where $\mathcal{F}_A$ was defined. Observe that the definitions of $(h,x)^\sim$ for $h \in E$ and of s-equivalence classes are exactly the same as before, the Hilbert space H being E under the norm $|\cdot|_K$.

Let $\mathcal{F}_K$ be the Banach algebra (under pointwise addition, complex scalar multiplication, and pointwise multiplication of elements) of all s-equivalence classes of functionals F of the form

$$F(x_1,x_2) = \int_E e^{i(\Pi_+ h,x_1)^\sim - i(\Pi_- h,x_2)^\sim} d\mu(h) \tag{4.2}$$

where

$$\mu \in \mathcal{M}(E) \tag{4.3}$$

PROPOSITION 4.1 The map

$$\mu \mapsto [F] \tag{4.4}$$

where F is defined by (4.2) is an algebra isomorphism between $\mathcal{M}(E)$ and $\mathcal{F}_K$. Moreover, $\mathcal{F}_K$ becomes a Banach algebra under the norm $\|F\| = \|\mu\|$, so that (4.4) is an isomorphism of Banach algebras.

Proof. The result follows from Proposition 3.2 if we set $A_1^{1/2} = \Pi_+$ and $A_2^{1/2} = -\Pi_-$. The verification that $\overline{\mathcal{R}(A_1 + A_2)} = E$ is immediate since every h in E can be expressed as $h = \Pi_+ h + \Pi_- h = A_1 h + A_2 h$.

The following result can now be deduced from Proposition 3.1. For the sake of brevity, we confine our attention to the Feynman integral.

PROPOSITION 4.2 The analytic Feynman integral over $B \times B$ with the parameter $q \in \mathbb{R}$ ($q_1 \neq 0$, $q_2 \neq 0$) exists for every

$F \in F_K$ and is given by

$$I_2^q(F) = \int_E e^{-(i/2)[q_1^{-1}(\Pi_+h,h)-q_2^{-1}(\Pi_-h,h)]}\, d\mu(h) \qquad (4.5)$$

$$I_2^{(1,-1)}(F) = \int_E e^{-(i/2)(h,h)}\, d\mu(h) \qquad (4.6)$$

Proof. Equations (4.5) and (4.6) are direct consequences of (3.5) and (3.6). To obtain (4.6), note that if $q_1 = 1$ and $q_2 = 1$,

$$(\Pi_+h,h) + (\Pi_-h,h) = ([\Pi_+ + \Pi_-]h,h) = (h,h)$$

REMARK 1. The expression for $I_2^{(1,-1)}(F)$ is of particular interest in that the right side of (4.6) involves only the measure μ in $M(E)$ and the given (indefinite) inner product in E. However, the class F_K depends on the choice of the fundamental symmetry K. Suppose that R is another fundamental symmetry for E. Since E is also a Hilbert space under the norm $|\cdot|_R$ it follows that the abstract Wiener space obtained is the same Banach space B as before and the abstract Wiener measure on it is Q with its covariance function determined by $(\cdot,\cdot)_R$ instead of $(\cdot,\cdot)_K$. The Banach algebra F_R is defined on (B,Q) when $[F] \in F_R$ is defined through (4.2), where Π_+ and Π_- refer to the fundamental projectors of the decomposition corresponding to R. The Feynman integral of F in this instance is also given by (4.5).

Analytic Feynman integrals for more general classes of functionals in the spirit of Secs. 3 and 5 can be defined in the indefinite metric space framework. We shall not pursue this matter further.

5. ANALYTIC WIENER AND EXTENDED FEYNMAN INTEGRALS

As in Secs. 2 and 3, (i,H,B) will denote an abstract Wiener space and P will be the probability on B induced by the weak

Gaussian distribution on H. The purpose of this section is to develop a definition of analytic Wiener integrals general enough to admit the expression

$$\int_H e^{-(i/2)[(D_1h,h)+i(D_2h,h)]}\, d\mu(h)$$

as an analytic Feynman integral when D_1 and D_2 are self-adjoint operators and $D_2 \leq 0$. Notice that this expression coincides with (3.7) when $D_2 = 0$. The example presented at the end of this section will show how this definition can be applied to the representation of time-invariant quasi-free states of the quantum mechanical harmonic oscillator given by Albeverio and Høegh-Krohn (sec. 8 of Ref. 1).

The definition to be given here will follow the general form described in Sec. 2. The number of complex variables involved will be k = 4. The first step is to select a domain $\Omega \subseteq \mathbb{C}^4$. It would seem natural to take Ω to be the region $\Omega_1 = \{(z_1, \cdots, z_4) \mid \mathrm{Re}(z_m) > 0 \text{ for all } 1 \leq m \leq 4\}$ in order to produce a theory analogous to that of Secs. 2 and 3. However, the application which will be presented below as well as our definition of the expression

$$\int_H e^{-(i/2)[(D_1h,h)+i(D_2h,h)]}\, d\mu(h)$$

as an analytic Feynman integral will require a larger domain permitting some of the variables z_m to have negative real parts. The principal constraint on our choice of Ω is that the function $J_F(\lambda_1,\lambda_2,\lambda_3,\lambda_4)$ (set k = 4 in the general discussion of Sec. 2) should have an analytic extension on all of Ω for each functional F that is of interest. In this section this requirement will be met by ensuring that

$$\sum_{m=1}^{4} \mathrm{Re}(z_m^{-1})A_m \geq 0$$

where A_m, $1 \leq m \leq 4$, are four bounded nonnegative self-adjoint operators on H, analogous to the two operators A_1 and A_2 of

Sec. 3. Another candidate for Ω is therefore the set

$$\Omega_2 = \{(z_1,z_2,z_3,z_4) \in \mathbb{C}^4 : \sum_1^4 \mathrm{Re}(z_m^{-1})A_m \geq 0\}$$

The problem here is that Ω_2, its interior, and its boundary, depend in a complicated way upon the interplay among the operators A_m. To avoid such difficulties we shall throughout this section impose the following conditions on the four bounded nonnegative self-adjoint operators A_m, $1 \leq m \leq 4$:

$$A_1 + A_2 \geq A_3 + A_4 \tag{5.1}$$

Let us now define

$$\Omega = \{(z_1,z_2,z_3,z_4) \in \mathbb{C}^4 : \mathrm{Re}(z_1) \wedge \mathrm{Re}(z_2) > 0, \; z_3 \neq 0, \; z_4 \neq 0 \text{ and } \mathrm{Re}(z_1^{-1}) \wedge \mathrm{Re}(z_2^{-1}) > -[\mathrm{Re}(z_3^{-1}) \wedge \mathrm{Re}(z_4^{-1})]\}$$

REMARK 1. (a) Clearly, $\Omega_1 \subsetneq \Omega$. On the other hand, for any $(z_1,z_2,z_3,z_4) \in \Omega$ there is a positive real number α such that

$$\mathrm{Re}(z_1^{-1}) \wedge \mathrm{Re}(z_2^{-1}) > \alpha > -[\mathrm{Re}(z_3^{-1}) \wedge \mathrm{Re}(z_4^{-1})]$$

It follows that $\Sigma_1^4 \mathrm{Re}(z_m^{-1})A_m \geq \alpha(A_1 + A_2 - A_3 - A_4) \geq 0$ according to (5.1). Hence $\Omega \subseteq \Omega_2$. Furthermore, this definition of Ω does not depend on the precise nature of the operators A_m, $1 \leq m \leq 4$.

(b) For future reference we also note here that (1) Ω is open connected, (2) any point of the form iq, $q \in \mathbb{R}^4$, belongs to the boundary of Ω (written $\partial\Omega$), and (3) any point of the form $(\zeta,\bar{\zeta},-\zeta,-\bar{\zeta})$, $\mathrm{Re}(\zeta) > 0$ belongs to $\partial\Omega$. The last assertion follows because this point does not belong to Ω but is the limit of some $(\zeta, \bar{\zeta}, -\zeta + \delta, -\zeta + \delta)$ as δ tends to zero.

For convenience in writing, $z = (z_1,z_2,z_3,z_4)$ will indicate a vector in $\mathbb{C}^4$, $\lambda = (\lambda_1,\lambda_2,\lambda_3,\lambda_4)$ will indicate a vector in $\mathbb{R}^4$, and Λ will denote the set of $\lambda \in \mathbb{R}^4$ having all coordinates $\lambda_m > 0$, $1 \leq m \leq 4$.

DEFINITION 5.1 Let F be a measurable complex-valued function on $B^4 = B \times B \times B \times B$. Suppose that the following conditions are satisfied:

(i) $J_F(\lambda) = \int_{B^4} F(\lambda_1^{-1/2}x_1, \lambda_2^{-1/2}x_2, \lambda_3^{-1/2}x_3, \lambda_4^{-1/2}x_4)$ $\times\, dP^4(x_1, x_2, x_3, x_4)$ exists for each vector $\lambda \in \Lambda$.

(ii) There is an analytic function $J_F^*(z)$, $z \in \Omega$, such that $J_F^*(\lambda) = J_F(\lambda)$ for $\lambda \in \Lambda$.

Then for any $z \in \Omega$ we shall call $I_4^z(F) = J_F^*(z)$ the analytic Wiener integral of F over B^4 with parameter z relative to Ω.

If $z \in \partial\Omega$ and $\lim_{z' \to z, z' \in \Omega} I_4^{z'}(F)$ exists, we shall denote the value of this limit by the symbol $I_4^q(F)$ when $z = (-iq_1, -iq_2, -iq_3, -iq_4)$, $q \in \mathbb{R}^4 \{0\}$, or by the symbol $I_4^z(F)$ when z is not of this form. In this case $I_4^q(F)$ will be called the analytic Feynman integral of F over B^4 with parameter q, whereas for $z \in \partial\Omega$ not of the form $(-iq_1, -iq_2, -iq_3, -iq_4)$ we shall call $I_4^z(F)$ the extended analytic Feynman integral of F over B^4 with complex parameter z relative to Ω.

REMARK 2. In Definitions 2.1, 3.1, and 5.1, the term "analytic Wiener integral" is used when the complex parameter z belongs to Ω. We have also followed the terminology of Refs. 3 and 5 in referring to $I^q(F)$ and $I_2^q(F)$ as "analytic Feynman integrals," corresponding to points $z \in \partial\Omega$ having all components purely imaginary. In the present section we will be particularly interested in parameters $z \in \partial\Omega$ which are not of this form. Since this situation does not appear to have been encountered before in the literature on the analytic continuation approach, we have adopted the term "extended analytic Feynman integral" to describe this case.

We proceed now just as in Secs. 2 and 3. Two functionals F and G on B^4 into $\mathbb{C}$ are called s-equivalent if

$$P^4(\{(x_1,x_2,x_3,x_4) \in B^4 : F(\alpha_1x_1,\alpha_2x_2,\alpha_3x_3,\alpha_4x_4)$$
$$\neq G(\alpha_1x_1,\alpha_2x_2,\alpha_3x_3,\alpha_4x_4)\}) = 0$$

for all $\alpha_m > 0$, $1 \leq m \leq 4$.

DEFINITION 5.2 Given any four bounded nonnegative self-adjoint operators A_m, $1 \leq m \leq 4$ on H, define $\mathcal{F}_{A_1,A_2,A_3,A_4}$ to be the space of s-equivalence classes of functionals

$$F(x_1,x_2,x_3,x_4) = \int_H e^{i\Sigma_1^4(A_m^{1/2}h,\lambda_m^{-1/2}x_m)^\sim}\, d\mu \tag{5.2}$$

where $\mu \in \mathcal{M}(H)$. For convenience we shall usually denote $\mathcal{F}_{A_1,A_2,A_3,A_4}$ by $\mathcal{F}_A$. When necessary [F] will represent the s-equivalence class determined by (5.2).

PROPOSITION 5.1 The space $\mathcal{F}_A$ is an algebra over the complex numbers under addition, multiplication, and scalar multiplication of s-equivalence classes. The map $\mu \mapsto [F]$ defined by equation (5.2) is a homomorphism of the complex algebra $\mathcal{M}(H)$ onto $\mathcal{F}_A$ which is one to one if and only if the range of $\Sigma_1^4 A_m$ is dense in H. In this case equation (5.2) establishes an isomorphism of Banach algebras, where $\|F\|$ is defined as $\|\mu\|$.

Proof. According to Remark 3 in Sec. 3, the range $\mathcal{R}(\Sigma_1^4 A_m)$ is dense if and only if the space $\Sigma_1^4 \mathcal{R}(A_m^{1/2})$ is dense. Then the one-to-one property is verified just as in Proposition 3.2. Either method of proof can be used. The other assertions are checked as in Proposition 3.2.

Until this point no use has been made of the condition (5.1) either in the definition of Ω or in the definition of $\mathcal{F}_A$ and the proposition above. We have to impose (5.1) in order to guarantee the existence of the integrals $I_4^z(F)$ when $z \in \Omega$ and $F \in \mathcal{F}_A$.

PROPOSITION 5.2 Let A_m, $1 \leq m \leq 4$, be bounded nonnegative self-adjoint operators on H satisfying condition (5.1) and let F be any functional in F_A.

(i) The analytic Wiener integral of F over B^4 with parameter $z \in \Omega$ exists and has the form

$$I_4^z(F) = \int_H e^{-(1/2)\Sigma_1^4 z_m^{-1}(A_m h,h)} d\mu(h) \tag{5.3}$$

where $\mu \in M(H)$ corresponds to F under (5.2).

(ii) The analytic Feynman integral F over B^4 exists for any parameter $q \in \mathbb{R}^4$ such that $q_m \neq 0$ for all $1 \leq m \leq 4$ and has the form

$$I_4^q(F) = \int_H e^{-(i/2)((\Sigma_1^4 q_m^{-1} A_m)h,h)} d\mu(h) \tag{5.4}$$

(iii) The extended analytic Feynman integral of F over B^4 exists for any complex parameter $z \in \partial\Omega$ provided that $z_m \neq 0$, $1 \leq m \leq 4$, and has the form (5.3).

These integrals depend only on the s-equivalence class of F in F_A and do not depend on the equivalent extension $(h,x)^\sim$ used in (5.2).

Proof. Using Fubini's theorem we can show for any $\lambda \in \Lambda$,

$$J_F(\lambda) = \int_H e^{-(1/2)\Sigma_1^4 \lambda_m^{-1}(A_m h,h)} d\mu(h)$$

Because $\Omega \subseteq \Omega_2 = \{z \in \mathbb{C}^4 : \Sigma_1^4 z_m^{-1} A_m \geq 0\}$ it follows that for $z \in \Omega$,

$$\left| e^{-(1/2)\Sigma_1^4 z_m^{-1}(A_m h,h)} \right| \leq 1 \tag{5.5}$$

Hence the function

$$J_F^*(z) = \int_H e^{-(1/2)\Sigma_1^4 z_m^{-1}(A_m h,h)} d\mu(h)$$

exists on the set $\bar{\Omega} \cap \{z \in \mathbb{C}^4 : z_m \neq 0, 1 \leq m \leq 4\}$. Continuity of J_F^* on this set follows by the dominated convergence theorem.

The easiest way to deduce the analyticity of $J_F^* = J_F^*|_\Omega$

is to use Morera's theorem. Fix $z^0 = (z_1^0, z_2^0, z_2^0, z_4^0) \in \Omega$ and let $\varepsilon > 0$ be chosen so that $\{z : |z_m - z_m^0| < \varepsilon,\ m = 1,\ldots,4\}$. Hold z_2^0, z_3^0, and z_4^0 fixed for the moment. Let $D = \{z_1 \in \mathbb{C} : |z_1 - z_1^0 < \varepsilon\}$ and $G_h(z_1) = e^{-(1/2z_1)(A_1h,h)}$ for each $h \in H$ and $z_1 \in D$. Every function G_h is analytic on D, so for any rectifiable closed curve C in D, $\int_C G_h(z_1)\, dz_1 = 0$. Letting $z_1(t)$ be any parameterization of C with $\int_{t_0}^{t_1} |z_1'(t)|\, dt < \infty$, Fubini's theorem gives

$$\int_C J_F^*(z_1, z_2^0, z_3^0, z_4^0)\, dz_1 = \int_{t_0}^{t_1} \int_H e^{-(1/2)\Sigma_2^4 (z_m^0)^{-1}(A_m h,h)} \times G_h(z_1(t)) z_1'(t)\, d\mu(h)\, dt$$

$$= \int_H [\int_C G_h(z_1)\, dz_1] \times e^{-(1/2)\Sigma_2^4 (z_m^0)^{-1}(A_m h,h)}\, d\mu(h)$$

$$= 0$$

where we have used again the bound (5.5). By Morera's theorem it follows that $J_F^*(z_1, z_2^0, z_3^0, z_4^0)$ is analytic on the disk $z_1 \in D$. Similar arguments work when the other sets of three variables are held fixed. Appealing once again to Theorem A1 of the Appendix, we see that J_F^* is analytic as a function of all four variables at any point of Ω.

Let us now see how Proposition 5.2 leads to extended Feynman integrals of the form

$$\int_H e^{-(i/2)[(D_1h,h)+i(D_2h,h)]}\, d\mu(h) \tag{5.6}$$

where D_1 and D_2 are bounded self-adjoint operators on H such that $D_2 \leq 0$.

Define

$$A_1 = \frac{1}{2}(D_1 - D_2)^+$$

$$A_2 = \frac{1}{2}(D_1 + D_2)^-$$

$$A_3 = \frac{1}{2}(D_1 - D_2)^-$$
$$A_4 = \frac{1}{2}(D_1 + D_2)^+$$

Then each A_m is a bounded nonnegative self-adjoint operator and $D_1 = A_1 - A_2 - A_3 + A_4$ and $-D_2 = (A_1 + A_2) - (A_3 + A_4)$. Hence $D_2 \leq 0$ implies condition (5.1). Take $F \in \mathcal{F}_{A_1,A_2,A_3,A_4}$ of the form (5.2) and fix $z^0 = (\zeta, \bar{\zeta}, -\zeta, -\bar{\zeta})$ where $\zeta = 1/(1 + i)$. Then $z^0 \in \partial\Omega$ [see Remark 1(b)] and the evaluation of (5.3) for the extended analytic Feynman integral of F over B^4 with complex parameter z^0 becomes

$$I_4^{z^0}(F) = J_F^*(z^0) = \int_H e^{-(i/2)[(D_1h,h)+i(D_2h,h)]} d\mu(h) \tag{5.8}$$

From another point of view, suppose that D denotes a bounded linear operator on the complexification $H^{\mathbb{C}} = H \oplus iH$ of H. (The norm on $H^{\mathbb{C}}$ is defined by $|h_1 \oplus ih_2| = [|h_1|^2 + |h_2|^2]^{1/2}$.) Suppose that D is dissipative, i.e., $\mathrm{Re}(Dh,h) \leq 0$ for all $h \in H^{\mathbb{C}}$. Writing $\mathrm{Re}(D) = (D + D^*)/2$ and $\mathrm{Im}(D) = (D - D^*)/2i$, we have

$$D = \mathrm{Re}(D) + i\ \mathrm{Im}(D)$$

Setting $D_2 = \mathrm{Re}(D)$ and $D_1 = -\mathrm{Im}(D)$, it follows that $D_2 \leq 0$ and (5.8) can be written

$$I_4^{z^0}(F) = \int_{H^{\mathbb{C}}} e^{(1/2)(Dh,h)}\ d\mu(h)$$

where z^0, F, and μ are as above, μ being concentrated on H.

We now discuss in some detail a physical problem involving the quantum mechanical harmonic oscillator with infinitely many degrees of freedom. The solution of this problem given in theorem 8.2 of Ref. 1 will be exhibited here as a limit of extended analytic Feynman integrals defined above. We will be concerned only with the mathematical problem, referring the reader to the discussion in Ref. 1 and the literature cited there for the physical interpretation.

EXAMPLE 1 [Time-Invariant Quasi-free States of the Harmonic Oscillator (Infinitely Many Degrees of Freedom)].

Let ν be a positive Borel measure on $\mathbb{R}^+ = (0,\infty)$ and let H be the real Hilbert space of functions $\alpha\colon \mathbb{R}\times\mathbb{R}^+ \to \mathbb{R}$ such that

$$\int\int_{\mathbb{R}\times\mathbb{R}^+} [(\frac{\partial\alpha}{\partial t})^2(t,\omega) + \omega^2\alpha^2(t,\omega)]\, dt\, d\nu(\omega) < \infty$$

This implies that for a.a. $\omega[\nu]$, $\alpha(t,\omega)$ is square integrable in the t variable and has a derivative $\frac{\partial\alpha}{\partial t}(t,\omega)$ at a.e. t which is also square integrable on $\mathbb{R}$. For α_1 and α_2 in H, the inner product is

$$(\alpha_1,\alpha_2)_H = \int\int_{\mathbb{R}\times\mathbb{R}^+} \left[\left(\frac{\partial\alpha_1}{\partial t}\cdot\frac{\partial\alpha_2}{\partial t}\right)(t,\omega) + \omega^2\alpha_1(t,\omega)\alpha_2(t,\omega)\right] \times dt\, d\nu(\omega) < \infty \qquad (5.9)$$

The Plancherel transform in the t variable,

$$\hat{\alpha}(p,\omega) = \frac{1}{\sqrt{2\pi}}\int_{\mathbb{R}} e^{ipt}\, \alpha(t,\omega)\, dt$$

then gives an isometric isomorphism of H onto the real subspace $\hat{H}$ of functions $\hat{\alpha}(p,\omega)$ in the complex Hilbert space $L^2(\mathbb{R}\times\mathbb{R}^+, (p^2+\omega^2)\, dp\, d\nu(\omega))$ satisfying $\hat{\alpha}(p,\omega) = \hat{\alpha}(-p,\omega)$. We shall write the inner product in H as

$$(\hat{\alpha}_1,\hat{\alpha}_2)_{\hat{H}} = \int\int_{\mathbb{R}\times\mathbb{R}^+} \alpha_1(t,\omega)\,\overline{\alpha_2(t,\omega)}(p^2+\omega^2)\, dp\, d\nu(\omega) \qquad (5.11)$$

The inner product on $L^2(\mathbb{R}\times\mathbb{R}^+, (p^2+\omega^2)\, dp\, d\nu(\omega))$ will be written $(\cdot,\cdot)_{L^2(\mathbb{R}\times\mathbb{R}^+)}$.

Let $t_1 \le t_2 \le \cdots \le t_n$ be real numbers and $u_j(\omega)$, $1 \le j \le n$, be real-valued functions for $\omega \in \mathbb{R}^+$ satisfying

$$\int_{\mathbb{R}^+} [u_j^2(\omega) + \frac{1}{\omega}\, u_j^2(\omega)]\, d\nu(\omega) < \infty \qquad (5.12)$$

for each j. Define

$$\hat{\alpha}_{u_j}^{t_j}(p,\omega) = \frac{1}{\sqrt{2\pi}}\,\frac{e^{ipt_j}}{p^2+\omega^2}\, u_j(\omega)$$

Since u_j is real and satisfies (5.12), it follows that each

$\hat{\alpha}_{u_j}^{t_j}$ belongs to $\hat{H}$ and hence corresponds to a unique element $\alpha_{u_j}^{t_j}$ in H under the inversion of (5.10).

Let $C(\omega)$ be a bounded measurable function on $\mathbb{R}^+$ such that $C(\omega) \geq 1$ for almost all $\omega[\nu]$. In theorem 8.2 of Ref. 1, the authors show that there is a one-to-one correspondence between such functions C and the time-invariant quasi-free states of the quantum mechanical harmonic oscillator. The state corresponding to C is determined by expressions of the form

$$\exp\left(-\frac{i}{2}\right)\left[\sum_{j,k=1}^{n} \int_{\mathbb{R}^+} G_C^{(t_j-t_k)}(\omega)u_j(\omega)u_k(\omega)\, d\nu(\omega)\right] \quad (5.13)$$

where t_j and u_j are as described above for $1 \leq j \leq n$ and

$$G_C^s(\omega) = -\left(\frac{1}{2\omega}\right)[\sin|s\omega| + iC(\omega)\cos(s\omega)]$$

for any real number s. The integrals in (5.13) exist because of (5.12). Using an extension of their definition of the Fresnel integral, called the Fresnel integral normalized with respect to a bilinear form, Albeverio and Høegh-Krohn then succeed in writing these expressions (5.13) in the form

$$F_{\Delta_C}(f) = \int_H^{\Delta_C} e^{iS(\alpha)} f(\alpha)\, d\alpha \quad (5.14)$$

where f is the Fourier transform of the probability measure μ_0 on H which has mass one at the point $\Sigma_{j=1}^n \alpha_{u_j}^{t_j}$ and $\Delta_C(\alpha_1,\alpha_2)$ is a bilinear form belonging to a certain dense linear subspace of H. For the definition of F_Δ, the normalized Fresnel integral, we refer to sec. 4 of Ref. 1. For our purposes, it suffices to note here that what Albeverio and Høegh-Krohn show is that the functions $\alpha_{u_j}^{t_j}$ defined above belong to the domain of Δ_C and that

$$\Delta_C\left(\alpha_{u_j}^{t_j}, \alpha_{u_k}^{t_k}\right) = \lim_{\varepsilon \downarrow 0}\left(\hat{D}^{\varepsilon}\alpha_{u_j}^{t_j}, \alpha_{u_k}^{t_k}\right)_{L^2(\mathbb{R}\times\mathbb{R}^+)}$$

$$= \int_{\mathbb{R}^+} G_C^{(t_j - t_k)}(\omega) u_j(\omega) u_k(\omega)\, d\nu(\omega) \qquad (5.15)$$

where $\hat{D}^{\varepsilon}$ is the bounded linear operator on $L^2(\mathbb{R} \times \mathbb{R}^+, (p^2 + \omega^2)\, dp\, d\nu(\omega))$ defined as multiplication by the function

$$\hat{D}^{\varepsilon}(p,\omega) = \frac{p^2 + \omega^2}{p^2 - \omega^2}\left[\operatorname{Re}\left(\frac{p^2 - \omega^2}{(p^2 - \omega^2) + i\varepsilon}\right) + iC(\omega)\operatorname{Im}\left(\frac{p^2 - \omega^2}{(p^2 - \omega^2) + i\varepsilon}\right) I_{(0,\varepsilon^{-1})}(\omega)\right]$$

[The factor $I_{(0,\varepsilon^{-1})}(\omega)$ is introduced here to make $\hat{D}^{\varepsilon}(p,\omega)$ a bounded function of (p,ω). This alteration does not affect (5.15) (see equations (8.36) and (8.42) of Ref. 1)].

Our goal now is to rewrite the expressions (5.13) in terms of the extended analytic Feynman integral introduced in this section. First, let $\hat{D}_1^{\varepsilon}(p,\omega) = \operatorname{Re}(\hat{D}^{\varepsilon}(p,\omega))$ and $\hat{D}_2^{\varepsilon}(p,\omega) = \operatorname{Im}(\hat{D}^{\varepsilon}(p,\omega))$. Note that these are both real and even functions in the variable p. Furthermore, $\hat{D}_2^{\varepsilon}(p,\omega) \leq 0$. It follows that multiplication by $\hat{D}_1^{\varepsilon}(p,\omega)$ and $\hat{D}_2^{\varepsilon}(p,\omega)$ produces two self-adjoint bounded linear operators on $\hat{H}$, the image of H under the Plancherel transform (5.10). Call these two operators $\hat{D}_1^{\varepsilon}$ and $\hat{D}_2^{\varepsilon}$ and denote the corresponding operators acting on H via (5.10) by D_1^{ε} and D_2^{ε}. In this way each operator $\hat{D}^{\varepsilon}$ gives rise to two self-adjoint bounded linear operators D_1^{ε} and D_2^{ε} on H having $D_2^{\varepsilon} \leq 0$.

We are now in a position to apply equation (5.8). For each $\varepsilon > 0$, let A_m^{ε}, $1 \leq m \leq 4$, be as in (5.7) where D_1 and D_2 are to be replaced by D_1^{ε} and D_2^{ε}. Let $F_{A^{\varepsilon}} = F_{A_1^{\varepsilon},A_2^{\varepsilon},A_3^{\varepsilon},A_4^{\varepsilon}}$ be the corresponding algebra. Take μ_0 to be the same complex measure on H as used by Albeverio and Høegh-Krohn, i.e., the probability measure concentrated at the point $\alpha_0 = \Sigma_{j=1}^{n} \alpha_{u_j}^{t_j}$. Fix a measurable norm on H and let (i,H,B) be the resulting abstract Wiener space. For $(x_1,x_2,x_3,x_4) \in B^4$ define

$$F_\varepsilon^C(x_1,x_2,x_3,x_4) = \int_H e^{i\Sigma_1^4((A_m^\varepsilon)^{1/2}\alpha,x_m)_H^{\sim}} d\mu_0(\alpha)$$

Then for each $\varepsilon > 0$, F_ε^C belongs to F_{A^ε} and for the parameter $z^0 = (\zeta,\bar{\zeta},-\zeta,-\bar{\zeta})$, $\zeta = 1/(1+i)$, we have by equation (5.8),

$$I_4^{z^0}(F_\varepsilon^C) = \int_H e^{-(i/2)[(D_1^\varepsilon\alpha,\alpha)+i(D_2^\varepsilon\alpha,\alpha)_H]} d\mu_0(\alpha)$$

$$= \exp\left(-\frac{1}{2}\right)[(D_1^\varepsilon\alpha_0,\alpha_0)_H + i(D_2^\varepsilon\alpha_0,\alpha_0)_H] \qquad (5.16)$$

To evaluate the limit of this expression as ε tends to zero, write

$$(D_1^\varepsilon\alpha_0,\alpha_0)_H + i(D_2^\varepsilon\alpha_0,\alpha_0)_H = (\hat{D}_1^\varepsilon\hat{\alpha}_0,\hat{\alpha}_0)_{\hat{H}} + i(\hat{D}_2^\varepsilon\hat{\alpha}_0,\hat{\alpha}_0)_{\hat{H}}$$

$$= ((\hat{D}_1^\varepsilon + i\hat{D}_2^\varepsilon)\hat{\alpha}_0,\hat{\alpha}_0)_{L^2(\mathbb{R}\times\mathbb{R}^+)}$$

$$= \sum_{j,k=1}^{n} (\hat{D}^\varepsilon\hat{\alpha}_{u_j}^{t_j},\hat{\alpha}_{u_k}^{t_k})_{L^2(\mathbb{R}\times\mathbb{R}^+)}$$

It follows from (5.15) that

$$\lim_{\varepsilon\downarrow 0} I^{z^0}(F_\varepsilon^C) = \exp\left(-\frac{i}{2}\right) \sum_{j,k=1}^{n} \int_{\mathbb{R}^+} G_C^{(t_j-t_k)}(\omega)$$

$$\times\ u_j(\omega)u_k(\omega)\, d\nu(\omega) \qquad (5.17)$$

The expressions (5.13) which determine the time-invariant quasi-free state associated with the function $C(\omega)$ are thus obtained in this way as limits of sequences of extended analytic Feynman integrals of Definition 5.1.

Since both (5.14) and (5.17) are two representations for (5.13), we have shown that

$$\lim_{\varepsilon \downarrow 0} I^{z^0}(F_\varepsilon^C) = F_{\Delta_C}(f)$$

The extended Fresnel integral appears to put the limiting operation inside the integral. The disadvantage of doing this is that the definition of the integral used, F_{Δ_C}, then depends on the function $C(\omega)$.

REMARK 3. The passage from (5.16) to (5.17) requires the second equality in (5.15). This can be verified directly without introducing the bilinear form Δ_C as follows. Write

$$(D^\varepsilon \alpha_{u_j}^{t_j}, \alpha_{u_k}^{t_k})_{L^2(\mathbb{R} \times \mathbb{R}^+)} = \int_0^\infty \int_{-\infty}^\infty \hat{D}^\varepsilon(p,\omega) \frac{1}{2\pi} e^{ip(t_j - t_k)} \times (p^2 + \omega^2)^{-1} u_j(\omega) u_k(\omega)\, dp\, d\nu(\omega)$$

$$= \int_0^\infty u_j(\omega) u_k(\omega) \times \left[\frac{1}{2\pi} \int_{\mathbb{R}} e^{ip(t_j - t_k)} \hat{D}^\varepsilon(p,\omega)\, dp\right] d\nu(\omega)$$

Directly computing the Fourier transform of $\hat{D}^\varepsilon(p,\omega)$ we can show that

$$\lim_{\varepsilon \downarrow 0} \frac{1}{2\pi} \int_{\mathbb{R}} e^{ip(t_j - t_k)} \hat{D}^\varepsilon(p,\omega)\, dp = -\left(\frac{1}{2\omega}\right)[\sin |t_j - t_k|\omega + iC(\omega) \cos (t_j - t_k)\omega]$$

for each $\omega > 0$ and that there are positive constants c_1 and c_2 such that

$$\left|\frac{1}{2\pi} \int_{\mathbb{R}} e^{ip(t_j - t_k)} \hat{D}^\varepsilon(p,\omega)\, dp\right| \le c_1 + c_2\omega$$

for all $\varepsilon > 0$ and $\omega > 0$. By (5.12) and the dominated convergence theorem it follows that

$$\lim_{\varepsilon \downarrow 0} (\hat{D}^\varepsilon \hat{\alpha}_{u_j}^{t_j}, \hat{\alpha}_{u_k}^{t_k})_{L^2(\mathbb{R} \times \mathbb{R}^+)} = \int_{\mathbb{R}^+} G_C^{(t_j - t_k)}(\omega) u_j(\omega) u_k(\omega)\, d\nu(\omega)$$

The other conditions that Albeverio and Høegh-Krohn assume

on u_j in order that $\alpha_{u_j}^{t_j}$ belong to the domain of Δ_C, namely continuity and boundedness, are not needed for this argument to go through. The class of functions u_j for which (5.17) holds is hence a bit wider.

Note

After completing this work, we saw the paper "The Equivalence of Two Approaches to the Feynman Integral" by G. W. Johnson, who has kindly sent us a preprint. Theorems 3 and 2 of sec. 2 of his paper are contained in Proposition 2.3 here. Propositions 3.2, 3.3, and 5.1 are further generalizations of Proposition 2.3 which cover the more general definitions of the analytic Feynman integral given here.

Acknowledgments

This work was begun when the first-named author was at Nagoya University in the summer of 1980. Some of the results were reported at the Conference on Measure Theory held at Oberwolfach in June 1981. The research was supported by AFOSR Grant 80-0080. This article was issued in October 1981 as a Technical Report of the Center for Stochastic Processes, University of North Carolina at Chapel Hill.

APPENDIX

THEOREM A1. Let $\Omega \subseteq \mathbb{C}^k$ be open. Suppose that $g: \Omega \to \mathbb{C}^k$ is continuous in Ω and analytic in each variable separately. That is, for each m, $i \leq m \leq k$, and each point $(z_1, z_2, \ldots, z_{m-1}, z_{m+1}, \ldots, z_k) \in \mathbb{C}^{k-1}$ such that $D = \{z_m \in \mathbb{C}: (z_1, \ldots, z_m, \ldots, z_k) \in \Omega\}$ is nonempty, the function $f(z_m) = g(z_1, \ldots, z_m, \ldots, z_k)$ is analytic in D. Then g is analytic as a function of k complex variables in Ω. If Ω is connected and contains the set $\Omega^+ = \{(z_1, \ldots, z_k): \mathrm{Re}(z_m) > 0, 1 \leq m \leq k\}$, then g is uniquely determined by its restriction to Ω^+.

For proofs, see Dieudonné [4], theorems 9.9.4 and 9.4.4.

REFERENCES

1. S. A. Albeverio and R. J. Høegh-Krohn, Mathematical Theory of Feynman Path Integrals, Lect. Notes Math. 523, Springer-Verlag, Berlin, 1976.
2. J. Bognár, Indefinite Inner Product Spaces, Ergebnisse der Mathematik und ihrer Grenzgebiete, Vol. 78, Springer-Verlag, Berlin, 1974.
3. R. H. Cameron and D. A. Storvick, Some Banach algebras of analytic Feynman integrable functionals, pp. 18-67 in Analytic Functions, Kozubnik, 1979, Lect. Notes Math. 798, Springer-Verlag, Berlin, 1980.
4. J. Dieudonné, Foundations of Modern Analysis, Academic Press, New York, 1960.
5. G. Kallianpur, A generalized Cameron-Feynman integral, in Essays in Honor of C. R. Rao, (G. Kallianpur et al., eds.), North-Holland, Amsterdam, 1981.
6. H-H. Kuo, Gaussian Measures in Banach Spaces, Lect. Notes Math. 463, Springer-Verlag, Berlin, 1975.
7. R. E. A. C. Paley, N. Wiener, and A. Zygmund, Notes on random functions, Math. Z. 33:647-668 (1933).
8. W. Pauli, On Dirac's new method of field quantization, Rev. Mod. Phys. 15:175-207 (1943).

9

Stochastic Differential Equations and Stochastic Flows of Homeomorphisms

HIROSHI KUNITA/Faculty of Engineering, Kyushu University, Fukuoka, Japan

1. INTRODUCTION

This chapter concerns the one-to-one correspondence between stochastic differential equations and stochastic flows of homeomorphisms. Recently, a lot of attention has been given to the problem of constructing stochastic flows of homeomorphisms generated by given stochastic differential equations (see Elworthy [3], Baxendale [1], Malliavin [10], Ikeda and Watanabe [5], Bismut [2], Kunita [7,8], etc.). Reference 8 gives a necessary and sufficient condition for a Stratonovich stochastic differential equation to generate a stochastic flow, assuming that the coefficients of the equation are of C^5-class.

In this chapter we relax the smoothness condition on the coefficients and consider the equation under the C^2-condition. Theorem 2.2 in Sec. 2 is a principal result. Assuming that the Stratonovich equation on a manifold

$$d\xi_t = \sum_{j=1}^{m} X_j(t,\xi_t)\circ dB_t^j + X_0(t,\xi_t)\,dt$$

is strongly complete (no explosion occurs whenever the initial state is), it generates a flow of homeomorphisms if and only if the adjoint equation

$$d\eta_t = -\sum_{j=1}^{m} X_j(t,\eta_t)\circ dB_t^j - X_0(t,\eta_t)\, dt$$

is also strongly complete. Here $B_t = (B_t^1,\ldots,B_t^m)$ is a standard Wiener process and the symbol $\circ$ denotes the Stratonovich integral. Our approach is based on some backward stochastic calculus. A key tool is a backward stochastic differential equation governing the inverse map of the solution.

In Sec. 3 we consider a converse problem: Given a stochastic flow of homeomorphisms defined on a Wiener space, we will find a stochastic differential equation governing the flow, assuming some smoothness condition to the mean and variances of the flow.

2. STOCHASTIC FLOWS OF HOMEOMORPHISMS GENERATED BY STOCHASTIC DIFFERENTIAL EQUATIONS

2.1 Preliminaries

Let $(\Omega,\underline{F},P)$ be a complete probability space, where an m-dimensional standard Wiener process $B_t = (B_t^1,\ldots,B_t^m)$, $t \in [0,T]$ is defined. We denote by $\underline{F}_s^t$ the least complete σ-field for which all $B_u - B_v$, $s \leq u \leq v \leq t$, are measurable. Then $\{\underline{F}_s^t;\ s \leq t\}$ is a family of σ-fields, which is continuous, increasing in t, and decreasing in s, i.e., $\underline{F}_s^t \subset \underline{F}_{s'}^{t'}$ is satisfied whenever $t \leq t'$ and $s' \leq s$. We will call $(\Omega,\underline{F},\underline{F}_s^t,P;B_t)$ a <u>Wiener space</u>.

Let M be a noncompact, paracompact, connected C^∞-manifold of dimension d. Let $\xi_{s,t}(x,\omega)$, $0 \leq s \leq t \leq T$, $x \in M$, be a random field with values in M. We will call it a <u>stochastic flow of homeomorphisms</u> acting on M if the following four conditions are satisfied.

It is continuous in (s,t,x) and $\lim_{t\downarrow s} \xi_{s,t}(x) = x$ holds for all s a.s. (2.1)

For almost all ω, the map $\xi_{s,t}(\cdot,\omega)$ is a homeomorphism from M onto M for any pair s, t with $s < t$. (2.2)

For almost all ω, the relation

$\xi_{s,u}(x,\omega) = \xi_{t,u}(\xi_{s,t}(x,\omega),\omega)$
is satisfied for any $s < t < u$. (2.3)

For any fixed (s,t,x), $\xi_{s,t}(x)$ is $\underline{F}_s^t$ measurable. (2.4)

If both maps $\xi_{s,t}$ and $\xi_{s,t}^{-1}$, $M \to M$, are of C^k-class for any $s < t$ a.s., then $\xi_{s,t}$ is called a stochastic flow of C^k-diffeomorphisms.

The purpose of this section is to show the existence of the stochastic flow of homeomorphisms generated by a given stochastic differential equation.

Our approach to the problem is based on some backward stochastic calculus. Since the forward (= usual) stochastic integral such as Itô's integral and Stratonovich's are known, we will define here backward integrals only.

Let t be a fixed time in [0,T] and let f(s), $s \in [0,t]$, be a real-valued continuous stochastic process which is $\underline{F}_s^t$-measurable for any s. The Itô backward integral is defined as

$$\int_s^t f(r)\, \hat{d}B_r = \lim_{|\Delta|\to 0} \sum_{k=0}^{n-1} f(t_{k+1})(B_{t_{k+1}} - B_{t_k})$$

where $\Delta = \{s = t_0 < \cdots < t_n = t\}$ are partitions of [s,t]. If $E[\int_s^t f(r)^2\, dr] < \infty$, then $M_s \equiv \int_s^t f(r)\, \hat{d}B_r$ is a backward $\underline{F}_s^t$-martingale. The Itô backward integral is of course defined for a more general process f(s): If f(s) is $\underline{F}_s^t$-measurable, progressively measurable, and $\int_s^t f(r)^2\, dr < \infty$ a.s., then Itô's backward integral is well defined, which is a local $\underline{F}_s^t$-martingale.

The Stratonovich backward integral is defined similarly. Let t be a fixed time and f(s), $s \in [0,t]$, be a continuous backward semimartingale relative to $\underline{F}_s^t$. Then the Stratonovich backward integral is defined by

$$\int_s^t f(r)\circ \hat{d}B_r \equiv \lim_{|\Delta|\to 0} \sum_{k=0}^{n-1} \frac{1}{2}[f(t_{k+1}) + f(t_k)](B_{t_{k+1}} - B_{t_k})$$

2.2 Main Results

We first consider a stochastic differential equation on a d-dimensional Euclidean space R^d. Let $X_0(t,x), \ldots, X_m(t,x)$, $t \in [0,T]$, $x \in R^d$, be R^d-valued continuous functions. Consider an Itô's stochastic differential equation

$$d\xi_t = \sum_{j=1}^{m} X_j(t,\xi_t)\, dB_t^j + X_0(t,\xi_t)\, dt \tag{2.5}$$

where the symbol dB_t^j denotes the Itô integral. A stochastic flow of homeomorphisms $\xi_{s,t}$ acting on R^d is said to be generated by the stochastic differential equation if it satisfies

$$\xi_{s,t}(x) = x + \sum_{j=1}^{m} \int_s^t X_j(r,\xi_{s,r}(x))dB_r^j + \int_s^t X_0(r,\xi_{s,r}(x))\, dr \tag{2.6}$$

for any (s,t,x), a.s.

THEOREM 2.1 Assume that $X_0, \ldots, X_m$ are of C^1-class in x and their first derivatives in x are bounded functions. Then there is a unique stochastic flow of homeomorphisms generated by the stochastic differential equation (2.5). Further, the inverse map $\xi_{s,t}^{-1}$ is governed by the following backward stochastic differential equation

$$\xi_{s,t}^{-1}(x) = x - \sum_{j=1}^{m} \int_s^t X_j(r,\xi_{r,t}^{-1}(x))\, \hat{d}B_r^j - \int_s^t X_0^*(r,\xi_{r,t}^{-1}(x))\, dr \tag{2.7}$$

where

$$X_0^*(t,x) = X_0(t,x) - \sum_{i,j} X_j^i(t,x) \frac{\partial}{\partial x^i} X_j(t,x) \tag{2.8}$$

The first part of Theorem 2.1 is shown in Ref. 7. Equation (2.7) is shown in Krylov and Rozovsky [6] for C^4-class coefficients and in Ref. 8 for C^5. The proof will be given in Sec. 2.3.

We next consider a stochastic differential equation on a manifold. Let M be a d-dimensional manifold mentioned in Sec. 2.1. Let $X_0(t,x), \ldots, X_m(t,x)$, $t \in [0,T]$, be vector fields on M with parameter t (first-order differential operators). With a local coordinate $(x^1,\ldots,x^d)$, these vector fields are represented as

$$X_j(t,x) = \sum_{i=1}^{d} X_j^i(t,x) \frac{\partial}{\partial x^i} \tag{2.9}$$

The Stratonovich stochastic differential equation on the manifold M is written as

$$d\xi_t = \sum_{j=1}^{m} X_j(t,\xi_t) \circ dB_t^j + X_0(t,\xi_t)\, dt \tag{2.10}$$

Precisely, a random field $\xi_{s,t}(x)$, $x \in M$, $0 < s < t < T(s,x,\omega)$, with values in M is called a solution of (2.10) if it satisfies

$$f(\xi_{s,t}(x)) = f(x) + \sum_{j=1}^{m} \int_s^t X_j(r) f(\xi_{s,r}(x)) \circ dB_r^j$$
$$+ \int_s^t X_0(r) f(\xi_{s,r}(x))\, dr \tag{2.11}$$

for any C^∞-function f. Here $T(s,x,\omega)$ is the explosion time of the sample function $\xi_{s,t}(x,\omega)$ with fixed (s,x,ω).

Assume that $X_1(t,x),\ldots,X_m(t,x)$ are continuous in (t,x), continuously differentiable in t, of C^2-class in x. Assume further that $X_0(t,x)$ is continuous in (t,x), of C^1-class in x. Then (2.11) has a unique solution $\xi_{s,t}(x)$, $t \in [s,T(s,x,\omega))$, which is continuous in (s,t,x) (see Refs. 7 and 8).

If the explosion time $T(s,x,\omega)$ is infinite (or the constant T) for all x a.s. for each s, (2.10) is called <u>strongly</u>

complete, following Elworthy [4].

REMARK 1. We introduced a stochastic differential equation on a manifold by means of a Stratonovich integral. A reason is that the Stratonovich equation is independent of the choice of local coordinates, whereas the Itô equation is not. In other words, let $\xi_{s,t}$ be the solution of (2.10) with local coordinate $(x^1,\ldots,x^d)$ and let $\tilde{\xi}_{s,t}$ be the solution with another local coordinate $(\tilde{x}^1,\ldots,\tilde{x}^d)$; then it holds that $\xi_{s,t} = \tilde{\xi}_{s,t}$. But if $\xi_{s,t}$ and $\hat{\xi}_{s,t}$ are solutions of Itô equations with these local coordinates, then $\xi_{s,t} = \tilde{\xi}_{s,t}$ is not satisfied in general.

THEOREM 2.2 Assume that vector fields $X_1(t,x), \ldots, X_m(t,x)$ are continuous in (t,x), continuously differentiable in t, and of C^2-class in x, and $X_0(t,x)$ is continuous in (t,x) and of C^1-class in x. Assume further that (2.10) is strongly complete. Then (2.10) generates a flow of homeomorphisms acting on M if and only if the following adjoint equation (2.12) is strongly complete:

$$d\eta_t = -\sum_{j=1}^{m} X_j(t,\eta_t)\circ dB_t^j - X_0(t,\eta_t)\,dt \tag{2.12}$$

Furthermore, under the above condition the inverse flow $\xi_{s,t}^{-1}$ satisfies the backward stochastic differential equation

$$\hat{d}\xi_{s,t}^{-1} = \sum_{j=1}^{m} X_j(s,\xi_{s,t}^{-1})\circ\hat{d}B_s^j + X_0(s,\xi_{s,t}^{-1})\,ds \tag{2.13}$$

i.e., for any C^∞-function f,

$$f(\xi_{s,t}^{-1}(x)) - f(x) = -\sum_{j=1}^{m}\int_s^t X_j(r)f(\xi_{r,t}^{-1}(x))\circ\hat{d}B_r^j$$

$$- \int_s^t X_0(r)f(\xi_{r,t}^{-1}(x))\,dr$$

The proof is given in Sec. 2.4.

Let us look at Theorem 2.2 from a slightly different standpoint. Forward equation (2.10) and backward equation

$$\hat{d}\xi_t = \sum_{j=1}^{m} X_j(t,\xi_t)\circ\hat{d}B_t^j + X_0(t,\xi_t)\, dt \tag{2.14}$$

are both written symbolically in the same form:

$$\frac{d\xi_t}{dt} = \sum_{j=1}^{m} X_j(t,\xi_t)\dot{B}_t^j + X_0(t,\xi_t) \tag{2.15}$$

where $\dot{B}_t^j$ denotes the formal derivative of B_t^j by t. By the forward solution of (2.15) starting at (s,x), we mean the process $\xi_{s,t}(x)$, $t \geq s$, satisfying (2.11) for any C^∞-function f. By the backward solution of (2.15) starting at (s,x), we mean the process $\xi_{s,t}(x)$, $t \leq s$, satisfying

$$f(\xi_{s,t}(x)) = f(x) - \sum_{j=1}^{m} \int_t^s X_j(r,\xi_{s,r}(x))\circ\hat{d}B_r^j$$
$$- \int_t^s X_0(r,\xi_{s,r}(x))\, dr$$

Theorem 2.2 states that if (2.15) is strongly complete both in the forward and the backward directions, then there is a modification of the forward and the backward solution denoted by $\xi_{s,t}(x)$, $s,t \in [0,T]$, which is a flow of homeomorphisms in the following sense: It satisfies (2.1) and (2.2') to (2.4') below:

For almost all ω, the map $\xi_{s,t}(\cdot,\omega)$; $M \to M$ is a homeomorphism for any s, t of $[0,T]$. (2.2')

For almost all ω, the relation $\xi_{s,u} = \xi_{t,u}\circ\xi_{s,t}$ is satisfied for any s, t, u of $[0,T]$. (2.3')

For any fixed (s,t,x), $\xi_{s,t}(x)$ is $F_{s\wedge t}^{s\vee t}$ measurable. (2.4')

Property (2.3') is obvious since $\xi_{s,t} = \xi_{t,s}^{-1}$ if $t < s$.

2.3 Proof of Theorem 2.1

The first assertion of Theorem 2.1 follows from theorem 1.2 in Ref. 7, because the coefficients $X_0, \ldots, X_m$ of the Itô stochastic differential equation are globally Lipschitz continuous.

For the proof of the backward equation governing $\xi_{s,t}^{-1}$, we require a lemma.

LEMMA 2.1 Let $g(r,x)$, $r \in [0,T]$, $x \in R^d$, be a continuous function, continuously differentiable in x. Then

$$\int_s^t g(r,\xi_{s,r}(y))\, dB_r^j \Big|_{y=\xi_{s,t}^{-1}(x)}$$

$$= \int_s^t g(r,\xi_{r,t}^{-1}(x))\, \hat{d}B_r^j - \int_s^t X_j(r) g(r,\xi_{r,t}^{-1}(x))\, dr \quad (2.16)$$

holds, where $X_j(r)$ is the first-order differential operator $\Sigma_{i=1}^d X_j^i(r,x)(\partial/\partial x^i)$.

REMARK 2. The above formula does not correspond to the usual formula because of the last member in (2.16). We will see at the end of this section that under additional conditions, the Stratonovich integral leads to the usual formula

$$\int_s^t g(r,\xi_{s,r}(y))\circ dB_r^j \Big|_{y=\xi_{s,t}^{-1}(x)} = \int_s^t g(r,\xi_{r,t}^{-1}(x))\circ \hat{d}B_r^j \quad (2.17)$$

Of course if we replace dB_r^j by dr, we have the usual formula

$$\int_s^t g(r,\xi_{s,r}(y))\, dr \Big|_{y=\xi_{s,t}^{-1}(x)} = \int_s^t g(r,\xi_{r,t}^{-1}(x))\, dr \quad (2.18)$$

as is easily verified.

Proof. We will show formula (2.16) for function g, which is continuously differentiable in t and of C^2-class in x. Once it is shown, the extension to the function g mentioned in the theorem is immediate, approximating g with a sequence of smooth functions g_n such that $g_n \to g$ and $X_j g_n \to X_j g$.

Now let $\Delta = \{s = t_0 < \cdots < t_n = t\}$ be partitions of the interval $[s,t]$ and let $|\Delta| = \max |t_{k+1} - t_k|$. Then

$$\int_s^t g(r,\xi_{s,r}(y))\, dB_r^j \Big|_{y=\xi_{s,t}^{-1}(x)}$$

$$= \lim_{|\Delta|\to 0} \sum_{k=0}^{n-1} g(t_{k+1},\xi_{s,t_{k+1}}(y))(B_{t_{k+1}}^j - B_{t_k}^j) \Big|_{y=\xi_{s,t}^{-1}(x)}$$

$$- \lim_{|\Delta|\to 0} \sum_{k=0}^{n-1} \{g(t_{k+1},\xi_{s,t_{k+1}}(y)) - g(t_k,\xi_{s,t_k}(y))\}$$

$$\times (B^j_{t_{k+1}} - B^j_{t_k})\Big|_{y=\xi^{-1}_{s,t}(x)} \tag{2.19}$$

holds. The first member of the right-hand side is equal to

$$\lim_{|\Delta|\to 0} \sum_{k=0}^{n-1} g(t_{k+1},\xi^{-1}_{t_{k+1},t}(x))(B^j_{t_{k+1}} - B^j_{t_k})$$

$$= \int_s^t g(r,\xi^{-1}_{r,t}(x))\ \hat{d}B^j_r$$

The second member equals

$$-2[\int_s^t g(r,\xi_{s,r}(y)]\circ dB^j_r - \int_s^t g(r,\xi_{s,r}(y))dB^j_r\Big|_{y=\xi^{-1}_{s,t}(x)}$$

$$= -\int_s^t X_j(r)g(r,\xi_{s,r}(y))\ dr\Big|_{y=\xi^{-1}_{s,t}(x)}$$

$$= -\int_s^t X_j(r)g(r,\xi^{-1}_{r,t}(x))\ dr$$

This proves (2.16).

Proof of equation (2.7). Let f be a C^2-class function. By Itô's formula,

$$f(\xi_{s,t}(x)) - f(x) = \sum_{j=1}^{m} \int_s^t X_j(r)f(\xi_{s,r}(x))\ dB^j_r$$

$$+ \int_s^t L(r)f(\xi_{s,r}(x))\ dr$$

holds, where L(r) is a second-order differential operator:

$$L(r)f = \frac{1}{2} \sum_{i,j=1}^{d} \left[\sum_{k=1}^{m} X^i_k(r)X^j_k(r)\right] \frac{\partial^2}{\partial x^i\ \partial x^j} f$$

$$+ \sum_{i=1}^{d} X^i_0(r) \frac{\partial}{\partial x^i} f$$

Then

$$f(x) - f(\xi^{-1}_{s,t}(x)) = \sum_{j=1}^{m} \int_s^t X_j(r)f(\xi_{s,r}(y))\ dB^j_r\Big|_{y=\xi^{-1}_{s,t}(x)}$$

$$+ \int_s^t L(r) f(\xi_{s,r}(y))\, dr \Big|_{y=\xi_{s,t}^{-1}(x)} \tag{2.20}$$

Apply formulas (2.16) and (2.18). Then we see that the right-hand side equals

$$\sum_{j=1}^m \int_s^t X_j(r) f(\xi_{r,t}^{-1}(x))\, \hat{d}B_r^j - \sum_{j=1}^m \int_s^t X_j(r)^2 f(\xi_{r,t}^{-1}(x))\, dr$$

$$+ \int_s^t L(r) f(\xi_{r,t}^{-1}(x))\, dr$$

Therefore, we have

$$f(\xi_{s,t}^{-1}(x)) - f(x) = - \sum_{j=1}^m \int_s^t X_j(r) f(\xi_{r,t}^{-1}(x))\, \hat{d}B_r^j$$

$$+ \int_s^t \left[\sum_{j=1}^m X_j(r)^2 - L(r) \right] f(\xi_{r,t}^{-1}(x))\, dr \tag{2.21}$$

A direct calculation yields

$$\sum_{j=1}^m X_j(r)^2 f - L(r) f = \frac{1}{2} \sum_{i,j=1}^d \left[\sum_{k=1}^m X_k^i(r,x) X_k^j(r,x) \right] \frac{\partial^2}{\partial x^i\, \partial x^j} f$$

$$- \sum_{i=1}^d X_0^{*i}(r,x) \frac{\partial}{\partial x^i} f$$

Setting $f(x) = x^i$, $i = 1, \ldots, d$ at (2.21), we see that $\xi_{s,t}^{-1}$ satisfies (2.7). The proof is complete.

COROLLARY TO THEOREM 2.1 Under the condition of Theorem 2.1, the inverse process $\xi_{s,t}^{-1}$, $s \in [0,t]$, is a backward semimartingale for each t. Let $g(t,x)$ be continuously differentiable in t and of C^2-class in x. Then formula (2.17) is valid.

Proof. It holds that

$$\int_s^t g(r, \xi_{s,r}(y)) \circ dB_r^j \Big|_{y=\xi_{s,t}^{-1}(x)}$$

$$= \int_s^t g(r, \xi_{s,r}(y))\, dB_r^j \Big|_{y=\xi_{s,t}^{-1}(x)}$$

$$+ \frac{1}{2} \int_s^t X_j(r) g(r, \xi_{s,r}(y))\, dr \Big|_{y=\xi_{s,t}^{-1}(x)}$$

$$= \int_s^t g(r,\xi_{r,t}^{-1}(x))\, \hat{d}B_r^j - \frac{1}{2}\int_s^t X_j(r)g(r,\xi_{r,t}^{-1}(x))\, dr$$

$$= \int_s^t g(r,\xi_{r,t}^{-1}(x))\circ\hat{d}B_r^j$$

2.4 PROOF OF THEOREM 2.2

We begin the discussion with the "only if" part of Theorem 2.2. Hence we suppose that there is a flow of homeomorphisms generated by (2.10). Then the inverse map $\xi_{s,t}^{-1}$ satisfies (2.21), by the same reasoning as in the proof of Theorem 2.1. Since $\xi_{s,t}$ is the solution of the Stratonovich equation (2.10), it holds that $L = (1/2)\Sigma_{j=1}^m X_j^2 + X_0$, so that $\Sigma_{j=1}^m X_j^2 - L = (1/2)\Sigma_{j=1}^m X_j^2 - X_0$. Therefore, (2.21) implies that

$$f(\xi_{s,t}^{-1}(x)) - f(x) = -\sum_{j=1}^m \int_s^t X_j(r)f(\xi_{r,t}^{-1}(x))\, \hat{d}B_r^j$$
$$+ \int_s^t [\tfrac{1}{2}\, \Sigma\, X_j(r)^2 - X_0(r)]$$
$$\times f(\xi_{r,t}^{-1}(x))\, dr \qquad (2.22)$$

Observing that $\xi_{r,t}^{-1}$ is a backward semimartingale, $X_j(r)f(\xi_{r,t}^{-1}(x))$ is also a backward semimartingale. Then the Stratonovich backward integral is well defined. It holds that

$$\int_s^t X_j(r)f(\xi_{r,t}^{-1})\circ\hat{d}B_r^j = \int_s^t X_j(r)f(\xi_{r,t}^{-1}(x))\, \hat{d}B_r^j$$
$$- \frac{1}{2}\int_s^t X_j(r)^2 f(\xi_{r,t}^{-1}(x))\, dr$$

Therefore, (2.22) implies that

$$f(\xi_{s,t}^{-1}(x)) - f(x) = -\sum_{j=1}^m \int_s^t X_j(r)f(\xi_{r,t}^{-1}(x))\circ\hat{d}B_r^j$$
$$- \int_s^t X_0(r)f(\xi_{r,t}^{-1}(x))\, dr \qquad (2.23)$$

This proves (2.13).

Now the above argument shows that the Stratonovich

backward equation (2.14) is strongly complete, which is clearly equivalent to the fact that adjoint equation (2.12) is strongly complete. Hence we have proved the "only if" part of the theorem.

We will next show the "if" part of the theorem. Let $\xi_{s,t}(x)$ be the solution of (2.10). Since the equation is strongly complete, we may choose a modification of it satisfying (2.1), (2.3), and (2.4). Instead of (2.2), a weaker assertion is satisfied.

For almost all ω, the map $\xi_{s,t}(\cdot,\omega)$ is a homeomorphism from M <u>into</u> M for any pair s, t with $s < t$. (2.2')

(see Ref. 7). We denote by $R_{s,t}(\omega)$ the image of the set M by the map $\xi_{s,t}(\cdot,\omega)$. Then $R_{s,t}(\omega)$ is open by the theorem of the invariance of the domain (see Meyer [11]). If we could show that $R_{s,t}(\omega)$ is closed, it would imply $R_{s,t}(\omega) = M$ since M is a connected manifold. The rest of this section is devoted to the proof of the closedness of the set $R_{s,t}(\omega)$.

Let us remark that the backward equation of the inverse $\xi^{-1}_{s,t}(x)$ is valid on the set $\{\omega \mid x \in R_{s,t}(\omega)\}$, namely:

PROPOSITION 2.1 Assume that vector fields $X_1(t,x), \ldots, X_m(t,x)$ of (2.10) are continuously differentiable in t and twice continuously differentiable in x. Assume further that (2.10) is strongly complete. Then the inverse $\xi^{-1}_{s,t}(x)$ satisfies Stratonovich backward equation (2.13) on the set $\{\omega \mid x \in R_{s,t}(\omega)\}$.

The proof is quite similar to that of Theorem 2.1. Consider now the backward stochastic differential equation (2.14). Since it is strongly complete by the assumption, it has a solution $\xi_{t,s}(x)$, $s \in [0,t]$, such that $\xi_{t,t}(x) = x$. The solution is continuous in (s,t,x) a.s. Clearly $\xi_{t,s}(x)$ is an extension of $\xi^{-1}_{s,t}(x)$ in the sense that $\xi_{t,s}(x) = \xi^{-1}_{s,t}(x)$ if $x \in R_{s,t}(\omega)$.

We will now prove that $R_{s,t}(\omega)$ is closed. We fix ω. Let $\{x_n\}$ be a Cauchy sequence belonging to $R_{s,t}(\omega)$. Then it holds that

$$x_n = \xi_{s,t} \circ \xi_{s,t}^{-1}(x_n) = \xi_{s,t} \circ \xi_{t,s}(x_n)$$

Let x be the limit of $\{x_n\}$. Then it holds that $x = \xi_{s,t} \circ \xi_{t,s}(x)$. This shows that x is an element of $R_{s,t}(\omega)$; i.e., $R_{s,t}(\omega)$ is closed. The proof of Theorem 2.2 is complete.

3. STOCHASTIC DIFFERENTIAL EQUATIONS GOVERNING STOCHASTIC FLOWS OF HOMEOMORPHISMS

3.1 PRELIMINARIES

In Sec. 2 we saw that a stochastic differential equation generates a stochastic flow of homeomorphisms if the corresponding vector fields satisfy some condition. In this section we consider a converse problem: Given a stochastic flow of homeomorphisms, we will find a stochastic differential equation which generates the flow.

Before we discuss the problem in detail, we state two preliminary results.

PROPOSITION 3.1 Let $\xi_{s,t}(x)$ be a stochastic flow of homeomorphisms on M. Then for each (s,x), the stochastic process $Y_t = \xi_{s,t}(x)$ is a Markov process with state space M.

Proof. Let f be a bounded continuous function on M. Noting that $\xi_{s,t+u} = \xi_{t,t+u} \circ \xi_{s,t}$ [property (2.3)] and that $\xi_{t,t+u}$ is independent of $\underline{F}_s^t$, we have

$$E[f(\xi_{s,t+u}(x)) | \underline{F}_s^t] = E[f(\xi_{t,t+u}(y))]_{y=\xi_{s,t}(x)}$$

This proves the Markov property.

Suppose now that $\xi_{s,t}(x)$ is a stochastic flow acting on R^d generated by Itô stochastic differential equation (2.5).

We will consider how coefficients $X_0, \ldots, X_m$ are computed from the data of input B_t and output $\xi_{s,t}$.

PROPOSITION 3.2 Let $\xi_{s,t}$ be a stochastic flow of Theorem 2.1. Then the following holds:

The mean $E[\xi_{s,t}(x)]$ is continuously differentiable in $t \in [s,T]$ for each s and satisfies

$$X_0(s,x) = \frac{\partial}{\partial t}E[\xi_{s,t}(x)]|_{t=s} = \lim_{h\to 0} \frac{1}{h}\{E[\xi_{s,s+h}(x)] - x\} \quad (3.1)$$

Covariances $E[\xi_{s,t}(x)B_t^j]$, $j = 1, \ldots, m$, are continuously differentiable in $t \in [s,T]$ for each s and satisfy

$$X_j(s,x) = \frac{\partial}{\partial t}E[\xi_{s,t}(x)B_t^j]|_{t=s} = \lim_{h\to 0} \frac{1}{h}E[\xi_{s,s+h}(x)B_{s+h}^j] \quad (3.2)$$

Here $E[\xi_{s,t}(x)]$ means the d-vector $(E[\xi_{s,t}^1(x)],\ldots,E[\xi_{s,t}^d(x)])$. The derivative at the endpoint $t = s$ stands for the right derivative.

Proof. Take the expectation of each term of (2.6). Then we have

$$E[\xi_{s,t}(x)] = x + \int_s^t E[X_0(r,\xi_{s,r}(x))]\, dr$$

Therefore, it is continuously differentiable in t and satisfies

$$\frac{\partial}{\partial t}E[\xi_{s,t}(x)] = E[X_0(t,\xi_{s,t}(x))]$$

This proves (3.1).

Consider next the covariance. We have

$$E[(\int_s^t X_i(r,\xi_{s,r}(x))\, dB_r^i)(B_t^j - B_s^j)]$$

$$= \delta_{ij}E[\int_s^t X_i(r,\xi_{s,r}(x))\, dr]$$

and

$$E[(\int_s^t X_0(r,\xi_{s,r}(x))\, dr)(B_t^j - B_s^j)]$$

$$= E[\int_s^t X_0(r,\xi_{s,r}(x))(B_r^j - B_s^j)\, dr]$$

Therefore,

$$E[\xi_{s,t}(x)B_t^j] = E[(\xi_{s,t}(x) - x)(B_t^j - B_s^j)]$$
$$= \int_s^t E[X_j(r,\xi_{s,r}(x))]$$
$$+ E[X_0(r,\xi_{s,r}(x))(B_r^j - B_s^j)]\,dr$$

It is differentiable in t and

$$\frac{\partial}{\partial t}E[\xi_{s,t}(x)B_t^j] = E[X_j(t,\xi_{s,t}(x))]$$
$$+ E[X_0(t,\xi_{s,t}(x))(B_t^j - B_s^j)]$$

The last member converges to 0 as $t \to s$. Thus we have the second assertion.

3.2 Stochastic Flow Acting on R^d

Let $\xi_{s,t}(x)$ be a stochastic flow of homeomorphisms acting on R^d defined on the Wiener space $(\Omega,\underline{F},\underline{F}_s^t,P;B_t)$. In order to find a stochastic differential equation governing the flow $\xi_{s,t}$, we have to find coefficients $X_0, \ldots, X_m$ of the equation. Proposition 3.2 suggests a method of finding these coefficients. We have, in fact, the following theorem.

THEOREM 3.1 Let $\xi_{s,t}(x)$ be a square-integrable stochastic flow of homeomorphisms on R^d defined on a Wiener space $(\Omega,\underline{F}_s^t,P;B_t)$. Assume that $\xi_{s,t}(x)$ satisfies the following two conditions.

For each (s,x), $E[\xi_{s,t}(x)]$ is continuously differentiable in $t \in [s,T]$, and the derivative is of linear growth in x.[†] (3.3)

For each (s,x), the covariances $E[\xi_{s,t}(x)B_t^j]$, $j = 1, \ldots, m$, are continuously differentiable in t and derivatives are of linear growth in x. (3.4)

Define d-vector functions X_0 and X_j, $j = 1, \ldots, m$, by

[†]The derivative $\varphi_{s,t}(x)$ satisfies $|\varphi_{s,t}(x)| \le K(1 + |x|)$, where K is a positive constant not depending on (s,t).

$$X_0(s,x) = \lim_{h\to 0} \frac{1}{h}\{E[\xi_{s,s+h}(x)] - x\} \tag{3.5}$$

$$X_j(s,x) = \lim_{h\to 0} \frac{1}{h} E[\xi_{s,s+h}(x)B^j_{s+h}] \qquad j = 1, \ldots, m \tag{3.6}$$

Then $\xi_{s,t}$ is governed by the Itô stochastic differential equation (2.5) with these functions $X_0, \ldots, X_m$ as coefficients.

The proof of the theorem is divided into two lemmas.

LEMMA 3.1 Set

$$M_{s,t}(x) = \xi_{s,t}(x) - x - \int_s^t X_0(r,\xi_{s,r}(x))\, dr \tag{3.7}$$

Then it is a square-integrable $\underline{F}_s^t$-martingale for each (s,x).

Proof. Let us first show that

$$E[\xi_{s,t}(x)] = x + \int_s^t E[X_0(r,\xi_{s,r}(x))]\, dr \tag{3.8}$$

Set $v_{s,t}(x) = E[\xi_{s,t}(x)]$. Then since $\xi_{s,t+h} = \xi_{t,t+h}\circ\xi_{s,t}$,

$$\frac{\partial}{\partial t}E[\xi_{s,t}(x)] = \lim_{h\to 0} \frac{1}{h} E[\xi_{t,t+h}\circ\xi_{s,t}(x) - \xi_{s,t}(x)]$$

$$= \lim_{h\to 0} E[\frac{1}{h}\{v_{t,t+h}(\xi_{s,t}(x)) - \xi_{s,t}(x)\}] \tag{3.9}$$

Let

$$\varphi_{t,u}(x) = \frac{\partial}{\partial u} v_{t,u}(x)$$

Then it holds that for some $\varepsilon > 0$,

$$\frac{1}{h}|v_{t,t+h}(x) - x| = |\varphi_{t,t+\theta h}(x)| \le K(1 + |x|)$$

$$\text{for any } |h| < \varepsilon$$

where $|\theta| \le 1$. Since $K(1 + |\xi_{s,t}(x)|)$ is integrable, we can change the order of lim and E in (3.9) by the Lebesgue convergence theorem. Hence we get

$$\frac{\partial}{\partial t} E[\xi_{s,t}(x)] = E[X_0(t,\xi_{s,t}(x))]$$

and (3.8) follows from this.

Now (3.8) implies that $E[M_{s,t}(x)] = 0$. Further, by the definition of $M_{s,t}(x)$, it is easily checked that

$$M_{s,u}(x) = M_{s,t}(x) + M_{t,u}(\xi_{s,t}(x)) \tag{3.10}$$

for $s < t < u$. Therefore,

$$E[M_{s,u}(x) | \underline{F}_s^t] = M_{s,t}(x) + E[M_{t,u}(y)]_{y=\xi_{s,t}(x)}$$
$$= M_{s,t}(x) \tag{3.11}$$

proving that $M_{s,t}$ is a martingale for each (s,x). The square integrability will be obvious.

LEMMA 3.2 It holds that

$$M_{s,t}(x) = \sum_{j=1}^{m} \int_s^t X_j(r, \xi_{s,r}(x))\, dB_r^j \tag{3.12}$$

Proof. For a fixed pair (s,x), $M_{s,t}(x)$ is a square-integrable d-vector martingale starting from 0 at time s. Therefore, by the representation theorem of martingales in the Wiener space (e.g., Kunita and Watanabe [9]) there are d-vector predictable processes $\varphi_j^{(s,x)}(t,\omega)$, $j = 1, \ldots, m$, such that

$$M_{s,t}(x) = \sum_{j=1}^{m} \int_s^t \varphi_j^{(s,x)}(r)\, dB_r^j$$

We shall prove that for each (s,x),

$$\varphi_j^{(s,x)}(t,\omega) = X_j(t, \xi_{s,t}(x)) \qquad \text{a.e. } (t,\omega) \text{ in } dt \otimes dP \tag{3.13}$$

holds. Let $A \in \underline{F}_0^t$. Then

$$E[(M_{s,t+h}(x) - M_{s,t}(x))(B_{t+h}^j - B_t^j); A]$$
$$= E[\int_t^{t+h} \varphi_j^{(s,x)}(r)\, dr; A]$$

The left-hand side is computed as

$$E[E[(\xi_{t,t+h}(y) - y)(B^j_{t+h} - B^j_t)]_{y=\xi_{s,t}(x)}; A]$$

$$- E[E[\int_t^{t+h} X_0(r,\xi_{t,r}(y))\, dr(B^j_{t+h} - B^j_t)]_{y=\xi_{s,t}(x)}; A]$$

$$= E[E[\xi_{t,t+h}(y)B^j_{t+h}]_{y=\xi_{s,t}(x)}; A]$$

$$- E[\int_t^{t+h} E[X_0(r,\xi_{t,r}(y))(B^j_r - B^j_t)]\, dr\big|_{y=\xi_{s,t}(x)}; A]$$

Obviously, the last member is $o(h)$ as $h \to 0$. Therefore, we get

$$\lim_{h\to 0} \frac{1}{h} E[\int_t^{t+h} \varphi_j^{(s,x)}(r)\, dr; A]$$

$$= \lim_{h\to 0} \frac{1}{h} E[E[\xi_{t,t+h}(y)B^j_{t+h}]_{y=\xi_{s,t}(x)}; A]$$

We can interchange the order of lim and E, and we get

$$E[\varphi_j^{(s,x)}(t); A] = E[X_j(t,\xi_{s,t}(x)); A] \qquad \text{for all } A \in \underline{F}_0^t$$

a.e. t. This proves (3.13). The proof is complete.

3.3 Stochastic Flow Acting on a Manifold

In this section we consider a stochastic flow of homeomorphisms acting on a manifold. The following proposition can be proved by an argument similar to that of Proposition 3.2.

PROPOSITION 3.3 Let $\xi_{s,t}(x)$ be a stochastic flow of homeomorphisms on the manifold M generated by the Stratonovich stochastic differential equation (2.10). Then we have the following†

For any $f \in C_0^\infty(M)$, $E[f(\xi_{s,t}(x))]$ is continuously differentiable in $t \in [s,T]$ for each (s,x) and satisfies

$$\lim_{h\to 0} \frac{1}{h}\{E[f(\xi_{s,s+h}(x))] - f(x)\}$$
$$= \frac{1}{2}\left[\sum_{j=1}^{m} X_j(s)^2 + X_0(s)\right] f(x) \tag{3.14}$$

For any $f \in C_0^\infty(M)$, $E[f(\xi_{s,t}(x))B^j_t]$, $j = 1, \ldots, m$, are

---

† $C_0^\infty(M)$ is the space of the C^∞-functions on M with compact support.

continuously differentiable in $t \in [s,T]$ for each (s,x) and satisfy

$$\lim_{h\to 0} \frac{1}{h} E[f(\xi_{s,s+h}(x))B^j_{s+h}] = X_j(s)f(x) \tag{3.15}$$

The purpose of this section is to prove the converse assertion.

THEOREM 3.2 Let $\xi_{s,t}(x)$ be a stochastic flow of homeomorphisms on the manifold M such that for any $f \in C_0^\infty(M)$, $E[f(\xi_{s,t}(x))]$ and $E[f(\xi_{s,t}(x))B^j_t]$ are differentiable in $t \in [s,T]$ for each (s,x) and their derivatives are bounded functions of (s,t,x), of C^1-class in t, and of C^2-class in x. Define

$$X_j(s)f(x) = \lim_{h\to 0} \frac{1}{h} E[f(\xi_{s,s+h}(x))B^j_{s+h}] \tag{3.16}$$

$$L(s)f(x) = \lim_{h\to 0} \frac{1}{h}\{E[f(\xi_{s,s+h}(x))] - f(x)\} \tag{3.17}$$

Then $X_1, \ldots, X_m$ and $X_0 = L - (1/2)\Sigma_{j=1}^m X_j^2$ are vector fields continuously differentiable in t and of C^2-class in x. Moreover, $\xi_{s,t}(x)$ satisfies the stochastic differential equation (2.10).

The proof is divided into three lemmas.

LEMMA 3.3 Let $f \in C_0^\infty(M)$ and let $L(r)f$ be the function defined by (3.17). Set

$$M^f_{s,t}(x) = f(\xi_{s,t}(x)) - f(x) - \int_s^t L(r)f(\xi_{s,r}(x))\, dr \tag{3.18}$$

Then it is a square-integrable martingale.

LEMMA 3.4 $M^f_{s,t}(x)$ is represented as

$$M^f_{s,t}(x) = \sum_{j=1}^m \int_s^t X_j(r)f(\xi_{s,r}(x))\, dB^j_r$$

where $X_j(r)f$ is the function defined by (3.16).

The proofs of Lemmas 3.3 and 3.4 are carried out similarly to those of Lemmas 3.1 and 3.2.

Now by Lemmas 3.3 and 3.4, $f(\xi_{s,t}(x))$ is represented as

$$f(\xi_{s,t}(x)) = f(x) + \sum_{j=1}^{m} \int_s^t X_j(r)f(\xi_{s,r}(x))\, dB_r^j$$
$$+ \int_s^t L(r)f(\xi_{s,r}(x))\, dr \tag{3.19}$$

We will show that $X_j(r)$, $j = 1, \ldots, m$, are first-order differential operators and $L(r)$ is a second-order operator.

LEMMA 3.5 X_j, $j = 1, \ldots, m$, defined by (3.16) are first-order differential operators. Furthermore, $X_0 \equiv L - (1/2) \Sigma X_j^2$ is also a first-order differential operator.

Proof. Let f, g be of class $C_0^\infty(M)$. Then $f(\xi_{s,t}(x))$, $g(\xi_{s,t}(x))$, and $f(\xi_{s,t}(x))g(\xi_{s,t}(x))$ are all semimartingales for any (s,x). By Lemmas 3.3 and 3.4,

$$f(\xi_{s,t}(x))g(\xi_{s,t}(x)) = f(x)g(x)$$
$$+ \sum_{j=1}^{m} \int_s^t X_j(r)(fg)(\xi_{s,r}(x))dB_r^j$$
$$+ \int_s^t L(r)(fg)(\xi_{s,r}(x))\, dr \tag{3.20}$$

holds. On the other hand, apply Itô's formula to the product of two semimartingales $f(\xi_{s,t}(x))$ and $g(\xi_{s,t}(x))$ which are written as (3.19), etc. Then

$$f(\xi_{s,t}(x))g(\xi_{s,t}(x)) =$$
$$= f(x)g(x) + \sum_{j=1}^{m} \int_s^t g(\xi_{s,r}(x))X_j(r)f(\xi_{s,r}(x))\, dB_r^j$$
$$+ \sum_{j=1}^{m} \int_s^t f(\xi_{s,r}(x))X_j(r)g(\xi_{s,r}(x))\, dB_r^j$$

$$+ \sum_{j=1}^{m} \int_s^t X_j(r)f(\xi_{s,r}(x))X_j(r)g(\xi_{s,r}(x))\ dr$$

$$+ \int_s^t g(\xi_{s,r}(x))L(r)f(\xi_{s,r}(x))\ dr$$

$$+ \int_s^t f(\xi_{s,r}(x))L(r)g(\xi_{s,r}(x))\ dr \tag{3.21}$$

The corresponding terms in (3.20) and (3.21) must coincide with each other: For example,

$$\int_s^t X_j(r)(fg)(\xi_{s,r}(x))\ dB_r^j$$
$$= \int_s^t g(\xi_{s,r}(x))X_j(r)f(\xi_{s,r}(x))\ dB_r^j$$
$$+ \int_s^t f(\xi_{s,r}(x))X_j(r)g(\xi_{s,r}(x))\ dB_r^j$$

holds for any $j = 1, \ldots, m$. This implies that

$$X_j(r)(fg) = gX_j(r)f + fX_j(r)g$$

Therefore, $X_j(r)$ are first-order differential operators.

By a similar consideration, we obtain from (3.20) and (3.21),

$$L(r)(fg) = fL(r)g + gL(r)f + \sum_{j=1}^{m} X_j(r)f\cdot X_j(r)g$$

Therefore, we have

$$L(r)(fg) - \frac{1}{2}\sum_{j=1}^{r} X_j(r)^2(fg)$$
$$= fL(r)g + gL(r)f - \frac{1}{2}\sum_j fX_j(r)^2 g - \frac{1}{2}\sum_j gX_j(r)^2 f$$

This implies that $X_0(r) \equiv L(r) - (1/2)\Sigma_j X_j(r)^2$ is also a first-order differential operator.

Now, since $X_j(r)f(\xi_{s,r}(x))$ is a semimartingale for each (s,x), the Stratonovich integral is well defined. It holds that

$$\int_s^t X_j(r)f(\xi_{s,r}(x))\circ dB_r^j = \int_s^t X_j(r)f(\xi_{s,r}(x))\, dB_r^j$$

$$+ \frac{1}{2}\int_s^t X_j(r)^2 f(\xi_{s,r}(x))\, dr$$

Therefore, equation (3.19) leads to equation (2.11). The proof of Theorem 3.2 is complete.

REFERENCES

1. P. Baxendale, Wiener processes on manifolds of maps Proc. Royal Soc. Edinburgh, 87A:127-152 (1980).
2. J. M. Bismut, A generalized formula of Itô and some other properties of stochastic flows, Z. Wahrscheinlichkeitsth. Verwend. Geb. 55:331-350 (1981).
3. K. D. Elworthy, Stochastic dynamical systems and their flows, pp. 79-95 in Stochastic Analysis, (A. Friedman and M. Pinsky, eds.), Academic Press, New York, 1978.
4. K. D. Elworthy, Stochastic flows and the C^0-diffusion property, Stochastics 6:233-238 (1982).
5. N. Ikeda and S. Watanabe, Stochastic Differential Equations and Diffusion Processes, North-Holland Kodansha, Amsterdam, 1981.
6. N. V. Krylov and B. L. Rozovsky, On the First Integrals and Liouville Equations for Diffusion Processes, pp. 117-125, Proc. 3rd Conf. Stochastic Differential Systems, Lect. Notes Control Inf. Sci. 36, Springer-Verlag, Berlin, 1983.
7. H. Kunita, On the decomposition of solutions of stochastic differential equations, pp. 213-255 in Stochastic Integrals, Proc. Durham Conf. Stochastic Integrals, Lect. Notes Math. 851, Springer-Verlag, Berlin, 1981.
8. H. Kunita, On backward stochastic differential equations, Stochastics 6:293-313 (1982).
9. H. Kunita and S. Watanabe, On square integrable martin-

gales, Nagoya Math. J. 30:209-245 (1967).

10. P. Malliavin, Un principe de transfert et son application au calcul des variations, C. R. Acad. Sci. Paris A284:187-189 (1977).

11. P. A. Meyer, Flot d'une équation différentielle stochastique, pp. 103-117 in Séminaire de probabilités XV, Lect. Notes Math. 850, Springer-Verlag, Berlin, 1981.

10

Approximation of Processes and Applications to Control and Communication Theory

HAROLD J. KUSHNER/Division of Applied Mathematics, Brown University, Providence, Rhode Island

1. INTRODUCTION

Diffusion models are of widespread use in many areas of control and communication theory. The models are frequently used for systems that are not quite diffusions but are, it is hoped, close to a diffusion in some sense. For example, the input noise might be "wide band" but not "white Gaussian." Many approximation techniques have been developed [1-10], under different sets of assumptions. The typical results are of a weak convergence nature. The physical process $x^\epsilon(\cdot)$ is parameterized by the parameter ϵ, and one tries to show that $\{x^\epsilon(\cdot)\}$ converges weakly to some diffusion $x(\cdot)$ as $\epsilon \to 0$. The limit process $x(\cdot)$ is then used to study various properties of $x^\epsilon(\cdot)$ for small ϵ.

If the system involves nonlinear functions of the input noise, or if the noise depends on the system state or if the system dynamics are not smooth, then the approximation problem is harder. Here we discuss a set of techniques which have proved to be quite useful for a variety of problems in control and communication. The main weak convergence theorems are stated and discussed in Sec. 2. Frequently, in applications, we are concerned with asymptotic properties, as $t \to \infty$ (for small ε), as well as with weak convergence. Such information is not normally provided by the weak convergence theory. In Sec. 2 we also discuss the problem of approximating functionals on the "tail" of $x^\varepsilon(\cdot)$ for small ε, by such functionals on the tail of $x(\cdot)$ [e.g., approximating the measures of $\{x^\varepsilon(t)$, large $t\}$, for small ε, by an invariant measure of $x(\cdot)$]. This is particularly useful in problems in communication theory where the (say) detection system is often supposed to be in operation for a very long time. The proofs of these theorems are given in the references.

In Secs. 3 and 4 we apply these theorems to two systems which are of frequent and basic use in communication theory, but which seem to be quite difficult to analyze without at least some sort of formal diffusion approximation technique. Other applications of these and related ideas to control and communication can be found in Refs. 9 and 11-14. While there has been widespread application of diffusion approximation methods to some topics in physics and biology, and operations research (e.g., to queueing theory), application to concrete nonlinear problems in control and communication theory is still in its infancy. There are numerous possibilities for other applications: e.g., to synchronization systems, robustified filters (where some nonlinear function of the observation is used in lieu of the observation), to adaptive modulation systems and filters, etc. These and related techniques and applications are dealt with in great detail in Ref. 20.

2. CONVERGENCE AND APPROXIMATION THEOREMS

2.1 Weak Convergence

Suppose that the system state $x^\varepsilon(\cdot)$ satisfies the differential equation $\dot{x}^\varepsilon = H^\varepsilon(\xi^\varepsilon, x^\varepsilon)$, where $\xi^\varepsilon(\cdot)$ is an input noise process whose bandwidth (BW) goes to ∞ as $\varepsilon \to 0$ (loosely speaking), and $x^{\varepsilon}(t) \in R^{r}$, euclidean r-space. We are first interested in showing that $\{x^{\varepsilon}(\cdot)\}$ converges weakly in $D^{r}[0,\infty)$ to some diffusion process $x(\cdot)$, defined by (2.1), where we assume that the martingale problem in $D^{r}[0,\infty)$ associated with (2.1) has a unique solution for each initial condition $x = x(0)$,

$$dx = \alpha(x)\, dt + \sigma(x)\, dB \qquad B(\cdot) = \text{standard Wiener process} \tag{2.1}$$

It is possible to treat problems where the limits are jump diffusions as in Refs. 7 and 8, but since the applications in Secs. 3 and 4 all have diffusion limits, the basic theorems here are specialized to this case.

Define the differential generator

$$A = \sum_{i} \alpha_i(x) \frac{\partial}{\partial x_i} + \frac{1}{2} \sum_{i,j} a_{ij}(x) \frac{\partial^2}{\partial x_i\, \partial x_j}$$

$$\text{where } a(x) = \{a_{ij}(x)\} = \sigma(x)\sigma'(x)$$

For each $N > 0$, let $b_N(\cdot)$ denote a continuously differentiable function satisfying $b_N(x) = 1$ for $x \in S_N = \{x : |x| \le N\}$, $0 \le b_N(x) \le 1$, and $b_N(x) = 0$ for $x \notin S_{N+1}$. Let A^N denote the differential generator of a diffusion $x^N(\cdot)$ with coefficients $\alpha^N(\cdot)$, $\sigma^N(\cdot)$ equal to $\alpha(\cdot)$, $\sigma(\cdot)$, respectively, in S_N. Define the truncated process $x^{\varepsilon,N}(\cdot)$ by $x^{\varepsilon,N} = H^{\varepsilon}(\xi^{\varepsilon}, x^{\varepsilon,N}) b_N(x^{\varepsilon,N})$. Then, if $x^{\varepsilon,N}(\cdot) \to x^N(\cdot)$ weakly in $D^{r}[0,\infty)$, as $\varepsilon \to 0$ for each N, we have [8] $x^{\varepsilon}(\cdot) \to x(\cdot)$ weakly also. The truncation is used as a technical device only, in that allowing us to work with bounded processes, it simplifies the proofs and the calculations in the applications of the theorem, although it makes for a slightly more complicated theorem statement.

Let $\hat{C}_0$ denote the space of real-valued continuous functions on R^r with compact support, and $\hat{C}_0^{\alpha,\beta}$ the subspace whose mixed α partial t-derivatives and β partial x-derivatives are continuous. Let $\{F_t^\varepsilon\}$ be a nondecreasing sequence of σ-algebras with F_t^ε measuring $\{\xi^\varepsilon(s), s \le t\}$. Let H denote the class of real-valued (progressively) measurable (ω,t) functions such that if $g(\cdot) \in H$, then $\sup_t E|g(t)| < \infty$, $E|g(t + \delta) - g(t)| \to 0$ as $\delta \to 0$, and $g(t)$ is F_t^ε measurable. Following Ref. 4, we say that $\text{p-lim}_{\delta\to 0} f^\delta(\cdot)$ if $\sup_{t\geq 0,\delta>0} E|f^\delta(t)| < \infty$ and $E|f^\delta(t)| \to 0$ as $\delta \to 0$, for each t. Define the operator $\hat{A}^\varepsilon$ with domain $\mathcal{D}(\hat{A}^\varepsilon)$ as follows: $g \in \mathcal{D}(\hat{A}^\varepsilon)$ and $\hat{A}^\varepsilon g(\cdot) = q(\cdot)$ if $g(\cdot)$ and $q(\cdot)$ are in H and $\text{p} - \lim_{\delta\to 0}[(E_t^\varepsilon g(\cdot + \delta) - g(\cdot))/\delta - q(\cdot)] = 0$. Thus $\hat{A}^\varepsilon$ resembles an infinitesimal operator. The following theorem is a specialization of a result in Ref. 8, which in turn is a development of ideas of Kurtz [4].

THEOREM 2.1 Let $\alpha(\cdot)$, $\sigma(\cdot)$ be continuous, and assume the above uniqueness condition on the solution of (2.1). Fix N. For each $f(\cdot) \in \hat{C}_0^{2,3}$, let there be a sequence $\{f^{\varepsilon,N}(\cdot)\} \in H$ such that

$$\begin{aligned}&\text{p-lim}_{\varepsilon\to 0} |f^{\varepsilon,N}(\cdot) - f(x^{\varepsilon,N}(\cdot),\cdot)| = 0\\&\text{p-lim}_{\varepsilon\to 0} |\hat{A}^\varepsilon f^{\varepsilon,N}(\cdot) - (A^N + \tfrac{\partial}{\partial t})f(x^{\varepsilon,N}(\cdot),\cdot)| = 0\end{aligned} \tag{2.2}$$

Then, if $\{x^{\varepsilon,N}(\cdot)\}$ is tight in $D^r[0,\infty]$ for each N, and $x^\varepsilon(0) \to x_0$ weakly, $\{x^\varepsilon(\cdot)\}$ converges weakly to $x(\cdot)$ [with $x(0) = x_0$] as $\varepsilon \to 0$.

REMARK 1. In the examples, it is shown how to get $\{f^{\varepsilon,N}(\cdot)\}$ in typical cases; the method is also discussed in Refs. 3, 7, 9, and 11-14. We use the form $f^{\varepsilon,N}(t) = f(x^{\varepsilon,N}(t),t) + \Sigma_{i=0 \text{ or } 1}^2 f_i^{\varepsilon,N}$ as discussed in Secs. 3 and 4. In the applications which we have in mind, it is convenient to prove tightness via Theorem 2.2 below. Let $\hat{C}_0$ denote the space of real-valued continuous functions on R^r, with compact support.

THEOREM 2.2 [8] Fix N. For each $f(\cdot)$ in a dense set $\mathcal{D}_1 \subset \hat{C}_0$ which contains the square of each function in it, let there be a sequence $\{f^{\varepsilon,N}(\cdot)\}$ in $\mathcal{H}$ such that $f^{\varepsilon,N}(\cdot) \in \mathcal{D}(\hat{A}^{\varepsilon})$, and for each real $T > 0$, let there be a random variable $M_T^{\varepsilon,N}(f)$ such that

$$P\{\sup_{t\leq T} |f^{\varepsilon,N}(t) - f(x^{\varepsilon,N}(t))| \geq \alpha\} \to 0 \qquad \text{as } \varepsilon \to 0, \text{ each } \alpha > 0$$

$$\sup_{t\leq T} |\hat{A}^{\varepsilon} f^{\varepsilon,N}(t)| \leq M_T^{\varepsilon,N}(f) \tag{2.3}$$

$$\sup_{\varepsilon>0} P\{M_T^{\varepsilon,N}(f) \geq K\} \to 0 \qquad \text{as } K \to \infty$$

Then $\{f(x^{\varepsilon,N}(\cdot))\}$ is tight in $D^1[0,\infty)$ for $f(\cdot) \in \mathcal{D}_1$, and $\{x^{\varepsilon,N}(\cdot)\}$ is tight in $D^r[0,\infty)$.

REMARK 2. Typically, $\mathcal{D}_1 = \hat{C}_0^{2,3}$, but with the time dependence suppressed, and the $f^{\varepsilon,N}(\cdot)$ are constructed as they are for Theorem 2.1. The conditions are not particularly restrictive, as will be seen in the examples below, or as can be seen from the applications in the references.

2.2 Discrete Parameter

Theorem 2.1 simplifies slightly in the discrete parameter case. Let $\{\tau_\varepsilon\}$ denote a sequence of positive numbers tending to zero as $\varepsilon \to 0$, and suppose that the random sequence $\{x_n^\varepsilon\}$ satisfies $x_{n+1}^\varepsilon = x_n^\varepsilon + h^\varepsilon(\xi_n^\varepsilon, x_n^\varepsilon)$, for some suitable measurable $h^\varepsilon(\cdot,\cdot)$ and random sequence $\{\xi_n^\varepsilon\}$. Define $\{x_n^{\varepsilon,N}\}$ by $x_{n+1}^{\varepsilon,N} = x_n^{\varepsilon,N} + h^\varepsilon(\xi_n^\varepsilon, x_n^{\varepsilon,N})b_N(x_n^{\varepsilon,N})$, and define $x^\varepsilon(\cdot)$, $x^{\varepsilon,N}(\cdot)$ by $x^\varepsilon(t) = x_n^\varepsilon$, $x^{\varepsilon,N}(t) = x_n^{\varepsilon,N}$ on $[n\tau_\varepsilon, n\tau_\varepsilon + \tau_\varepsilon)$. Write E_n^ε for $E_{n\tau_\varepsilon}^\varepsilon$. Then for $f(\cdot) \in \mathcal{H}$, we now define $\hat{A}^\varepsilon f(n\tau_\varepsilon) = [E_n^\varepsilon f(n\tau_\varepsilon + \tau_\varepsilon) - f(n\tau_\varepsilon)]/\tau_\varepsilon$. In the application of Sec. 4, $\tau_\varepsilon = \varepsilon$. Theorem 2.1 can be now rewritten as:

THEOREM 2.3 [8] Assume the conditions of Theorem 2.1, but where (2.4) replaces (2.2), and the $f^{\varepsilon,N}(\cdot)$ are constant on the $[n\tau_\varepsilon, n\tau_\varepsilon + \tau_\varepsilon)$ intervals.

$$\sup_{n,\varepsilon} E|f^{\varepsilon,N}(n\tau_\varepsilon)| < \infty \qquad \sup_{n,\varepsilon} E|\hat{A}^\varepsilon f^{\varepsilon,N}(n\tau_\varepsilon)| < \infty$$

$$E|f^{\varepsilon,N}(n\tau_\varepsilon) - f(x^{\varepsilon,N}(n\tau_\varepsilon),n\tau_\varepsilon)| \to 0 \tag{2.4}$$

$$E|\hat{A}^\varepsilon f^{\varepsilon,N}(n\tau_\varepsilon) - (\frac{\partial}{\partial t} + A^N)f(x^{\varepsilon,N}(n\tau_\varepsilon),n\tau_\varepsilon)| \to 0$$

for each t and $n\tau_\varepsilon \to t$ as $\varepsilon \to 0$. Then the conclusions of Theorem 2.1 continue to hold.

(There is also the obvious discrete parameter simplification of Theorem 2.2.)

REMARK ON A SPECIAL CASE

A useful method for constructing the $f^{\varepsilon,N}(\cdot)$ for both discrete and continuous parameter cases is illustrated in the examples and references. When we are concerned only with averaging and the limit process is the solution to an ordinary differential equation, then by use of a slightly different perturbed test function, the conditions for convergence can often be improved. We simply remark on a discrete parameter case where the $\{\xi_n\}$ is bounded, stationary, and strong mixing with arbitrary mixing rate. Write $x^\varepsilon_{n+1} = x^\varepsilon_n + \varepsilon\bar{h}(x^\varepsilon_n) + \varepsilon\delta h(x^\varepsilon_n,\xi_n)$, where $\bar{h}(\cdot)$ is continuous, $\dot{x} = \bar{h}(x)$ has a unique solution for each initial condition, $E\delta h(x,\xi) \equiv 0$, and $\delta h(x,\xi)$ is bounded on bounded x-sets. The $(1 - \varepsilon)^{j-n}$ weighing allows the use of an arbitrary mixing rate.

For $f(\cdot) \in \hat{C}_0^{2,3}$, define $f^\varepsilon(n\varepsilon) = f(x^\varepsilon_n) + f^\varepsilon_1(n\varepsilon)$, where $f^\varepsilon_1(n\varepsilon) = f^\varepsilon_1(x^\varepsilon_n,n\varepsilon)$ and

$$f^\varepsilon_1(x,n\varepsilon) = \varepsilon\sum_{j=n}^{\infty} (1 - \varepsilon)^{j-n} f'_x(x,n\varepsilon)E_n\delta h(x,\xi_j)$$

Assume that for each $\delta > 0$, there is an $N_\delta < \infty$ and $\varepsilon_0 > 0$ such that $\varepsilon \leq \varepsilon_0$ and $j - n \geq N_\delta$ implies that

$\frac{1}{\varepsilon} E|E_n f_x'(x_{n+1}^\varepsilon, n\varepsilon + \varepsilon)\delta h(x_{n+1}^\varepsilon,\xi_j) - f_x'(x_n^\varepsilon,n\varepsilon)\delta h(x_n^\varepsilon,\xi_j)| \le \delta$

Then the untruncated sequence $\{x^\varepsilon(\cdot)\}$ converges weakly to a solution of $\dot{x} = \bar{h}(x)$ with initial condition $x(0)$ if $x^\varepsilon(0)$ converges weakly to $x(0)$.

OUTLINE OF PROOF

By boundedness of $h(x,\xi)$ on bounded x-sets, the (N-truncated) $\{x^\varepsilon(\cdot)\}$ is tight if $\{x^\varepsilon(0)\}$ is. Thus we only need check (2.4). First note that if for real $\{b_n\}$, $b_n \to 0$, then $\varepsilon \Sigma_0^\infty (1 - \varepsilon)^j b_j \to 0$ as $\varepsilon \to 0$. By the mixing condition $E|E_n \delta h(x_n^\varepsilon,\xi_j)| \to 0$ as $n - j \to \infty$ (uniformly in n). Next, for $f(\cdot) \in \hat{C}_0^{2,3}$, we have the expansion

$$E_n f^\varepsilon(n\varepsilon + \varepsilon) - f^\varepsilon(n\varepsilon) = \varepsilon f_t(x_n^\varepsilon,n\varepsilon) + \varepsilon f_x'(x_n^\varepsilon,n\ )\bar{h}(x_n^\varepsilon) + o(\varepsilon) + \varepsilon \sum_{j=n+1}^{\infty} \varepsilon(1 - \varepsilon)^{j-n-1}$$

$$E_n[f_x'(x_{n+1}^\varepsilon, n\varepsilon + \varepsilon)\delta h(x_{n+1}^\varepsilon,\xi_j) - f_x'(x_n^\varepsilon,n\varepsilon)\delta h(x_n^\varepsilon,\xi_j)]\ \varepsilon^{-1}$$

$$+ \varepsilon \sum_{j=n+1}^{\infty} \varepsilon(1 - \varepsilon)^{j-n-1} E_n \delta h(x_n^\varepsilon,\xi_j)$$

By our hypotheses, the coefficients of ε in the last two terms satisfy $\text{p-lim}_{\varepsilon\to 0}(\cdot) = 0$, and (2.4) is satisfied for the appropriate operator $\hat{A}^\varepsilon$ or $\hat{A}^{\varepsilon,N}$.

2.3 Approximation of Asymptotic and Invariant Measures

Theorems 2.1 to 2.3 yield weak convergence of $\{x^\varepsilon(\cdot)\}$ to $x(\cdot)$ in $D^r[0,\infty)$. Frequently, in applications we are more interested in the measures of $x^\varepsilon(t)$ for large t and small ε. Suppose that $x(\cdot)$ has a unique invariant measure $\mu(\cdot)$. In practice this measure is often used as an approximation to the "asymptotic" measures of $x^\varepsilon(t)$. In particular, it is of interest to show at least that the measure of $x^\varepsilon(t)$

converges in some sense to $\mu(\cdot)$ as $t \to \infty$, $\varepsilon \to 0$ in any way at all. Weak convergence does not yield such information directly. But weak convergence together with an "averaged Liapunov function" technique and some assumptions on $x(\cdot)$ can yield the desired result [15]. Here we cite some results which will be used in the examples. The basic assumptions are (A1) to (A4), and Theorem 2.4 is the basic convergence theorem. Conditions guaranteeing (A4) will be given after the statement of Theorem 2.4.

Theorem 2.4 requires:

A1. $x(\cdot)$ is a Feller diffusion process with continuous coefficients and a unique weak sense solution on $[0,\infty)$ for each $x(0) = x$.

A2. $x(\cdot)$ has a unique invariant measure $\mu(\cdot)$ and $P_x\{x(t) \in \cdot\}$ converges weakly to $\mu(\cdot)$ for each x, as $t \to \infty$. The convergence is uniform in compact x-sets; i.e., for each $f(\cdot)$ in the space of continuous bounded functions on R^r, $E_x f(x(t)) \to E_\mu f(x(0))$ uniformly in x in compact sets, as $t \to \infty$.

A3. $x^\varepsilon(\cdot) \to x(\cdot)$ [initial condition $x(0)$] weakly if $x^\varepsilon(0) \to x(0)$ weakly in $D^r[0,\infty)$.

A4. There is an $\varepsilon_0 > 0$ such that $\{x^\varepsilon(t),\ 0 < \varepsilon \leq \varepsilon_0,\ t \geq 0\}$ is tight.

THEOREM 2.4 [15] Assume (A1) to (A4). Then for each integer m, real $T < \infty$, and bounded continuous $f(\cdot)$ and $\delta > 0$, there are $t_0(f,\delta) < \infty$ and $\varepsilon_0(f,\delta) > 0$ such that for $t \geq t_0(f,\delta)$ and $\varepsilon \leq \varepsilon_0(f,\delta)$, and any sequence $\{x^\varepsilon(\cdot)\}$ which converges weakly to $x(\cdot)$ and $T \geq \Delta_i \geq 0$, $i = 1, \ldots, m$,

$$|Ef(x^\varepsilon(t + \Delta_i),\ i \leq m) - E_\mu f(x(\Delta_i),\ i \leq m)| < \delta$$

Let $(x^\varepsilon(\cdot),\xi^\varepsilon(\cdot))$ be Markov and have an invariant measure $v^\varepsilon(\cdot)$ with x-marginal $\mu^\varepsilon(\cdot)$. Replace (A4) by: There is a sequence T_ε (which might tend to ∞ as $\varepsilon \to 0$, and T_ε can depend on the initial condition) such that (A4) holds for $t \geq T_\epsilon$. Then $\{\mu^\epsilon(\cdot)\}$ is tight. Let $\sup_{t>0,\epsilon>0} E|\xi^\epsilon(t)| < \infty$, where the expectations are with respect to the stationary measure $v^\epsilon(\cdot)$, and suppose that (A3) holds if $\xi^\epsilon(0)$ varies in any compact set not depending on ϵ. Then $\{\mu^\epsilon(\cdot)\}$ converges weakly to $\mu(\cdot)$ as $\epsilon \to 0$.

REMARK 3. In the proof of the last assertion of the theorem, we only need prove that $\{\mu^\epsilon(\cdot)\}$ is tight, and the given tightness condition is used for this. In applications, it is often possible to prove a result such as $\overline{\lim}_{t\to\infty} E_{x,\xi}|x^\epsilon(t)| \leq K$, where x, $\xi = \xi(0)$ are bounded random variables and K does not depend on x, ξ for small ϵ. This implies the first sentence of the replacement for (A4) in the theorem (see also Theorem 3.1).

A CRITERION FOR (A4)

The tightness (A4) can be proved if certain stability assumptions (of the "recurrence" type) are made on the limit $x(\cdot)$; in particular, if there is a Liapunov function $V(\cdot)$ of the sort frequently used to prove recurrence for $x(\cdot)$. An "averaged" form of $V(\cdot)$ is used; the averaged function is obtained similarly to the way the perturbation $f^\epsilon(\cdot)$ to $f(\cdot)$ is obtained in applications of Theorem 2.1 to 2.3. Owing to the mixing condition (B1) and to the smoothness conditions (B2) and (B4) below, Theorem 2.5 is too strong for many applications. The conditions are weakened in Theorem 3.1. Further comments and developments appear in Ref. 15.

Let us specialize to the form

$$\dot{x}^\epsilon = \frac{F(x^\epsilon,\xi^\epsilon)}{\epsilon} + G(x^\epsilon,\xi^\epsilon) + \overline{G}(x^\epsilon) \tag{2.5}$$

where $\xi^\epsilon(t) = \xi(t/\epsilon^2)$ for some stationary process $\xi(\cdot)$. Let $EF(x,\xi) = EG(x,\xi) = 0$. This noise scaling is a common way of getting a wide-band noise process, with "bandwidth" parameter ϵ. Under various sets of conditions [5,7,14] the differential generator of $x(\cdot)$ is given by

$$Af(x) = f_x'(x)\overline{G}(x) + \int_0^\infty E[f_x'(x)F(x,\xi(t))]_x' F(x,\xi(0))\, dt \quad (2.6a)$$

If $\xi(\cdot)$ is Markov and $F(\cdot,\cdot)$ not smooth, then we use the representation

$$Af(x) = f_x'(x)\overline{G}(x) + \int_0^\infty E[E_{\xi(0)} f_x'(x)F(x,\xi(t))]_x' \times F(x,\xi(0))\, dt \quad (2.6b)$$

if the derivatives exist for $t > 0$ and are integrable.

We will require (B1) to (B4) in Theorem 2.5. The constant K might vary from usage to usage. These conditions as well as the extensions to be discussed in Theorem 3.1 fit many common applications.

B1. $\xi(\cdot)$ is a bounded, right continuous, stationary φ-mixing process [16] with $\int_0^\infty \varphi^{1/2}(t)\, dt < \infty$.

B2. $F(\cdot,\cdot)$, $G(\cdot,\cdot)$, and $\overline{G}(\cdot)$ are continuous, R^r-valued functions whose growth (as $|x| \to \infty$) is $O(|x|)$. The partial derivatives of $F(\cdot,\xi)$ up to order 2 [and of $G(\cdot,\xi)$ up to order 1] are bounded uniformly in x, ξ, and $EF(x,\xi) \equiv 0 \equiv EG(x,\xi)$.

B3. There is a diffusion process $x(\cdot)$ with differential generator A defined by (2.6), and which satisfies (A1) to (A2). There is a continuous Liapunov function $0 \le V(x) \to \infty$ as $|x| \to \infty$, and a λ_0 and $\alpha_0 > 0$ such that $AV(x) \le -\alpha_0$ for $x \notin Q_0 = \{x : V(x) \le \lambda_0\}$. The partial derivatives of $V(\cdot)$ up to order 3 are continuous.

B4. There are constants K such that, uniformly in x, ξ,

$$|V_x'(x)G(x,\xi)| + |V_x'(x)F(x,\xi)| \le K(1 + V(x))$$

$$|(V_x'(x)F(x,\xi))_x' F(x,\xi)| \le K(1 + V(x))$$

$$|(V_x'(x)G(x,\xi))_x' U(x,\xi)| \le K(1 + |AV(x)|)$$

$$\text{for } U = F,\ G,\ \overline{G}$$

$$|(V_x'(x)F(x,\xi))_x'U(x,\xi)| \le K(1 + |AV(x)|)$$
$$U = G, \bar{G}$$
$$|((V_x'(x)F(x,\xi))_x'F(x,\xi))_x'U(x,\xi)| \le K(1 + |AV(x)|)$$
$$U = F, G, \bar{G}$$

THEOREM 2.5 Under (B1) to (B4) and the tightness of $\{x^{\varepsilon}(0)\}$, condition (A4) holds.

3. EXAMPLE 1. A PHASE-LOCKED LOOP WITH A LIMITER

The phase-locked loop (PLL) is a basic device used for synchronization in communications systems; in particular, to estimate the phase, frequency, and certain "timing" data of the incoming signal. In this example we treat a simple PLL, but with a commonly used (and hard to analyze) nonlinearity. The circuit and definitions of some terms are given by Figure 1. The VCO (voltage-controlled oscillator) is an oscillator which oscillates at a frequency whose deviation from a given central frequency is proportional to its input voltage. First, we describe the input noise and the parametrization, and then simplify the model slightly. Then Theorems 2.1 to 2.4 will be applied. The noise model used is chosen for ease of presentation, and because the technical details of proof [14] are not too hard. Further details of this example can be found in Ref. 14, which also contains analyses of several other phase-locked-loop devices, with either nonlinearities or other features which seem to require a robust diffusion approximation method such as the one presented here. See Refs. 17, 18 for a general background.

Refer to Figure 1. The object is to track the unknown and time-varying phase $\theta(\cdot)$ of the input signal. Let $\theta(\cdot)$ be continuously differentiable here. With not much extra

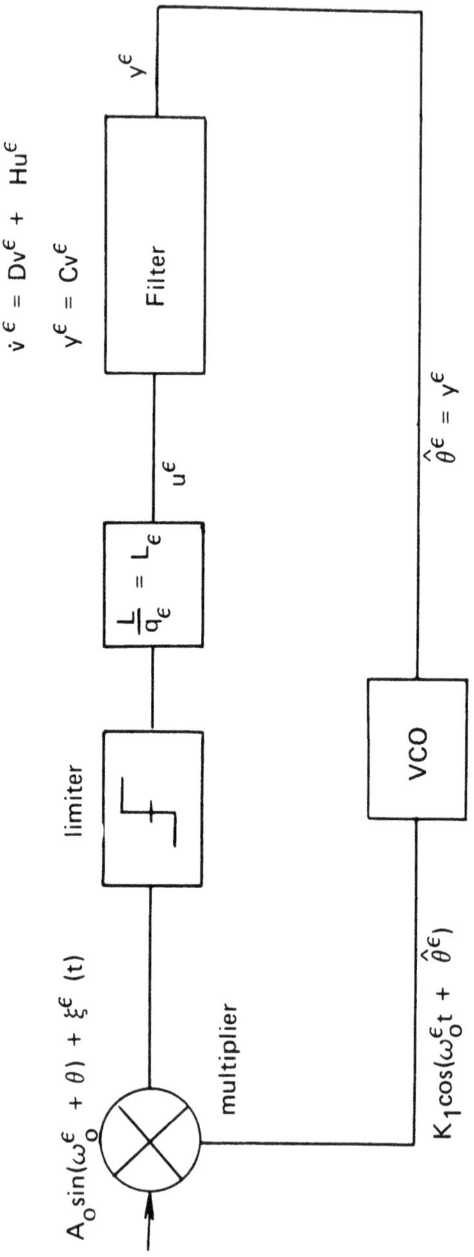

Figure 1. Phase-locked-loop circuit.

trouble, various random process models can be used for both $\theta(\cdot)$ and also for A_0 (which is a constant here). The system operates with a high central frequency and a noise bandwidth which is wide but small relative to the central frequency. Let $\{q_\varepsilon\}$ be a sequence of positive real numbers tending to zero such that $\varepsilon/q_\varepsilon \to 0$ as $\varepsilon \to 0$. Let $\{\varphi_i, z_i(\cdot); i = 1, 2\}$ be mutually independent with the φ_i uniformly distributed on $[0,2\pi]$, and the $z_i(\cdot)$ Gaussian processes with spectral density $S(\omega)$ and unit variance. Define $z_i^\varepsilon(t) = z_i(t/q_\varepsilon^2)$ and

$$\xi^\varepsilon(t) = \frac{\sigma[z_1^\varepsilon(t)\cos(\omega_0^\varepsilon t + \varphi_1) + z_2^\varepsilon(t)\sin(\omega_0^\varepsilon t + \varphi_2)]}{q_\varepsilon} \tag{3.1}$$

where ω_0^ε is the input (and central VCO) frequency. Then $(1/2)[S(q_\varepsilon^2(\omega - \omega_0^\varepsilon)) + S(q_\varepsilon^2(\omega + \omega_0))] = S_\varepsilon(\cdot)$ is the spectral density of $\xi^\varepsilon(\cdot)$. Let $\omega_0^\varepsilon = \omega_0/\varepsilon^2$. Then $S_\varepsilon(\cdot)$ is of the desired parametrized wide-band form. With no loss of generality, we set $\varphi_i = 0$. The matrix $\dot{D}$ in the filter in Figure 1 is stable.

A gain of roughly the form $L_\varepsilon = L/q_\varepsilon$ in Figure 1 (either before or after the filter) is needed because if $q_\varepsilon L_\varepsilon \to 0$ as $\varepsilon \to 0$, then the input to the VCO will go to zero weakly as $\varepsilon \to 0$. If $q_\varepsilon L_\varepsilon \to \infty$ as $\varepsilon \to 0$, then the output will diverge to infinity. The scaling is consistent with heuristic methods used for analyzing such systems. Let sign (0) be any number in $[-1,1]$. Let $z_i(\cdot)$ be Gauss-Markov; in particular, let $z_i(t) = MZ_i(t)$, where $dZ_i = AZ_i\,dt + B\,dw_i$, and $\{w_i(\cdot); i = 1,2\}$ are independent standard Wiener processes and A a stable matrix.

We now simplify the model slightly. The output of the multiplier can be written in the form

$$\frac{\sigma}{2q_\varepsilon}\{z_1^\varepsilon(t)[\cos\hat{\theta}^\varepsilon + \cos(2\omega_0^\varepsilon t + \hat{\theta}^\varepsilon)]$$
$$+ z_2^\varepsilon(t)[\sin(-\hat{\theta}^\varepsilon) + \sin(2\omega_0^\varepsilon t + \hat{\theta}^\varepsilon)]\}$$

$$+ \frac{A_0}{2} [\sin (\theta^\varepsilon - \hat{\theta}^\varepsilon) + \sin (\theta^\varepsilon + \hat{\theta}^\varepsilon + 2\omega_0^\varepsilon t)] \qquad (3.2)$$

If we proceeded with this "output," the terms involving $2\omega_0^\varepsilon t$ would have no effect on the limit, and for notational simplicity here we drop them at this point. In fact, it is common engineering practice to assume that the multiplier acts as a filter which eliminates these same terms involving $2\omega_0^\varepsilon t$. While one must be careful in using this "filtering" idea, since the system is nonlinear and time varying and the properties of $\hat{\theta}^\varepsilon$ are not known a priori for small ε, these terms do not, in fact, affect the limit process (as can be shown by retaining them throughout the analysis). Let L = 1 (absorb it into the other parameters).

The main result is the following: $\{v^\varepsilon(\cdot), \hat{\theta}^\varepsilon(\cdot)\}$ converges weakly to the diffusion $v(\cdot)$, $\hat{\theta}(\cdot)$ defined by (3.3), where $B(\cdot)$ is a standard Wiener process and $\rho(t) = Ez_i(t)z_i(0)/Ez_i^2(t)$ (the stationary measure is to be used here). If $\rho(t) = \exp(-a|t|)$, $a > 0$, then $\sigma_0^2 = (2 \ln 2)/a$ (Ref. 11, sec. 6).

$$dv = [Dv + \frac{A_0}{\sigma} H\sqrt{\frac{2}{\pi}} \sin (\theta - \hat{\theta})] dt + H\sigma_0 \, dB \qquad (3.3a)$$

$$d\hat{\theta} = Cv \, dt \qquad \dot{\theta} \text{ given} \qquad (3.3b)$$

$$\sigma_0^2 = 4 \int_0^\infty [P\{z_i(u) > 0, z_i(0) > 0\} - P\{z_i(u) < 0, z_i(0) > 0\} du$$

$$= \frac{4}{\pi} \int_0^\infty \sin^{-1} \rho(u) \, du$$

If we consider the standard PLL, where the limiter and gain L_ε are replaced by simply a unity gain, then the limit process is defined by

$$dv = [Dv + H \frac{A_0}{2} \sin (\theta - \hat{\theta})] dt + \sigma_1 H \, dB$$

$$d\hat{\theta} = Cv \, dt \qquad (3.4)$$

where

$$\sigma_1^2 = \frac{\sigma^2}{2} \int_0^\infty \rho(u)\, du$$

In the latter case, the analysis is much simpler and the Gaussian assumption can be weakened.

For small σ, the system with the limiter is preferable to the system without the limiter, and conversely for large σ. This interesting and very useful result has also been verified by simulations.

In order to use the limit results (3.3) or (3.4) to get probability estimates in any particular application, we must estimate the values of ω_0, σ^2, $\rho(\cdot)$ for that application. This can often be done via measurements on the system, and is discussed briefly in Ref. 14.

APPLICATION OF THEOREMS 2.1 AND 2.2

Dropping the terms in (3.2) which contain $2\omega_0^\varepsilon t$, the input to the filter in Figure 1 is $u^\varepsilon(t,\hat{\theta}^\varepsilon(t),\theta(t))$, where

$$u^\varepsilon(t,\hat{\theta},\theta) = \frac{1}{q_\varepsilon} \operatorname{sign} \{\frac{\sigma}{2q_\varepsilon} [z_1^\varepsilon(t) \cos \hat{\theta} - z_2^\varepsilon(t) \sin \hat{\theta}] + \frac{A_0}{2} \sin (\theta - \hat{\theta})\}$$

Also,

$$\begin{aligned} \dot{v}^\varepsilon &= Dv^\varepsilon + HEu^\varepsilon + H(u^\varepsilon - Eu^\varepsilon) \\ \dot{\hat{\theta}}^\varepsilon &= Cv^\varepsilon, \qquad (v^\varepsilon,\hat{\theta}^\varepsilon) = x^\varepsilon \\ Eu^\varepsilon(t,\hat{\theta},\theta) &= \sqrt{\frac{2}{\pi}}\, \frac{A_0}{\sigma} \sin (\theta - \hat{\theta}) + O(q_\varepsilon) \end{aligned} \tag{3.5}$$

Throughout expectations and conditional expectations are over the $z_i^\varepsilon(\cdot)$ only. The $\hat{\theta}^\varepsilon(t)$, $\theta(t)$ are considered to be parameters when taking expectations.

Now, apply Theorem 2.1 to a test function $f(\cdot) \in \hat{C}_0^{2,3}$. To be consistent with Theorems 2.1 and 2.2, we should work with $\{x^{\varepsilon,N}(\cdot)\}$, for large N. We will give only a rough outline of the procedure and the notation is simpler if we drop the N-superscript and the $b_N(\cdot)$ (unless mentioned otherwise) The given estimates are actually valid for $\{x^\varepsilon(\cdot)\}$, but this is unimportant for application of Theorems 2.1 and 2.2. Now, applying $\hat{A}^\varepsilon$ to $f(\cdot)$ and writing $x^\varepsilon(t) = x^\varepsilon$ yields

$$\begin{aligned}\hat{A}^\varepsilon f(x^\varepsilon,t) &= f_t(x^\varepsilon,t) + f_{\hat{\theta}}(x^\varepsilon,t)Cv^\varepsilon \\ &\quad + f_v'(x^\varepsilon,t)[Dv^\varepsilon + HEu^\varepsilon(t,\hat{\theta}^\varepsilon,\theta(t))] \\ &\quad + f_v'(x^\varepsilon,t)H\,\delta u^\varepsilon(t,\hat{\theta}^\varepsilon,\theta(t))\end{aligned}$$

where $\delta u^\varepsilon(t,\hat{\theta},\theta) = u^\varepsilon(t,\hat{\theta},\theta) - Eu^\varepsilon(t,\hat{\theta},\theta)$. Only the "noise term" $f_v'H\,\delta u^\varepsilon$ needs to be averaged out. These so-called averaging operations follow the scheme in Refs. 3 and 7 and they yield the perturbations f_i^ε to f required by Theorems 2.1 and 2.2. (See also Refs. 9 and 12 to 15 for applications and illustrations of this averaging idea.)

Define the first perturbation $f_1^\varepsilon(t) = f_1^\varepsilon(x^\varepsilon(t),t)$ by

$$\begin{aligned}f_1^\varepsilon(x^\varepsilon,t) &= \int_0^\infty ds\ E_t^\varepsilon f_v'(x^\varepsilon,\ t+s)H\,\delta u^\varepsilon(t+s,\hat{\theta}^\varepsilon,\theta(t)) \\ &= q_\varepsilon^2 \int_0^\infty ds\ E_t^\varepsilon f_v'(x^\varepsilon,\ t+q_\varepsilon^2 s)H \\ &\quad \times \delta u^\varepsilon(t+q_\varepsilon^2 s,\ \hat{\theta}^\varepsilon,\theta(t))\end{aligned} \tag{3.6}$$

Note that the integrand of the first integral of (3.6) at $s = 0$ is just the noise term which we wish to average out. Owing to the sign function in δu^ε and to the fact that the operation $\hat{A}^\varepsilon f_1^\varepsilon$ involves a differentiation, it is convenient to write $E_t^\varepsilon\ \delta u^\varepsilon$ as follows. Define

$$[t+s] \equiv z_1^\varepsilon(t+s)\cos\hat{\theta}^\varepsilon - z_2^\varepsilon(t+s)\sin\hat{\theta}^\varepsilon + \frac{q_\varepsilon A_0}{2}\sin(\theta - \hat{\theta}^\varepsilon)$$

The distribution of $[t+s]$ conditioned on $\{z_i^\varepsilon(u),\ u \le t,\ i = 1, 2\}$ is $N(m_\varepsilon(s),\sigma_\varepsilon^2(s))$, where

$$m_\varepsilon(s) = \rho\left(\frac{s}{q_\varepsilon^2}\right)[z_1^\varepsilon(t)\cos\hat{\theta}_\varepsilon - z_2^\varepsilon(t)\sin\hat{\theta}^\varepsilon] + \frac{A_0 q_\varepsilon}{2}\sin(\theta - \hat{\theta}^\varepsilon)$$

$$\sigma_\varepsilon^2(s) = 1 - \rho^2\left(\frac{s}{q_\varepsilon^2}\right)$$

Then, using the notation $dN(m,\sigma^2) = d\xi \exp[-(\xi - m)^2/2\sigma^2]/\sqrt{2\pi}\,\sigma$, we have

$$E_t^\varepsilon\, \delta u^\varepsilon(t + s, \hat{\theta}^\varepsilon, \theta(t))$$
$$= \frac{1}{q_\varepsilon}\int_{-\infty}^{\infty} (\text{sign } \xi)\left\{(dN[m_\varepsilon(s),\sigma_\varepsilon^2(s)] - dN\left[\frac{A_0 q_\varepsilon}{\sigma}\sin\left(\theta(t) - \hat{\theta}^\varepsilon\right)\ 1\right]\right\} \tag{3.7}$$

It can be shown that $f_1^\varepsilon(\cdot) \in \mathcal{D}(\hat{A}^\varepsilon)$, and that

$$|f_1^\varepsilon(t)| = O(q_\varepsilon)[1 + |z^\varepsilon(t)|] \tag{3.8}$$

and [writing $x^\varepsilon(t) = x^\varepsilon$, and letting I denote the integrand of the first integral in (3.6)]

$$\hat{A}^\varepsilon f_1^\varepsilon(x^\varepsilon,t) = -f_v'(x^\varepsilon,t)H\,\delta u^\varepsilon(t,\hat{\theta}^\varepsilon,\theta(t)) + \int_0^\infty ds(E_t^\varepsilon I)_{\hat{\theta}}\dot{\hat{\theta}}^\varepsilon + \int_0^\infty ds(E_t^\varepsilon I)_\theta\dot{\theta}(t) + \int_0^\infty ds(E_t^\varepsilon I)_v'\dot{v}^\varepsilon \tag{3.9}$$

In (3.9), the representation (3.7) is used for $E_t^\varepsilon \delta u^\varepsilon$, so all derivatives are taken with respect to parameters in the normal density function. The first two integrals on the right of (3.9) are bounded by the right side of (3.8) (to see this, change variables $s/q_\varepsilon^2 \to s$), and the last term equals (3.10) modulo a term which is bounded by the right side of (3.8). In (3.10), [t + s] denotes the term defined below (3.6):

$$\frac{1}{q_\varepsilon^2}\int_0^\infty ds\, H' f_{vv}(x^\varepsilon,\ t+s)H(E_t^\varepsilon \text{ sign } [t+s]$$
$$- E \text{ sign } [t+s])(\text{sign } [t] - E \text{ sign } [t])$$
$$\equiv k^\varepsilon(t,x^\varepsilon) \tag{3.10}$$

The term k^ε is neither negligible nor part of $Af(x^\varepsilon,t)$ and must be averaged out. To do this, we define the second perturbation function $f_2^\varepsilon(t) = f_2^\varepsilon(t,x^\varepsilon(t))$, where

$$f_2^\varepsilon(t,x^\varepsilon) = \int_0^\infty d\tau[E_t^\varepsilon k^\varepsilon(t+\tau,\ x^\varepsilon) - Ek^\varepsilon(t+\tau,\ x^\varepsilon)]$$

It can be shown that $|f_2^\varepsilon(t)| = O(q_\varepsilon)[1 + |z^\varepsilon(t)|^2]$, that $f_2^\varepsilon(\cdot) \in \mathcal{D}(\hat{A}^\varepsilon)$, and that [again writing $x^\varepsilon = x^\varepsilon(t)$]

$$\hat{A}^\varepsilon f_2^\varepsilon(t,x^\varepsilon) = -k^\varepsilon(t,x^\varepsilon) + Ek^\varepsilon(t,x^\varepsilon)$$
$$+ \text{ error term} \tag{3.11}$$

where $|\text{error term}| = O(q_\varepsilon)[1 + |z^\varepsilon(t)|^3]$. In applying $\hat{A}^\varepsilon$ to $f_2^\varepsilon(\cdot)$ we use a representation of the type used in (3.7) for the conditional expectations and expectations, so that the derivatives are taken only with respect to parameters in the normal density functions. Finally, it can be shown via the technique in Ref. 11, sec. 6 that $Ek^\varepsilon(t,x) - H'f_{vv}(x,t)H\sigma_0^2/2 \to 0$ uniformly in bounded (x,t) sets as $\varepsilon \to 0$.

Now, let us summarize the above calculations, and reintroduce the superscript N on $x^{\varepsilon,N}(\cdot)$ only. The stated bounds on $|f_i^\varepsilon(\cdot)|$, $|f^\varepsilon(\cdot) - f(\cdot)|$, and on the error term in (3.11) and on the first two integrals on the right side of (3.9) all hold for each N. Now Theorem 2.2 implies tightness for $\{x^{\varepsilon,N}(\cdot)\}$ for each N, and $\text{p-lim}_{\varepsilon\to 0}\, [f^\varepsilon(\cdot) - f(x^{\varepsilon,N}(\cdot),\cdot)] = 0$. The expressions given for $\hat{A}^\varepsilon f_i^\varepsilon(x^\varepsilon,t)$ above equal $\hat{A}^\varepsilon f_i^\varepsilon(x^{\varepsilon,N},t)$, if $x^{\varepsilon,N}$ and x^ε are in S_N. Thus

$$\text{p-lim}_{\varepsilon\to 0}[\hat{A}^\varepsilon f^\varepsilon(x^{\varepsilon,N}(\cdot),\cdot) - (\frac{\partial}{\partial t} + A)f(x^{\varepsilon,N}(\cdot),\cdot)]$$
$$\times\, b_N(x^{\varepsilon,N}(\cdot)) = 0 \tag{3.12}$$

where A is the differential generator of the process (3.3), and we conclude that (3.3) is the weak limit of $\{v^{\varepsilon}(\cdot),\hat{\theta}^{\varepsilon}(\cdot)\}$ if $\{v^{\varepsilon}(0),\hat{\theta}^{\varepsilon}(0)\}$ converge weakly to $\{v(0),\hat{\theta}(0)\}$. Further details on this and on similar problems are given in Refs. 11, 12 and 14.

APPROXIMATE INVARIANT MEASURES FOR (3.5)

Since the noise $z^{\varepsilon}(\cdot)$ is unbounded, Theorem 2.5 cannot be applied, but there is a result similar to Theorem 2.4 which is useful. Let $\theta(t) = \theta_0$, a constant. We interpret both $\hat{\theta}^{\varepsilon}(\cdot)$ and $\hat{\theta}(\cdot)$ modulo 2π, so they are always bounded. Also, conditions (A1) to (A3) hold. The main problem is in showing that $\{v^{\varepsilon}(\cdot);\ t \geq 0,\ \varepsilon \text{ small}\}$ is tight. In particular, we want to show that $(\hat{\theta}^{\varepsilon}(\cdot),v^{\varepsilon}(\cdot)) = x^{\varepsilon}(\cdot)$ has an invariant measure $m^{\varepsilon}(\cdot)$ for small ε, and that the v-marginals $\{\mu^{\varepsilon}(\cdot)\}$ of $\{m^{\varepsilon}(\cdot)\}$ are tight.

There is a useful "averaged Liapunov method" which is similar to Theorem 2.5, but where the noise is allowed to be unbounded, and the conditions are tailored to a class of cases which occur frequently. For motivation, we first consider the present case. For some positive definite and symmetric P, let $V(x) = v'Pv$ be a Liapunov function for $\dot{v} = Dv$, and let L denote the differential generator of (3.3a). Then there are $\gamma > 0$, $K < \infty$ such that $LV(x) \leq -\gamma V(x) + K$. An averaging method can now be applied to $V(\cdot)$, to yield a perturbed Liapunov function $V^{\varepsilon}(\cdot)$, which in turn is useful for proving the required tightness, via a method which resembles those used for "Liapunov function" proofs of recurrence of strong Markov processes. Actually, since $\hat{A}^{\varepsilon}$ cannot be applied to unbounded functions, the averaging is done on and the technical conditions are introduced on the truncated Liapunov function defined by $V_M(x) = V(x)b_M(v)$, where $b_M(\cdot)$

is the truncation function defined earlier.

The perturbed or averaged truncated Liapunov function $V_M^\varepsilon(t) = V_M^\varepsilon(x^\varepsilon(t),t)$ is obtained by perturbing $V_M(v^\varepsilon(t))$ in exactly the same way that we perturbed $f(x^{\varepsilon,N}(t),t)$ to get $f^\varepsilon(x^{\varepsilon,N}(t),t) = f^\varepsilon(t)$, except that here we use the original process $x^\varepsilon(\cdot)$ and not the truncation $x^{\varepsilon,N}(\cdot)$.

In more generality, the technique, assumptions, and result are formalized as follows. We suppose that $\dot{x}^\varepsilon = H^\varepsilon(x^\varepsilon,\xi^\varepsilon)$ takes the form (2.5) and that $\xi^\varepsilon(t) = \xi(t/\varepsilon^2)$. (In the example of this section, q_ε is used in lieu of ε.) Suppose that for some $\varepsilon_0 > 0$, $V_M^\varepsilon(\cdot) \in \mathcal{D}(\hat{A}^\varepsilon)$ for $\varepsilon \leq \varepsilon_0$. Suppose that there are random variables $\tilde{\xi}^\varepsilon(t)$, $\bar{\xi}^\varepsilon(t)$, integers p, q, and functions $\tilde{V}(\cdot)$, $\bar{V}(\cdot)$ satisfying (3.13). For the first two lines of (3.13), let $x \in S_M = \{x : |x| \leq M\}$. Let A be defined by (2.6):

$$\hat{A}^\varepsilon V_M^\varepsilon(x,t) = AV(x) + O(\varepsilon)[1 + V(x)] + O(\varepsilon)\tilde{\xi}^\varepsilon(t)\tilde{V}(x,t)$$

$$V_M^\varepsilon(x,t) = V_M(x) + O(\varepsilon)[1 + V(x)] + O(\varepsilon)\bar{\xi}^\varepsilon(t)\bar{V}(x,t)$$

$$\sup_{t,\varepsilon} E|\tilde{\xi}^\varepsilon(t)|^p < \infty \qquad \sup_{t,\varepsilon} E|\bar{\xi}^\varepsilon(t)|^q < \infty \tag{3.13}$$

$$|\tilde{V}(x,t)|^{p/p-1} = O(V(x)) \qquad |\bar{V}(x,t)|^{q/q-1} = O(V(x))$$

for large $|x|$

We also need (C1) to (C4).

C1. $\bar{G}(\cdot)$, $F(\cdot,\cdot)$, and $G(\cdot,\cdot)$ are measurable and are $O(|x|)$ for large $|x|$, uniformly in bounded ξ-sets.

C2. For the given sequence $\{x^\varepsilon(0)\}$, $\sup_{\varepsilon\leq\varepsilon_0} EV(x^\varepsilon(0)) < \infty$.

C3. For some initial condition, $\sup E|\xi(t)| < \infty$. For small ε, $\{x^\varepsilon(\cdot), \xi^\varepsilon(\cdot)\}$ and $\xi(\cdot)$ are Markov-Feller processes with right-continuous paths and homogeneous transition functions.

C4. $0 \leq V(x) \to \infty$ as $|x| \to \infty$ and $AV(x) \leq -\gamma V(x)$ for some $\gamma > 0$ and large $|x|$.

THEOREM 3.1 [15] Assume (C1) to (C4), (A3), (B3), (3.13), and the conditions above (3.13). Then there is an $\varepsilon_1 > 0$ such that for $\varepsilon \leq \varepsilon_1$, $(x^\varepsilon(\cdot),\xi^\varepsilon(\cdot))$ has an invariant measure $m^\varepsilon(\cdot)$. The x-marginals $\{\mu^\varepsilon(\cdot)\}$ of any such sequence of invariant measures are tight. If (A3) holds when $(x^\varepsilon(0),\xi^\varepsilon(0))$ has measure $m^\varepsilon(\cdot)$, then $\{\mu^\varepsilon(\cdot)\}$ converges weakly to $\mu(\cdot)$.

APPLICATION OF THEOREM 3.1 TO EXAMPLE 1

Let $\theta(t) \equiv \theta(0)$. Since $\hat{\theta}(\cdot)$ and $\hat{\theta}^\varepsilon(\cdot)$ are bounded by 2π, we need only let $V(\cdot)$ depend on v. Use $V(x) = v'Pv$, and q_ε in lieu of ε. Then we can get the form (3.13), where $p = q = 2$, $\bar{V}(x,t) = O(|v| + 1)$, $\tilde{V}(x,t) = O(|v| + 1)$, and both $\tilde{\xi}^\varepsilon(t)$ and $\bar{\xi}^\varepsilon(t)$ are bounded by $O(1 + |z^\varepsilon(t)|^3)$. Then the theorem implies that $\{x^\varepsilon(\cdot),z_1^\varepsilon(\cdot),z_2^\varepsilon(\cdot)\}$ has an invariant measure for small $\varepsilon > 0$, and that the x-marginals converge to $\mu(\cdot)$, the unique invariant measure of $x(\cdot) = (v(\cdot),\hat{\theta}(\cdot))$, as $\varepsilon \to 0$. This justifies the use of the stationary measure of (3.3) as an approximation to the stationary measures of the system of Figure 1, for small enough ε, when $\theta(t) \equiv \theta_0$.

4. EXAMPLE 2. A DIGITAL PHASE-LOCKED LOOP (DPLL)

Consider the digital signaling problem where a random sequence $\{s_n\}$ is transmitted. The received signal is $s(t) + \text{noise}(t) \equiv y_T(t)$, where $s(t) = s_n$ in $[nT + \delta_0, nT + T + \delta_0)$. The time T is the symbol interval and δ_0 is unknown. For detection with an acceptably low error rate, a very good estimate of δ_0 is needed, and a DPLL system is used to provide a sequence of estimates of δ_0 (see, e.g., the systems in Ref. 17). Here we deal with a specific simple (but important) case. Let $s_n = \pm A_0$, $A_0 > 0$, where $\{s_n\}$ is independent, identically distributed, and

$P\{s_n = A_0\} = 1/2$. Let $\hat{\delta}_n$ denote the n th estimate of δ_0, and define $\lambda_n = (\hat{\delta}_0 - \delta_0)/T$. In the study of the algorithm, the exact value of δ_0 is unimportant, so without loss of generality, set $\delta_0 = 0$.

The particular DPLL studied will now be defined. It was chosen because variations of it are used frequently [17], and rigorous analysis via currently used techniques in communication theory is very hard. The diffusion approximation method used here involves scalings which are consistent with those used currently in communication theory (explicitly or implicitly), and seems to be a robust and systematic tool for this type of problem. Let noise$(\cdot) = \psi_T(\cdot)$, where $w_T(t) \equiv \int_0^t \psi_T(s)\, ds$ is a Wiener process with variance $\sigma_T^2 t$. The variance σ_T^2 must depend on T (for small T) if the problem is to be meaningful. In any particular case, of course, σ_T^2 would have to be determined from the problem data. To get the scaling of σ_T^2, suppose that $\hat{\delta}_n = 0$ and that the statistic $y = s_0 T + \int_0^T \psi_T(s)\, ds$ (the integral of the observation, over an interval of the same width as the symbol interval) is used to estimate $s_0 (= \pm A_0)$ via a likelihood ratio test. But

$$P\{s_0 = A_0 \text{ chosen} \mid s_0 = -A_0\} = \int_{A_0 T/\sigma_T \sqrt{T}}^{\infty} dN(0,1)$$

Thus $\sigma_T^2 = \sigma^2 T$ for some $\sigma^2 > 0$ is the natural scaling if the decision problem is not to degenerate for small $T > 0$.

The algorithm DPLL is defined as follows. Let $0 < \Delta < 1/4$. Define $e_n(\lambda_{n-1}, \lambda_n)$ by

$$e_n(\lambda_{n-1},\lambda_n) = \left| \int_{(n+\Delta)T+\hat{\delta}_{n-1}}^{(n+1+\Delta)T+\hat{\delta}_n} y_T(s)\, ds \right| - \left| \int_{(n+1-\Delta)T+\hat{\delta}_{n-1}}^{(n+2-\Delta)T+\hat{\delta}_n} y_T(s)\, ds \right| \tag{4.1}$$

Thus we can write, for a Wiener process $W(\cdot)$ with variance $\sigma^2 t$,

$$g_n(\lambda_{n-1},\lambda_n) \equiv \frac{e_n(\lambda_{n-1},\lambda_n)}{T} = |(1 - \Delta - \lambda_{n-1})s_n + (\Delta + \lambda_n)s_{n+1} + W(n + 1 + \Delta + \lambda_n) - W(n + \Delta + \lambda_{n-1})| - |(\Delta - \lambda_{n-1})s_n + (1 - \Delta + \lambda_n)s_{n+1} + W(n + 2 - \Delta + \lambda_n) - W(n + 1 - \Delta + \lambda_{n-1})| \tag{4.2}$$

We can use $W(t)$ instead of $w_T(tT)/T$, since the latter process is a Wiener process with variance $\sigma^2 t$. Finally, with <u>adjustment parameter</u> $\varepsilon > 0$, we write λ_n as λ_n^ε and let $\{\lambda_n^\varepsilon\}$ be defined by the recursive algorithm (the DPLL):

$$\lambda_{n+1}^\varepsilon = \lambda_n^\varepsilon + \varepsilon g_n(\lambda_{n-1}^\varepsilon,\lambda_n^\varepsilon) \qquad \varepsilon > 0 \tag{4.3}$$

Normally, the adjustment parameter ε depends on T. Since we are concerned with tracking over real time intervals $\{n : nT \le t\}$, all real t, ε should be $\approx$ (constant)T for small T for otherwise (smaller order of T), the sequence $\{\lambda_n\}$ moves too slowly, or (larger order of T) is excessively sensitive to the noise. Let $\lambda_0^\varepsilon = O(\sqrt{\varepsilon})$, for otherwise, the system (4.3) will not be able to improve the estimate for small ε.

Define $U_n^\varepsilon = \lambda_n^\varepsilon/\sqrt{\varepsilon}$, and define $U^\varepsilon(t) = U_n^\varepsilon$ on $[n\varepsilon, n\varepsilon + \varepsilon)$. Using parameters λ, λ' (in lieu of λ_{n-1}, λ_n), define $\bar{g}(\lambda,\lambda') = Eg_n(\lambda,\lambda')$, and set $\xi_n^\varepsilon(\lambda_{n-1}^\varepsilon,\lambda_n^\varepsilon) = g_n(\lambda_{n-1}^\varepsilon,\lambda_n^\varepsilon) - \bar{g}(\lambda_{n-1}^\varepsilon,\lambda_n^\varepsilon)$. Then

$$U_{n+1}^\varepsilon = U_n^\varepsilon + \varepsilon\,\frac{\bar{g}(\lambda_{n-1}^\varepsilon,\lambda_n^\varepsilon)}{\sqrt{\varepsilon}} + \sqrt{\varepsilon}\,\xi_n^\varepsilon(\lambda_{n-1}^\varepsilon,\lambda_n^\varepsilon)$$

$$\equiv U_n^\varepsilon + \varepsilon q_\varepsilon(\lambda_{n-1}^\varepsilon,\lambda_n^\varepsilon)$$

Define $U_n^{\varepsilon,N}, U^{\varepsilon,N}(\cdot)$ as $x_n^{\varepsilon,N}$ and $x^{\varepsilon,N}(\cdot)$ were defined in Sec. 2. In particular, $U_{n+1}^{\varepsilon,N} = U_n^{\varepsilon,N} + \varepsilon q_\varepsilon(\lambda_{n-1}^{\varepsilon,N},\lambda_n^{\varepsilon,N})b_N(U_n^{\varepsilon,N})$, where $\lambda_n^{\varepsilon,N} = \sqrt{\varepsilon}\,U_n^{\varepsilon,N}$ is used. There is a $\theta > 0$ such that $-\theta = (d/d\lambda)\bar{g}(\lambda,\lambda)|_{\lambda=0}$. Equation (4.4) can be written as

$$U_{n+1}^{\varepsilon} = U_n^{\varepsilon} - \theta U_n^{\varepsilon} + \sqrt{\varepsilon}\,\xi_n^{\varepsilon}(\lambda_{n-1}^{\varepsilon}, \lambda_n^{\varepsilon}) + \varepsilon v_n^{\varepsilon} \tag{4.5}$$

where $v_n^{\varepsilon} = O(|\lambda_n^{\varepsilon}|^2 + |\lambda_{n-1}^{\varepsilon}|^2 + |U_n^{\varepsilon} - U_{n-1}^{\varepsilon}|)$, and similarly for the $\{U_n^{\varepsilon,N}\}$ equation. The main result is that if U_0^{ε} converges weakly to a random variable U_0, then $\{U^{\varepsilon}(\cdot)\}$ converges weakly to $U(\cdot)$, where

$$dU = -\theta U\,dt + v\,dB \qquad U(0) = U_0$$

$$v^2 = 2E[g_{n+1}(0,0) - \bar{g}(0,0)][g_n(0,0) - \bar{g}(0,0)] + E[g_n(0,0) - \bar{g}(0,0)]^2 \qquad \text{for any } n \geq 1 \tag{4.6}$$

If the $\{v_n^{\varepsilon}\}$ were carried through in the calculations, they would contribute nothing to the limit, so for notational simplicity here, we drop them from (4.5) (and from the equation for $\{U_n^{\varepsilon,N}\}$) now.

The representation (4.6) is useful for calculating the statistics of the DPLL (4.3) when the adjustment rate ε is small, say $\varepsilon \approx T$, where the symbol interval is small. In any case, rough qualitative properties and parametric dependencies can be obtained. We do not know how small ε must be for (4.6) to be a good (scaled) approximation, although simulations on related systems suggest that the range is within reason for practical systems.

APPLICATION OF THEOREM 2.3

To get the limit (4.6), use Theorem 2.3. Fix $f(\cdot) \in \hat{C}_0^{2,3}$. The method of getting and using the $f^{\varepsilon,N}(\cdot)$ [written here as $f^{\varepsilon}(\cdot)$, for notational simplicity] is very similar to that for the continuous parameter case. We start by calculating $\hat{A}^{\varepsilon} f(U^{\varepsilon,N}(n\varepsilon))$, and successively average out the noise terms, except that the f_i^{ε} are defined by sums rather than integrals. More details on the method for this and related DPLLs are in Ref. 13, and details of similar approximations for other scaled and interpolated discrete parameter processes can be found in Refs. 9 and 19.

First, we calculate $\hat{A}^{\epsilon}f(U_n^{\epsilon,N},n\epsilon)$ (discrete parameter form). Let $E_n^{\epsilon}(\cdot)$ = expectation conditioned on $W(t), s(t)$, $t \leq$ time at which λ_n^{ϵ} is first known.

$$\begin{aligned}\epsilon\hat{A}^{\epsilon}f(U_n^{\epsilon,N},n\epsilon) &= \epsilon f_t(U_n^{\epsilon,N},n\epsilon) + O(\epsilon^2)\\ &\quad - \epsilon f_u(U_n^{\epsilon,N},n\epsilon)\theta U_n^{\epsilon,N} b_N(U_n^{\epsilon,N})\\ &\quad + \sqrt{\epsilon}\, f_u(U_n^{\epsilon,N},n\epsilon) b_N(U_n^{\epsilon,N}) E_n^{\epsilon}\xi_n^{\epsilon}(\lambda_{n-1}^{\epsilon,N},\lambda_n^{\epsilon,N})\\ &\quad + \epsilon \frac{f_{uu}(U_n^{\epsilon,N},n\epsilon)}{2} b_N^2(U_n^{\epsilon,N})\\ &\qquad E_n^{\epsilon}(\xi_n^{\epsilon}(\lambda_{n-1}^{\epsilon,N},\lambda_n^{\epsilon,N}))^2 + O_{1n}^{\epsilon,N} \end{aligned} \tag{4.7}$$

The $O_{1n}^{\epsilon,N}$, a remainder in the truncated Taylor expansion, satisfies

$$O_{1n}^{\epsilon,N} = O(\epsilon^{3/2})E_n^{\epsilon}[1 + |\xi_n^{\epsilon}(\lambda_{n-1}^{\epsilon,N},\lambda_n^{\epsilon,N})|^3] \tag{4.8}$$

Using the properties of the Wiener process, for each $t < \infty$,

$$\begin{aligned}&\lim_{\epsilon\to 0}\sup_{n\epsilon\leq t}\left|\frac{O_{1n}^{\epsilon,N}}{\epsilon}\right| = 0\\ &\lim_{\epsilon\to 0}\sup_{n\epsilon\leq t}\left|\frac{O_{1n}^{\epsilon,N}}{\epsilon}\right| = 0 \quad \text{with probability one}\end{aligned} \tag{4.9}$$

All the $\{O_{1n}^{\epsilon}\}$ introduced below satisfy (4.9) for each N.

Only the first and third terms on the right of (4.7) are part of the limit operator $\epsilon(\partial/\partial t + A)f(U_n^{\epsilon,N},n\epsilon)$ (for $U_n^{\epsilon,N} \in S_N$). The rest are either negligible or must be averaged out. Henceforth, drop the N superscript on $U_n^{\epsilon,N}$, $\lambda_n^{\epsilon,N}$. The bounds on the error terms are valid for each $\{U_n^{\epsilon,N}, \epsilon > 0\}$. We use the form $f^{\epsilon}(n\epsilon) = f(U_n^{\epsilon},n\epsilon) + \Sigma_{i=0}^{2} f_i^{\epsilon}(n\epsilon)$, where it will be true that (each $t > 0$)

$$\lim_{\epsilon\to 0}\sup_{n\epsilon\leq t} |f_i^{\epsilon}(n\epsilon)| = 0 \quad \text{with probability one}$$

$$\lim_{\varepsilon\to 0} \sup_{n\varepsilon \le t} E|f_i^\varepsilon(n\varepsilon)| = 0 \tag{4.10}$$

To average out the f_{uu} term in (4.7), introduce the first perturbation $f_0^\varepsilon(n\varepsilon) = f_0^\varepsilon(U_n^\varepsilon, n\varepsilon)$, where

$$f_0^\varepsilon(U,n\varepsilon) = \frac{\varepsilon f_{uu}(U,n\varepsilon)}{2} b_N^2(U) \sum_{j=n}^{\infty} [E_n^\varepsilon(\xi_j^\varepsilon(\lambda_{n-1}^\varepsilon,\lambda_n^\varepsilon))^2 - E(\xi_j^\varepsilon(\lambda_{n-1}^\varepsilon,\lambda_n^\varepsilon))^2]$$

The $\lambda_{n-1}^\varepsilon, \lambda_n^\varepsilon$ are taken as parameters in calculating all expectations. Due to $\Delta \le 1/4$, and to the properties of the Wiener process $W(\cdot)$ and of $\{s_n\}$, all the terms in the sum are zero except the lowest [since $E_n^\varepsilon(\cdot) = E(\cdot)$ for the other terms]. Now

$$\varepsilon\hat{A}^\varepsilon f_0^\varepsilon(U_n^\varepsilon,n\varepsilon) = -[\text{fifth term of (4.7)}] + \text{terms satisfying (4.9)} + \frac{\varepsilon f_{uu}(U_n^\varepsilon,n\varepsilon)}{2} b_N^2(U_n^\varepsilon) E\xi_n^\varepsilon(\lambda_{n-1}^\varepsilon,\lambda_n^\varepsilon)^2$$

The last component yields the last term of the variance v^2 in (4.6).

To average out the $\sqrt{\varepsilon}$ term in (4.7) use

$$f_1^\varepsilon(n\varepsilon) = \sqrt{\varepsilon}\, f_u(U_n^\varepsilon,n\varepsilon) b_N(U_n^\varepsilon) \sum_{j=n}^{\infty} E_n^\varepsilon \xi_j^\varepsilon(\lambda_{n-1}^\varepsilon,\lambda_n^\varepsilon)$$
$$= \sqrt{\varepsilon}\, f_u(U_n^\varepsilon,n\varepsilon) b_N(U_n^\varepsilon) E_n^\varepsilon \xi_n^\varepsilon(\lambda_{n-1}^\varepsilon,\lambda_n^\varepsilon)$$

We can write

$$\varepsilon\hat{A}^\varepsilon f_1^\varepsilon(n\varepsilon) = -f_1^\varepsilon(n\varepsilon) + 0_{2n}^\varepsilon + \sqrt{\varepsilon}\, E_n^\varepsilon b_N(U_{n+1}^\varepsilon) f_u(U_{n+1}^\varepsilon,n\varepsilon)\xi_{n+1}^\varepsilon(\lambda_n,\lambda_{n+1})$$

The last term can be written as

$$\sqrt{\varepsilon}\, E_n^\varepsilon(f_u(U_n^\varepsilon,n\varepsilon)b_N(U_n^\varepsilon))_u(U_{n+1}^\varepsilon - U_n^\varepsilon)\xi_{n+1}^\varepsilon(\lambda_n^\varepsilon,\lambda_{n+1}^\varepsilon) + 0_{3n}^\varepsilon$$
$$= \varepsilon(f_n(U_n^\varepsilon,n\varepsilon)b_N(U_n^\varepsilon))_u E_n^\varepsilon \xi_{n+1}^\varepsilon(\lambda_n^\varepsilon,\lambda_n^\varepsilon)\xi_n^\varepsilon(\lambda_{n-1}^\varepsilon,\lambda_n^\varepsilon) + 0_{4n}^\varepsilon \tag{4.11}$$

This term is neither negligible nor part of $\varepsilon Af(U_n^\varepsilon)$, so it needs to be averaged further. This is done via the last perturbation $f_2^\varepsilon(n\varepsilon)$:

$$\varepsilon(f_u(U_n^\varepsilon,n\varepsilon)b_N(U_n^\varepsilon))_u \sum_{j=n}^{\infty} [E_n^\varepsilon \xi_{j+1}^\varepsilon(\lambda_n^\varepsilon,\lambda_n^\varepsilon)\xi_j^\varepsilon(\lambda_{n-1},\lambda_n)$$

$$- E\xi_{j+1}^\varepsilon(\lambda_n^\varepsilon,\lambda_n^\varepsilon)\xi_j^\varepsilon(\lambda_{n-1}^\varepsilon,\lambda_n^\varepsilon)] = f_2^\varepsilon(n^\varepsilon)$$

$\varepsilon\hat{A}^\varepsilon f_2^\varepsilon(n\varepsilon)$ can be shown to equal O_{5n}^ε - (4.11) + $(\varepsilon/2)(f_u(U_n^\varepsilon,n\varepsilon)b_N(U_n^\varepsilon))$ × [first component of variance v^2 in (4.6)]. The constructed $f^\varepsilon(n\varepsilon)$ can be shown to satisfy the requirements of Theorem 2.3, where the operator A is that of (4.6). Also, the $f^\varepsilon(n\varepsilon)$ can be used in Theorem 2.2 to yield tightness.

ACKNOWLEDGMENT

This research was supported in part by the Air Force Office of Scientific Research under AFOSR 76-3063D, in part by the National Science Foundation under NSF-Eng 77-12946-A02, and in part by the Office of Naval Research under N00014-76-C-0279-P0004.

REFERENCES

1. E. Wong and M. Zakai, On the convergence of ordinary integrals to stochastic integrals, Ann. Math. Stat. 36: 1560-1564 (1965).
2. G. C. Papanicolaou and W. Kohler, Asymptotic theory of mixing stochastic ordinary differential equations, Commun. Pure Appl. Math. 29:641-668 (1974).
3. G. Blankenship and G. C. Papanicolaou, Stability and control of stochastic systems with wide-band noise disturbances, SIAM J. Appl. Math. 34:437-476 (1978).
4. T. G. Kurtz, Semigroups of conditional shifts and

approximation of Markov processes, Ann. Prob. 4:618-642 (1975).

5. T. G. Kurtz, Approximation of Population Processes, SIAM, Philadelphia, 1981.

6. R. Z. Khazminskii, A limit theorem for solutions of differential equations with random right-hand sides, Theory Prob. Appl. 11:390-406 (1966).

7. H. J. Kushner, Jump-diffusion approximations for ordinary differential equations with wide-band random right-sides, SIAM J. Control Optimization 17:729-744 (1979).

8. H. J. Kushner, A martingale method for the convergence of a sequence of processes to a jump-diffusion process, Z. Wahrscheinlichtkeitsth. Verwend. Geb. 53:207-219 (1980).

9. H. J. Kushner and Hai Huang, Averaging methods for the asymptotic analysis of learning and adaptive systems, with small adjustment rate, SIAM J. Optimization Control 19:635-650 (Sept. 1981).

10. G. C. Papanicolaou, D. Strook, and S. R. S. Varadham, Martingale approach to some limit theorems, Duke Univ. Conf. Turbulence, 1976.

11. H. J. Kushner, Diffusion approximations to output processes of nonlinear systems with wide inputs, IEEE Trans. Inf. Theory IT-26:715-725 (1980).

12. H. J. Kushner and Y. Bar-Ness, Analysis of nonlinear systems with wide-band inputs, IEEE Trans. Autom. Control, AC-25:1072-1078 (1980).

13. H. J. Kushner and Hai Huang, Diffusion approximations for the analysis of digital phase locked loops, LCDS Tech. Rep. 80-2, Brown University, 1980; IEEE Trans. Inf. Theory IT-28, 1982, 384-390.

14. H. J. Kushner and W. T. Y. Ju, Diffusion approximations for nonlinear phase locked loop type systems with wide-band inputs, LCDS Tech. Report 80-2, Brown University, 1980; J. Math. Anal. Appl. 82, 1982, 518-541.

15. H. J. Kushner, Asymptotic distributions of solutions of ordinary differential equations with wide band noise inputs; approximate invariant measures, LCDS Tech. Rep. 81-2, Jan. 1981, Stochastics 6, 1982, 259-278.

16. P. Billingsley, Convergence of Probability Measures, Wiley, New York, 1968.

17. W. C. Lindsey, Synchronization Systems in Communications and Control, Prentice-Hall, Englewood Cliffs, N. J., 1972.

18. F. L. Gardner, Phaselock Techniques, 2nd ed., Wiley, New York, 1979.

19. H. J. Kushner and Hai Huang, On the weak convergence of general stochastic difference equations to a diffusion, SIAM J. Appl. Math. 40:528-542 (1981).

20. H. J. Kushner, Approximation and Weak Convergence Methods for Random Processes, with Applications to Stochastic Systems Theory, to be published by MIT Press, Cambridge, Mass, March, 1984.

11

An Analogue to the Stochastic Integral for $\partial/\partial t = -\Delta^2$

MINORU MOTOO/Faculty of Science, Tokyo Institute of Technology, Tokyo, Japan

1. INTRODUCTION

In this chapter we define the stochastic integral on the signed measure space corresponding to $\partial/\partial t = -\partial^4/\partial x^4$, which is an extension of the one introduced by Hochberg [1]. Following the idea by K. Ito [2] and T. Ueno [3], we shall consider the space of measures on $L = L_4[0,T]$. Each measure $F(d\lambda)$ in the space can be regarded as corresponding to a function $\tilde{F}(W)$ on the paths W_t: $0 \leq t \leq T$ such that

$$\tilde{F}(W) = \int_L \exp\left[i\int_0^T \lambda(t)\, \frac{dW_t}{dt}\, dt \right] F(d\lambda)$$

which has only a symbolical meaning. Then, stochastic integrals can be realized on a certain completion of the above space. As an application, we show that solutions of the equation

$$\frac{\partial u}{\partial t} = -\frac{\partial^4 u}{\partial x^4} - 4f_3(x)\frac{\partial^3 u}{\partial x^3} - 12f_2(x)\frac{\partial^2 u}{\partial x^2} - 24f_1(x)\frac{\partial u}{\partial x} + f_0(x)u$$

with smooth $f_p(x)$ ($p = 0, 1, 2, 3$) can be given by a Girsanov-type transformation.

2. PRELIMINARIES

Throughout, time $T > 0$ is fixed. Let L be the set of all measurable functions λ on $[0,T]$ such that $\int_0^T \lambda(t)^4\, dt < \infty$. $\underline{B}(L)$ is the topological Borel field of L with respect to usual L^4-topology. Set $L(J) = \{\lambda \in L : \lambda(t) = 0 \text{ for } t \notin J\}$, for an interval J in $[0,T]$. $\underline{N}$ is the set of all complex measures F on $\underline{B}(L)$ with bounded total variations $|F|$. On $\underline{N}$ we introduce multiplication by convolution; that is,

$$F\cdot G(A) = \int_{\mu+\nu\in A} F(d\mu)G(d\nu) \tag{2.1}$$

for F and G in $\underline{N}$ and A in $B(L)$. It is easily seen that

$$|FG| \le |F||G| \tag{2.2}$$

Set

$$\|F\| = \sup_{\mu\in L} \int \exp\, [-\int_0^T (\lambda(t) + \mu(t))]^4\, F(d\lambda) \tag{2.3}$$

Then

$$\|F\| \le |F| \qquad \text{and} \qquad \|F\cdot G\| \le |F|\|G\| \tag{2.4}$$

Identifying F and G in $\underline{N}$ with $\|F - G\| = 0$ we can define the normed space:

$$\underline{M} = \{\{F\} : F \in \underline{N}\} \qquad \|\{F\}\| = \|F\|$$

In the following, we use the same notation for $\{F\}$ and F. $\overline{\underline{M}}$ is the completion of $\underline{M}$ by the norm $\|\cdot\|$. For $G \in \underline{N}$ and $X \in \overline{\underline{M}}$, set

$$GX = \lim_{n\to\infty} GF_n$$

if F_n is in $\underline{M}$ and $F_n \to X$. By (4) GX is independent of the approximating sequence F_n. Moreover, if $\|G\| = 0$, then $GF_n = 0$ and $GX = 0$. Therefore, we can define GX for any G in $\underline{M}$ and X in $\overline{\underline{M}}$. Let f be a function on R^n such that

$$f(x_1, \ldots, x_n) = \int \exp\, (i \sum_{k=1}^{n} x_k\xi_k)\hat{f}(d\xi_1, \ldots, d\xi_n) \tag{2.5}$$

Then set, for $0 \le t_1 < t_2 < \cdots < t_n \le T$ if $\int d|\hat{f}| < \infty$,

$$f(x + W_{t_1}, \ldots, x + W_{t_n}) = \int \exp\, i\,x \sum_{k=1}^{n} \xi_k I_{t_k} \hat{f}(d\xi_1, \ldots, d\xi_n) \tag{2.6}$$

where $\delta(\lambda)$ in $\underline{M}$ is the measure concentrated at the point λ in L and I_t in L is the indicator of set $[0,t)$. The left side of the equality has only a symbolical meaning.

3. EXPECTATIONS OF ELEMENTS IN $\overline{\underline{M}}$

For $F \in \underline{M}$, set

$$E(F) = \int \exp\left[-\int_0^T \lambda(s)^4 \right] ds\ F(d\lambda)$$

Then $E(F)$ is linear in $\underline{M}$ and $|E(F)| \leq \|F\|$. For $X \in \overline{\underline{M}}$, if $F_n \in \underline{M}$ and $\|F_n - X\| \to 0$, set

$$E(X) = \lim_{n\to\infty} E(F_n)$$

$E(X)$ is independent of the approximating sequence F_n.

In particular,

$$E[f(x + W_{t_1}, \ldots, x + W_{t_n})] = \int \exp\{ix \sum_{k=1}^{n} \xi_k - t_1(\xi_1 + \cdots + \xi_n)^4 - \cdots - (t_n - t_1)\xi_n^4\}\hat{f}(d\xi_1, \ldots, d\xi_n) \tag{3.1}$$

where f is given by (2.5) and (2.6). For X in $\overline{\underline{M}}$ and λ in L, set

$$X(\lambda) = E(\delta(\lambda)X)$$

In particular, for F in $\underline{M}$,

$$F(\lambda) = \int \exp\left[-\int_0^T (\lambda + \mu)^4 \right] ds\ F(d\lambda)$$

and

$$\|F\| = \sup_\lambda |F(\lambda)|$$

In general, for $X \in \overline{\underline{M}}$ it holds that

$$\|X\| = \sup_\lambda |X(\lambda)| \tag{3.2}$$

Therefore, a sequence $\{X_n\}$ in $\overline{\underline{M}}$ converges to X if and only if $\{X_n(\lambda)\}$ converges to $X(\lambda)$ uniformly in λ. For F in $\underline{M}$, X in $\underline{M}$, and μ in L,

$$E(FX) = \int X(\lambda)F(d\lambda)$$
$$FX(\mu) = \int X(\lambda + \mu)F(d\lambda) \tag{3.3}$$

Set $W_\varepsilon = (1/i\varepsilon)(\delta(\varepsilon I_J) - \delta(0))$, where I_J is the indicator of an interval J in $[0,T]$. Noting that

$$W_\varepsilon^p(\lambda) = \frac{1}{(i\varepsilon)^p} \sum_{k=0}^{p} (-1)^k \binom{p}{k} \exp -\int_0^t [\lambda + (p-k)\varepsilon I_J]^4 \, ds$$

we can see that

$$W^p(J) = \lim_{\varepsilon \to 0} W_\varepsilon^p \qquad p = 1, 2, 3, 4$$

exist in $\overline{\underline{M}}$, and

$$W(J)(\lambda) = 4i(\int_J \lambda^3 \, ds)e^{-\Lambda}$$
$$W^2(J)(\lambda) = [12\int_J \lambda^2 \, ds - 16(\int_J \lambda^3 \, ds)^2]e^{-\Lambda}$$
$$W^3(J)(\lambda) = i[-24\int_J \lambda \, ds + 144(\int_J \lambda^3 \, ds)(\int_J \lambda^2 \, ds) - 64(\int_J \lambda^3 \, ds)^3]e^{-\Lambda}$$
$$W^4(J)(\lambda) = [-24|J| + 432(\int_J \lambda^2 \, ds)^2 + 384(\int_J \lambda^3 \, ds)(\int_J \lambda \, ds) - 1152(\int_J \lambda^3 \, ds)(\int_J \lambda^2 \, ds) + 256(\int_J \lambda^3 \, ds)^4]e^{-\Lambda}$$

where $\Lambda = \int_0^T \lambda^4(s) \, ds$.

4. STOCHASTIC INTEGRALS

For an interval J in $[0,T]$, we define X to be in $\overline{\underline{M}}_J$ if and only if there exists a sequence $\{F_n\}$ in $\underline{M}$ which converges to X and support of each F_n is contained in $L(J)$. In particular, we set $\overline{\underline{M}}_t = \overline{\underline{M}}_{[0,t]}$. If the support of F is in $L(J)$, then

$$F(\lambda) = \int \exp[-\int_0^T (I_J\mu + \lambda)^4 \, ds]F(d\mu)$$
$$= F(I_J\lambda) \exp(-\int_{J^c} \lambda^4 \, ds) \tag{4.1}$$

Therefore, for X in $\overline{\underline{M}}_J$, we have

$$X(\lambda) = X(I_J\lambda) \exp\left(-\int_{J^c} \lambda^4 \, ds\right) \tag{4.2}$$

Let X be in $\underline{\overline{M}}_J$ and Y in $\underline{\overline{M}}_K$ and $J \cap K = \emptyset$; then there exists an element XY in $\underline{\overline{M}}$ with the following property:

$$XY(\lambda) = X(I_J\lambda)Y(I_K\lambda) \exp\left(-\int_{(J\cup K)^c} \lambda^4 \, ds\right) \tag{4.3}$$

In fact, let $\{F_n\}$ and $\{G_n\}$ be sequences in $\underline{\overline{M}}$ whose supports are contained in L(J) and L(K), respectively, and $F_n \to X$ and $G_n \to Y$. Then

$$F_nG_n(\lambda) = F_n(I_J\lambda)G_n(I_K\lambda) \exp\left(-\int_{(J\cup K)^c} \lambda^4 \, ds\right)$$

Therefore, $XY = \lim F_nG_n$ exists and (4.3) holds. By (4.3) XY is independent of special choices of $\{F_n\}$ and $\{G_n\}$.

LEMMA 4.1 Let λ be in L and m be any nonnegative function, and

$$n(t) = \left\{\int_0^t m(s)\,|\lambda^p(s)| \exp\left[-\int_s^t \lambda^4(u)\, du\right] ds\right\}$$

Then

$$n(t)^4 \leq C_p\left[\int_0^t m(s)^4 \, ds\right]t^{3-p}$$

where C_p is an absolute constant (p = 0, 1, 2, 3).

Proof. Using Hölder's inequality, we have for p = 0,

$$n(t) \leq \int_0^t m \, ds \leq \left(\int_0^t m^4 \, ds\right)^{1/4} t^{3/4}$$

and for p = 1, 2, 3,

$$n(t) \leq \left(\int_0^t m^4 \, ds\right)^{1/4} \left[\int_0^t |\lambda|^{4p/3} \, ds \exp\left(-\frac{4}{3}\int_s^t \lambda^4 \, du\right)\right]^{3/4}$$

And for p = 3,

$$\int_0^t |\lambda|^{12/3} \, ds \exp\left(-\frac{4}{3}\int_s^t \lambda^4 \, du\right) \leq \frac{3}{4}$$

For p = 1, 2, using Hölder's inequality again, we have

$$\left[\int_0^t |\lambda|^{4p/3} \, ds \exp\left(-\frac{4}{3}\int_s^t \lambda^4 \, du\right)\right]$$

$$\leq \left[\int_0^t \lambda^4 \, ds \exp\left(-\frac{4}{3}\int_s^t \lambda^4 \, du\right)\right]^{p/3} t^{(3-p)/3}$$

$$\leq \left(\frac{p}{4}\right)^{p/3} t^{(3-p)/3}$$

and

$$n(t) \le \int_0^t m^4 \, ds \cdot (\tfrac{p}{4})^{p/3} t^{(3-p)/4}$$

Let $0 = t_0 < t_1 < \cdots < t_n = t$ be a subdivision of an interval $[0,t]$ $(t \le T)$. If X_t in $\underline{\overline{M}}$ is bounded and continuous in the $\|\cdot\|$-norm, we can easily see that

$$Y_t^{(0)} = \lim_{\max(t_{k+1}-t_k)\to 0} \sum_{k=0}^{n-1} X_{t_k}(t_{k+1} - t_k)$$

exists in $\underline{\overline{M}}$ and

$$Y_t^{(0)}(\lambda) = \int_0^t X_s(\lambda) \, ds$$

THEOREM 4.1 If X_t in $\underline{\overline{M}}_t$ is bounded and continuous in the $\|\cdot\|$-norm, then

$$Y_t^{(p)} = \lim_{\max(t_{k+1}-t_k)\to 0} \sum_{k=0}^{n-1} X_{t_k} W^p([t_k,t_{k+1})) \tag{4.4}$$

$(p = 1, 2, 3, 4)$ exist in $\underline{M}_t$ and

$$Y_t^{(p)}(\lambda) = -\frac{4!}{(4-p)!} \frac{1}{i^p} \int_0^t \lambda^{4-p}(s) X_s(\lambda) \, ds \tag{4.5}$$

In particular, $Y_t^{(4)} = -24\int_0^t X_s \, ds$. We shall write $Y_t^{(p)} = \int_0^t X_s \, dW_s^p$.

Proof. For example, we shall prove the theorem for $p = 2$. The other cases are proved in a similar way. By (4.1) and (4.3), we have, for $J_k = [t_k,t_{k+1})$,

$$X_{t_k} W^2(J_k)(\lambda)$$
$$= X_{t_k}(I_{t_k}\lambda) \; 12\int_{J_k} \lambda^2 \, ds - 16 \int_{J_k} \lambda^3 \, ds^{\;2}$$
$$\times \exp \; -\int_{t_k}^T \lambda^4 \, ds$$

Noting, for $u < s$, that

$$|X_s(\lambda) - X_u(\lambda)| \le |X_s(I_s\lambda) - X_u(I_s\lambda)| \exp\left(-\int_s^T \lambda^4 \, dv\right)$$
$$\le \|X_s - X_u\| \exp\left(-\int_s^T \lambda^4 \, dv\right)$$

we have by Lemma 4.1,

$$\int_0^t X_s(\lambda)\lambda^2(s)\,ds - \sum_{k=0}^{n-1} X_{t_k}(I_{t_k}\lambda)\int_{J_k}\lambda^2\,ds\,\exp\left(-\int_{t_k}^T \lambda^4\,ds\right)$$

$$\leq \sup_{k,s,u\in J_k}\|X_s - X_u\| \int_0^t \lambda^2(s)\exp\left(-\int_s^T \lambda^4\,dv\right)ds$$

$$\leq \sup_{k,s,u\in J_k}\|X_s - X_u\| C_1 t^{1/4} \tag{4.6}$$

On the other hand,

$$X_{t_k}(I_{t_k}\lambda)\left(\int_{J_k}\lambda^3\,ds\right)^2 \exp\left(-\int_{t_k}^T \lambda^4\,ds\right)$$

$$\leq \sup_t\|X_t\|(t_{k+1} - t_k)^{1/2}\left(\int_{J_k}\lambda^4\,ds\right)^{3/2}\exp\left(-\int_{t_k}^T\lambda^4\,ds\right)$$

Set $a_k = \int_{J_k}\lambda^4\,ds$; then

$$\int_{J_k}\lambda^4\,ds \quad \exp \quad -\int_{t_k}^T \lambda^4\,ds$$

$$\leq \sup_{a>0} ae^{-a/2}\cdot a_k \exp\left(-\frac{1}{2}\sum_{\ell=k}^n a_\ell\right) \leq C_2$$

Then

$$\sum_k X_{t_k}(I_{t_k}\lambda)\left(\int_{J_k}\lambda^3\,ds\right)^2\exp\left(-\int_{t_k}^T\lambda^4\,ds\right)$$

$$\leq C_3 \max_k (t_{k+1} - t_k)^{1/2} \tag{4.7}$$

Since both of the right sides of (4.6) and (4.7) converge to 0 as max $(t_{k+1} - t_k) \to 0$ uniformly in λ, we have proved the theorem.

For X_s in $\overline{M}_s$,

$$|X_s(\lambda)| \leq |X_s(I_s\lambda)| \exp\left(-\int_s^T \lambda^4\,du\right)$$

$$\leq \|X_s\| \exp\left(-\int_s^T \lambda^4\,du\right)$$

Therefore, applying Lemma 4.1 to (4.4) and (4.5), we can show that

$$\|Y_t^{(0)}\|^4 \leq C_0\left(\int_0^t \|X_s\|^4\, ds\right) t^3 \tag{4.8}$$

$$\|Y_t^{(p)}\|^4 \leq C_p\left(\int_0^t \|X_s\|^4\, ds\right) t^{p-1} \qquad p = 1, 2, 3$$

Similarly, we can show that

$$\|Y_t^{(0)} - Y_s^{(0)}\| \leq C_0\left(\int_s^t \|X\|_s^4\, ds\right) t^3 \tag{4.9}$$

$$\|Y_t^{(p)} - Y_s^{(p)}\| \leq C_0\left(\int_s^t \|X\|_s^4\, ds\right) t^{p-1} \qquad p = 1, 2, 3$$

REMARK 1. If X_t $(0 \leq t \leq T)$ is in $\underline{\overline{M}}_t$ and there exists a sequence $\{X_t^{(n)}\}$ $(0 \leq t \leq T)$ in $\underline{\overline{M}}_t$ such that each $X_t^{(n)}$ is continuous in t and

$$\lim_{n\to\infty} \int_0^T \|X_s^{(n)} - X_s\|^4\, ds = 0$$

then by (4.8) we can define the stochastic integrals for X_t, which also satisfy (4.5), (4.8), and (4.9).

5. LINEAR STOCHASTIC DIFFERENTIAL EQUATIONS

In this section we solve linear stochastic differential equations if their coefficients are in $\underline{M} \cap \underline{\overline{M}}_t$.

LEMMA 5.1 Let X_n $(n = 0, 1, 2, \ldots)$ be in $\underline{\overline{M}}$ and F_n $(n = 0, 1, 2, \ldots)$ be in $\underline{M}$. Assume that $\sup_{n\geq 0}\|X_n\|$ and $\sup_n |F_n|$ are finite, and $\lim_{n\to\infty} \|F_n - F_0\| = 0$ holds. Then it holds that

$$\lim_{n\to\infty} \|F_n X_n - F_0 X_0\| = 0$$

Proof. $\overline{\lim}_{n\to\infty} \|F_n X_n - F_n X_0\| \leq \lim |F_n|\|X_n - X_0\| = 0.$ On the other hand, for any $\varepsilon > 0$, take G in $\underline{M}$ such that $\|G - X_0\| < \varepsilon$. Then

$$\begin{aligned}\overline{\lim_{n\to\infty}} \|F_n X_0 - F_0 X_0\| &\leq 2 \sup |F_n|\|G - X_0\| + \lim_{n\to\infty} \|F_n G - F_0 G\| \\ &\leq 2\varepsilon \sup |F_n| + \lim_{n\to\infty} |G|\|F_n - F_0\| \\ &\leq 2\varepsilon \sup |F_n|\end{aligned}$$

Since ε may be arbitrarily small, we have proved the lemma.

LEMMA 5.2 Let F be in $\underline{M}$. Define F_t by

$$F_t(A) = F(\{\lambda: I_t\lambda \in A\})$$

for $0 \leq t \leq T$. Then F is in $\underline{M} \cap \overline{\underline{M}}_t$, $|F_t| \leq |F|$, and F_t is continuous in the $\|\cdot\|$-norm.

Proof. We shall prove the continuity of F_t in t. The other parts are obvious. Since

$$\lim_{\Delta\to 0+} F(\{\mu: \int_t^{t+\Delta} \mu^4 \, ds \geq \delta\}) = 0$$

for any $\delta > 0$, for each n, we can choose $\{\Delta_n\}$ such that $F(A_n^c) \leq 1/n^4$, where

$$A_n = \int_t^{t+\Delta_n} \mu^4 \, ds \leq \frac{1}{n^4}$$

For given $\varepsilon > 0$, take n and K such that

$$e^{-(K^4-1)} < \varepsilon \qquad \text{and} \qquad \frac{1}{n^4} + \frac{4}{n}K^3 + \frac{6}{n^2}K^2 + \frac{4}{n^3}K < \min\{\varepsilon, 1\}$$

For $0 < \Delta < \Delta_n$,

$$|F_{t+\Delta}(\lambda) - F_t(\lambda)| \leq |\int| \exp[-\int_t^{t+\Delta} (\lambda + \mu)^4 \, ds] - \exp(-\int_t^{t+\Delta} \lambda^4 \, ds)|F(d\mu)|$$

If $\int_t^{t+\Delta} \lambda^4 \, ds \leq K^4$ and $\mu \in A_n$, then by Hölder's inequality,

$$|\int_t^{t+\Delta} \{(\lambda + \mu)^4 - \lambda^4\} \, ds| \leq \frac{4K^3}{n} + \frac{6K^2}{n^2} + \frac{4K}{n^3} + \frac{1}{n^4} < \varepsilon$$

$$|F_{t+\Delta}(\lambda) - F_t(\lambda)| < \frac{1}{n^4} + (e^{\varepsilon} - 1)|F| < \varepsilon$$

If $\int_t^{t+\Delta} \lambda^4 \, ds > K^4$ and $\mu \in A_n$ then

$$\int_t^{t+\Delta} (\lambda + \mu)^4 \, ds \geq K^4 - \frac{4K^3}{n} - \frac{6K^2}{n^2} - \frac{4K}{n^3} - \frac{1}{n^4} > K^4 - 1$$

$$|F_{t+\Delta}(\lambda) - F_t(\lambda)| \leq \frac{1}{n^4} + e^{-(K^4-1)}|F| < \varepsilon(1 + F)$$

Therefore, $\lim_{\Delta\to 0+} \|F_{t+\Delta} - F_t\| = 0$ has been proved. In the same way, we can prove the left continuity of F_t.

THEOREM 5.1 Let X_t be an element in $\overline{\underline{M}}_t$, and $F_t^{(p)}$ ($p = 0, 1, 2, 3$) be in $\underline{M} \cap \overline{\underline{M}}_t$ for $0 \leq t \leq T$. Assume that $F_t^{(p)}$ and X_t are continuous in the $\|\cdot\|$-norm and that $\sup_t |F_t^{(p)}| < \infty$ and $\sup_t \|X_t\| < \infty$ hold. Then the solution Z_t $(0 \leq t \leq T)$ of the equation

$$Z_t = X_t + \int_0^t Z_s(F_s^{(0)}\, ds + \sum_{p=1}^{3} F_s^{(p)}\, dW^p) \tag{5.1}$$

exists, and Z_t is in $\overline{\underline{M}}_t$ and bounded and continuous in the $\|\cdot\|$-norm. Conversely, such a solution Z_t of (5.1) is unique.

Proof. Set $Z_t^{(0)} = 0$, $Z_t^{(1)} = X_t$, and

$$Z_t^{(n+1)} = X_t + \int_0^t Z_s^{(n)}(F_s^{(0)}\, ds + \sum_{p=1}^{3} F_s^{(p)}\, dW^p)$$

Note that the $Z_s^{(n)}F_s^{(p)}$'s are continuous in the $\|\cdot\|$-norm by Lemma 5.1. Then, by (4.8),

$$\|Z_t^{(n+1)} - Z_t^{(n)}\|^4 \leq C \int_0^t \|Z_s^{(n)} - Z_s^{(n-1)}\|^4 (\sum_{p=0}^{4} |F_s^p|)^4\, ds$$
$$\leq K \int_0^t \|Z_s^{(n)} - Z_s^{(n-1)}\|^4\, ds$$

Therefore,

$$\|Z^{(n+1)} - Z^{(n)}\| \leq (\frac{1}{n!} K^n t^n)^{1/4} \sup_t \|X_t\|$$

By the usual arguments of successive approximation, we can prove the existence and uniqueness of Z_t.

Let $a(t)$, $b(t)$, $C(t)$, and $d(t)$ be continuous functions on $[0,T]$. Define Z_t by

$$Z_t = 1 + \int_0^t z_s(a\, dW_s + b\, dW_s^2 + c\, dW_s^3 + d\, ds)$$

Then

$$z_t(\lambda) = \exp(-\int_0^T \lambda^4\, ds) + \int_0^t z_s(\lambda)(4i\lambda^3 a + 12\lambda^2 b - 24i\lambda c + d)\, ds \tag{5.2}$$

We can easily find the unique solution of (5.2):

$$Z_t(\lambda) = \exp\,[-\int_0^t (4i\lambda^3 a + 12\lambda^2 b - 24i\lambda c + d - \lambda^4)\,ds - \int_t^T \lambda^4\,ds]$$

In particular, setting $a = i\alpha(t)$, $b = (1/2)[i\alpha(t)]^2$, $c = (1/6)[i\alpha(t)]^3$, and $d = -\alpha^4(t)$, we have

$$z_t(\lambda) = \exp\,[-\int_0^t (\alpha + \lambda)^4\,ds - \int_t^T \lambda^4\,ds] = \delta(I_t\alpha)(\lambda)$$

Then $z_t = \delta(I_t\alpha)$ and

$$\delta(I_t\alpha) = 1 + \int_0^t \delta(I_s\alpha)[i\alpha\,dW_s + \frac{1}{2}(i\alpha)^2 dW_s^2 \pm \frac{1}{6}(i\alpha)^3\,dW_s^3 - (i\alpha)^4\,ds]$$

Take a function

$$f(x) = \int_{-\infty}^{\infty} \exp(x\xi)\hat{f}(d\xi) \qquad \text{with } \int \xi^4|\hat{f}|(d\xi) < \infty$$

Then

$$\begin{aligned} f(W_t) &= \int \delta(I_t\xi)\hat{f}(d\xi) \\ &= f(0) + \int_0^t \{\int \delta(I_s\xi)[i\xi\,dW_s + \frac{1}{2}(i\xi)^2\,dW_s^2 \\ &\qquad + \frac{1}{6}(i\xi)^3\,dW_s^3 - (i\xi)^4]\hat{f}(d\xi)\}\,ds \\ &= f(0) + \int_0^t (f'(W_s)\,dW_s + \frac{1}{2}f''(W_s)\,dW_s^2 \\ &\qquad + \frac{1}{6}f'''(W)\,dW_s^3 - f''''(W_s)\,ds) \end{aligned} \tag{5.3}$$

In fact, multiplying both sides by $\delta(\lambda)$ and taking expectations, we can easily prove (5.3). Formula (5.3) is given in Ref. 1.

6. A GIRSANOV-TYPE THEOREM

As an application of the stochastic integral defined in Sec. 4, we prove the following theorem.

THEOREM 6.1 $f_p(x) = \int e^{ix\xi}\hat{f}_p(d\xi)$ $(p = 0, 1, 2, 3)$ and $\varphi(x) = \int e^{ix\xi}\hat{\psi}(d\xi)$ are functions of real variable with

$$\int (1 + \xi^4)|\hat{f}_p|(d\xi) < \infty \text{ and } \int |\hat{\psi}|(d\xi) < \infty.$$

Let Z_t^x in $\bar{M}_t$ be a solution of

$$Z_t^x = 1 + \int_0^t f_0(x + W_s)Z_s^x\, ds + \sum_{p=1}^{3} f_p(x + W)Z_s^x\, dW_s^{4-p} \tag{6.2}$$

Then $u(t,x) = E(\psi(x + W_t)Z_t^x)$ is continuously differentiable with respect to $t > 0$ and four times continuously differentiable with respect to x, and satisfies

$$\frac{\partial u}{\partial t} = -\frac{\partial^4 u}{\partial x^4} - 4f_3(x)\frac{\partial^3 u}{\partial x^3} - 12f_2(x)\frac{\partial^2 u}{\partial x^2} - 24f_1(x)\frac{\partial u}{\partial x} + f_0(x)u \tag{6.3}$$

with $u(0,x) = \psi(x)$.

Proof. Since $f_p(x + W_s)$ $(p = 0, 1, 2, 3)$ are continuous in the $\|\cdot\|$-norm by Lemma 5.1, Z_t^x is well defined by Theorem 5.1. Also, by Lemmas 5.1 and 5.2, and by (5.19), we easily see that

$$\lim_{t\to 0} u(t,x) = E \lim_{t\to 0} \psi(x + W_t)Z_t^x = \psi(x)$$

Set $v_x(t,\xi) = e^{ix\xi}Z^x(\xi I_t)$; then by (3.3),

$$u(t,x) = \int e^{ix\xi}Z_t^x(\xi\ I_t)\hat{\psi}(d\xi)$$
$$= \int v_x(t,\xi)\hat{\psi}(d\xi)$$

If $v_x(t,\xi)$ is differentiable with respect to t and four times with respect to x, the derivatives are bounded and continuous, and $v_x(t,\xi)$ satisfies (6.3) for each ξ, then $u(t,x)$ satisfies (6.3). Then we prove that $v_x(t,\xi)$ has the above properties. By the same argument as above, we also see that

$$e^{ix\xi}[f_p(x + W_s)Z_s^x](\xi I_t)$$
$$= E[e^{ix\xi}Z_s^x f_p(x + W_s)\ \delta(\xi I_s)]\ \exp\,[-\xi^4(t - s)]$$
$$= \int \exp\,[-\xi^4(t - s)]v_x(s,\ \xi + \eta)\hat{f}_p(d\eta)$$

Therefore, from (6.2) it follows that

$$v_x(t,\xi) = \exp\,(ix\xi - \xi^4 t) + \int_0^t \exp\,[-\xi^4(t - s)]\, ds \int v_x(s,\ \xi + \eta) \sum_{p=0}^{3} c_p(i\xi)^p\hat{f}_p(d\eta) \tag{6.4}$$

where $c_0 = 1$ and $c_p = -4!/(4-p)!$ $(p = 1, 2, 3)$. For a function $g(t,\xi)$ which is continuous in t and ξ, set

$$\|g\|_t = \sup_{\xi, 0\le s\le t} |g(s,\xi)|$$

$$\|g\|_p = \sup_{\xi,t} \frac{|g(t,\xi)|}{1+|\xi|^p} \qquad p = 0, 1, 2, 3, 4$$

and $\|g\| = \|g\|_T = 2\|g\|_0$. Set

$$Kg(t,\xi) = \int g(t, \xi+\eta) \sum_{p=0}^{3} c_p(i\xi)^p \hat{f}_p(d\eta)$$

$$Lg(t,\xi) = \int_0^t \exp[-\xi^4(t-s)]Kg(s,\xi)\,ds$$

and $\sigma(d\eta) = \Sigma|c_p|\,|\hat{f}_p|(d\eta)$. Then

$$\sum_{p=0}^{3} |c_p(i\xi)^p|\,|\hat{f}_p|)(d\eta) \le (1+|\xi|^3)\sigma(d\eta)$$

and by (6.1),

$$\int (1+\eta^4)\sigma(d\eta) < \infty \tag{6.5}$$

LEMMA 6.1 (i) If $\|g\|_p < \infty$ $(p = 0, 1, 2, 3, 4)$, then $Lg(t,\xi)$ is continuous in t and ξ, and

$$\|Lg\| \le C\|g\| \qquad \text{if } p = 0$$

$$\|Lg\|_{p-1} \le C\|g\|_p \qquad \text{if } p > 0$$

(ii) If $\|g\| < \infty$, then

$$\|L^n g\| \le \frac{C^n}{n!}\|g\|$$

(iii) If $\|g\|_4 < \infty$, then $\Sigma_{n=4}^{\infty} \|L^n g\| < \infty$. In particular, $\Sigma_{n=0}^{\infty} L^n g(t,\xi)$ is convergent uniformly in t and ξ.

Proof. Since

$$Lg(t,\xi) = \int_0^t I_t(s) \exp[-\xi^4(t-s)]\,ds \int g(s, \xi+\eta) \times \sum_{p=0}^{3} c_p(i\xi)^p \hat{f}_p(d\eta)$$

and

$$|g(s, \xi+\eta)| \le \|g\|_4(1+\xi+\eta|^4) \le C\|g\|_4(1+|\xi|^4+|\eta|^4)$$

and $\int (1 + \xi^4 + \eta^4)\sigma(d\eta)$ is bounded if ξ is in a compact set, by the dominated convergence theorem continuity of $Lg(t,\xi)$ follows. For p = 1, 2, 3, 4, noting (6.5), we have

$$\frac{|Lg(t,\xi)|}{1 + |\xi|^{p-1}} \le C_1 \int_0^t \exp[-\xi^4(t - s)]\, ds$$

$$\times \int \|g\|_p \frac{1 + |\xi + \eta|^p}{1 + |\xi|^{p-1}}(1 + |\xi|^3)\sigma(d\eta)$$

$$\le C_2 \int_0^t \exp[-\xi^4(t - s)]\|g\|_p(1 + \xi^4)\, ds$$

$$\le C\|g\|_p$$

then $\|Lg\|_{p-1} \le C\|g\|_p$ (p = 1, 2, 3, 4) is proved. If $\|g\| < \infty$,

$$|Lg(t,\xi)| \le C_1 \int_0^t \exp[-\xi^4(t - s)]\|g\|_s(1 + |\xi|^3)\, ds$$

and by Lemma 4.1,

$$\|Lg\|_t^4 \le C^2 \int_0^t \|g\|_s^4\, ds \qquad t \le T$$

In particular, $\|Lg\| \le C\|g\|$. By induction we also have

$$\|L^n g\|_t^4 \le \frac{C_2^n}{n!} t^n \|g\|^4$$

By (i), $\|L^4 g\| < \infty$ if $\|g\|_4 < \infty$. Therefore, (iii) follows from (ii).

LEMMA 6.2 Assume that $\frac{\partial g}{\partial t}(t,\xi)$ exists and is continuous, and $\|g\| < \infty$ and $\|\partial g/\partial t\|_4 < \infty$ holds. Then

(i) $\frac{\partial}{\partial t} Lg(t,\xi)$ exists, and

$$\frac{\partial}{\partial t} Lg(t,\xi) = L\frac{\partial g}{\partial t}(t,\xi) + Kg(0,\xi)\exp(-\xi^4 t)$$

$$\left\|\frac{\partial}{\partial t} Lg\right\|_3 \le C\left(\left\|\frac{\partial g}{\partial t}\right\|_4 + \|g\|\right)$$

(ii) $\frac{\partial}{\partial t} L^n g$ exists, and

$$\frac{\partial}{\partial t} L^n g(t,\xi) = L^n g(t,\xi) + L^{n-1}h(t,\xi)$$

where $h(t,\xi) = Kg(0,\xi)\exp(-\xi^4 t)$.

(iii) $\frac{\partial}{\partial t} \sum_{n=0}^{\infty} L^n g$ exists and

$$\frac{\partial}{\partial t} \Sigma L^n g = \Sigma L^n \frac{\partial g}{\partial t} + \Sigma L^n h$$

Proof. Since $\frac{\partial g}{\partial t}(t,\xi)$ is continuous and $\|\partial g/\partial t\|_4 < \infty$, we can apply formal differentiation to

$$Lg(t,\xi) = \int_0^t \exp(-\xi^4 s)\, ds \int g(t-s, \xi+\eta) \times \sum_{p=0}^{3} c_p (i\xi)^p \hat{f}_p(d\eta)$$

and we have

$$\frac{\partial}{\partial t} Lg = L \frac{\partial g}{\partial t} + Kg(0,\xi) \exp(-\xi^4 t)$$

Noting that $\|h\|_3 \le C_1\|g\|$, by Lemma 6.1(i) we have

$$\|L \frac{\partial g}{\partial t} + h\|_3 \le C(\|\frac{\partial g}{\partial t}\|_4 + \|g\|)$$

Since $L^n g(0,\xi) = 0$ for $n \ge 1$, we have (ii) by induction. Since $\Sigma L^n \frac{\partial g}{\partial t}(t,\xi)$ and $\Sigma L^n h(t,\xi)$ are convergent uniformly, by Lemma 6.1(iii), (iii) is obvious.

The following lemma can be proved in a similar way.

LEMMA 6.3 Let $g_x(t,\xi)$ be a function with real parameter x. Assume that $\partial g_x/\partial x\,(t,\xi)$ exists, $g_x(t,\xi)$ and $(\partial g/\partial x)(t,\xi)$ are continuous in t, ξ, and x, and $\sup_x \|g_x\|_4 < \infty$ and $\sup_x \|\partial g_x/\partial x\|_4 < \infty$ hold. Then:

(i) $Lg_x(t,\xi)$ is continuous in t, ξ, and x.

(ii) $\frac{\partial}{\partial x} Lg_x(t,\xi)$ exists and $\frac{\partial}{\partial x} Lg_x(t,\xi) = L \frac{\partial g_x}{\partial x}(t,\xi)$.

(iii) $\frac{\partial}{\partial x} \sum_{n=0}^{\infty} L^n g_x(t,\xi)$ exists, and $\frac{\partial}{\partial x} \Sigma L^n g_x = \Sigma L^n \frac{\partial g_x}{\partial x}$.

Proof. Now, we can prove that $v_x(t,\xi)$ satisfies (6.3) for each ξ. Set $\psi_x(t,\xi) = \exp(ix\xi - \xi^4 t)$. Since

$$v_x(t,\xi) = \psi_x(t,\xi) + Lv_x(t,\xi)$$

by Lemma 6.1(iii),

$$v_x(t,\xi) = \sum_{n=0}^{\infty} L^n \psi_x(t,\xi)$$

$$\frac{\partial \psi_x}{\partial t}(t,\xi) = -\xi^4 \psi_x(t,\xi)$$

$$\begin{aligned} h_x(t,\xi) &\equiv K\psi_x(0,\xi) \exp(-\xi^4 t) \\ &= \sum_{p=0}^{3} c_p (i\xi)^p \exp(-\xi^4 t) \int e^{ix(\xi+\eta)} \hat{f}_p(d\eta) \\ &= \sum_{p=0}^{3} c_p (i\xi)^p f_p(x) \psi_x(t,\xi) \end{aligned}$$

$$\frac{\partial^p \psi_x}{\partial x^p}(t,\xi) = (i\xi)^p \psi_x(t,\xi) \qquad p = 1, 2, 3, 4$$

and $\sup_x \|\partial\psi_x/\partial t\|_4 < \infty$, $\sup_x \|h_x\|_4 < \infty$, and $\sup_x \|\partial^p \psi_x/\partial x^p\|_4 < \infty$ $(p = 1, 2, 3, 4)$ hold. Therefore, by Lemmas 6.2(iii) and 6.3(iii), we can easily see that

$$\frac{\partial v_x}{\partial t}(t,\xi) = \frac{\partial^4 v_x}{\partial x^4}(t,\xi) + \sum_{p=0}^{3} c_p f_p(x) \frac{\partial^p v_x}{\partial x^p}(t,\xi)$$

Noting that each derivative is continuous and bounded for each fixed ξ, we complete the proof.

REFERENCES

1. K. J. Hochberg, A signed measure on path space related to Wiener measure, Ann. Prob. 6:433-458 (1978).
2. K. Itô, Generalized uniform complex measures in the Hilbertian metric space with their application to the Feynman integral, pp. 163-182 in Proceedings of the Fifth Berkeley Symposium, Vol. 2, Part 1, University of California Press, Berkeley, Calif., 1967.
3. T. Ueno, Continual integral II, pp. 89-105 in Report of Research Group of Mathematical Sciences (in Japanese).

12

A Complex Measure Related to the Schrödinger Equation

KUNIO NISHIOKA/Tokyo Metropolitan University, Tokyo, Japan

1. INTRODUCTION

Our purpose in this chapter is to apply a stochastic method to the Schrödinger equation. More precisely, we treat the fundamental solution $q(t,x,y)$, defined by (2.2), of the Schrödinger equation (2.1) as if it were a complex-valued transition probability density of a "Markov process $X(t)$". Since $q(t,x,y)$ is not of bounded variation, we cannot define a bounded variation measure of $X(t)$ in a function space. But we can study some properties of a Markov process $X(t)$.

For $X(t)$, we show that its "path" is not necessarily continuous. We define a "first hitting time" of $X(t)$ by solving the Dirichlet problem of equation (2.1), and show that it behaves like that of a Brownian motion. By using a "subordination" by $X(t)$, we give a solution of a multidimensional Schrödinger equation with variable coefficients. Also using a subordination by a Brownian motion for a Schrödinger equation, we obtain a solution of a fourth-order and multidimensional parabolic equation with variable coefficients.

We note that Krylov [4], Hochberg [2] and Motoo (see Chap. 11) treat their signed fundamental solutions for fourth and higher-order parabolic equations as if they are signed transition probability densities of Markov processes.

2. The transition probability

Consider the Schrödinger equation for a free particle:

$$\begin{aligned} u_t &= \frac{i}{2} u_{xx} \qquad x \in R^1,\ t > 0 \\ u(0,x) &= f(x) \end{aligned} \tag{2.1}$$

The fundamental solution of (2.1) is given by

$$q(t,x,y) = \frac{1}{\sqrt{2\pi t i}} \exp \frac{i(x-y)^2}{2t} \tag{2.2}$$

and the solution of (2.1) for the reasonable initial function is obtained by

$$u(t,x) = T_t f(x) = \lim_{R\to\infty} \int_{|y|<R} q(t,x,y) f(y)\, dy$$

Denote by D the linear space of constants and functions of the Schwartz class $\mathcal{S}$. We note the properties of $\{T_t\}$:

THEOREM 2.1 (i) $T_t f(x)$ is defined for a function f such that $\sup_s |f(s)| < \infty$ and $\int |f(s) - f(\theta(s)\sqrt{s^2 + 2t\pi})|\, ds < \infty$, where $\theta(s) \equiv$ sign s.

(ii) $T_t T_s f = T_{t+s} f$ if f is a step function or belongs to D.

(iii) $\lim_{t\to 0} T_t f(x) = f(x)$ if f is a step function or belongs to D and if f is continuous at x.

(iv) $T_t : D \to D$.

<u>Proof</u>. (i) $|T_t f(x)| = \int_{-\infty}^{\infty} \frac{1}{\sqrt{2\pi t i}} \exp\, i \frac{s^2}{2t}\, f(x+s)\, ds$

$$= (\sqrt{2\pi t})^{-1} \sum_{k=-\infty}^{\infty} \int_{2kt\pi}^{(2k+2)t\pi} \left(\cos \frac{s'}{2t} + \sin \frac{s'}{2t}\right) \frac{f(x + \theta(s')\sqrt{|s'|})}{\sqrt{|s'|}}\, ds'$$

$$\le (\sqrt{2\pi t})^{-1} \int_{-\infty}^{\infty} \left| \cos \frac{s'}{2t} + \sin \frac{s'}{2t} \right|$$

$$\times \left| \frac{f(x+\theta(s')\sqrt{|s'|})}{2\sqrt{|s'|}} - \frac{f(x+\theta(s')\sqrt{|s'|+2t\pi})}{2\sqrt{|s'|+2t\pi}} \right| ds'$$

$$\leq (\sqrt{2\pi t})^{-1} \int_{-\infty}^{\infty} \left(1 - \frac{|s|}{\sqrt{s^2+2t\pi}}\right) |f(x+s)|\, ds$$

$$+ (2\pi t)^{-1} \int_{-\infty}^{\infty} \frac{|s|}{\sqrt{s^2+2t\pi}} |f(x+s) - f(x+\theta(s)\sqrt{s^2+2t\pi})|\, ds$$

Since $|s|/\sqrt{s^2+2t\pi} \leq 1$ and $1 - (|s|/\sqrt{s^2+2t\pi})$ is integrable, we obtain (i).

(ii) to (iv) are obtained by a calculation analogous to (i).

3. The Markov process X(t)

In order to define a Markov process X(t) related to (2.1), we define the complex-valued transition probability $q(t,x,dy) = q(t,x,y)\, dy$, which satisfies

i. $\int_R q(t,x,y)\, dy = 1$ (Fresnel integral).

ii. $\int_R q(t,x,y)q(s,y,z)\, dy = q(t+s, x, z)$.

iii. $\int |y|^{\alpha} q(t,x,y)\, dy < \infty$ for $|\alpha| < 1$.

where the integrals always mean the improper Riemann integrals. Let Ω be a set of functions $X(\cdot)$: $[0,\infty) \to X(t)$ with $X(0) = x$, and let C be a cylinder set $C \equiv \{X;\ a_i < X(t_i) < b_i,\ i = 1, \ldots, n\}$ for $0 = t_1 < t_2 < \cdots < t_n$. Then, in the usual way, a complex-valued finitely additive measure P_x is defined for C by

$$P_x(C) = \int_{a_1}^{b_1} q(0,x,y)\, dy_1 \int_{a_2}^{b_2} q(t_2 - t_1, y_1, y_2)\, dy_2 \cdots \int_{a_n}^{b_n} q(t_n - t_{n-1}, y_{n-1}, y_n)\, dy_n$$

E_x denotes the "expectation" by P_x. It is very well if the finitely additive measure P_x can be extended to a bounded variation measure on Ω, because we may use tools of the Lebesgue integral, i.e., Lebesgue's theorem, etc. But unfortunately, it is not true by the results of Daletskii [1]. Moreover, P is not concentrated on continuous functions.

THEOREM 3.1 There is a constant $\delta > 0$ and a subsequence $\{n_j\}$ of $\{n\}$, such that

$$\lim_{j\to\infty} \left| P_0\left\{ \sup_{1\le k\le n_j} \left| X\left(\frac{k}{n_j}\right) - X\left(\frac{k-1}{n_j}\right) \right| > \delta \right\} \right| = \infty$$

$$\lim_{j\to\infty} \left| P_0\left\{ \sup_{1\le k\le n_j} \left| X\left(\frac{k}{n_j}\right) - X\left(\frac{k-1}{n_j}\right) \right| \le \delta \right\} \right| = \infty$$

$$\lim_{j\to\infty} \left| \frac{P_0\left\{ \sup \left| X\left(\frac{k}{n_j}\right) - X\left(\frac{k-1}{n_j}\right) \right| > \delta \right\}}{P_0\left\{ \sup \left| X\left(\frac{k}{n_j}\right) - X\left(\frac{k-1}{n_j}\right) \right| \le \delta \right\}} \right| = 1$$

Proof.

$$P_0\left\{ \sup_{0\le k\le n} X\left(\frac{k}{n}\right) - X\left(\frac{k-1}{n}\right) > \delta \right\}$$

$$= \int_{|x_1|>\delta} q\left(\frac{1}{n},0,x_1\right) dx_1$$

$$+ \int_{|x_1|\le\delta} q\left(\frac{1}{n},0,x_1\right) dx_1 \int_{|x_2-x_1|>\delta} q\left(\frac{1}{n},x_1,x_2\right) dx_2$$

$$+ \cdots + \int_{|x_1|\le\delta} q\left(\frac{1}{n},0,x_1\right) dx_1$$

$$\times \int_{|x_2-x_1|\le\delta} q\left(\frac{1}{n},x_1,x_2\right) dx_2$$

$$\cdots \int_{|x_n-x_{n-1}| > \delta} q\left(\frac{1}{n},x_{n-1},x_n\right) dx_n$$

$$= \sum_{k=0}^{n} a(n,\delta)^k b(n,\delta) = 1 - [1 - b(n,\delta)]^n$$

where $a(n,\delta) = \int_{x\le\delta} q(1/n,0,x)\,dx$ and $b(n,\delta) = 1 - a(n,\delta)$. If we take $\delta^2 = 3\pi/2M$ for an integer $M > 0$ and a subsequence $n_j = (8j + 1)M$, $j = 0, 1, 2, \ldots$, then we obtain that $-\alpha/\sqrt{j} \ge \mathrm{Re}(b(n_j,\delta)) \ge -\beta/\sqrt{j}$ for some positive constants α and β. Thus $-n_j\,\mathrm{Re}(b(n_j,\delta)) \to \infty$ as $j \to \infty$, and $|1 - b(n_j,\delta)|^{n_j j} \to \infty$ by Lemma 3.1. Noting that

$$P_0\left\{\sup_{0\le k\le n_j}\left|X\left(\frac{k}{n_j}\right)-X\left(\frac{k-1}{n_j}\right)\right|\le\delta\right\}=[1-b(n_j,\delta)]^{n_j}$$

we complete the proof.

LEMMA 3.1 $f(x)$ is a positive-valued function and $g(x)$ is a complex-valued smooth function. Let $f(x)\to\infty$ and $|g(x)|\to 0$ as $x\to\infty$. Then, if $-f(x)\,\mathrm{Re}(g(x))\to-\infty,\ c,\ +\infty$ as $x\to\infty$, it follows that $|1-g(x)|^{f(x)}\to 0,\ e^c,\ +\infty$, respectively.

Proof. By using the Taylor expansion of $1-g(x)$, we obtain the lemma.

REMARK 1. In the proof of Theorem 3.1, if we take $\delta^2=7\pi/2M$ for an integer $M>0$ and $n_j=(8j+1)M$, $j=0,1,2,\ldots,$ then it holds that

$$\lim_{j\to\infty}\left|P_0\left\{\sup_{1\le k\le n_j}\left|X\left(\frac{k}{n_j}\right)-X\left(\frac{k-1}{n_j}\right)\right|>\delta\right\}\right|=1$$

$$2\ge\limsup,\ \liminf\left|P_0\left\{\sup\left|X\left(\frac{k}{n_j}\right)-X\left(\frac{k-1}{n_j}\right)\right|\le\delta\right\}\right|\ge 0$$

4. The first hitting time

Let $X^\varepsilon(m)$ be a Markov chain with a transition probability density $q(\varepsilon,x,y)$. The first hitting time σ_0^ε for $x=0$ by $X^\varepsilon(m)$ is defined as follows:

$$\sigma_0^\varepsilon\equiv\inf\{n\varepsilon;\ X^\varepsilon(m)>0\text{ for }m<n\text{ and }X^\varepsilon(n)<0\}\qquad(4.1)$$

For a positive constant λ, we define $v_\lambda^\varepsilon(x)$ by

$$v_\lambda^\varepsilon(x)\equiv\sum_{n=0}^{\infty}\exp(-\lambda n\varepsilon)Q_\varepsilon^n\cdot H_0(x)\varepsilon$$

Here

$$Q_\varepsilon\cdot g(x)=\int H_0(x)q(\varepsilon,x,y)H_0(y)g(y)\,dy$$

with $H_0(x)=1$ for $x>0$ and 0 for $x\le 0$. Thanking Motoo [6], we have by some modifications:

LEMMA 4.1 There is a function $G_\lambda(x)$ such that

$\lim_{\varepsilon\downarrow 0} v_\lambda^\varepsilon(x) = G_\lambda(x)$. Moreover, $G_\lambda(x)$ is a weak solution of

$$\lambda G_\lambda(x) - \frac{i}{2}\frac{d^2}{dx^2} G_\lambda(x) = 1 \qquad x > 0$$
$$G_\lambda(0) = 0 \tag{4.2}$$

On the other hand, $u(t,x) = (2/\sqrt{2\pi ti}) \int_0^x \exp[i(y^2/2t)]\, dy$ $[\equiv P_x(\sigma_0 > t)]$ is the solution of (4.3) (see Theorem 6.1):

$$u_t = \frac{i}{2} u_{xx} \qquad t > 0 \text{ and } x > 0$$
$$u(t,0) = 0 \qquad t > 0 \tag{4.3}$$
$$u(0,x) = 1 \qquad x > 0$$

The Laplace transform of $u(t,x)$ satisfies equation (4.2). Since the bounded solution of (4.2) is unique, we may say that $P_x(\sigma_0 > t)$ is the limit of the distribution of the first hitting time for the ε-Markov chain $X^\varepsilon(t)$. We have the following property for $P_x(\sigma_0 > t)$:

THEOREM 4.1 Let f be a C^2-function with a compact support or a step function. Let the support of f be included in $(-\infty,0]$. Then, for $x > 0$,

$$E_x f(X(t)) = \int_0^t P_x(\sigma_0 \in ds) E_0 f(X(t-s)) \tag{4.4}$$

Proof. (a) First we show that the right side of (4.4) exists. Let f be a step function, say $f(x) = 1$ for $x \leq -a < 0$ and 0 for $x > -a$.

The right side of (4.4)

$$= \lim_{\varepsilon\downarrow 0} \int_\varepsilon^t ds \frac{x}{\sqrt{2\pi s^3 i}} \exp \frac{ix^2}{2s} \int_{-\infty}^{-a} \frac{1}{\sqrt{2\pi(t-s)i}} \times \exp\left[\frac{iy^2}{2(t-s)}\right] dy$$

$$= \lim_{\varepsilon\downarrow 0} \int_\varepsilon^t ds \frac{x}{\sqrt{2\pi s^3 i}} \exp\left(\frac{ix^2}{2s}\right) \int_{a/\sqrt{t-s}}^\infty \frac{1}{\sqrt{2\pi i}} \exp\left(\frac{iy^2}{2}\right) dy$$

$$= \lim_{\varepsilon\downarrow 0} \int_\varepsilon^t ds \left[\frac{x}{\sqrt{2\pi s^3 i}} \exp\left(\frac{ix^2}{2s}\right)\right] \left\{1 - \left(\int_0^{a/\sqrt{t}} + \int_{a/\sqrt{t}}^{a/\sqrt{t-s}}\right)\right.$$

$$\left[\frac{1}{\sqrt{2\pi i}} \exp\left(\frac{iy^2}{2}\right) dy\right]\Bigg\} \equiv I_1 - \{I_2 + I_3\}$$

$$|I_1 - 1| = \left|\lim_{\varepsilon \downarrow 0} \int_{x/\sqrt{t}}^{x/\sqrt{\varepsilon}} \frac{2}{\sqrt{2\pi i}} \exp\left(\frac{is^2}{2}\right) ds - 1\right| \leq \frac{K\sqrt{t}}{x}$$

with a positive constant K.

$$|I_2| = |I_1| \left|\int_0^{a/\sqrt{t}} \frac{1}{\sqrt{2\pi i}} \exp\left(\frac{iy^2}{2}\right) dy\right| \leq |1 + K\sqrt{t}|^2$$

Note that

$$\left|\int_{a/\sqrt{t}}^{a/\sqrt{t-s}} \frac{1}{\sqrt{2\pi i}} \exp\left(\frac{iy^2}{2}\right) dy\right| \leq \frac{1}{\sqrt{2\pi}} \int_{a^2/t}^{a^2/t-s} \left(\frac{1}{\sqrt{y}} - \frac{1}{\sqrt{y + 2\pi}}\right) dy$$

$$+ \frac{2}{\sqrt{2\pi}}\left[\left(\frac{a}{\sqrt{t - s}} - \frac{a}{\sqrt{t}}\right) \wedge 2\pi\sqrt{t}\right]$$

which is obtained by a similar calculation as that for the proof of Theorem 2.1. Thus we obtain

$$|I_3| \leq \int_0^t \frac{ds}{\sqrt{2\pi s^3}} \left|\int_{a/\sqrt{t}}^{a/\sqrt{t-s}} \frac{1}{\sqrt{2\pi i}} \exp\left(\frac{iy^2}{2}\right) dy\right| \leq K\left(\frac{1}{\sqrt{t}} \vee 1\right)$$

We have

$$[\text{the right side of (4.4)}] \leq K\left(\frac{1}{\sqrt{t}} \vee t\right)$$

(b) We take the Laplace transform of (4.4):

$$\int_0^\infty e^{-\lambda t} dt \lim_{\varepsilon \downarrow 0} \chi_\varepsilon(t) \int_\varepsilon^t P_x(\sigma_0 \in ds) E_0[f(X(t - s))]$$

$$= \lim_{\varepsilon \downarrow 0} \int_\varepsilon^\infty dt\, e^{-\lambda t} \int_\varepsilon^t P_x(\sigma_0 \in ds) E_0[f(X(t - s))]$$

$$= \lim_{\varepsilon \downarrow 0} \int_\varepsilon^\infty P_x(\sigma_0 \in ds) \int_s^\infty dt\, e^{-\lambda t} E_0[f(X(t - s))]$$

$$= \lim_{\varepsilon \downarrow 0} \int_\varepsilon^\infty ds\, e^{-\lambda s} \frac{x}{\sqrt{2\pi s^3 i}} \exp\left(\frac{is^2}{2s}\right)$$

$$\times \int \frac{1}{\sqrt{2i\lambda}} \exp[(i - 1)\sqrt{\lambda}\, |y|] f(y)\, dy$$

$$= \exp[(i - 1)\sqrt{\lambda}\, |x|] \int \frac{1}{\sqrt{2i\lambda}} \exp[(i - 1)\sqrt{\lambda}\, |y|] f(y)\, dy$$

where $\chi_\varepsilon(t) = 0$ for $t \leq \varepsilon$, 1 for $\varepsilon < t$. To verify the first equality, we use the result of step (a), and for the second equality, we use Fubini's theorem.

On the other hand, note that

$$\lim_{R\to\infty} \int_{-R}^{-a} \frac{1}{\sqrt{2\pi t i}} \exp\left(\frac{iy^2}{2t}\right) dy$$

is uniformly bounded and converges uniformly for t in a finite interval. We have that

[the Laplace transform of the left side of (4.4)]

$$= (\sqrt{2i\lambda})^{-1} \int_{-\infty}^{\infty} dy \exp [(i-1)\sqrt{\lambda}\ |y-x|]f(y)$$

The proof is completed for a step function. For a compact support function, the proof is similar.

COROLLARY 4.1 Define $E_x[f(X(t)); \sigma_0 > t] \equiv$ [a solution of (6.1) and (6.2)], where f is a step function. Then it holds that

$$E_x[f(X(t))] = E_x[f(X(t)); \sigma_0 > t] + \int_0^t P_x(\sigma_0 \in ds)E_0[f(X(t-s))] \tag{4.5}$$

In other words, we may conclude as follows. Define $P_x(\sigma_0 > t) \equiv (2/\sqrt{2\pi t i}) \int_0^x \exp(iy^2/2t)\, dy$, which is an analogy of André's reflection principle (see [3]). Then $F_x\{\sigma_0 > t)$ satisfies (4.3), like a Brownian motion does for the heat equation. That $P_x(\sigma_0 > t)$ is the limit of the distribution of σ_0^{ε}, given by (4.1), for a Markov chain $X^{\varepsilon}(t)$. Moreover, (4.5) holds for a step function. But it seems to be difficult to require that (4.5) hold for more general functions, because of Theorem 3.1.

The next properties of $P_x(\sigma_0 > t)$ are clear:

$\lim_{t\to\infty} P_x(\sigma_0 > t) = 0$ for each x

$\lim_{x\to\infty} P_x(\sigma_0 > t) = 1$ for each $t > 0$

5. Subordination

We define a "subordination" by $X(t)$ in order to solve a Schrödinger equation with variable coefficients.

Define a second-order elliptic operator A:

$$Au(x) = \sum_{i,j=1}^{d} a_{ij}(x) \frac{\partial^2 u}{\partial x_i \, \partial x_j} + \sum_{i=1}^{d} b_i(x) \frac{\partial u}{\partial x_i} + c(x)u \quad (5.1)$$

where the matrix $(a_{ij}(x))$ is nonnegative definite.

Let $u(t,x)$ be a smooth solution of the following second-order hyperbolic equation.

$$\begin{aligned} u_{tt} &= Au \qquad \text{for } t > 0 \text{ and } t < 0,\ x \in R^d \\ u(0,x) &= f(x) \qquad \text{and} \qquad u_t(0,x) = g(x) \end{aligned} \quad (5.2)$$

THEOREM 5.1 Let $u(t,x)$ satisfy, for each $x \in R^d$,

(i) $\sup_{y \in U(x)} |tu(t,y)|$ and $\sup_{y \in U(x)} |tu_t(t,y)|$ vanish as $|t| \to \infty$

(ii) $\sup_{y \in U(x)} \int |t^2 u(t,x)| \, dt < \infty$ and $\sup_{y \in U(x)} \int |Au(t,x)| \, dt < \infty$ (5.3)

where $U(x)$ is a compact neighborhood of x. Then.

$$v(t,x) \equiv E_0 u(X(t),x) \equiv \int_{-\infty}^{\infty} q(t,0,s) u(s,x) \, ds$$

is a solution of

$$\begin{aligned} v_t &= \frac{i}{2} Av \qquad \text{for } t > 0 \text{ and } t < 0,\ x \in R^d \\ v(0,x) &= f(x) \end{aligned} \quad (5.4)$$

Proof. Note that $u(s,x)$ and $u_s(s,x)$ are continuous for $-\infty < s < \infty$. Thus we can verify (5.4) easily, by using (5.3).

THEOREM 5.2 Let $v(t,x)$ be a smooth solution of (5.4). Assume that $\sup_{y \in U(x)} |v(t,y)| \le F(t)$ for each $x \in R^d$, where $F(t)$ is a function such that $\int_{-\infty}^{\infty} F(t) \exp(-\alpha t^2) \, dt < \infty$ for any $\alpha > 0$. Then

$$w(t,x) = \int_{-\infty}^{\infty} (2\pi t)^{-1/2} \exp\left(\frac{-s^2}{2t}\right) u(s,x) \, ds$$

is a solution of

$$w_t = -\frac{1}{8}A^2 w \qquad \text{for } t > 0 \text{ and } x \in R^d$$

$$w(0,x) = f(x)$$

Proof. Note that $v(s,x)$ and $v_s(s,x)$ are continuous for $-\infty < s < \infty$, and the proof is clear.

REMARK 1. Condition (5.3) is studied as a lacuna or energy decay of equation (5.2) (see Morawetz [5] and Murata [7]).

In Theorem 5.1 we may consider a first-order differential equation as a special case of (5.2). Let $Y(t; x)$ be a solution of

$$\frac{dY}{dt} = a(Y) \qquad Y(0) = x \in R^1$$

where $a(x)$ is a C^1-function with a bounded derivative. The next results are obtained from the behavior of $Y(t; x)$.

Case 1. (Transformation of a Variable). Let $a(x) \geq c_1$, with a constant $c_1 > 0$. Let $f \in D$; then

$$v(t,x) = \int q(t,0,s)f(Y(s;x))\,ds \tag{5.5}$$

satisfies (5.4) with $d = 1$ and

$$A = a(x)\frac{\partial}{\partial x}\left(a(x)\frac{\partial}{\partial x}\right)$$

We assume that $f \in S$ and that

$$c_2(1 + |x|^{\alpha}) \geq a(x) \geq c_1 \qquad \text{for large } |x|$$

with constants $c_2 > 0$ and $0 \leq \alpha < 1$. If $\alpha = 1/2$, then

$$\int |v(t,x)|^2 \frac{1}{a(x)}\,dx < \infty \qquad \text{for each } t$$

If $0 \leq \alpha < 1/2$, then

$$\int |v(t,x)|^2 \frac{1}{a(x)}\,dx = \int |f(x)|^2 \frac{1}{a(x)}\,dx \qquad \text{for each } t$$

Case 2. Let $a(x_1) = a(x_2) = 0$ for $\infty < x_1 < x_2 < +\infty$ and $a(x) > 0$ for $x \in (x_1,x_2)$. If $f(x)$ is a C^2-function whose support is in (x_1,x_2), then $v(t,x)$, given by (5.5), satisfies (5.4) with the same A as case (1). Moreover, $v(t,x) \equiv 0$ for each t and $x \in (x_1,x_2)^c$

6. Boundary value problems

We treat a Dirichlet and a Neumann problem for the Schrödinger equation. In quantum mechanics, a Dirichlet problem for the Schrödinger equation appears when the potential blows up to infinity at a boundary of a region. Our problem is

$$\begin{aligned} v_t &= \frac{i}{2} v_{xx} \qquad t > 0,\ x > 0 \\ v(0,x) &= f(x) \end{aligned} \tag{6.1}$$

with the boundary condition

$$v(t,0) = 0 \qquad t > 0 \tag{6.2}$$

or

$$v_x(t,0) = 0 \qquad t > 0 \tag{6.3}$$

Direct calculations prove the next theorem.

THEOREM 6.1 Assume that $f \in D$.
(i) Let $v^D(t,x) \equiv \int_0^\infty [g(t,x,s) - q(t,-x,s)]f(s)\, ds$. Then v^D satisfies (6.1) with (6.2).
(ii) Let $v^N(t,x) \equiv \int_0^\infty [q(t,x,s) + q(t,-x,s)]f(s)\, ds$. Then v^N satisfies (6.1) with (6.3).

Let v be a solution of (6.1).

$$\int_0^R |v(t,x)|^2\, dx - \int_0^R |f(x)|^2\, dx = \int_0^R dx \int_0^t d\tau\, \frac{\partial}{\partial\tau}|v(\tau,x)|^2$$

$$= \frac{i}{2}\int_0^t d\tau \{[v_x(\tau,R)\,\overline{v(\tau,R)} - v(\tau,R)\,\overline{v_x(\tau,R)}]$$

$$- [v_x(\tau,0)\,\overline{v(\tau,0)} - v(\tau,0)\, v_x(\tau,0)]\}$$

If $f \in S$, then the first term of the last equation vanishes as $R \to \infty$, and if (6.2) or (6.3) is satisfied, then the second term vanishes. We have the next remark.

REMARK 1. Let v be a solution of (6.1) with (6.2) or (6.3). If $f \in S$, then it holds that $\int_0^\infty |v(t,x)|^2\, dx = \int_0^\infty |f(x)|^2\, dx$.

Now we consider a d-dimensional case with variable coefficients.

$$v_t = \frac{i}{2} Av \qquad t > 0,\ x \in G$$
$$v(0,x) = f(x) \tag{6.4}$$

where A is of the form (5.1) and let $G = \{x \in R^d;\ x_1 > 0\}$ and $\partial G = \{x \in R^d;\ x_1 = 0\}$. We extend the operator A and the initial function f to the whole space R^d, as follows:

$$a^*_{ij}(x_1,\ldots,x_d) = a_{ij}(|x_1|,\ldots,x_d), \qquad i, j = 1, \ldots, d$$
$$b^*_i(x_1,\ldots,x_d) = b_i(|x_1|,\ldots,x_d), \qquad i = 1, \ldots, d$$
$$c^*(x_1,\ldots,x_d) = c(|x_1|,\ldots,x_d)$$
$$f^D(x) = \begin{cases} f(x) & \text{for } x \in G \\ 0 & \text{for } x \in G^c \end{cases}$$
$$f^N(x_1,\ldots,x_d) = \frac{1}{2} f(|x_1|,\ldots,x_d)$$

Let $u^D(t,x)$ [respectively, $u^N(t,x)$] be a solution of

$$u_{tt} = A^*u \qquad t > 0 \text{ and } t < 0,\ x \in R^d$$
$$u(0,x) = f^D(x) \qquad [\text{respectively, } f^N(x)] \tag{6.5}$$
$$u_t(0,x) = 0$$

Denote by x^* a point $(-x_1,x_2,\ldots,x_d) \in R^d$ for $x = (x_1,x_2,\ldots,x_d) \in R^d$.

THEOREM 6.2 If $u^D(t,x)$ [respectively, $u^N(t,x)$] satisfies (5.3), then

$$v^D(t,x) \equiv E_0\{u^D(X(t),x) - u^D(X(t), x^*)\}$$

[respectively, $v^N(t,x) \equiv E_0\{u^N(X(t),x) + u^N(X(t),x^*)\}$]

satisfies (6.4) with $v(t,x) = 0$ for $x \in \partial G$ [respectively, $\frac{\partial v}{\partial x_1}(t,x) = 0$ for $x \in \partial G$].

Proof. The assumption of the theorem justifies the following formal calculations:

$$v^D_{tt} = \frac{i}{2} \int q(t,0,s)[u^D_{ss}(s,x) - u^D_{ss}(s,x^*)]\, ds$$
$$= \frac{i}{2} \int q(t,0,s)[A^*u^D(s,x) - A^*u^D(s,x^*)]\, ds$$

$$= \frac{i}{2} A^*v = \frac{i}{2} Av \qquad \text{for } x \in G$$

$$v^D(t,x) = 0 \qquad \text{for } x \in \partial G, \text{ because } x = x^* \text{ for } x \in \partial G$$

$$\lim_{t \downarrow 0} v^D(t,x) = u^D(0,x) - u^D(0,x^*) = f(x) \qquad \text{for } x \in G$$

The calculations are similar for v^N.

REMARK 2. If we obtain a solution of a hyperbolic equation with a boundary condition and if that solution satisfies (5.3), then the "subordination" method, like Theorem 5.1, is useful for the Schrödinger equation with the same A and with the same boundary condition.

REFERENCES

1. Yu. L. Daletskii, Functional integrals connected with certain differential equations and systems, Sov. Math. Dokl. 2 (1961).
2. K. J. Hochberg, A signed measure on path space related to Wiener measure, Ann. Prob. 6 (1978).
3. K. Itô and H. P. McKean, Diffusion Processes and Their Sample Paths, Springer-Verlag, Berlin, 1965.
4. V. Yu. Krylov, Some properties of the distribution corresponding to the equation $\partial u/\partial t = (-1)^q(\partial^q u/\partial x^q)$, Sov. Math. Dokl. 1 (1960).
5. C. S. Morawetz, Note on time decay and scattering for some hyperbolic problems, Reg. Conf. Ser. Appl. Math. 19, SIAM, Philadelphia, 1975.
6. M. Motoo, Lecture at Tokyo Metropolitan Univ., 1980.
7. M. Murata, Asymptotic expansions in time for solutions of Schrödinger-type equations, to appear.

13

On the Nearness of Two Solutions in Comparison Theorems for One-Dimensional Stochastic Differential Equations

YUKIO OGURA/Faculty of Science and Engineering, Saga University, Saga, Japan

1. INTRODUCTION

In a previous paper [3], we have studied strong (or strict) comparison theorems for the solutions of one-dimensional stochastic differential equations (SDE). There we obtained strict inequalities $x_1(t) < x_2(t)$, $t \geq 0$, a.s. for two solutions $x_1(t)$ and $x_2(t)$ of SDE under relevant assumptions, whereas most of the preceding authors on this subject gave inequalities $x_1(t) \leq x_2(t)$, $t \geq 0$, a.s. under some weaker assumptions.

The object of this chapter is to investigate the relation of the above two kinds of comparison theorems. More precisely, we will measure the "nearness" of two solutions $x_1(t)$ and $x_2(t)$ by means of two quantities, and study how the assumptions in the above theorems affect the quantities and clarify when two solutions become separated from each other. In Sec. 2 we adopt the minimum of the distance $|x_2(t) - x_1(t)|$, $0 \leq t \leq T \wedge \sigma_L$, as the scale of nearness and estimate the order of decay of its distribution in a neighborhood of 0. In Sec. 3 we choose the sojourn time in $[0,x]$ for the process $|x_2(t) - x_1(t)|$, $0 \leq t \leq T \wedge \sigma_L$, and dominate the order of its decay as $x \downarrow 0$.

Throughout, we denote by $(\Omega, \mathcal{F}, P; \{\mathcal{F}_t\})$ a complete probability space with right-continuous increasing family $\{\mathcal{F}_t\}_{t\geq 0}$ of complete sub-σ-fields of $\mathcal{F}$ and by $B(t)$ a $\mathcal{F}_t$-Brownian motion satisfying $B(0) = 0$. For $i = 1, 2$, let $x_i(t)$ be a continuous $\mathcal{F}_t$-adapted solution of the SDE

$$x_i(t) = x_i + \int_0^t a(s, x_i(s))\, dB(s) + \int_0^t b_i(s, x_i(s))\, ds \quad (1.1)$$

a.s. on $\{t < \zeta_i\}$, where $a(t,x)$ and $b_i(t,x)$ are real continuous functions on $[0,\infty) \times R$ and $\zeta_i = \sup\{t; \sup_{0\leq s\leq t} |x_i(s)| < \infty\}$. Further, assume that there is a continuous increasing function $r(u)$ on $[0,\infty)$ such that $r(0) = 0$ and

$$|a(t,x) - a(t,y)| \leq r(|x - y|) \qquad x,y \in R,\ t \geq 0$$

We also assume that $x_2 - x_1 > 0$ and

$$b_2(t,x) - b_1(t,x) > 0 \qquad x \in R,\ t \geq 0$$

$$\int_0^1 r^{-2}(u)\, du = \infty$$

Due to the usual comparison theorem, this ensures that

$$P(x_1(t) \leq x_2(t),\ 0 \leq t \leq \zeta) = 1$$

where $\zeta = \zeta_1 \wedge \zeta_2$, the minimum of ζ_1 and ζ_2.

2. MINIMUM DISTANCE OF TWO SOLUTIONS

In this section we estimate the distribution function of the minimum distance

$$m(T) = \min\{x_2(t) - x_1(t);\ 0 \leq t \leq T\}$$

For each $L > 0$, let

$$\sigma_L = \sigma_L^1 \wedge \sigma_L^2 \qquad \sigma_L^i = \inf\{t \geq 0;\ x_i(t) \geq L\}$$

$(\inf \emptyset = \infty)$. Then $\sigma_L^i \uparrow \zeta_i$ as $L \uparrow \infty$, so $\sigma_L \uparrow \zeta$ as $L \uparrow \infty$, a.s. Let also, for each $T, L > 0$,

$$c(T,L) = \min\{b_2(t,x) - b_1(t,x);\ (t,x) \in [0,T] \times [-L,L]\}$$

and for each $c > 0$,

$$f_c(x) = \int_x^1 \exp[cR(y)]\, dy \qquad x > 0$$

where $R(y) = \int_y^1 r^{-2}(u)\, du$.

THEOREM 2.1 Suppose that the assumptions in Sec. 1 are fulfilled. Then for each T, L > 0 and $c \in (0,c(T,L))$, there exists a positive constant $K_1 = K_1(c,T,L)$ such that

$$P(m(T \wedge \sigma_L) \leq x) \leq \frac{K_1}{f_{2c}(x)} \qquad 0 < x < 1 \tag{2.1}$$

Proof. The basic idea is same as that in Ref. 3, proof 2 of theorem 4.1. Indeed, by choosing a $\delta > 0$ so that

$$|b_1(t,x) - b_1(t,y)| < c(T,L) - c$$
$$|x - y| < \delta,\ x,y \in [-L,L],\ 0 \leq t \leq T$$

we have

$$b_2(t,x) - b_1(t,y) \geq c \qquad |x - y| < \delta,\ x,y \in [-L,L],\ 0 \leq t \leq T \tag{2.2}$$

For $x \geq 0$, let

$$\sigma_x' = \inf \{0 \leq t < \zeta;\ x_2(t) - x_1(t) = x\}$$

($\inf \emptyset = \zeta$) and $T' = T \wedge \sigma_x' \wedge \sigma_L$ ($< \zeta$). Then by the method of Ref. 3, equation (4.21), we get

$$\begin{aligned} &Ef_{2c}(x_2(T') - x_1(T')) \\ &\quad \leq f_{2c}(x_2 - x_1) + e^{2cR(\delta)} \sup_{\substack{0 \leq t \leq T \\ |x| \leq L}} \{(|b_1(t,x)| + |b_2(t,x)| \\ &\qquad + c\} ET \wedge \sigma_L \\ &\quad \equiv K_1(c,T,L) \end{aligned}$$

But the function $f_{2c}(x)$ is decreasing on $(0,\infty)$. Hence

$$\begin{aligned} P(m(T \wedge \sigma_L) \leq x) &= P(\sigma_x' \leq T \wedge \sigma_L) \\ &\leq f_{2c}(x)^{-1} Ef_{2c}(x_2(T') - x_1(T')) \end{aligned}$$

and the inequality (2.1) follows.

The next small variation of strong comparison theorem 4.1 of Ref. 3 is a direct consequence of Theorem 2.1.

COROLLARY 2.1 Suppose that the assumptions in Sec. 1 are fulfilled, and let T, L > 0. If there is a $c \in (0,c(T,L))$ such that $f_{2c}(0+) = \infty$, then

$$P(x_1(t) < x_2(t),\ 0 \leq t \leq T \wedge \sigma_L) = 1$$

Further, if the above conditions are satisfied for all T,

$L > 0$, then

$$P(x_1(t) < x_2(t),\ 0 \leq t < \zeta) = 1 \qquad (2.3)$$

In particular, if $f_{2c}(0+) = \infty$ for some c with

$$0 < c < c_0 \equiv \inf_{T,L>0} c(T,L)$$

then the assertion (2.3) follows.

REMARK 1. If $f_{2c_0}(0+)$ is finite, then one cannot expect the strict comparison theorem in general, because there is a case in which all the assumptions in Sec. 1 are satisfied and the left-hand side of (2.1) remains positive as $x \downarrow 0$. Indeed, let $a(t,x) \equiv r(x)$, $b_1(t,x) \equiv 0$, $b_2(t,x) \equiv c_0 > 0$, $x_1 = 0$, and $x_2 > 0$. Then it is obvious that $x_1(t) \equiv 0$ and $x_2(t)$ is a realization of the diffusion process $X = (x(t), P_x)$ on $[0,\infty)$ corresponding to the generator $r^2(x)d^2/2\ dx^2 + c_0 d/dx$ starting at x_2. The scale $s(x)$ of this diffusion coincides with $-f_{2c_0}(x)$, and the boundary 0 is regular if

$$f_{2c_0}(0+) = \int_0^1 \exp\,[2c_0 R(y)]\,dy < \infty$$

Hence, for these $x_1(t)$ and $x_2(t)$,

$$P(m(T \wedge \sigma_L) \leq x) \geq P_{x_2}(\sigma_0 \leq T \wedge \sigma_L) > 0$$
$$0 < x < 1 \wedge L$$

where σ_x stands for the first hitting time for the state x with respect to the diffusion X.

REMARK 2. An inspection of the proof of Theorem 2.1 produces a more general assertion: In addition to the assumptions in Sec. 1, suppose that there is a $\delta > 0$ for which the inequality (2.2) holds for all $|x - y| < \delta$, $x,y \in [-L,L]$, $t \geq 0$, and suppose that $E\sigma_L < \infty$. Then there is a constant $K_2 = K_2(c,L)$ such that

$$P(m(\sigma_L) \leq x) \leq \frac{K_2}{f_{2c}(x)} \qquad 0 < x < 1$$

REMARK 3. The estimate (2.1) is not so bad in the sense that we have an example for which all the assumptions in Sec. 1

are satisfied and

$$P(m(T \wedge \sigma_L) \leq x) \geq \frac{K_3}{f_{2c(T,L)}(x)} \qquad 0 < x < \frac{1}{2} \tag{2.4}$$

for an appropriate positive K_3, if $L \geq 1$ and T is large enough. Indeed, consider the processes in Remark 1. If $f_{2c_0}(0+)$ is finite, then (2.4) is obvious by Remark 1. Here is a proof of (2.4) for the case $f_{2c_0}(0+) = \infty$, which is a joint product with M. Tomisaki.

First note that

$$\begin{aligned} P(m(T \wedge \sigma_L) \leq x) &= P_{x_2}(\sigma_x \leq T \wedge \sigma_L) \\ &\geq P_{x_2}(\sigma_x \leq T \leq \sigma_L) \\ &= E_{x_2}[P_x(T - U \leq \sigma_L)|_{U=\sigma_x}; \sigma_x \leq T] \\ &\geq P_{x_2}(\sigma_x \leq T)P_{1/2}(T \leq \sigma_L) \qquad 0 < x < \frac{1}{2} \end{aligned}$$

and for $\alpha > \beta > 0$,

$$\begin{aligned} E_{x_2} \exp(-\alpha\sigma_x) &= E_{x_2}[\exp(-\alpha\sigma_x); \sigma_x \leq T] \\ &\quad + E_{x_2}[\exp(-\alpha\sigma_x); \sigma_x > T] \\ &\leq P_{x_2}(\sigma_x \leq T) + e^{-(\alpha-\beta)T}E_{x_2} \exp(-\beta\sigma_x) \end{aligned}$$

Hence, denoting $g(x; \lambda) = 1/E_{x_2} \exp(-\lambda\sigma_x)$, we have

$$P_{x_2}(\sigma_x \leq T) \geq g(x;\alpha)^{-1} - e^{-(\alpha-\beta)T}g(x;\beta)^{-1}$$

But $-g^+(x;\lambda) \equiv -dg(x;\lambda)/ds(x)$ is decreasing and $g^+(0+;\lambda)$ is finite because the boundary 0 is now entrance (Ref. 1, p. 130). Hence we can find positive constants K_4 and K_5 satisfying

$$g(x;\alpha) \leq K_4\, s(x) \qquad g(x;\beta) \geq K_5 s(x) \qquad 0 < x < \frac{1}{2}$$

where the scale is now normalized as $s(x) = -f_{2c_0}(x)$. This with the above inequalities implies (2.4), provided that T is large enough.

Of course the only interesting examples for Theorem 2.1 fall in the case when $f_{2c}(0+) = \infty$. We will give two of them.

EXAMPLE 1. Let $r(u) = r_0 u^{1/2}$ $(r_0 > 0)$ and $c(T,L) > r_0^2/2$. Then $R(y) = -\ell n\, y^{1/r_0^2}$ and

$$f_{2c}(x) = \frac{x^{-(2c/r_0^2-1)} - 1}{2c/r_0^2 - 1}$$

provided that $2c/r_0^2 \neq 1$. Hence, for each $c \in (r_0^2/2, c(T,L))$, we have

$$P(m(T \wedge \sigma_L) \leq x) \leq K_1 x^{2c/r_0^2-1} \qquad 0 < x < 1$$

EXAMPLE 2. Let $r(u) = r_0 u^{(1+p)/2}$ $(r_0 > 0,\ 0 < p \leq 1)$. Then $R(y) = (y^{-p} - 1)/pr_0^2$ and

$$f_{2c}(x) = \exp\left(\frac{-2c}{pr_0^2}\right)\int_x^1 \exp\left(\frac{2cy^{-p}}{pr_0^2}\right) dy$$

But for $0 < x < 1/2$,

$$\int_x^1 \exp\left(\frac{2cy^{-p}}{pr_0^2}\right) dy \geq \int_x^{2x} \exp\left(\frac{2cy^{-p}}{pr_0^2}\right) dy$$
$$\geq x \exp\left(\frac{2c(2x)^{-p}}{pr_0^2}\right)$$

Hence, for some positive K_6,

$$P(m(T \wedge \delta_L) \leq x) \leq K_1 x^{-1} \exp(-K_6 x^{-p}) \qquad 0 < x < \frac{1}{2}$$

In contrast with the next "scale of nearness", the relevant one makes sense for two solutions of one SDE with different initial points. Let $x_i(t) = x(t,x_i)$ be a continuous $\mathcal{F}_t$-adapted solution of the SDE

$$x_i(t) = x_i + \int_0^t a(s,x_i(s))\, dB(s) + \int_0^t b(t,x_i(s))\, ds$$

a.s. on $\{t < \zeta_i\}$, where $a(t,x)$ and $b(t,x)$ satisfy the corresponding conditions in Sec. 1. Further, let $k(u)$ be a continuous increasing function on $[0,\infty)$ such that

$$|b(t,x) - b(t,y)| \leq k(|x - y|) \qquad x,y \in R,\ t \geq 0$$

Let also

$$f(x) = \int_x^1 \int_y^1 r^{-2}(u)\, du\, dy$$

and assume that $f(0+) = \infty$ as well as

$$\lim_{x \downarrow 0} \sup_{x \leq y \leq 1} \frac{k(y) \int_y^1 r^{-2}(u)\, du}{f(x)} = 0$$

Then we have the following:

THEOREM 2.2 Under the above assumptions, for each T, L > 0, there is a positive constant $K_7 = K_7(T,L)$ such that

$$P(m(T \wedge \sigma_L) \leq x) \leq \frac{K_7}{f(x)} \qquad 0 < x < 1 \qquad (2.5)$$

The proof is just the combination of the arguments in the proof of Ref. 3, theorem 1.1 and that of Theorem 2.1.

EXAMPLE 3. Let $r(u) = r_0 u$ ($r_0 > 0$) and $b(t,x)$ satisfy the conditions of Theorem 2.2. Then

$$f(x) \sim r_0^{-2} \ln\left(\frac{1}{x}\right) \qquad \text{as } x \downarrow 0$$

and so, for each positive T and L,

$$P(m(T \wedge \sigma_L) \leq x) \leq \frac{K_8}{\ln(1/x)} \qquad 0 < x < 1$$

for some constant $K_8 = K_8(T,L)$.

EXAMPLE 4. Let $r(u) = r_0 u(1 - \ln u)^{1/2}$ and $b(t,x)$ satisfy the conditions of Theorem 2.2. In this case

$$\int_y^1 r^{-2}(u)\, du = \frac{1}{r_0^2 e} \int_1^{1-\ln y} \frac{e^v}{v}\, dv$$

$$\geq \frac{1 - y}{r_0^2 y(1 - \ln y)} \qquad 0 < y < 1$$

Hence, for some positive constant K_9,

$$f(x) \geq K_9 \ln(1 - \ln x) \qquad 0 < x < 1$$

and so, for each positive T and L,

$$P(m(T \wedge \sigma_L) \leq x) \leq \frac{K_{10}}{\ln(1 - \ln x)} \qquad 0 < x < 1$$

Finally, we note that Theorem 2.2 can be easily extended to the multidimensional case in relevant form if one wishes.

3. SOJOURN TIME FOR THE DISTANCE OF TWO SOLUTIONS

In this section we dominate the Lebesgue measure $|S(x,T)|$ of the sojourn time

$$S(x,T) = \{0 \leq t \leq T;\ 0 \leq x_2(t) - x_1(t) \leq x\}$$

Using the same notations as in Sec. 2, we let

$$g_c(x) = \exp[-cR(x)] \qquad c,\ x > 0$$

In the following, the description

$$Y(x,\omega) < g(x) \qquad \text{as } x \downarrow 0, \text{ a.s.}$$

means that there is a null set N such that for each $\omega \in \Omega \backslash N$ there is a positive $h = h(\omega)$ satisfying $Y(x,\omega) < g(x)$ for all $x \in (0,h)$.

THEOREM 3.1 Suppose that the assumptions in Sec. 1 are fulfilled. Then for each T, L > 0 and $c \in (0,c(T,L))$, it holds that

$$|S(x,T \wedge \sigma_L)| < g_{2c}(x) \qquad \text{as } x \downarrow 0, \text{ a.s.} \tag{3.1}$$

For the proof we need one lemma.

LEMMA 3.1 Under the assumptions in Sec. 1,

$$E|S(0,\zeta)| = 0 \tag{3.2}$$

Further, for each T, L > 0 and $c \in (0,c(T,L))$ it holds that

$$\begin{aligned} K_{11} &= K_{11}(c,T,L) \\ &\equiv E\int_{S(1,T\wedge\sigma_L)\backslash S(0,T\wedge\sigma_L)} \exp[2cR(x_2(s) - x_1(s))]\, ds \\ &< \infty \end{aligned} \tag{3.3}$$

Proof. Fix T, L > 0 and $c \in (0,c(T,L))$, and set

$$f_{(m)}(x) = f_{2c}\left(x + \frac{1}{m}\right) \qquad m = 1, 2, 3, \ldots$$

Then, for the $\delta > 0$ in (2.2), let

$$S_0 = S(0,T') \qquad S_1 = S(\delta,T') \qquad S_2 = S(L,T')\backslash S(\delta,T')$$

where $T' = T \wedge \sigma_L$. Since $f_{(m)} \in C^2(-1/m,\infty)$ and $x_2(t) - x_1(t)$ is nonnegative on $\{t < \zeta\}$, one can apply Itô's formula to obtain

$$\begin{aligned}
Ef_{(m)}&(x_2(T') - x_1(T')) - f_{(m)}(x_2 - x_1) \\
&= E\int_{S_1} f'_{(m)}(x_2(s) - x_1(s))[b_2(s,x_2(s)) \\
&\quad - b_1(s,x_1(s))]\,ds \\
&\quad + E\int_{S_2} f'_{(m)}(x_2(s) - x_1(s))[b_2(s,x_2(s)) \\
&\qquad\qquad - b_1(s,x_1(s))]\,ds \\
&\quad + \frac{1}{2}E\int_{S_1} f''_{(m)}(x_2(s) - x_1(s))[a(s,x_2(s)) \\
&\qquad\qquad - a(s,x_1(s))]^2\,ds \\
&\quad + \frac{1}{2}E\int_{S_2} f''_{(m)}(x_2(s) - x_1(s))[a(s,x_2(s)) \\
&\qquad\qquad - a(s,x_1(s))]^2\,ds \\
&\equiv I_1 + I_2 + I_3 + I_4
\end{aligned} \tag{3.4}$$

But

$$f'_{(m)}(x) = -\exp\left[2cR\left(x + \frac{1}{m}\right)\right]$$

$$f''_{(m)}(x) = \frac{2c\exp[2cR(x + 1/m)]}{r^2(x + 1/m)}$$

Hence

$$I_1 \le -c(T,L)E\int_{S_1} \exp\left[2cR\left(x_2(s) - x_1(s) + \frac{1}{m}\right)\right]ds$$

$$I_3 \le cE\int_{S_1} \exp\left[2cR\left(x_2(s) - x_1(s) + \frac{1}{m}\right)\right] \times \frac{r^2(x_2(s) - x_1(s))}{r^2(x_2(s) - x_1(s) + 1/m)}\,ds$$

Noting that $r(x)$ is increasing, we obtain

$$I_1 + I_3 \le -(c(T,L) - c)\,E\int_{S_1} \exp\left[2cR\left(x_2(s) - x_1(s) + \frac{1}{m}\right)\right]ds$$

Similarly,

$$I_2 + I_4 \leq \exp\,[2cR(\delta + \tfrac{1}{m})]$$

$$\times \sup_{\substack{0 \leq t \leq T \\ |x| \leq L}} \{(|b_1(t,x)| + |b_2(t,x)|) + c\} ET \wedge \sigma_L$$

Since $R(x)$ is decreasing on $(0,1)$, this implies that

$$I_2 + I_4 \leq K_{12}(c,T,L) \qquad m = 1, 2, 3, \ldots$$

Now, substitution of the above inequalities into (3.4) shows

$$(c(T,L) - c)E \int_{S_1} \exp\,[2cR(x_2(s) - x_1(s) + \tfrac{1}{m})]\, ds$$

$$\leq f_{(m)}(x_2 - x_1) - Ef_{(m)}(x_2(T') - x_1(T')) + K_{12}$$

$$\leq f_{2c}(x_2 - x_1) + |f_{2c}(2L + 1)| + K_{12} \equiv K_{13}$$

$$m = 1, 2, 3, \ldots \tag{3.5}$$

The first term in (3.5) is not less than

$$(c(T,L) - c) \exp\,[2cR(\tfrac{1}{m})]\, E|S_0|$$

But $R(1/m) \uparrow \int_0^1 r^{-2}(u)\, du$ $(m \uparrow \infty)$, which is equal to infinity, as we have assumed in Sec. 1. Hence

$$E\, |S(0,T \wedge \sigma_L)| = E\, |S_0| = 0$$

Now let $T \to \infty$ and $L \to \infty$ to obtain (3.2).

In order to see (3.3), note that (3.2) and (3.5) ensure that

$$(c(T,L) - c)E \int_{S_1 \setminus S_0} \exp\,[2cR(x_2(s) - x_1(s) + \tfrac{1}{m})]\, ds$$

$$\leq K_{13} \qquad m = 1, 2, 3, \ldots$$

Hence by letting $m \uparrow \infty$, we have

$$E \int_{S_1 \setminus S_0} \exp\,[2cR(x_2(s) - x_1(s))]\, ds < \infty$$

Since it is clear that

$$E \int_{S_2} \exp\,[2cR(x_2(s) - x_1(s))]\, ds \leq \exp\,[2cR(\delta)] ET \wedge \sigma_L < \infty$$

the desired assertion (3.3) follows.

REMARK 1. If one needs (3.2) only, there is a simpler proof. In fact, we have only to trace the proof of Ref. 2, lemma 2.

Proof of Theorem 3.1. Take a $c_1 \in (c, c(T,L))$, and define the sequence $1 = x_1 > x_2 > x_3 > \cdots \to 0$ by

$$(c_1 - c) \int_{x_n}^{1} r^{-2}(u)\, du = \ell n\ n$$

This is possible because $R(x)$ is continuous and $R(0+) = \infty$. From (3.2) and (3.3), it follows that, for each positive T and L,

$$\begin{aligned} E\ |S(x_n, T \wedge \sigma_L)| &= E\ |S(x_n, T \wedge \sigma_L) \backslash S(0, T \wedge \sigma_L)| \\ &\leq \exp\ [-2c_1 R(x_n)] K_{11}(c_1, T, L) \\ &\qquad n = 1, 2, 3, \ldots \end{aligned}$$

Hence

$$\begin{aligned} &P(|S(x_n, T \wedge \sigma_L)| \geq g_{2c}(x_{n+1})) \\ &\quad \leq K_{11}(c_1, T, L) \exp\ [2cR(x_{n+1}) - 2c_1 R(x_n)] \\ &\qquad\quad n = 1, 2, 3, \ldots \end{aligned}$$

But by our choice of $\{x_n\}$,

$$2cR(x_{n+1}) - 2c_1 R(x_n) = -2\ \ell n\ n + \frac{2c}{c_1 - c}\ \ell n\ (1 + \frac{1}{n})$$

Thus we have

$$P(|S(x_n, T \wedge \sigma_L)| \geq g_{2c}(x_{n+1})) \leq \frac{K_{14}}{n^2}$$

and by the Borel-Cantelli theorem,

$$P(|S(x_n, T \wedge \sigma_L)| \geq g_{2c}(x_{n+1}),\ \text{i.o.}) = 0 \qquad (3.6)$$

Further, by the monotonicity of $S(x, T \wedge \sigma_L)$ and $g_{2c}(x)$ in x,

$$\begin{aligned} &\{|S(x, T \wedge \sigma_L)| \geq g_{2c}(x) \text{ for some } x \in (x_{n+1}, x_n]\} \\ &\quad \subset \{|S(x_n, T \wedge \sigma_L)| \geq g_{2c}(x_{n+1})\} \end{aligned}$$

This with (3.6) verifies (3.1).

REMARK 2. The above proof gives us a slight generalization of Theorem 3.1: In addition to the assumptions in Sec. 1, suppose that $E\sigma_L < \infty$ and that there are $\delta > 0$ and $c_1 > 0$ for which the inequality (2.2) with $c = c_1$ holds for all $|x - y| < \delta$, $x,y \in [-L,L]$. Then we have, for each $c \in (0,c_1)$,

$$|S(x,\sigma_L)| < g_{2c}(x) \qquad \text{as } x \downarrow 0, \quad \text{a.s.} \tag{3.7}$$

REMARK 3. The estimate (3.1) is not so bad in the sense that we have an example in which the assumptions in Sec. 1 are satisfied and

$$P\left(\lim_{x\downarrow 0} \frac{|S(x, T \wedge \sigma_L)|}{g_{2c(T,L)}(x)} > 0\right) = 1 \tag{3.8}$$

Indeed, we can take the same example as in Remark 1 of Sec. 2. Since the boundary 0 is regular for the diffusion $X = (x(t),P_x)$, it has a positive local time

$$\ell(T) = \lim_{x\downarrow 0} \frac{|S(x,T)|}{m[0,x]}$$

where $m[0,x]$ is the mass of the speed measure of the interval $[0,x]$ (Ref. 1, pp. 174-176). But for our diffusion, it is given by

$$m[0,x] = 2\int_0^x \exp[-2c_0R(y)]r^{-2}(y)\,dy = \frac{g_{2c_0}(x)}{c_0}$$

Since it is obvious that $\ell(T \wedge \sigma_L)$ is positive for $L > x_2$, (3.8) follows.

REMARK 4. If $f_{2c}(0+)$ is infinite for all positive c, then due to the strong comparison theorem in Ref. 3 there is a positive random variable $x(\omega)$ for which $S(x(\omega), T \wedge \sigma_L) = \emptyset$. Hence our assertion (3.1) holds trivially for all $g(x)$ which are positive on $(0,\infty)$.

Thus the only interesting examples for Theorem 3.1 fall in the case that $f_{2c}(0+)$ is finite for some positive c.

EXAMPLE 1. Let $r(u) = r_0u^{1/2}$ $(r_0 > 0)$. Then $R(x) = -\ell n\, x^{1/r_0^2}$ Hence for each positive T, L and $c \in (0,c(T,L))$,

$$|S(x, T \wedge \sigma_L)| < x^{2c/r_0^2} \qquad \text{as } x \downarrow 0, \quad \text{a.s.}$$

EXAMPLE 2. Let $r(u) = r_0[u(1 - \log u)]^{1/2}$ $(r_0 > 0)$. Then $R(x) = r_0^{-2}\ \ell n(1 - \ell n\ x)$. Hence for each positive T, L, and $c \in (0,c(T,L))$, it holds that

$$|S(x, T \wedge \sigma_L)| < (1 - \ell n\ x)^{-2c/r_0^2} \qquad \text{as } x \downarrow 0, \quad \text{a.s.}$$

EXAMPLE 3. Let $r(u) = r_0 u^{1/2}(1 - \ell n\ u)^{p/2}$ $(r_0 > 0,$ $0 < p < 1)$. Then

$$R(x) = r_0^{-2}(1 - p)^{-1}[(1 - \ell n\ x)^{1-p} - 1]$$

Hence, for each positive T, L, and $c \in (0,c(T,L))$,

$$|S(x, T \wedge \sigma_L)| < K_{14} \exp [-K_{15}(1 - \ell n\ x)^{1-p}] \qquad \text{as } x \downarrow 0, \text{ a.s.}$$

where $K_{14} = \exp [2c/r_0^2(1 - p)]$ and $K_{15} = 2c/r_0^2(1 - p)$.

ACKNOWLEDGMENT

I would like to thank Professor T. Yamada and Dr. M. Tomisaki for their helpful discussions.

REFERENCES

1. K. Itô and H. P. McKean, Jr., Diffusion Processes and Their Sample Paths, Springer-Verlag, Berlin, 1965.
2. S. Manabe and T. Shiga, On one-dimensional stochastic differential equations with non-sticky boundary conditions, J. Math. Kyoto Univ. 13:595-603 (1973).
3. T. Yamada and Y. Ogura, On the strong comparison theorems for solutions of stochastic differential equations, Z. Wahrscheinlichkeitsth. Verwend. Geb. 56:3-19 (1981).

14
Estimation Theory in Hilbert Spaces and Its Applications

SHIGERU OMATU/University of Tokushima, Tokushima, Japan

1. INTRODUCTION

A number of important physical phenomena may be modeled as discrete-time distributed parameter systems. When estimation problems are encountered in such systems, the measurements are also frequently discrete in time. A great deal of work has been carried out on estimation problems for continuous-time distributed parameter systems [1-4]. Tzafestas [5,6] and Nagamine, Omatu, and Soeda [7] have derived optimal estimators for discrete-time distributed parameter systems. Tzafestas employed a Bayesian approach, whereas Nagamine, Omatu, and Soeda considered only the filtering problem based on the Wiener-Hopf theory. Recently, Bencala and Seinfeld [3] have derived the optimal filter for continuous-time distributed parameter systems with discrete-time observations by the Wiener-Hopf approach.

The object of this chapter is twofold. First, we seek to derive optimal filtering and smoothing algorithms for discrete-time distributed parameter systems by a unified Wiener-Hopf approach. Fixed-point, fixed-interval, and fixed-lag smoothers are considered. Second, we wish to apply the results to the estimation of atomospheric sulfur dioxide concentrations in the Tokushima prefecture of Japan.

2. DESCRIPTION OF THE DISTRIBUTED PARAMETER SYSTEM

Let D be a bounded open domain of an r-dimensional euclidean space with smooth boundary ∂D. The spatial coordinate vector will be denoted by $x = (x_1, \ldots, x_r) \in D$. Consider a linear distributed parameter system described by

$$u(k+1, x) = L_x u(k,x) + G(k,x)w(k,x) \qquad x \in D \qquad (2.1)$$

where $u(k+1, x)$ is an n-dimensional vector function of the system, $w(k,x)$ is a vector-valued Gaussian process, L_x is a linear spatial matrix differential operator, and $G(k,x)$ is a known matrix function.

The initial and boundary conditions are given by

$$u(0,x) = u_0(x) \qquad (2.2)$$

$$\Gamma_\xi u(k+1, \xi) = S(k+1, \xi) \qquad \xi \in \partial D \qquad (2.3)$$

$$\Gamma_\xi[\cdot] = \alpha\, \xi[\cdot] + \frac{[1 - \alpha(\xi)]\partial[\cdot]}{\partial \underset{\sim}{n}} \qquad (2.4)$$

where $\underset{\sim}{n}$ is an exterior normal vector to the boundary ∂D at a point $\xi \in \partial D$ and $\alpha(\xi)$ is a function of class c^2 on ∂D satisfying $0 \leq \alpha(\xi) \leq 1$. $S(k+1, \xi)$ denotes a source function at the boundary and is assumed to be known.

Assume that $u_0(x)$ is a Gaussian random function the mean and covariance functions of which are given by

$$E[u_0(x)] = 0 \qquad (2.5)$$

$$E[u_0(x)u_0'(y)] = P_0(x,y) \qquad (2.6)$$

where $E[\cdot]$ and the prime denote the expectation and transpose operators, respectively.

Let the observed data be taken at m points, $x^1, \ldots, x^m \in \bar{D} = D \cup \partial D$ and let an mn-dimensional column vector $u_m(k)$ be defined by

$$u_m(k) = \mathrm{Col}[u(k,x^1), \ldots, u(k,x^m)] \qquad (2.7)$$

Let the observations be related to the states by

$$z(k) = H(k)u_m(k) + v(k) \tag{2.8}$$

where $z(k)$ is a p-dimensional observation vector at the m observation points, $x^1,\ldots,x^m \in \bar{D}$, $H(k)$ is a known $p \times mn$ matrix, and $v(k)$ is a p-dimensional vector-valued white Gaussian process. Assume that the white Gaussian process $w(k,x)$ in (2.1) and $v(k)$ in (2.8) are statistically independent of each other and also independent of the initial condition $u_0(x)$. Their mean and covariance functions are given by

$$E[w(k,x)] = 0 \qquad E[v(k)] = 0 \tag{2.9}$$

$$E[w(k,x)w'(s,y)] = Q(k,x,y)\delta_{ks} \qquad x,y \in D \tag{2.10}$$

$$E[v(k)v'(s)] = R(k)\delta_{ks} \tag{2.11}$$

where δ_{ks} is the Kronecker delta function and $Q(k,x,y)$ and $R(k)$ are symmetric positive semidefinite and positive definite matrices, respectively.

3. DESCRIPTION OF THE ESTIMATION PROBLEMS

The general problem considered here is to find an estimate $\hat{u}(\tau,x/k)$ of the state $u(\tau,x)$ at time τ based on the measurement data z_0^k, denoting a family of $z(\sigma)$ from $\sigma = 0$ up to the present time k. Specifically, for $\tau > k$ we have the prediction problem, for $\tau = k$ the filtering problem, and for $\tau < k$ the smoothing problem. As in the Kalman-Bucy approach, an estimate $\hat{u}(\tau,x/k)$ of $u(\tau,x)$ is sought through a linear operation on the past and present observation values z_0^k as follows:

$$\hat{u}(\tau,x/k) = \sum_{\sigma=0}^{k} \tilde{F}(\tau,x,\sigma)z(\sigma) \tag{3.1}$$

where $\tilde{F}(\tau,x,\sigma)$ is an $n \times p$ matrix kernel function.

To differentiate between the prediction, filtering, and smoothing problems, we replace (3.1) with different notation for each problem:

1. Prediction ($\tau > k$)

$$\hat{u}(\tau,x/k) = \sum_{\sigma=0}^{k} A(\tau,x,\sigma)z(\sigma) \tag{3.2}$$

2. Filtering ($\tau = k$)

$$\hat{u}(k,x/k) = \sum_{\sigma=0}^{k} F(\tau,x,\sigma)z(\sigma) \tag{3.3}$$

3. Smoothing ($\tau < k$)

$$\hat{u}(\tau,x/k) = \sum_{\sigma=0}^{k} B(\tau,k,x,\sigma)z(\sigma) \tag{3.4}$$

The estimation error is denoted by $\tilde{u}(\tau,x/k)$,

$$\tilde{u}(\tau,x/k) = u(\tau,x) - \hat{u}(\tau,x/y) \tag{3.5}$$

The estimate $\hat{u}(\tau,x/k)$ that minimizes

$$J(\hat{u}) = E[\|\tilde{u}(\tau,x/k)\|^2] \tag{3.6}$$

is said to be optimal, where $\|\cdot\|$ denotes the euclidean norm.

THEOREM 3.1 (Wiener-Hopf Theory) A necessary and sufficient condition for the estimate $\hat{u}(\tau,x/k)$ to be optimal is that the following Wiener-Hopf equation hold for $\alpha = 0, 1,\ldots,k$ and $x \in \bar{D}$:

$$\sum_{\sigma=0}^{k} \tilde{F}(\tau,x,\sigma)E[z(\sigma)z'(\alpha)] = E[u(\tau,x)z'(\alpha)] \tag{3.7}$$

Furthermore, (3.7) is equivalent to

$$E[\tilde{u}(\tau,x/k)z'(\alpha)] = 0 \tag{3.8}$$

for $\alpha = 0, 1,\ldots,k$ and $x \in \bar{D}$.

Proof. Let $F_\Delta(\tau,x,\sigma)$ be an $n \times p$ matrix function and let ε be a scalar-valued parameter. The trace of the covariance of the estimate

$$\hat{u}_\varepsilon(\tau,x/k) = \sum_{\sigma=0}^{k} [F(\tau,x,\sigma) + \varepsilon F_\Delta(\tau,x,\sigma)]z(\sigma)$$

is given by

$$J(\hat{u}_\varepsilon) = E[\|u(\tau,x) - \hat{u}(\tau,x/k) - \varepsilon \sum_{\sigma=0}^{k} F_\Delta(\tau,x,\sigma)z(\sigma)\|^2]$$

$$= E[\|\tilde{u}(\tau,x/k)\|^2] - 2\varepsilon E[\tilde{u}'(\tau,x/k) \sum_{\sigma=0}^{k} F_\Delta(\tau,x,\sigma)z(\sigma)]$$

$$+ \epsilon^2 E[\| \sum_{\sigma=0}^{k} F_\Delta(\tau,x,\sigma)z(\sigma)\|^2]$$

A necessary and sufficient condition for $\hat{u}(\tau,x/k)$ to be optimal is that

$$\left.\frac{\partial J(\hat{u}_\epsilon)}{\partial \epsilon}\right|_{\epsilon=0} = 0$$

that is,

$$E[\tilde{u}'(\tau,x/k) \sum_{\sigma=0}^{k} F_\Delta(\tau,x,\sigma)z(\sigma)] = 0$$

for any $n \times p$ matrix $F_\Delta(\tau,x,\sigma)$. Using the relation between the trace and inner product yields

$$E[\tilde{u}'(\tau,x/k) \sum_{\sigma=0}^{k} F_\Delta(\tau,x,\sigma)z(\sigma)]$$

$$= \mathrm{tr}\, E[\tilde{u}(\tau,x/k) \sum_{\sigma=0}^{k} z'(\sigma)F_\Delta'(\tau,x,\sigma)]$$

$$= \sum_{\sigma=0}^{k} \mathrm{tr}\, E[\tilde{u}(\tau,x/k)z'(\sigma)]F_\Delta'(\tau,x,\sigma) = 0$$

Setting $F_\Delta(\tau,x,k) = E[\tilde{u}(\tau,x/k)z'(\sigma)]$ in the above equation, it follows that (3.8) is a necessary condition for $\hat{u}(\tau,x/k)$ to be optimal. Sufficiency of (3.8) also follows from the above equation.

COROLLARY 3.1 (Orthogonal Projection Lemma) The following orthogonality condition holds:

$$E[\tilde{u}(\tau,x/k)\hat{u}'(\zeta,y/k)] = 0 \qquad x,y \in \overline{D} \tag{3.9}$$

where ζ is any time instant, for example, $\zeta < k$, $\zeta = k$, or $\zeta > k$.

Proof. Multiplying each side of (3.8) by $\tilde{F}(\zeta,y,\alpha)$ and summing from $\sigma = 0$ to $\sigma = k$ yields

$$E[\tilde{u}(\tau,x/k)\sum_{\alpha=0}^{k} z'(\alpha)\tilde{F}'(\zeta,y,\alpha)] = 0$$

Substituting (3.1) into the above equation yields (3.9).

Then the following lemma can be proved.

LEMMA 3.1 (Uniqueness of the Optimal Kernel) Let $\tilde{F}(\tau,x,\sigma)$ and $\tilde{F}(\tau,x,\sigma) + N(\tau,x,\sigma)$ be optimal matrix kernel functions satisfying the Wiener-Hopf equation (3.7). Then it follows that

$$N(\tau,x,\sigma) \equiv 0 \qquad \sigma = 0, 1, \ldots, k \text{ and } x \in \bar{D} \tag{3.10}$$

and the optimal matrix kernel function $\tilde{F}(\tau,x,\sigma)$ is unique.

Proof. From (3.7) we have

$$\sum_{\sigma=0}^{k} \tilde{F}(\tau,x,\sigma)E[z(\sigma)z'(\alpha)] = E[u(\tau,x)z'(\alpha)]$$

$$= \sum_{\sigma=0}^{k} (\tilde{F}(\tau,x,\sigma) + N(\tau,x,\sigma))E[(z(\sigma)z'(\alpha)]$$

Thus

$$\sum_{\sigma=0}^{k} N(\tau,x,\sigma)E[z(\sigma)z'(\alpha)] = 0$$

Multiplying each side of the above equation by $N'(\tau,x,\alpha)$ and summing from $\alpha = 0$ to $\alpha = k$ yields

$$\sum_{\sigma=0}^{k}\sum_{\alpha=0}^{k} N(\tau,x,\sigma)E[z(\sigma)z'(\alpha)]N'(\tau,x,\alpha) = 0$$

On the other hand, from (2.8) and (2.11) we have

$$E[z(\sigma)z'(\alpha)] = H(\sigma)E[u_m(\sigma)u_m'(\alpha)]H'(\alpha) + R(\sigma)\delta_{\sigma\alpha}$$

Then it follows that

$$\sum_{\sigma=0}^{k}\sum_{\alpha=0}^{k} N(\tau,x,\sigma)H(\sigma)E[u_m(\sigma)u_m'(\alpha)]H'(\alpha)N'(\tau,x,\alpha)$$

$$+ \sum_{\sigma=0}^{k} N(\tau,x,\sigma)H(\sigma)R(\sigma)H'(\sigma)N'(\tau,x,\sigma) = 0$$

Since both terms on the left-hand side of the above

equation are positive semidefinite because of the positive definiteness of $R(\sigma)$, a necessary and sufficient condition for the above equation to hold is $N(\tau,x,\sigma) \equiv 0$, $\sigma = 0, 1, \ldots, k$ and $x \in \bar{D}$. Thus the proof of the lemma is complete.

To facilitate the derivation of the optimal estimators, we rewrite (3.7) in terms of the following corollary.

COROLLARY 3.2 The Wiener-Hopf equation (3.7) is rewritten for the prediction, filtering, and smoothing problems as follows:

(i) Prediction ($\tau > k$)

$$\sum_{\sigma=0}^{k} A(\tau,x,\sigma)E[z(\sigma)z'(\sigma)] = E[u(\tau,x)z'(\alpha)] \tag{3.11}$$

for $\alpha = 0, 1, \ldots, k$ and $x \in \bar{D}$.

(ii) Filtering ($\tau = k$)

$$\sum_{\sigma=0}^{k} F(k,x,\sigma)E[z(\sigma)z'(\alpha)] = E[u(k,x)z'(\alpha)] \tag{3.12}$$

for $\alpha = 0, 1, \ldots, k$ and $x \in \bar{D}$.

(iii) Smoothing ($\tau < k$)

$$\sum_{\sigma=0}^{k} B(\tau,k,x,\sigma)E[z(\sigma)z'(\alpha)] = E[u(\tau,x)z'(\alpha)] \tag{3.13}$$

for $\alpha = 0, 1, \ldots, k$ and $x \in \bar{D}$.

In what follows, let us denote the estimation error covariance matrix function by $P(\tau,x,y/k)$,

$$P(\tau,x,y/k) = E[\tilde{u}(\tau,x/k)\tilde{u}'(\tau,y/k)] \tag{3.14}$$

4. DERIVATION OF THE OPTIMAL PREDICTOR

In this section we derive the optimal prediction estimator by using the Wiener-Hopf theory in the preceding section.

THEOREM 4.1 The optimal prediction estimator is given by

$$\hat{u}(k+1, x/k) = L_x\hat{u}(k,x/k) \tag{4.1}$$

$$\Gamma_\xi\hat{u}(k+1, \xi/k) = S(k+1, \xi) \qquad \xi \in \partial D \tag{4.2}$$

Proof. From (3.11) and (2.1) we have

$$\sum_{\sigma=0}^{k} A(k+1, x, \sigma)E[z(\sigma)z'(\alpha)] = L_x E[u(k,x)z'(\alpha)]$$

Since w(k,x) is independent of $z(\alpha)$, $\alpha = 0, 1, \ldots, k$. From the Wiener-Hopf equation (3.12) for the optimal filtering problem we have

$$\sum_{\sigma=0}^{k} \{A(k+1, x, \sigma) - L_x F(k,x,\sigma)\}E[z(\sigma)z'(\alpha)] = 0$$

Defining $\bar{N}(k,x,\sigma)$ by

$$\bar{N}(k+1, x, \sigma) = A(k+1, x, \sigma) - L_x F(k,x,\sigma)$$

it is clear that $A(k+1, x, \sigma) + \bar{N}(k+1, x, \sigma)$ also satisfies the Wiener-Hopf equation (3.11). From the uniqueness of $A(k+1, x, \sigma)$ by Lemma 3.1 to follows that $\bar{N}(k+1, x, \sigma) \equiv 0$, that is,

$$A(k+1, x, \sigma) = L_x F(k,x,k) \tag{4.3}$$

Thus, from (3.2) and (3.3) we have

$$\hat{u}(k+1, x/k) = L_x \sum_{\sigma=0}^{k} F(k,x,\sigma)z(\sigma) = L_x \hat{u}(k,x/k)$$

Since the forms of Γ_ξ and $S(k+1,\xi)$ are known, the predicted estimate $\hat{u}(k+1, \xi/k)$ also satisfies the same boundary condition as (2.3), $\Gamma_\xi \hat{u}(k+1, \xi/k) = S(k+1, \xi)$, $\xi \in \partial D$. Thus the proof of the theorem is complete.

THEOREM 4.2 The optimal prediction error covariance matrix function $P(k+1, x, y/k)$ is given by

$$P(k+1, x, y/k) = L_x P(k, x, y/k) L_y' + \tilde{Q}(k,x,y) \tag{4.4}$$

$$\Gamma_\xi P(k+1, \xi, y/k) = 0 \qquad \xi \in \partial D \tag{4.5}$$

where

$$\tilde{Q}(k,x,y) = G(k,x)Q(k,x,y)G'(k,y) \tag{4.6}$$

Proof. From (2.1), (3.5), and (4.1) it follows that

$$\tilde{u}(k+1, x/k) = L_x \tilde{u}(k,x/k) + G(k,x)w(k,x) \tag{4.7}$$

and from (2.3), (3.5), and (4.2)

$$\Gamma_\xi \tilde{u}(k+1, \xi/k) = 0 \qquad \xi \in \partial D \tag{4.8}$$

Then we have from (4.6) $P(k+1, x, y/k) =$
$E[\tilde{u}(k+1, x/k)\tilde{u}'(k+1, y/k)] = {}_xP(k,x,y/k)\,{}'_y + \tilde{Q}(k,x,y)$
and from (4.8), $E[\Gamma_\xi \tilde{u}(k+1, \xi/k)\tilde{u}'(k+1, y/k)] =$
$\Gamma_\xi P(k+1, \xi, y/k) = 0$. Thus the proof of the theorem is complete.

5. DERIVATION OF THE OPTIMAL FILTER

Let us derive the optimal filter by using the Wiener-Hopf theorem for the filtering problem. From (3.12) it follows that

$$F(k+1, x, k+1)E[z(k+1)z'(\alpha)] + \sum_{\sigma=0}^{k} F(k+1,x,\sigma)E[Z(\sigma)Z'(\alpha)]$$
$$= E[u(k+1,x)z'(\alpha)] \tag{5.1}$$

for $\alpha = 0, 1, \ldots, k+1$.

From (2.1) and the independence of z_0^{k+1} and $w(k+1, x)$, it follows that $E[u(k+1, x)z'(\alpha)] = L_x E[u(k,x)z'(\alpha)]$. Applying the Wiener-Hopf equation (3.12) to the right-hand side of the above equation yields

$$E[u(k+1, x)z'(\alpha)] = L_x \sum_{\sigma=0}^{k} F(k,x,\sigma)E[z(\sigma)z'(\alpha)] \tag{5.2}$$

Furthermore, from (2.8) and the whiteness of $v(k+1)$ we have

$$E[z(k+1)z'(\alpha)] = H(k+1)E[u_m(k+1)z'(\alpha)]$$

Let us introduce $L_*[\cdot]$ and $[\cdot]L'_*$ as follows:

$$L_*[\cdot] = \begin{bmatrix} L_{x^1}[\cdot] & & 0 \\ & \ddots & \\ 0 & & L_{x^m}[\cdot] \end{bmatrix} \tag{5.3}$$

and

$$[\cdot]L'_* = (L_*[\cdot])' \tag{5.4}$$

Then from (2.1) and (2.7) it follows that

$$u_m(k+1) = L_* u_m(k) + \tilde{w}_m(k) \tag{5.5}$$
$$\tilde{w}_m(k) = \mathrm{Col}[G(k,x^1)w(k,x^1), \ldots, G(k,x^m)w(k,x^m)] \tag{5.6}$$

Then we have for $\alpha < k+1$, $E[z(k+1)z'(\alpha)] =$

$H(k+1)L_* E[u_m(k)z'(\alpha)]$. Applying the Wiener-Hopf equation (3.12) to the right-hand side of the above equation yields

$$E[z(k+1)z'(\alpha)] = H(k+1)L_* \sum_{\sigma=0}^{k} F_m(k,\sigma)E[z(\sigma)z'(\alpha)] \tag{5.7}$$

where

$$F_m(k,\sigma) = \begin{bmatrix} F(k,x^1,\sigma) \\ \vdots \\ F(k,x^m,\sigma) \end{bmatrix} \tag{5.8}$$

Substituting (5.2) and (5.7) into (5.1) yields

$$\sum_{\sigma=0}^{k} N_\Delta(k,x,\sigma)E[z(\sigma)z'(\alpha)] = 0 \qquad \alpha = 0, 1, \ldots, k$$

where

$$N_\Delta(k,x,\sigma) = F(K+1, x, k+1)H(k+1)L_* F_m(k,\sigma) - L_x(k,x,\sigma) + F(k+1, x, \sigma)$$

Since it is clear that $F(k,x,\sigma) + N_\Delta(k,x,\sigma)$ also satisfies the Wiener-Hopf equation (3.12), it follows from Lemma 3.1 that $N_\Delta(k,x,\sigma) \equiv 0$. Thus we have the following lemma.

LEMMA 5.1 The optimal matrix kernel function $F(k,x,\sigma)$ of the filter is given by

$$F(k+1, x, \sigma) = L_x F(k,x,\sigma) - F(k+1, x, k+1)H(k+1)L_* F_m(k,\sigma) \qquad \sigma = 0, 1, \ldots, k \tag{5.9}$$

THEOREM 5.1 The optimal filtering estimate $\hat{u}(k,x/k)$ is given by

$$\hat{u}(k+1, x/k+1) = L_x \hat{u}(k,x/k) + F(k+1, x, k+1)\nu(k+1) \tag{5.10}$$

$$\nu(k+1) = z(k+1) - H(k+1)L_* \hat{u}_m(k/k) \tag{5.11}$$

$$\hat{u}(0,x/0) = 0 \tag{5.12}$$

$$\Gamma_\xi \hat{u}(k+1, \xi/k+1) = S(k+1,\xi) \qquad \xi \in \partial D \tag{5.13}$$

where

$$\hat{u}_m(k/k) = \mathrm{Col}[\hat{u}(k,x^1/k), \ldots, \hat{u}(k,x^m/k)] \tag{5.14}$$

Proof. Using (3.3) and (5.9) yields

$$\hat{u}(k+1, x/k+1) = F(k+1, x, k+1)z(k+1)$$
$$+ L_x \sum_{\sigma=0}^{k} F(k,x,\sigma)z(\sigma)$$
$$- F(k+1, x, k+1)H(k+1)L_* \sum_{\sigma=0}^{k} F_m(k,\sigma)z(\sigma)$$

Again from (3.3) we have

$$\hat{u}(k+1, x/k+1) = L_x\hat{u}(k,x/k) + F(k+1, x, k+1)\nu(k+1)$$

Since we have no information at the initial time, it is suitable to assume an initial value of $\hat{u}(k+1, x/k+1)$ as $\hat{u}(0,x/0) = E[u_0(x)] = 0$. Furthermore, since we know the exact forms of Γ_ξ and $S(k+1, \xi)$, the boundary value $\hat{u}(k+1, \xi/k+1)$ also satisfies the same boundary condition as $u(k+1, \xi)$. Thus we have $\Gamma_\xi\hat{u}(k+1, \xi/k+1) = S(k+1, \xi)$, $\xi \in \partial D$, and the proof of the theorem is complete.

Note that $\nu(k+1)$ defined by (5.11) is rewritten by using the prediction value of (4.1) as follows:

$$\nu(k+1) = z(k+1) - H(k+1)\hat{u}_m(k+1/k) \tag{5.15}$$

or

$$\nu(k+1) = H(k+1)\tilde{u}_m(k+1/k) + v(k+1) \tag{5.16}$$

where

$$\hat{u}_m(k+1/k) = \mathrm{Col}[\hat{u}(k+1, x^1/k), \ldots, \hat{u}(k+1, x^m/k)] \tag{5.17}$$

and

$$\tilde{u}_m(k+1/k) = u_m(k+1) - \hat{u}_m(k+1/k) \tag{5.18}$$

$\nu(k+1)$ is termed the innovation process [8,9].

In order to find the optimal matrix kernel function $F(k+1, x, k+1)$ for the filtering problem, we introduce the following notation:

$$p_m(\tau,x/k) = [p(\tau,x,x^1/k),\ldots,p(\tau,x,x^m/k)] \tag{5.19}$$

and

$$p_{mm}(\tau/k) = \begin{bmatrix} p(\tau,x^1/k) & p(\tau,x^1,x^1/k),\ldots,p(\tau,x^1,x^m/k) \\ \vdots & \vdots \quad \ddots \quad \vdots \\ p(\tau,x^m/k) & p(\tau,x^m,x^1/k),\ldots,p(\tau,x^m,x^m/k) \end{bmatrix} \tag{5.20}$$

Note from the definitions of $p_m(\tau, x/k)$ and $p_{mm}(\tau/k)$ that

$$p_m(\tau, x/k) = E[\tilde{u}(\tau, x/k)\tilde{u}_m'(\tau/k)] \tag{5.21}$$

and

$$p_{mm}(\tau/k) = E[\tilde{u}_m(\tau/k)\tilde{u}_m'(\tau/k)] \tag{5.22}$$

Furthermore, we define the covariance matrix of the innovation process $\nu(k+1)$ by $\Gamma(k+1/k)$,

$$\Gamma(k+1/k) = E[\nu(k+1)\nu'(k+1)] \tag{5.23}$$

Then from (5.16) it follows that

$$\Gamma(k+1/k) = H(k+1)p_{mm}(k+1/k)H'(k+1) + R(k+1) \tag{5.24}$$

Then the following theorem holds.

THEOREM 5.2 The optimal filtering gain matrix function $F(k+1, x, k+1)$ is given by

$$F(k+1, x, k+1) = p_m(k+1, x/k)H'(k+1)\Gamma^{-1}(k+1/k) \tag{5.25}$$

or

$$F(k+1, x, k+1) = p_m(k+1, x/k)\psi(k+1/k)H'(k+1)R^{-1}(k+1) \tag{5.26}$$

where

$$\psi(k+1/k) = (I + \tilde{R}(k+1)p_{mm}(k+1/k))^{-1} \tag{5.27}$$

and

$$\tilde{R}(k+1) = H'(k+1)R^{-1}(k+1)H(k+1) \tag{5.28}$$

Proof. From the Wiener-Hopf equation (3.12) it follows that

$$F(k+1, x, k+1)E[z(k+1)z'(k+1)] + \sum_{\sigma=0}^{k} F(k+1, x, \sigma)E[z(\sigma)z'(k+1)] = E[u(k+1, x)z'(k+1)]$$

Substituting (5.9) into the above equation yields

$$F(k+1, x, k+1)(E[\{z(k+1) - H(k+1)L_* \sum_{\sigma=0}^{k} F_m(k,\sigma)z(\sigma)\} \times z'(k+1)]) = E[\{u(k+1, x) - L_x \sum_{\sigma=0}^{k} F(k,x,\sigma)z(\sigma)\} z'(k+1)]$$

Substituting (3.3) into the right-hand side of the above equation and using (4.1) and the orthogonality condition of (3.9) yields

$$E[\{u(k+1, x) - L_x \sum_{\sigma=0}^{k} F(k,x,\sigma)z(\sigma)\}z'(k+1)]$$

$$= E[\tilde{u}(k+1, x/k)z'(k+1)] = E[\tilde{u}(k+1, x/k)$$

$$u_m'(k+1)]H'(k+1) = p_m(k+1,x/k)H'(k+1)$$

Using the orthogonality condition of (3.9) gives

$$E[\nu(k+1)z'(k+1)] = H(k+1)E[\tilde{u}_m(k+1/k)u_m'(k+1)]H'(k+1)$$

$$+ R(k+1) = H(k+1)p_{mm}(k+1/k)H'(k+1) + R(k+1)$$

$$= \Gamma(k+1/k) \quad (5.29)$$

Then we have

$$F(k+1, x, k+1)\Gamma(k+1/k) = p_m(k+1, x/k)H'(k+1) \quad (5.30)$$

Thus, (5.25) is derived. In order to show the equivalence between (5.25) and (5.26), we use the following matrix inversion lemma:

$$PH'(HPH' + R)^{-1} = P(I + H'R^{-1}HP)^{-1}H'R^{-1} \quad (5.31)$$

From (5.25) and (5.31) we have

$$F(k+1, x, k+1) = p_m(k+1, x/k)\psi(k+1/k)H'(k+1)R^{-1}(k+1)$$

Then (5.26) is derived, and the proof of the theorem is complete.

The equation for the optimal filtering error covariance matrix function $p(k+1, x, y/k+1)$ now must be derived.

THEOREM 5.3 The optimal filtering error covariance matrix function $p(k+1, x, y/k+1)$ is given by

$$p(k+1, x, y/k+1) = p(k+1, x, y/k) - p_m(k+1, x/k)H'(k+1)\Gamma^{-1}(k+1/k) \times H(k+1)p_m'(k+1, y/k) \quad (5.32)$$

or

$$p(k+1, x, y/k+1) = p(k+1, x, y/k) - p_m(k+1, x/k) \times \psi(k+1/k)\tilde{R}(k+1)p_m'(k+1, y/k) \quad (5.33)$$

where

$$p(0,x,y/0) = p_0(x,y) \tag{5.34}$$

and

$$\Gamma_\xi p(k+1,\ \xi,\ y/k+1) = 0 \qquad \xi \in \partial D \tag{5.35}$$

Proof. From (2.1) and (5.10) we have

$$\tilde{u}(k+1,\ x/k+1) = \tilde{u}(k+1,\ x/k) - F(k+1,\ x,\ k+1)\nu(k+1) \tag{5.36}$$

and from (2.3) and (5.13),

$$\Gamma_\xi \tilde{u}(k+1,\ \xi/k+1) = 0 \qquad \xi \in \partial D \tag{5.37}$$

Using the independence property between $v(k+1)$ and $\tilde{u}(k+1,\ y/k)$ yields, from (5.36),

$$\begin{aligned} p(k+1,\ x,\ y/k+1) &= E[\tilde{u}(k+1,\ x/k+1)\tilde{u}'(k+1,\ y/k+1)] \\ &= p(k+1,\ x,\ y/k) + F(k+1,\ x,\ k+1)E[\nu(k+1) \\ &\quad \times \nu'(k+1)]F'(k+1,\ y,\ k+1) - F(k+1,\ x,\ k+1) \\ &\quad \times H(k+1)E[\tilde{u}_m(k+1/k)\tilde{u}'(k+1,\ y/k)] - E[\tilde{u}(k+1,\ x/k) \\ &\quad \times \tilde{u}_m'(k+1/k)]H'(k+1)F'(k+1,\ y,\ k+1) \end{aligned}$$

Using (5.25) and (5.30) it follows that

$$\begin{aligned} p(k+1,\ x,\ y/k+1) &= p(k+1,\ x,\ y/k) \\ &\quad - p_m(k+1,\ x/k)H'(k+1)F''(k+1,\ y,\ k+1) \\ &= p(k+1,\ x,\ y/k) - p_m(k+1,\ x/k)H'(k+1) \\ &\quad \times \Gamma^{-1}(k+1/k)H(k+1)p_m'(k+1,\ y/k) \end{aligned}$$

Thus, (5.32) is derived. The equivalence between (5.32) and (5.33) is easily shown by using (5.31). Since the initial value $\hat{u}(0,x/0)$ of $\hat{u}(k+1,\ x/k+1)$ is zero from (5.12), it is clear that $P(0,x,y/0) = E[\tilde{u}(0,x/0)\tilde{u}'(0,y/0)] = P_0(x,y)$. Multiplying each side of (5.37) by $\tilde{u}'(k+1,\ y/k+1)$ and taking the expectation yields $\Gamma_\xi P(k+1,\ \xi,\ y/k+1) = 0$, $\xi \in \partial D$. Thus the proof of the theorem is complete.

COROLLARY 5.1 $\hat{u}_m(k+1/k+1)$ and $p_m(k+1,\ x/k+1)$ satisfy the following relations:

$$\hat{u}_m(k+1/k+1) = \hat{u}_m(k+1/k) + F_m(k+1,\ k+1)\nu(k+1) \tag{5.38}$$

$$F_m(k+1,k+1) = p_{mm}(k+1/k)\psi(k+1/k)H'(k+1)R^{-1}(k+1) \tag{5.39}$$

or

$$F_m(k+1, \ k+1) = p_{mm}\frac{k+1}{k+1}H'(k+1)R^{-1}(k+1) \tag{5.40}$$

$$p_{mm}(k+1/k+1) = p_{mm}(k+1/k) - p_{mm}(k+1/k)$$
$$\times \ \psi(k+1/k)\tilde{R}(k+1)p_{mm}(k+1/k) \tag{5.41}$$

or

$$p_{mm}(k+1/k+1) = p_{mm}(k+1/k)\psi(k+1/k) \tag{5.42}$$

Proof. From the definitions (5.8) and (5.17) of $F_m(k+1, k+1)$ and $\hat{u}_m(k+1/k)$, it is clear that (5.38), (5.39), and (5.41) hold. From (5.27) and (5.41) it follows that

$$\begin{aligned} p_{mm}(k+1/k+1) &= p_{mm}(k+1/k)\psi(k+1/k)\{\psi^{-1}(k+1/k) \\ &\quad - \tilde{R}(k+1)p_{mm}(k+1/k)\} \\ &= p_{mm}(k+1/k)\psi(k+1/k)\{I+\tilde{R}(k+1)p_{mm}(k+1/k) \\ &\quad - \tilde{R}(k+1)p_{mm}(k+1/k)\} \\ &= p_{mm}(k+1/k)\psi(k+1/k) \end{aligned}$$

Thus (5.42) is derived and (5.40) is clear from (5.39) and (5.42).

The present result corresponds to that of Santis, Saeks, and Tung [17], which is an abstract form of the filter.

6. DERIVATION OF THE EQUATIONS FOR THE OPTIMAL SMOOTHING ESTIMATOR

In this section we derive the basic equations for the optimal smoothing estimator by using the Wiener-Hopf theory.

LEMMA 6.1 The optimal matrix kernel function $B(\tau,k+1,x,\sigma)$ of the smoothing estimator is given by

$$\begin{aligned} B(\tau,k+1,\ x,\ \sigma) &= B(\tau,k,x,\sigma) \\ &\quad - B(\tau,k+1,\ x,\ k+1)H(k+1)L_*F_m(k,\sigma) \\ &\qquad \sigma = 0,\ 1,\ \dots,\ k \end{aligned} \tag{6.1}$$

Proof. From the Wiener-Hopf equation (3.13) we have

$$\sum_{\sigma=0}^{k+1} B(\tau, K+1, x, \sigma)E[z(\sigma)z'(\alpha)] = E[u(\tau,x)z'(\alpha)] \qquad \alpha = 0, \ldots, k+1$$

and

$$\sum_{\sigma=0}^{k} B(\tau,k,x,\sigma)E[z(\sigma)z'(\alpha)] = E[u(\tau,x)z'(\alpha)], \qquad \alpha = 0, \ldots, k$$

Subtracting the latter equation from the former yields

$$B(\tau, k+1, x, k+1)E[z(k+1)z'(\alpha)] + \sum_{\sigma=0}^{k} [B(\tau,k+1, x, \sigma) - B(\tau,k,x,\sigma)]E[z(\sigma)z'(\alpha)] = 0$$

From (2.8) and (3.12) we have

$$E[z(k+1)z'(\alpha)] = H(k+1)L_* E[u_m(k)z'(\alpha)] = H(k+1)L_* \sum_{\sigma=0}^{k} F_m(k,\sigma)E[z(\sigma)z'(\alpha)]$$

Then it follows that

$$\sum_{\sigma=0}^{k} \tilde{N}(\tau,k,x,\sigma)E[z(\sigma)z'(\alpha)] = 0$$

where

$$\tilde{N}(\tau,k,x,\sigma) = B(\tau,k+1, x, \sigma) - B(\tau,k,x,\sigma) + B(\tau, k+1, x, k+1)H(k+1)L_* F_m(k,\sigma)$$

Since it is easily seen that $B(\tau,k,x,\sigma) + \tilde{N}(\tau,k,x,\sigma)$ also satisfies the Wiener-Hopf equation (3.13), from Lemma 3.1 we have $\tilde{N}(\tau,k,x,\sigma) \equiv 0$, and the proof of the lemma is complete.

THEOREM 6.1 The optimal smoothing estimate $\hat{u}(\tau, x/k+1)$ is given by

$$\hat{u}(\tau, x/k+1) = \hat{u}(\tau,x/k) + B(\tau, k+1, x, k+1)\nu(k+1) \tag{6.2}$$

$$\Gamma_\xi \hat{u}(\tau, \xi/k+1) = S(\tau,\xi) \qquad \xi \in \partial D,\ k=\tau,\ \tau=1, \ldots \tag{6.3}$$

Proof. From (3.4) it follows that

$$\hat{u}(\tau, x/k+1) = B(\tau, k+1, x, k+1)z(k+1) + \sum_{\sigma=0}^{k} B(\tau, k+1, x, \sigma)z(\sigma)$$

Substituting (6.1) into the above equation yields

$$\hat{u}(\tau, x/k+1) = B(\tau, k+1, x, k+1)(z(k+1) - H(k+1)L_* \times \sum_{\sigma=0}^{k} F_m(k,\sigma)z(\sigma)) + \sum_{\sigma=0}^{k} B(\tau,k,x,\sigma)z(\sigma)$$

Substituting (3.3) and (3.4) into the above equation yields

$$\hat{u}(\tau, x/k+1) = \hat{u}(\tau,x/k) + B(\tau, k+1, x, k+1)\nu(k+1)$$

Since we have no additional information about the boundary value of $u(\tau,x)$, except for $S(\tau,\xi)$ and the exact form Γ_ξ, we have $\Gamma_\xi \hat{u}(\tau, \xi/k + 1) = S(\tau,\xi)$, $\xi \in \partial D$, and the proof of the theorem is complete.

THEOREM 6.2. The optimal smoothing gain matrix function $B(\tau, k+1, x, k+1)$ is given by

$$B(\tau, k+1, x, k+1) = L_m(\tau,x/k)L_*'H'(k+1)\Gamma^{-1}(k+1/k) \quad (6.4)$$

or

$$B(\tau, k+1, x, k+1) = J(\tau, x/k+1)H'(k+1)R^{-1}(k+1) \quad (6.5)$$

where

$$J(\tau, x/k+1) = L_m(\tau, x/k)L_*'(I+\tilde{R}(k+1)p_{mm}(k+1/k))^{-1} \quad (6.6)$$

$$L_m(\tau,x/k) = [L(\tau,x,x^1/k),\ldots,L(\tau,x,x^m/k)] \quad (6.7)$$

and

$$L(\tau,x,y/k) = E[\tilde{u}(\tau,x/k)\tilde{u}'(k,y/k)] \quad (6.8)$$

Proof. From the Wiener-Hopf equation (3.13) it follows that

$$B(\tau, k+1, x, k+1)E[z(k+1)z'(k+1)] + \sum_{\sigma=0}^{k} B(\tau, k+1, x, \sigma)E[z(\sigma)z'(k+1)] = E[u(\tau,x)z'(k+1)]$$

Substituting (6.1) into the above equation yields

$$B(\tau, k+1, x, k+1)E[\nu(k+1)z'(k+1)] = E[\tilde{u}(\tau,x/k)z'(k+1)]$$

On the other hand, from (5.15) and (5.16) we have

$$E[\nu(k+1)z'(k+1)] = E[\nu(k+1)(\nu(k+1)+H(k+1)\hat{u}_m(k+1/k))'] = E[\nu(k+1)\nu'(k+1)] = \Gamma(k+1/k) \quad (6.9)$$

From (2.8) and the independence of $v(k+1)$ and $\tilde{u}(\tau,x/k)$, we have

$$E[\tilde{u}(\tau, x/k)z'(k+1)] = E[\tilde{u}(\tau,x/k)\tilde{u}_m'(k+1/k)]H'(k+1)$$

But from (4.1) and (5.5) it follows that

$$\tilde{u}_m(k+1/k) = L_*\tilde{u}_m(k/k) + \tilde{w}_m(k) \tag{6.10}$$

Then we have

$$B(\tau, k+1, x, k+1)\Gamma(k+1/k) = L_m(\tau,x/k)L_*'H'(k+1)$$

The equivalence between (6.4) and (6.5) is easily seen by using the matrix inversion lemma (5.31). Thus the proof of the theorem is complete.

Let us now derive the equation for $L(\tau, x, y/k+1)$. Using the orthogonality condition (3.9) yields

$$L(\tau, x, y/k+1) = E[u(\tau,x)\tilde{u}'(k+1, y/k+1)]$$

Substituting (5.36) into the above equation yields

$$L(\tau, x, y/k+1) = L(\tau,x,y/k)L_y' - L_m(\tau,x/k)L_*'H'(k+1)F'(k+1, y, k+1) \tag{6.11}$$

From (2.3) and (6.3) it follows that $\Gamma_\xi\tilde{u}(\tau, \xi/k+1) = 0$, $\xi \in \partial D$. Multiplying each side by $\tilde{u}'(k+1, y/k+1)$ and taking the expectation yields

$$\Gamma_\xi L(\tau, \xi, y/k+1) = 0 \qquad \xi \in \partial D \tag{6.12}$$

Then the following theorem holds.

THEOREM 6.3 $J(\tau, x/k+1)$ in (6.5) is given by

$$J(\tau, x/k+1) = J(\tau,x/k)L_*'\psi(k+1/k) \tag{6.13}$$

$$J(\tau,x/\tau) = p_m(\tau,x/\tau) \tag{6.14}$$

$$\Gamma_\xi J(\tau, \xi/k+1) = 0 \qquad \xi \in \partial D \tag{6.15}$$

Proof. From (6.11) and (5.26) it follows that

$$L_m(\tau, x/k+1) = L_m(\tau,x)L_*'\{I - \tilde{R}(k+1)\psi'(k+1/k)p_{mm}(k+1/k)\}$$

But we have

$$I - \tilde{R}(I + P\tilde{R})^{-1}P = \tilde{R}(I + P\tilde{R})^{-1}[(I + P\tilde{R})\tilde{R}^{-1} - P]$$
$$= [(I + P\tilde{R})\tilde{R}^{-1}]^{-1}\tilde{R}^{-1} = [\tilde{R}(\tilde{R}^{-1} + P)]^{-1} = (I + \tilde{R}P)^{-1}$$

Thus

$$L_m(\tau, x/k+1) = L_m(\tau,x)L_*'(I + \tilde{R}(k+1)p_{mm}(k+1/k))^{-1}$$

Therefore, from (6.6) it follows that

$$J(\tau, x/k+1) = L_m(\tau, x/k+1) \tag{6.16}$$

and from (6.6) we have

$$J(\tau,\ x/k+1) = J(\tau,x/k)L_*'(I + \tilde{R}(k+1)p_{mm}(k+1/k))^{-1}$$

Then it follows that

$$J(\tau,x/\tau) = L_m(\tau,x/\tau) = p_m(\tau,x/\tau)$$

Since (6.15) is clear from (6.12) and (6.16), the proof of the theorem is complete.

Let us now derive the equation for the optimal smoothing error covariance matrix function $p(\tau,x,y/k)$ defined by

$$p(\tau,x,y/k) = E[\tilde{u}(\tau,x/k)\tilde{u}(\tau,y/k)] \tag{6.17}$$

From (6.2) and (6.3) it follows that

$$\tilde{u}(\tau,\ x/k+1) = \tilde{u}(\tau,x/k) - B(\tau,k+1),\ x,\ k+1)\nu(k+1) \tag{6.18}$$

$$\Gamma_\xi\tilde{u}(\tau,\ \xi/k+1) = 0 \qquad \xi \in \psi D \tag{6.19}$$

Then the following theorem holds.

THEOREM 6.4 The optimal smoothing error covariance matrix function $p(\tau,\ x,\ y/k+1)$ is given by

$$p(\tau,\ x,\ y/k+1) = p(\tau,x,y/k) - L_m(\tau,x/k)L_*'H'(k+1)$$
$$\Gamma^{-1}(k+1/k)H(k+1)L_*L_m'(\tau,y/k) \tag{6.20}$$

or

$$p(\tau,\ x,\ y/k+1) = p(\tau,x,y/k) - J(\tau,\ x/k+1)$$
$$\psi^{-1}(k+1/k)\tilde{R}(k+1)J'(\tau,\ y/k+1) \tag{6.21}$$

$$\Gamma_\xi p(\tau,\ \xi,\ y/k+1) = 0 \qquad \xi \in \partial D \tag{6.22}$$

Proof. From (6.18) it follows that

$$p(\tau,\ x,\ y/k+1) = p(\tau,x,y/k) + B(\tau,\ k+1,\ x,\ k+1)$$
$$E[\nu(k+1)\nu'(k+1)]B'(\tau,\ k+1,\ y,\ k+1)$$
$$- B(\tau,\ k+1,\ x,\ k+1)E[\nu(k+1)\tilde{u}'(\tau,y/k)]$$
$$- E[\tilde{u}(\tau,x/k)\nu'(k+1)]B'(\tau,k+1,\ y,\ k+1)$$

$$E[\tilde{u}(\tau,x/k)\nu'(k+1)] = E[\tilde{u}(\tau,x/k)\tilde{u}_m'(k/k)]L_*'H'(k+1)$$
$$= L_m(\tau,x/k)L_*'H'(k+1)$$

and

$$E[\nu(k+1)\tilde{u}'(\tau,y/k)] = H(k+1)L_*L_m'(\tau,y/k)$$

Thus, we have

$$p(\tau, x, y/k+1) = p(\tau,x,y/k) + B(\tau, k+1, x, k+1)$$
$$\times \Gamma(k+1/k)B'(\tau, k+1, y, k+1) - B(\tau,kk+1, x, k+1)$$
$$\times H(k+1) L_* L_m'(\tau,y/k) - L_m(\tau,x/k) L_*' H'(k+1)$$
$$\times B'(\tau, k+1, y, k+1)$$

Substituting (6.4) into the above equation yields

$$p(\tau, x, y/k+1) = p(\tau,x,y/k) - L_m(\tau,x/k)$$
$$\times L_*' H'(k+1)\Gamma^{-1}(k+1/k)H(k+1) L_* L_m'(\tau,y/k)$$

In order to derive (6.21), note that from (6.6),

$$L_m(\tau,x/k) L_*' = J(\tau, x/k+1)\psi^{-1}(k+1/k)$$

and from the matrix inversion lemma (5.31),

$$H'(HPH' + R)^{-1}H = (I + H'R^{-1}HP)^{-1}H'R^{-1}H$$

Then we have

$$H'(k+1)\Gamma^{-1}(k+1/k)h(k+1) = \psi(k+1/k)\tilde{R}(k+1)$$

and

$$p(\tau, x, y/k+1) = p(\tau,x,y/k) - J(\tau, x/k+1)$$
$$\times \psi^{-1}(k+1/k)\tilde{R}(k+1)J'(\tau, y/k+1)$$

Multiplying each side of (6.19) by $\tilde{u}(\tau, y/k + 1)$ and taking the expectation yields $\Gamma_\xi p(\tau, \xi, y/k+1) = 0$, $\xi \in \partial D$. Thus the proof of the theorem is complete

COROLLARY 6.1 $J(\tau,x/k)$ satisfies the following relations:

$$J(\tau, x/k + 1) = A(\tau,x)J_m(\tau + 1/k + 1) \tag{6.23}$$

and

$$J(\tau + 1, x/k) = D(\tau,x)J_m(\tau/k) \tag{6.24}$$

where

$$J_m(\tau/k) = \begin{bmatrix} J(\tau,x^1/k) \\ \vdots \\ J(\tau,x^m/k) \end{bmatrix} \tag{6.25}$$

$$A(\tau,x) = p_m(\tau,x/\tau) L_*' p_{mm}^{-1}(\tau + 1/\tau) \tag{6.26}$$

$$D(\tau,x) = p_m(\tau + 1, x/\tau)[p_{mm}(\tau/\tau) L_*']^{-1} \tag{6.27}$$

Proof. Letting $\Phi(k + 1)$ be given by $\Phi(k + 1) = L_*'(I + \tilde{R}(k + 1)p_{mm}(k + 1/k))^{-1}$, from (6.13) and (6.14) it follows that $J(\tau, x/k + 1) = p_m(\tau,x/\tau)\Phi(\tau + 1)\Phi(\tau + 2)\cdots \Phi(k + 1)$ and $J_m(\tau + 1/k + 1) = p_{mm}(\tau + 1/\tau + 1)\Phi(\tau + 2)\cdots$

$\Phi(k+1)$. From the above equations and (5.42) we have

$$J(\tau, x/k+1) = p_m(\tau,x/\tau)\Phi(\tau+1)p_{mm}^{-1}(\tau+1/\tau+1)J_m(\tau+1/k+1)$$
$$= p_m(\tau,x/\tau)L_*'\psi(\tau+1/\tau)\psi^{-1}(\tau+1/\tau)p_{mm}^{-1}(\tau+1/\tau)J_m$$
$$\times J_m(\tau+1/k+1) = A(\tau,x)J_m(\tau+1/k+1)$$

From (6.13) and (6.14) it follows that $J(\tau+1, x/k) = p_m(\tau+1, x/\tau+1)\Phi(\tau+2)\cdots\Phi(k)$ and $J_m(\tau/k) = p_{mm}(\tau/\tau)\Phi(\tau+1)\Phi(\tau+2)\cdots\Phi(k)$. Thus we have from the above equations,

$$J(\tau+1, x/k) = p_m(\tau+1, x/\tau+1)[I + \tilde{R}(\tau+1)p_{mm}(\tau+1/\tau)]$$
$$\times [p_{mm}(\tau/\tau)L_*']^{-1}J_m(\tau/k)$$
$$= p_m(\tau+1, x/\tau)[p_{mm}(\tau/\tau)L_*']^{-1}J_m(\tau/k)$$

where the following equality derived from (5.33) has been used:

$$p_m(\tau+1, x/\tau+1) = p_m(\tau+1, x/\tau)[I + \tilde{R}(\tau+1)$$
$$p_{mm}(\tau+1/\tau)]^{-1} \tag{6.28}$$

Thus the proof of the corollary is complete.

THEOREM 6.5 The optimal smoothing estimator is given by

$$\hat{u}(\tau,x/k) = \hat{u}(\tau,x/\tau) + \sum_{\ell=\tau+1}^{k} J(\tau,x/\ell)\tilde{\nu}(\ell) \tag{6.29}$$

$$\Gamma_\xi \hat{u}(\tau,\xi/k) = S(\tau,\xi) \qquad \xi \in \partial D \tag{6.30}$$

where

$$\tilde{\nu}(\ell) = H'(\ell)R^{-1}(\ell)\nu(\ell) \tag{6.31}$$

Furthermore, the optimal smoothing error covariance matrix function $p(\tau,x,y/k)$ is given by

$$p(\tau,x,y/k) = p(\tau,x,y/\tau)$$
$$- \sum_{\ell=\tau+1}^{k} J(\tau,x/\ell)\psi^{-1}(\ell/\ell-1)\tilde{R}(\ell)J'(\tau,y/\ell) \tag{6.32}$$

$$\Gamma_\xi p(\tau,\xi,y/k) = 0 \qquad \xi \in \partial D \tag{6.33}$$

Proof. From (6.2) and (6.5), (6.29) can be directly obtained and from (6.21), (6.32) is clear. Thus the proof of the theorem is complete.

7. SUMMARY OF THE OPTIMAL SMOOTHING ESTIMATORS

7.1 Fixed-Point Smoother (τ = fixed, $k = \tau + 1, \tau + 2, \ldots$)

THEOREM 7.1 The optimal fixed-point smoothing estimator is given by

$$\hat{u}(\tau, x/k+1) = \hat{u}(\tau, x/k) + J(\tau, x/k+1)\tilde{\nu}(k+1) \tag{7.1}$$

$$J(\tau, x/k+1) = J(\tau, x/k)\, L_*'\psi(k+1/k) \tag{7.2}$$

$$\psi(k+1/k) = (I + \tilde{R}(k+1)p_{mm}(k+1/k))^{-1} \tag{7.3}$$

$$J(\tau, x/\tau) = p_m(\tau, x/\tau) \tag{7.4}$$

$$\Gamma_\xi \hat{u}(\tau, \xi/k+1) = S(\tau, \xi) \qquad \xi \in \partial D \tag{7.5}$$

$$\Gamma_\xi J(\xi, x/k+1) = 0 \qquad \xi \in \partial D \tag{7.6}$$

Furthermore, the optimal fixed-point smoothing error covariance matrix function $p(\tau, x, y/k+1)$ is given by

$$p(\tau, x, y/k+1) = p(\tau, x, y/k) - J(\tau, x/k+1)\psi^{-1}(k+1/k) \times \tilde{R}(k+1)J'(\tau, y/k+1) \tag{7.7}$$

$$\Gamma_\xi p(\tau, \psi, y/k+1) = 0 \qquad \xi \in \partial D \tag{7.8}$$

7.2 Fixed-Interval Smoothing Estimator (k = fixed, $\tau = k - 1, k - 2, \ldots$)

From Theorem (6.5) it follows that

$$\hat{u}(\tau+1, x/k) = \hat{u}(\tau+1, x/\tau+1) + \sum_{\ell=\tau+2}^{k} J(\tau+1, x/\ell)\tilde{\nu}(\ell) \tag{7.9}$$

and

$$p(\tau+1, x, y/k) = p(\tau+1, x, y/\tau+1) - \sum_{\ell=\tau+2}^{k} J(\tau+1, x/\ell)\, \psi^{-1}(\ell/\ell-1)\tilde{R}(\ell)J'(\tau+1, y/\ell) \tag{7.10}$$

Then the following theorem holds.

THEOREM 7.2 The optimal fixed-interval smoothing estimator is given by

$$\hat{u}(\tau+1, x/k) = \hat{u}(\tau+1, x/\tau+1) + A(\tau+1, x) \times [\hat{u}_m(\tau+2/k) - \hat{u}_m(\tau+2/1)] \tag{7.11}$$

$$\Gamma_\xi \hat{u}(\tau+1, \xi/k) = S(\tau+1, \xi) \qquad \xi \in \partial D \tag{7.12}$$

Furthermore, the optimal fixed-interval smoothing error covariance matrix function is given by

$$p(\tau+1, x, y/k) = p(\tau+1, x, y/\tau+1) - A(\tau+1, x)$$
$$\times (p_{mm}(\tau+1/k) - p_{mm}(\tau+1/\tau))A'(\tau+1, y) \tag{7.13}$$
$$\Gamma_{\xi} p(\tau+1, \xi, y/k) = 0 \qquad \xi \in \partial D \tag{7.14}$$

Proof. From (6.23) and (7.9) we have

$$\hat{u}(\tau+1, x/k) = \hat{u}(\tau+1, x/\tau+1) + A(\tau+1, x) \times \sum_{\ell=\tau+2}^{k} J_m(\tau+1/\ell)\tilde{\nu}(\ell)$$

But from Theorem 6.5,

$$\hat{u}(\tau+2, x/k) = \hat{u}(\tau+2, x/\tau+2) + \sum_{\ell=\tau+3}^{k} J(\tau+2, x/\ell)\tilde{\nu}(\ell)$$

and from (5.10) and (5.26),

$$\hat{u}(\tau+2, x/\tau+2) = \hat{u}(\tau+2, x/\tau+1) + F(\tau+2, x, \tau+2) \times \nu(\tau+2)$$
$$= \hat{u}(\tau+2, x/\tau+1) + J(\tau+2, x/\tau+2) \times \tilde{\nu}(\tau+2)$$

Thus we have

$$\hat{u}_m(\tau+2/k) - \hat{u}_m(\tau+2/\tau+1) = \sum_{\ell=\tau+2}^{k} J_m(\tau+2/\ell)\tilde{\nu}(\ell)$$

Then we have

$$\hat{u}(\tau+1, x/k) = \hat{u}(\tau+1, x/\tau+1) + A(\tau+1, x) \times [\hat{u}_m(\tau+2/k) - \hat{u}_m(\tau+2/\tau+1)]$$

From (6.23) and (7.10),

$$p(\tau+1, x, y/k) = p(\tau+1, x, y/\tau+1) - A(\tau+1, x)$$
$$\sum_{\ell=\tau+2}^{k} J_m(\tau+2/\ell)\psi^{-1}(\ell/\ell-1)\tilde{R}(\ell)J_m'(\tau+2/\ell)A'(\tau+1, y)$$

From Theorem 6.5,

$$p(\tau + 2, x, y/k) = p(\tau + 2, x, y/\tau + 2) - \sum_{\ell=\tau+3}^{k} J(\tau + 2, x/\ell)\psi^{-1}(\ell/\ell - 1)\tilde{R}(\ell)J'(\tau + 2, y/\ell)$$

and from (5.33),

$$p(\tau + 2, x, y/\tau + 2) = p(\tau + 2, x, y/\tau + 1) - p_m(\tau + 2, x/\tau + 1) \psi(\tau + 2/\tau + 1)\tilde{R}(\tau + 2)p_m'(\tau + 2, y/\tau + 1)$$

Taking into consideration that, from (6.28), $J(\tau + 2, x/\tau + 2) = p_m(\tau + 2, x/\tau + 1)\psi(\tau + 2/\tau + 1)$ and

$$p_{mm}(\tau + 2/k) - p_{mm}(\tau + 2/\tau + 1) = - \sum_{\ell=\tau+2}^{k} J_m(\tau + 2/\ell) \times \psi^{-1}(\ell/\ell - 1)\tilde{R}(\ell)J_m'(\tau + 2/\ell)$$

we have

$$p(\tau + 1, x, y/k) = p(\tau + 1, x, y/\tau + 1) - A(\tau + 1, x)[p_{mm}(\tau + 2/k) - p_{mm}(\tau + 2/\tau + 1)]A'(\tau + 1, y)$$

Since the boundary conditions (7.12) and (7.14) are clear from (6.30) and (6.33), respectively, the proof of the theorem is complete.

7.3 Fixed-Lag Smoothing Estimator ($\tau = k + 1$, $k = k + 1 + \Delta$, Δ fixed)

From Theorem 6.5, we have

$$\hat{u}(k + 1, x/k + 1 + \Delta) = \hat{u}(k + 1, x/k + 1) + \sum_{\ell=k+2}^{k+1+\Delta} J(k + 1, x/\ell)\tilde{\nu}(\ell) \tag{7.15}$$

$$p(k + 1, x, y/k + 1 + \Delta) = p(k + 1, x, y/k + 1) - \sum_{\ell=k+2}^{k+1+\Delta} J(k + 1, x/\ell)\psi^{-1}(\ell/\ell - 1)\tilde{R}(\ell)J'(k + 1, y/\ell) \tag{7.16}$$

Then the following theorem holds.

THEOREM 7.3 The optimal fixed-lag smoothing estimator is given by

$$\hat{u}(k + 1, x/k + 1 + \Delta) = L_x\hat{u}(k, x/k + \Delta) + C(x, k + 1, \Delta) \times F_m(k + 1 + \Delta/k + 1 + \Delta)\nu(k + 1 + \Delta) + \tilde{Q}_m(k,x)[p_{mm}(k/k)L_*')^{-1}$$

$$\times\,(\hat{u}_m(k/k+\Delta) - \hat{u}_m(k/k)] \tag{7.17}$$

$$\Gamma_\xi \hat{u}(k+1,\ \xi/k+1+\Delta) = S(k+1,\ \xi) \qquad \xi \in \partial D \tag{7.18}$$

where

$$C(x,\ k+1,\ \Delta) = A(k+1,\ x)A_m(k+1) \quad \ldots \quad A_m(k+\Delta) \tag{7.19}$$

and

$$A_m(k) = \begin{bmatrix} A(k,x^1) \\ \vdots \\ A(k,x^m) \end{bmatrix}$$

Furthermore, the optimal fixed-lag smoothing error covariance matrix function $p(k+1,\ x,\ y/k+1+\Delta)$ is given by

$$p(k+1,\ x,\ y/k+1+\Delta) = p(k+1,\ x,\ y/k) - C(x,\ k+1,\ \Delta)$$
$$F_m(k+1+\Delta/k+1+\Delta)H(k+1+\Delta)p_{mm}(k+1+\Delta/k+\Delta)C'(y,k+1,\Delta)$$
$$- D(k,x)[p_{mm}(k/k) - p_{mm}(k/k+\Delta)]D'(k,y) \tag{7.20}$$
$$\Gamma_\xi p(k+1,\ \xi,\ y/k+1+\Delta) = 0 \qquad \xi \in \partial D \tag{7.21}$$

Proof. From (5.10) and (5.26) we have

$$\hat{u}(k+1,\ x/k+1) = L_x\hat{u}(k,x/k) + J(k+1,\ x/k+1)\tilde{\nu}(k+1) \tag{7.22}$$

From (7.15) and (7.22) it follows that

$$\hat{u}(k+1,\ x/k+1+\Delta) = L_x\hat{u}(k,x/k) + \sum_{\ell=k+1}^{k+\Delta} J(k+1,\ x/\ell)\tilde{\nu}(\ell)$$
$$+ J(k+1,\ x/k+1+\Delta)\tilde{\nu}(k+1+\Delta)$$

From (6.13) it follows that

$$J(k+1,\ x/k+1+\Delta) = p_m(k+1,\ x/k+1)L_*'\psi(k+2/k+1)$$
$$\times\ L_*'\psi(k+3/k+2)\ \cdots$$
$$\times\ L_*'\psi(k+2+\Delta/k+1+\Delta)$$

Substituting (5.42) into the right-hand side yields

$$J(k+1,\ x/k+1+\Delta) = p_m(k+1,\ x/k+1)L_*'p_{mm}^{-1}(k+1/k+1)$$
$$\times\ p_{mm}(k+2/k+2)L_*'\psi(k+3/k+2)\ \cdots$$
$$\times\ L_*'\psi(k+2+\Delta/k+1+\Delta)$$

Repeating the same procedure and using (6.26) yields

$$J(k+1,\ x/k+1+\Delta) = A(k+1,\ x)A_m(k+2)\ \cdots$$
$$\times\ A_m(k+\Delta)p_{mm}\frac{k+1+\Delta}{k+1+\Delta}$$

Thus we have

$$J(k+1,\ x/k+1+\Delta)\tilde{\nu}(k+1+\Delta) = C(x,\ k+1,\ \Delta) \times p_{mm}(k+1+\Delta/k+1+\Delta) \tag{7.23}$$

From (6.24) it follows that

$$\sum_{\ell=k+1}^{k+\Delta} J(k+1,\ x/\ell)\tilde{\nu}(\ell) = \sum_{\ell=k+1}^{k+\Delta} p_m(k+1,\ x/k) \times [p_{mm}(k/k)L_*']^{-1} J_m(k/\ell)\tilde{\nu}(\ell)$$

But from (4.4) we have

$$p_m(k+1,\ x/k) = L_x p_m(k,x/k)L_*' + Q_m(k,x)$$

From (6.23) and (6.24) we have

$$J(k,x/\ell) = A(k,x)J_m(k+1/\ell) = p_m(\tau,x/\tau) \times [p_{mm}(k/k)L_*']^{-1} J_m(k/\ell)$$

Then it follows that

$$\sum_{\ell=k+1}^{k+\Delta} J(k+1,\ x/\ell)\tilde{\nu}(\ell) = L_x \sum_{\ell=k+1}^{k+\Delta} J(k,x/\ell)\tilde{\nu}(\ell) + \tilde{Q}_m(k,x) \times \sum_{\ell=k+1}^{k+\Delta} [p_{mm}(k/k)L_*']^{-1} J_m(k/\ell)\tilde{\nu}(\ell)$$

and

$$\hat{u}(k+1,\ x/k+1+\Delta) = L_x \hat{u}(k,\ x/k+\Delta) + C(x,\ k+1,\ \Delta) \times F_m(k+1+\Delta/k+1+\Delta)\nu(k+1+\Delta) + \tilde{Q}_m(k,x)(p_{mm}(k/k)L_*')^{-1} \times \sum_{\ell=k+1}^{k+\Delta} J_m(k/\ell)\tilde{\nu}(\ell)$$

But from Theorem 6.5 we have

$$\hat{u}_m(k/k+\Delta) - \hat{u}_m(k/k) = \sum_{\ell=k+1}^{k+\Delta} J_m(k/\ell)\tilde{\nu}(\ell)$$

Thus, we have (7.17). From (5.32) and (7.16) it follows that

$$p(k+1,\ x,\ y/k+1+\Delta) = p(k+1,\ x,\ y/k) - J_1 - J_2$$

where

$$J_1 = \sum_{\ell=k+1}^{k+\Delta} J(k+1,\ x/\ell)\psi^{-1}(\ell/\ell - 1)\tilde{R}(\ell)J'(k+1,\ y/\ell)$$

$$J_2 = j(k+1,\ x/k+1+\Delta)\psi^{-1}(k+1+\Delta/k+\Delta)\tilde{R}(k+1+\Delta)$$
$$\times J'(k+1,\ y/k+1+\Delta)$$

From (5.42) and (7.22) we have

$$J_2 = C(x,\ k+1,\ \Delta)p_{mm}(k+1+\Delta/k+1+\Delta)\tilde{R}(k+1+\Delta)$$
$$\times p_{mm}(k+1+\Delta/k+\Delta)C'(y,k+1,\ \Delta)$$
$$= C(x,\ k+1,\ \Delta)F_m(k+1+\Delta/k+1+\Delta)H(k+1+\Delta)$$
$$\times p_{mm}(k+1+\Delta/k+\Delta)C'(y,\ k+1,\ \Delta)$$

Substituting (6.24) into J_1 yields

$$J_1 = D(k,x) \sum_{\ell=k+1}^{k+\Delta} J_m(k/\ell)\psi^{-1}(\ell/\ell - 1)\tilde{R}(\ell)J_m'(k/\ell)D'(k,y)$$

But from Theorem 6.5 we have

$$p(k,\ x,\ y/k+\Delta - p(k,x,y/k) = -\sum_{\ell=k+1}^{k+\Delta} J(k,x/\ell)\psi^{-1}(\ell/\ell - 1)$$
$$\times \tilde{R}(\ell)J'(k,y/\ell)$$

and

$$p_{mm}(k/k+\Delta) - p_{mm}(k/k) = -\sum_{\ell=k+1}^{k+\Delta} J_m(k/\ell)\psi^{-1}(\ell/\ell - 1)$$
$$\times \tilde{R}(\ell)J_m'(k/\ell)$$

Then we have

$$p(k+1,\ x,\ y/k+1+\Delta) = p(k+1,\ x,\ y/k) - C(x,\ k+1,\ \Delta)$$
$$\times F_m(k+1+\Delta/k+1+\Delta)H(k+1+\Delta)p_{mm}(k+1+\Delta/k+\Delta)C'(y,k+1,\Delta)$$
$$- D(k,x)[p_{mm}(k/k) - p_{mm}(k/k+\Delta)]D'(k,y)$$

Since the boundary conditions (7.18) and (7.21) are clear from (6.30) and (6.33), respectively, the proof of the theorem is complete.

Kelly and Anderson [18] proved that the fixed-lag smoothing algorithm of Theorem 7.3 may be unstable, but

Chirarattananon and Anderson [19] derived a stable version of the algorithm. It is possible to derive a comparable version here, although stability problems should not arise in our use of the algorithm of Theorem 7.3 as long as it is used over a finite time interval.

8. APPLICATION TO ESTIMATION OF AIR POLLUTION

Distributed parameter estimation theory has recently been applied to simulated air pollution data to demonstrate the capability of estimating atmospheric concentration levels from routine monitoring data [10,11]. A problem identified in these early studies was how to specify the statistical properties of the assumed system and observation noise. In this section we expand upon the prior studies in two respects. First, we consider actual monitoring data for sulfur dioxide (SO_2), in particular those measured each hour during the period December 1-31, 1975, at four locations in Tokoshima Prefecture, Japan (see Figure 1). Second, we apply the method of Sage and Husa [12] to estimate the unknown noise covariances in the system equation and measurements.

Hourly sulfur dioxide data are available at the four locations shown in Figure 1 for the period December 1-31, 1975. The data for day k at location i may be denoted by $e_k(x^i,t)$. It is useful to average the data for December 1-30 to produce

$$\langle e(x^i,t)\rangle = \frac{1}{30}\sum_{k=1}^{30} e_k(x^i,t) \tag{8.1}$$

where we will consider December 31 as a day to test the algorithms.

If it can be assumed that the wind flows are such that there are no north-south variations of concentration and that vertical mixing is rapid enough to eliminate variations of concentration with altitude, then the region can be

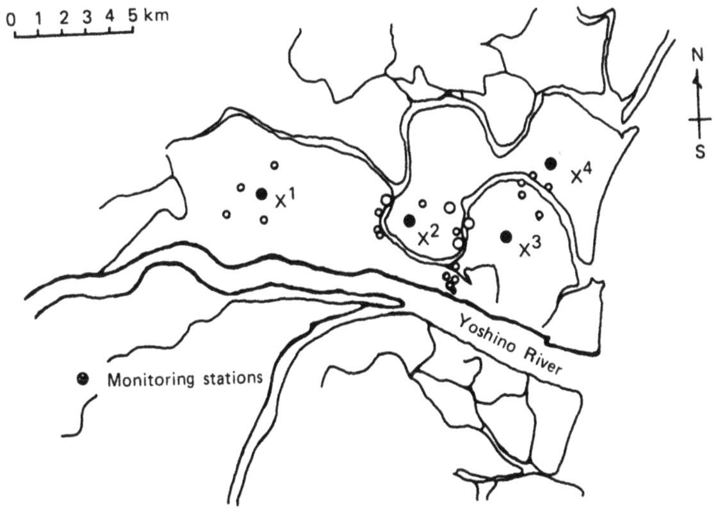

Figure 1. Map of Tokushima Prefecture, Japan. The four air pollution monitoring stations shown are located as follows: x^1, Aizumi; x^2, Kitajima; x^3, Kawauchi; x^4, Matsushige. Sources of sulfur dioxide have been lumped according to the following three source sizes, as indicated by the open circles, in order of descending size: 30-50 = m^3/h, 10-30 = m^3/h, and < 10 = m^3/h.

considered to be one-dimensional along the east-west coordinate. The SO_2 concentration at any particular time can be assumed to be described by the atmospheric diffusion equation [13],

$$\frac{\partial c}{\partial t} + \zeta \frac{\partial c}{\partial x} = a \frac{\partial^2 c}{\partial x^2} + S(x,t) \tag{8.2}$$

where ζ is the wind velocity, a is a diffusion coefficient, and S is the rate of emission of SO_2 as a function of location and time.

Equation (8.2) holds at any instant of time, but we desire an equation governing the monthly mean concentration $\langle c \rangle$. Although no such equation exists, we can formally average (8.2) over the 30 realizations (days) to produce

$$\frac{\partial \langle c \rangle}{\partial t} + \langle \zeta \frac{\partial c}{\partial x} \rangle = \langle a \frac{\partial^2 c}{\partial x^2} \rangle + S \tag{8.3}$$

One object will be to estimate the diffusion parameter a. This parameter will in general vary with location and time of day, although for simplicity we seek a constant value for the month. Thus, the first term on the right-hand side of (8.3) becomes a $\partial^2 \langle u \rangle / \partial x^2$. We can form the residuals, $u = c - \langle c \rangle$ amd $z = e - \langle e \rangle$. By subtracting (8.3) from (8.2) we obtain

$$\frac{\partial u}{\partial t} + \frac{\partial c}{\partial x} - \langle \zeta \frac{\partial c}{\partial x} \rangle = a \frac{\partial^2 u}{\partial x^2} \tag{8.4}$$

Since wind data are not available with which to evaluate the second and third terms on the left-hand side of (8.4) let us rewrite (8.4) as

$$\frac{\partial u}{\partial t} = a \frac{\partial^2 u}{\partial x^2} + w(x,t) \tag{8.5}$$

where w(x,t) includes those unknown features associated with the velocity terms.

The boundary conditions on (8.2) are

$$\frac{\partial c}{\partial x} = 0 \qquad x = 0,\ 1 \tag{8.6}$$

expressing the assumption that there is no diffusive flux of SO_2 into or out of the region at the boundaries. After averaging and forming the residual, (8.6) becomes

$$\frac{\partial u}{\partial x} = 0 \qquad x = 0,\ 1 \tag{8.7}$$

The problem is now to estimate u(x,t) based on the data,

$$z(x^i,t) = u(x^i,t) + v_i(t) \qquad i = 1,\ 2,\ 3,\ 4 \tag{8.8}$$

Since hourly data are available, (8.5) can be cast into the discrete-time form (2.1),

$$u(k + 1,\ x) = L_x u(k,x) + w(x,t) \tag{8.9}$$

with $L_x = 1 + a\partial^2/\partial x^2$. Observation error is estimated from the mean-square error of predicted values and observed data,

$$PA(i) = \frac{1}{24} \sum_{k=1}^{24} [z_i(k) - \hat{z}_i(k/k - 1)]^2 \qquad i = 1, 2, 3, 4 \tag{8.10}$$

An index of overall estimation error is

$$J = \sum_{i=1}^{4} PA(i) \tag{8.11}$$

To apply discrete-time distributed parameter estimation theory to predict air pollution levels, we must consider three problems. The first problem is how to simulate the distributed parameter system. The second is how to determine the covariances of system and observation noise. The last is how to determine the diffusion coefficient a. For the first problem we use the Fourier expansion method and approximate the original distributed parameter system by a finite-dimensional system. For the second problem, we apply the algorithm of Sage and Husa [12], which necessitates the simultaneous application of the optimal filtering and smoothing algorithms. For the third problem we apply the maximum likelihood approach in the smoothing form [14]. We now consider these problems in more detail.

8.1 Fourier Expansion Method

It is well known that the state u(k,x) of the distributed parameter system (8.9) with boundary condition (8.7) can be represented by using the eigenfunctions $\varphi_i(x)$ as follows:

$$u(k,x) = \sum_{i=1}^{\infty} u_i(k)\varphi_i(x) \tag{8.12}$$

where

$$L_x\varphi_i(x) = \lambda_i\varphi_i(x) \qquad x \in (0,1)$$

$$\frac{\partial\varphi_i(\xi)}{\partial\xi} = 0 \qquad \xi = 0,\ 1 \tag{8.13}$$

and

$$\int_0^1 \varphi_i(x)\varphi_j(x)\,dx = \delta_{ij}$$

λ_i is the eigenvalue of L_x corresponding to $\varphi_i(x)$. In this case, it is easily seen that the eigenfunction $\varphi_i(x)$ and the eigenvalue λ_i are given by

$$\varphi_1(x) = 1 \quad \varphi_i(x) = \sqrt{2}\cos\pi ix \qquad i = 1,\ 2,\ \ldots$$

and (8.14)

$$\lambda_i = 1 - a\pi^2(i-1)^2 \qquad i = 1,\ 2,\ \ldots$$

Then $\hat{u}(\tau,x/k)$, $p(\tau,xy/k)$, and $A(\tau,x)$ can be represented as follows:

$$\hat{u}(t,x/k) = \sum_{i=1}^{\infty} \hat{u}_{ij}(\tau/k)\varphi_i(x)$$

$$p(\tau,x,y/k) = \sum_{i,j=1}^{\infty} \tilde{p}_{ij}(\tau/k)\varphi_i(x)\varphi_j'(y) \tag{8.15}$$

$$A(\tau,x) = \sum_{i=1}^{\infty} a_i(\tau)\varphi_i(x)$$

Let us approximate these infinite expansions by the first N terms and define the following matrices and vectors:

$$\hat{u}(\tau/k) = \text{Col}[\hat{u}_1(\tau/k), \ldots, \hat{u}_N(\tau/k)]$$
$$A(\tau) = \text{Col}[a_1(\tau), \ldots, a_N(\tau)]$$
$$A = \text{diag}[\lambda_1, \ldots, \lambda_N]$$
$$p(\tau/k) = \begin{bmatrix} \tilde{p}_{11}(\tau/k) & \cdots & \tilde{p}_{1N}(\tau/k) \\ \vdots & & \vdots \\ \tilde{p}_{N1}(\tau/k) & \cdots & \tilde{p}_{NN}(\tau/k) \end{bmatrix}$$
$$Q(k) = \begin{bmatrix} q_{11}(k) & \cdots & q_{1N}(k) \\ \vdots & & \vdots \\ q_{N1}(k) & \cdots & q_{NN}(k) \end{bmatrix}$$
$$\Phi = \begin{bmatrix} \varphi_1(x^1) & \cdots & \varphi_N(x^1) \\ \vdots & & \vdots \\ \varphi_1(x^m) & \cdots & \varphi_N(x^m) \end{bmatrix}$$

where $q_{ij}(k)$ denotes the (i,j)th Fourier coefficient of $\tilde{Q}(k,x,y)$.

Then, from Theorems 4.2, 5.1, and 5.2 we have

$$\begin{aligned} \hat{u}(k+1/k+1) &= \Lambda\hat{u}(k/k) + F(k+1)\nu(k+1) \\ F(k+1) &= P(k+1/k)\Phi'H'(k+1)[H(k+1) \\ &\quad \times \Phi P(k+1/k)\Phi'H'(k+1) + R(k+1)]^{-1} \\ P(k+1/k) &= \Lambda P(k/k)\Lambda' + Q(k) \\ P(k+1/k+1) &= [I - F(k+1)H(k+1)\Phi]P(k+1/k) \end{aligned} \tag{8.16}$$

Furthermore, from Theorem 7.1 we have

$$\begin{aligned} \hat{u}(\tau+1/k) &= \hat{u}(\tau+1/\tau+1) + A(\tau+1)\Phi(\hat{u}(\tau+2/k) - \hat{u}(\tau+2/\tau+1)) \\ A(\tau+1) &= P(\tau+1/\tau+1)\Lambda P^{-1}(\tau+1/\tau)\Phi^{-1} \\ P(\tau+1/k) &= P(\tau+1/\tau+1) - A(\tau+1)\Phi(P(\tau+1/k) \\ &\quad - P(\tau+1/\tau))\Phi'A'(\Phi+1) \end{aligned} \tag{8.17}$$

Note that the fixed-interval smoothing estimator does not depend on the matrix Φ which reflects the effect of sensor location.

8.2 Determination of the Noise Covariances

In order to determine the unknown covariance matrices of the system and observation noises, we adopt Sage and Husa's algorithm [12] given by

$$\hat{Q}(k) = \frac{1}{k}\sum_{j=1}^{k} [\hat{u}(j/k) - \Lambda\hat{u}(j-1/k)][\hat{u}(j/k) - \Lambda\hat{u}(j-1/k)] \tag{8.18}$$

and

$$\hat{R}(k) = \frac{1}{k}\sum_{j=1}^{k} [z(j) - H(j)\Phi\hat{u}(j/k)][z(j) - H(j)\Phi\hat{u}(j/k)] \tag{8.19}$$

where $\hat{Q}(k)$ and $\hat{R}(k)$ denotes the estimated values of $Q(k)$ and $R(k)$, respectively. Note that in the identification algorithm the fixed-interval smoothing estimate $\hat{u}(j/k)$ is used.

8.3 Identification of the Unknown Parameter a

To determine the unknown parameter a we use the maximum likelihood approach in smoothing form [14]. The log-likelihood function $\gamma(k;a)$ is given from Ref. 14 by

$$\gamma(k;a) = \frac{1}{2}(\gamma_{bias} + \gamma_{obs}) \tag{8.20}$$

where

$$\gamma_{bias} = -kp\ \ln(2\pi) - \sum_{j=1}^{k} \ln\det \Sigma_\nu(j/j-1;\ a)$$

$$\gamma_{obs} = -\sum_{j=1}^{k} \{\nu'(j;a)R^{-1}\nu(j;a) + (\hat{u}(j/k,a) - \hat{u}(j-1/k,\ a))'Q^{-1}(\hat{u}(j/k,a) - \hat{u}(j-1/k,\ a))\}$$

$$\nu(j;a) = z(j) - H(j)\hat{u}(j/j-1,\ a)$$

$$\Sigma_\nu(i/j-1;\ a) = E[\nu(j;a)\nu'(j;a)]$$

where p is the dimension of $z(k)$, and $\hat{u}(j/k-1,\ a)$ denotes $\hat{u}(j/k-1)$ under the condition that the unknown parameter is assumed to be a.

To maximize $\gamma(k;a)$ we use the following gradient method:

$$a_{i+1} = a_i + G(i)\gamma_g(k;a_i)$$

$$\gamma_g(k;a_i) = \left.\frac{\partial\gamma(k;a)}{\partial a}\right|_{a=a_i} \tag{8.21}$$

where G(i) is a suitable matrix. Therefore, we adopt the following recursive algorithm to identify the unknown parameters Q, R, and a:

1. Make an initial guess a_0 of a.
2. Compute $\hat{Q}(a_0)$ and $\hat{R}(a_0)$ by using (8.18) and (8.19).
3. Compute $\hat{a}_i$ by using (8.20).
4. Compute $\hat{Q}(\hat{a}_i)$ and $\hat{R}(\hat{a}_i)$ by using (8.18) and (8.19).
5. Return to step 3 by changing i to i + 1 and repeat until these values do not change.

8.4 Numerical Results

We use the observed data from December 1-30 to identify the unknown parameter a and noise covariances Q and R. After four iterations the algorithm for determining a converged to the value, $\hat{a} = 0.001$. The Fourier expansion has been truncated at N = 4. The estimated diagonal elements of noise covariance matrices are:

$Q_{11} = 6.44$ $\quad R_{11} = 0.29$

$Q_{22} = 1.40$ $\quad R_{22} = 0.61$

$Q_{33} = 5.75$ $\quad R_{33} = 1.96$

$Q_{44} = 3.68$ $\quad R_{44} = 1.34$

To consider the effect of the number and location of monitoring stations, we assume that we have data at only one monitoring station. In this case from the previous results of Kumar and Seinfeld [15] and Omatu, Koide, and Soeda [16] we expect that the optimal sensor location is closest to the boundary. Thus either x^1 or x^4 is the optimal single sensor location among the four monitoring stations, x^1, x^2, x^3, x^4. In Table 1 we show the values of PA(i) and J for several monitoring stations. We see that Aizumi or Matsushige is optimal for the one-point sensor location case. Similar conclusions hold for two or three monitoring stations. Finally, we illustrate the actual observation data and 1-

hour-ahead predicted values for December 31 in Figures 2 to 5 for Aizumi, Kitajiima, Kawauchi, and Matsushige, respectively.

8.5 Comparison with Other Approaches

It is of interest to compare results of the present filtering and smoothing approaches with others available for air pollution estimation. We consider, therefore, the same SO_2 estimation problem by the following methods: (1) AR model, (2) persistence, and (3) weighted ensemble.

The AR model method is based on the following AR(p) model:

$$u_k^{(i)} = a_1 u_{k-1}^{(i)} + \cdots + a_p u_{k-p}^{(i)} + e_k^{(i)}$$
$$i = 1, 2, 3, 4 \tag{8.22}$$

where the $u_k^{(i)}$'s are the concentration levels at time k and at monitoring station x^i, $a_1, a_2, \ldots, a_p$ are the corresponding AR parameters, and the $e_k^{(i)}$'s are residuals. We used the Levinson algorithm to determine the AR parameters, while the optimal order p of the AR process is determined by using the minimum AIC (Akaike's information criterion) [20]. Then the 1-hour-ahead predicted concentration is given by

$$\hat{u}_{k/k-1}^{(i)} = a_1 u_{k-1}^{(i)} + \cdots + a_p u_{k-p}^{(i)} \tag{8.23}$$

and the prediction error variance is

$$J = \sum_{i=1}^{4} \frac{1}{24} \sum_{k=1}^{24} \left(u_k^{(i)} - \hat{u}_{k/k-1}^{(i)} \right)^2 \tag{8.24}$$

Table 2 shows the AR parameters and minimum AIC value at each monitoring station.

The weighted ensemble method uses the mean of the past observation data at each time k weighted by a linear function of the source strength as the prediction value at time k. Based on the number of emission sources, the weighting functions are assumed here to be 0.15, 0.41, 0.26, and 0.18

Table 1. Effect of the Number of Observation Locations on the Overall Error, J

Number of sensor locations	Sensor location	J
1	Aizumi (x^1)	39.9
	Kitajima (x^2)	83.1
	Kawauchi (x^3)	46.3
	Matsushige (x^4)	39.1
2	x^1, x^2	26.9
	x^1, x^3	25.8
	x^1, x^4	23.9
	x^2, x^3	41.3
	x^2, x^4	37.7
	x^3, x^4	29.2
3	x^1, x^2, x^3	22.0
	x^1, x^2, x^4	22.9
	x^1, x^3, x^4	23.4
	x^2, x^3, x^4	29.0
4	x^1, x^2, x^3, x^4	10.6

at x^1, x^2, x^3, and x^4, respectively. Table 3 shows the performance criteria of the four methods. From Table 3 we can see that the present method possesses almost the same accuracy as the AR model method. By multiplying each eigenfunction coefficient by the corresponding eigenfunction and summing them, however, the present method enables us to estimate concentrations over the entire region. Therefore, the present method is more powerful than the AR model method.

9. CONCLUSIONS

Optimal estimators for discrete-time distributed parameter systems have been derived based on Wiener-Hopf theory. A notable point of the present work is that the smoothing esti-

Table 2. AR Parameters and Minimum AIC (MAIC)

	x^1	x^2	x^3	x^4
MAIC	3.24	6.86	9.89	7.52
(optimal p)	(p = 5)	(p = 1)	(p = 10)	(p = 6)
a_1	-0.870	-0.811	-0.721	-0.841
a_2	0.058		-0.002	-0.001
a_3	0.073		-0.074	-0.040
a_4	-0.052		0.003	0.017
a_5	-0.056		0.007	-0.079
a_6			0.016	0.133
a_7			-0.029	
a_8			0.048	
a_9			0.014	
a_{10}			0.069	

mators have been derived by the same approach as the filter, thus providing a unified approach for this class of distributed parameter estimation problems. The estimation algorithms have been applied to the problem of predicting atmospheric sulfur dioxide levels in Tokushima,Prefecture, Japan.

Table 3. Comparison of the Four Methods at the Four Monitoring Sites for 1-Hour-Ahead Predicted Values-Prediction Error Squared

Method	x^1	x^2	x^3	x^4	Total
Current	1.02	2.32	4.21	3.08	10.63
AR	0.99	2.36	4.31	3.11	10.77
Persistence	0.92	2.87	6.08	2.71	12.58
Weighted ensemble	2.10	6.55	10.46	5.48	24.59

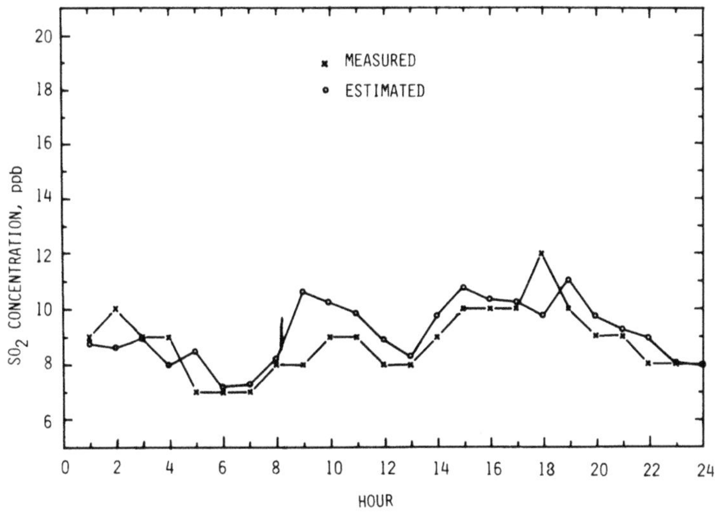

Figure 2. Measured and estimated sulfur dioxide concentrations on December 31, 1975, at Aizumi monitoring station (x^1).

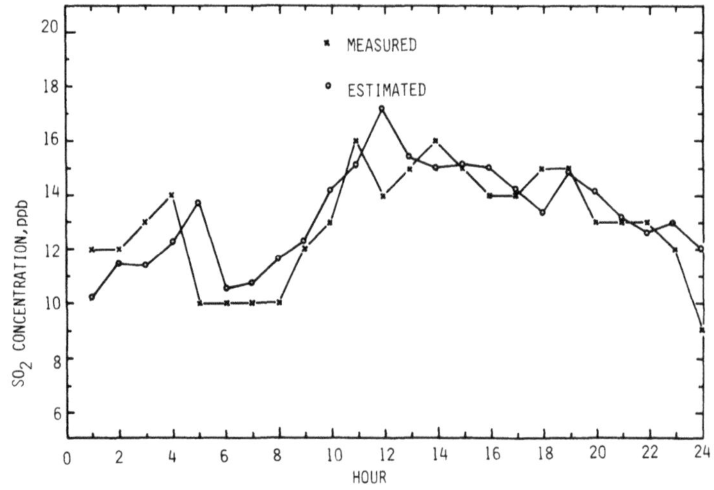

Figure 3. Measured and estimated sulfur dioxide concentrations on December 31, 1975, at Kitajima monitoring station (x^2).

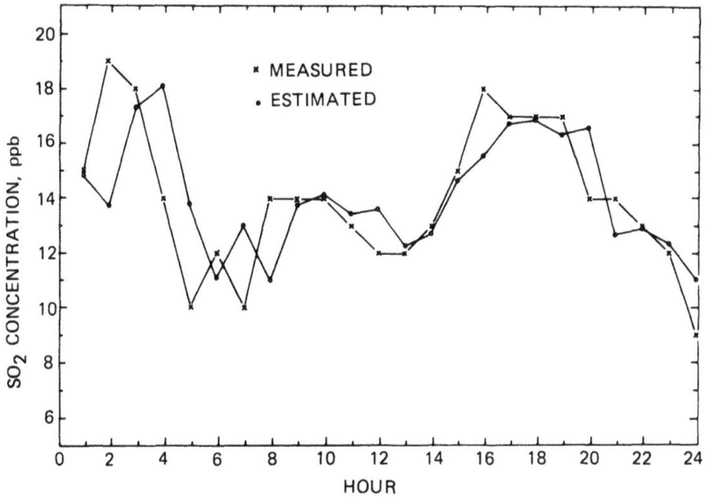

Figure 4. Measured and estimated sulfur dioxide concentrations on December 31, 1975, at Kawauchi monitoring station (x^3).

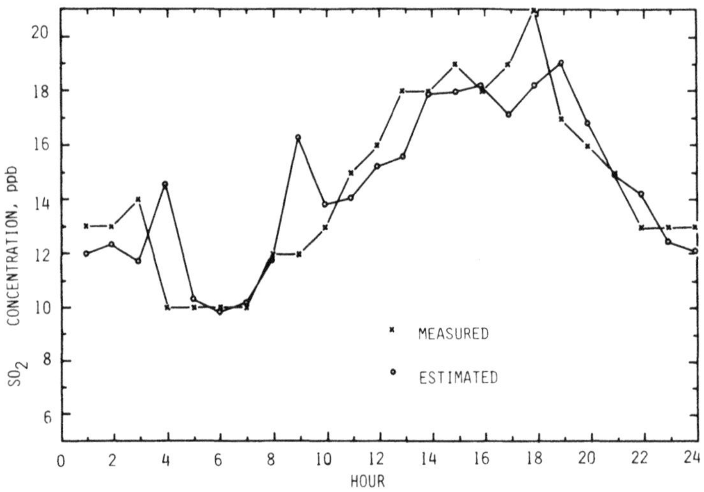

Figure 5. Measured and estimated sulfur dioxide concentrations on December 31, 1975, at Matsushige monitoring station (x^4).

REFERENCES

1. R. F. Curtain, A survey of infinite dimensional filtering, SIAM Rev. 17(3):395-411 (1975).
2. Y. Sawaragi, T. Soeda, and S. Omatu, Modeling, Estimation, and Their Applications for Distributed Parameter Systems, Lect. Notes Control Inf. Sci. 11, Springer-Verlag, Berlin, 1978.
3. K. E. Bencala and J. H. Seinfeld, Distributed parameter filtering: boundary noise and discrete observations, Int. J. Syst. Sci. 10(5):493-512 (1979).
4. J. S. Meditch, A survey of data smoothing for linear and nonlinear dynamic systems, Automatica, 9(2):151-162 (1973).
5. S. G. Tzafestas, Bayesian approach to distributed-parameter filtering and smoothing, Int. J. Control 15(2): 273-295 (1972).
6. S. G. Tzafestas, On optimal distributed-parameter filtering and fixed-interval smoothing for colored noise, IEEE Trans. Autom. Control AC-17(4):448-458 (1972).
7. H. Nagamine, S. Omatu, and T. Soeda, The optimal filtering problem for a discrete-time distributed parameter system, Int. J. Syst. Sci. 10(7):735-749 (1979).
8. T. Kailath, An innovation approach to least-squares estimation, Part I: Linear filtering in additive white noise, IEEE Trans. Autom. Control AC-13(6):646-655 (1968).
9. T. Kailath and P. Frost, An innovation approach to least-squares estimation, Part II: Linear smoothing in additive white noise, IEEE Trans. Autom. Control AC-13 (6):655-660 (1968).
10. A. A. Desalu, L. A. Gould, and F. C. Schweppe, Dynamic estimation of air pollution, IEEE Trans. Autom. Control AC-19(6):904-910 (1974).

11. M. Koda and J. H. Seinfeld, Estimation of urban air pollution, Automatica 14(6):583-595 (1978).
12. A. P. Sage and G. W. Husa, Adaptive filtering with unknown prior statistics, JACC:760-769 (1969).
13. J. H. Seinfeld, Air Pollution: Physical and Chemical Fundamentals, McGraw-Hill, New York, 1975.
14. F. C. Schweppe, Uncertain Dynamic Systems, Prentice-Hall, Englewood Cliffs, N. J., 1973.
15. S. Kumar and J. H. Seinfeld, Optimal location of measurements for distributed parameter estimation, IEEE Trans. Autom. Control AC-23(4):690-698 (1978).
16. S. Omatu, S. Koide, and T. Soeda, Optimal sensor location problem for a linear distributed parameter system, IEEE Trans. Autom. Control AC-23(4):665-673 (1978).
17. R. M. D. Santis, R. Saeks, and L. J. Tung, Basic optimal estimation and control problems in Hilbert space, Math. Syst. Theory 12:175-203 (1978).
18. C. N. Kelly and B. D. O. Anderson, On the stability of fixed-lag smoothing algorithms, J. Franklin Inst. 291 (4):271-281 (1971).
19. S. Chirarattananon and B. D. O. Anderson, The fixed-lag smoother as a stable finite-dimensional linear system, Automatica 7:657-669 (1971).
20. R. K. Mehra and D. G. Lainiotis, System Identification Advances and Case Studies, pp. 27-96, Academic Press, New York, 1976.

15

Homogenization of Diffusion Processes with Boundary Conditions

HIROSHI TANAKA/Faculty of Science and Technology, Keio University, Yokohama, Japan

1. INTRODUCTION

Let L_ε be a uniformly elliptic second-order partial differential operator of the form (2.1) of Sec. 2 indexed by a parameter $\varepsilon > 0$. The homogenization problem has been investigated by many authors; usually, the problem is to analyze the behavior, as $\varepsilon \downarrow 0$, of the solution $u_\varepsilon(x)$ of $L_\varepsilon u_\varepsilon = f$ in a domain D of $\mathbb{R}^d$ subject to an appropriate boundary condition under the assumption that the coefficients $a^{ij}(x)$, $b^i(x)$, and $c^i(x)$ are periodic [1-7], almost periodic [8], or more generally, stationary random fields [9,10]. In a probabilistic approach a problem of the following type is discussed. To what diffusion process does the diffusion process $X_\varepsilon(t)$ with generator L_ε converge in the law sense? The result in this direction for diffusion processes in the whole of $\mathbb{R}^d$ was obtained by Bensoussan, Lions, and Papanicolaou [4,5] in the case of periodic coefficients and by Papanicolaou and Varadhan [10] in the case when the coefficients are stationary random fields. In the case of the presence of boundary conditions in $X_\varepsilon(t)$, some results were obtained by Papanicolaou, Stroock and Varadhan [6].

In this chapter we consider the homogenization problem in the case when $X_\varepsilon(t)$ is a diffusion process on the half-space $\overline{D}$ with generator L_ε inside D and subject to the boundary condition $L_\varepsilon u = 0$ on ∂D, where $D = \{(x^1, \ldots, x^d) \in \mathbb{R}^d; x^1 > 0\}$ with boundary ∂D, $d \geq 2$, and L_ε is given by (2.2). The main assumptions on the coefficients are that they be smooth (say, C^∞) functions periodic in each variable (period 1), that $b^i(x)$ and $\beta^i(x)$ satisfy certain centering conditions, and that $\gamma^1(x) = 1$ (precise assumptions on the coefficients are stated in Sec. 2). The situation here may be somewhat simple because of the periodicity assumption, but we emphasize that the coefficients in the boundary condition are also rapidly oscillating as $\varepsilon \downarrow 0$. Obviously, the behavior inside D of the limit diffusion process $X_0(t)$ (if it exists) is known from a result of Bensoussan, Lions, and Papanicolaou [4], so the crucial point in our discussions is to check some tightness condition for the family of processes $X_\varepsilon(t)$, $\varepsilon > 0$, and then to find the boundary condition describing $X_0(t)$. The following two cases are considered separately. Let $L^{(0)}$ be given by (2.4). Case (1): $L^{(0)}$ is uniformly elliptic on ∂D. Case (2): $L^{(0)} = 0$ (oblique reflection). Even if γ is a constant in case (2), the limit diffusion process will have an oblique reflection with coefficients different from γ in general. In each case we shall arrive at our result by describing the sample paths of the process X_ε as the solutions of the stochastic differential equation with boundary condition due to S. Watanabe [11] and then performing stochastic calculus as in Refs. 4 and 5. The C^∞ assumption on the coefficients might be too much to be needed, but we do so because our main interest is in the determination of an explicit form of the boundary condition of $X_0(t)$.

An exact statement of the problem and the main results are given in Sec. 2. After preparing some known results and preliminary lemmas in Secs. 3 and 4, the proof of the main results are given in Secs. 5 and 6.

2. STATEMENTS OF THE PROBLEM AND THE MAIN RESULTS

Let $D = \{(x^1,\ldots,x^d) \in \mathbb{R}^d : x^1 > 0\}$ and let

$$\mathcal{L}_\varepsilon = \sum_{i,j=1}^{d} a^{ij}\left(\frac{x}{\varepsilon}\right)\partial_i\partial_j + \frac{1}{\varepsilon}\sum_{i=1}^{d} b^i\left(\frac{x}{\varepsilon}\right)\partial_i + \sum_{i=1}^{d} c^i\left(\frac{x}{\varepsilon}\right)\partial_i \tag{2.1}$$

$$L_\varepsilon = \sum_{i,j=2}^{d} \alpha^{ij}\left(\frac{x}{\varepsilon}\right)\partial_i\partial_j + \frac{1}{\varepsilon}\sum_{i=2}^{d} \beta^i\left(\frac{x}{\varepsilon}\right)\partial_i + \sum_{i=1}^{d} \gamma^i\left(\frac{x}{\varepsilon}\right)\partial_i \tag{2.2}$$

be given ($\varepsilon > 0$), where $\partial_i = \partial/\partial x^i$. We set

$$\mathcal{L}^{(0)} = \sum_{i,j=1}^{d} a^{ij}(x)\partial_i\partial_j + \sum_{i=1}^{d} b^i(x)\partial_i \qquad x \in \mathbb{R}^d \tag{2.3}$$

$$L^{(0)} = \sum_{i,j=2}^{d} \alpha^{ij}(x)\partial_i\partial_j + \sum_{i=2}^{d} \beta^i(x)\partial_i \qquad x \in \partial D \tag{2.4}$$

and require the following:

1. $\mathcal{L}^{(0)}$ is uniformly elliptic and the matrix $a(x) = \{a^{ij}(x)\}$ is factored as $\sigma(x)^t\sigma(x)/2$.
2. The functions

   $\sigma: \mathbb{R}^d \to \mathbb{R}^d \otimes \mathbb{R}^d$
   $b: \mathbb{R}^d \to \mathbb{R}^d$
   $c: \mathbb{R}^d \to \mathbb{R}^d$
   are smooth (C^∞) and periodic (period = 1) in each variable.
3. The matrix $\alpha(x) = \{\alpha^{ij}(x)\}_{i,j=2}^d$ is factored as $\tau(x)^t\tau(x)/2$.
4. The functions

   $\tau: \partial D(\cong \mathbb{R}^{d-1}) \to \mathbb{R}^{d-1} \otimes \mathbb{R}^{d-1}$
   $\beta: \partial D(\cong \mathbb{R}^{d-1}) \to \mathbb{R}^{d-1}$
   $\gamma: \partial D(\cong \mathbb{R}^{d-1}) \to \mathbb{R}^d$
   are smooth (C^∞) and periodic (period = 1) in each

variable, and $\gamma^1(x) = 1$.

The differential operator L_ε inside D together with the boundary condition $L_\varepsilon u = 0$ on ∂D determines a unique diffusion process on $\bar{D}$ which is called the $(L_\varepsilon, L_\varepsilon)$-diffusion for simplicity. A precise description of the $(L_\varepsilon, L_\varepsilon)$-diffusion is given in the next section by means of stochastic differential equations. Now our problem is stated as follows: To what diffusion process does the $(L_\varepsilon, L_\varepsilon)$-diffusion converge in the law sense as $\varepsilon \downarrow 0$? This problem will be solved under assumptions on the coefficients (Assumptions 1 and 2). By requirements 1 and 2 there exists the $L^{(0)}$-diffusion on $\mathbb{R}^d$ (the unique diffusion process on $\mathbb{R}^d$ with generator $L^{(0)}$). By the periodicity assumption on the coefficients this process induces a diffusion process on the d-dimensional torus $\underline{T}^d$, which is called the $L^{(0)}$-diffusion on $\underline{T}^d$; let m be its invariant probability measure.

ASSUMPTION 1 (Centering Condition). $\int_{\underline{T}^d} b \, dm = 0$.

The following two cases are considered separately.

Case 1. $L^{(0)}$ is uniformly elliptic.

Case 2. $L^{(0)} = 0$ $(\tau = 0, \beta = 0)$.

The result in case (1): There exists the $L^{(0)}$-diffusion on $\partial D (\cong \mathbb{R}^{d-1})$ which, again by the periodicity assumption, induces the $L^{(0)}$-diffusion on $\underline{T}^{d-1}$; let m_0 be the invariant probability measure of the $L^{(0)}$-diffusion on $\underline{T}^{d-1}$. In addition to Assumption 1 we make the following assumption in case 1.

ASSUMPTION 2 (Centering Condition). $\int_{\underline{T}^d} \beta \, dm_0 = 0$.

First we notice that under Assumption 1 there exists a periodic solution χ^k of $L^{(0)}\chi^k = -b^k$ for each $k = 1, \ldots, d$. For example, we have a solution of the form $\chi^k(x) =$

$\int_0^\infty E\{b^k(X(t))\}\,dt$ where X(t) is the $L^{(0)}$-diffusion starting at x. We set

$$\chi = \begin{bmatrix} \chi^1 \\ \vdots \\ \chi^d \end{bmatrix} \qquad \nabla\chi = \begin{bmatrix} \partial_1\chi^1 & \cdots & \partial_d\chi^1 \\ \vdots & & \vdots \\ \partial_1\chi^d & \cdots & \partial_d\chi^d \end{bmatrix} \tag{2.5}$$

$$a_0 = \{a_0^{ij}\} = \int_{\underline{T}^d} (I + \nabla\chi)a(x)^t(I + \nabla\chi)\,dm \tag{2.6}$$

$$c_0 = \{c_0^i\} = \int_{\underline{T}^d} (I + \nabla\chi)c(x)\,dm \tag{2.7}$$

$$L_0 = \sum_{i,j=1}^{d} a_0^{ij}\partial_i\partial_j + \sum_{i=1}^{d} c_0^i\partial_i \tag{2.8}$$

Next, under Assumption 2 there exists a periodic solution χ_0^k of $L^{(0)}\chi_0^k = -\beta^k$ on ∂D for each k ($2 \le k \le d$), and a bounded (and periodic in $x^2, \ldots, x^d$) solution ξ^k of

$$\begin{aligned} L^{(0)}\xi^k &= -b^k && \text{in } D \\ \xi^k &= \chi_0^k && \text{on } \partial D \end{aligned} \tag{2.9}$$

for each k = 1, ..., d (we set $\chi^1 = 0$). We set

$$\chi_0 = \begin{bmatrix} \chi_0^2 \\ \vdots \\ \chi_0^d \end{bmatrix} \quad \nabla\chi_0 = \begin{bmatrix} \partial_2\chi_0^2 & \cdots & \partial_d\chi_0^2 \\ \vdots & & \vdots \\ \partial_2\chi_0^d & \cdots & \partial_d\chi_0^d \end{bmatrix} \quad \xi = \begin{bmatrix} \xi^1 \\ \vdots \\ \xi^d \end{bmatrix}, \text{ etc.} \tag{2.10}$$

$$\alpha_0 = \int_{\underline{T}^{d-1}} (I + \nabla\chi_0)\alpha(x)^t(I + \nabla\chi_0)\,dm_0 \tag{2.11}$$

$$\gamma_0 = \int_{\underline{T}^{d-1}} (I + \nabla\xi)\gamma\,dm_0 \tag{2.12}$$

$$L_0 = \sum_{i,j=2}^{d} \alpha_0^{ij}\partial_i\partial_j + \sum_{i=1}^{d} \gamma_0^i\partial_i \tag{2.13}$$

THEOREM 2.1 In case 1, under Assumptions 1 and 2 the $(L_\varepsilon, L_\varepsilon)$-diffusion converges in the law sense to the (L_0, L_0)-diffusion as $\varepsilon \downarrow 0$.

<u>The result in case 2</u>: We write $u = H\varphi$ for the solution u of

$$\begin{aligned} L^{(0)}u &= 0 && \text{in } D \\ u &= \varphi && \text{on } \partial D \end{aligned} \tag{2.14}$$

Then H sends functions on ∂D to functions on $\bar{D}$, while $\partial_\gamma H$ sends functions on ∂D to functions on ∂D where $\partial_\gamma = \Sigma_{i=1}^d \gamma^i \partial_i$. It is known that there exists a unique Markov process on ∂D with generator $\partial_\gamma H$. By the periodicity assumption this Markov process induces a Markov process on the torus $\underline{T}^{d-1}$; let $\tilde{m}$ be the invariant measure of the induced Markov process. We set

$$\tilde{\gamma} = \{\tilde{\gamma}^i\} = \int_{\underline{T}^{d-1}} (I + \nabla\chi)\gamma \, d\tilde{m} \qquad \tilde{L} = \sum_{i=1}^{d} \tilde{\gamma}^i \partial_i \tag{2.15}$$

THEOREM 2.2 In case 2, under Assumption 1 the $(L_\varepsilon, L_\varepsilon)$-diffusion converges in the law sense to the $(L_0, \tilde{L})$-diffusion as $\varepsilon \downarrow 0$.

REMARKS 1. It can be proved that $\gamma_0^1 > 0$, $\tilde{\gamma}^1 > 0$, so $\tilde{L}$ is actually an oblique derivative (Lemma 4.5).

2. The convergence in the law sense in the theorems means, of course, as follows. Let $W(\bar{D})$ be the space of continuous functions from $[0,\infty)$ to $\bar{D}$, and $X_\varepsilon(t)$, $X_0(t)$ be processes with sample paths in $W(\bar{D})$. Let P_ε and P_0 be the probability measures on $W(\bar{D})$ induced by these processes. Then the process $X_\varepsilon(t)$ is said to converge in the law sense to $X_0(t)$ as $\varepsilon \downarrow 0$ if $P \to P_0$ (weakly) as $\varepsilon \downarrow 0$.

3. STOCHASTIC DIFFERENTIAL EQUATIONS AND SOME KNOWN RESULTS

3.1 SDE with Boundary Condition

Let L_ε and L_ε be given as in Sec. 2. Then, by the result of S. Watanabe [11], the $(L_\varepsilon, L_\varepsilon)$-diffusion $X_\varepsilon(t)$ can be obtained by solving a certain stochastic differential equation (SDE) with boundary condition. For simplicity we set $\varepsilon = 1$.

PROPOSITION 3.1 [11] (see also Ref. 12). Let $x \in \bar{D}$. On a suitable probability space (Ω, F, P) with an increasing family

$\{\mathcal{F}_t\}_{t\geq 0}$ of sub-σ-fields of $\mathcal{F}$, we can construct four processes $B(t)$, $M(t)$, $\Phi(t)$, $X(t)$ satisfying the following conditions.

(i) $B(t)$ and $M(t)$ are d-dimensional processes such that their component processes $B^i(t)$ and $M^i(t)$ are square-integrable continuous $\mathcal{F}_t$-martingale with mean 0 satisfying

$$B^i(0) = M^i(0) = 0 \quad (1 \leq i \leq d) \qquad M^1(t) = 0$$

$$\langle B^i(t), B^j(t)\rangle = \delta^{ij} t \qquad 1 \leq i,j \leq d$$

[in particular, $B(t)$ is a d-dimensional Brownian motion]

$$\langle B^i(t), M^j(t)\rangle = 0 \qquad 1 \leq i,j \leq d$$

$$\langle M^i(t), M^j(t)\rangle = \delta^{ij}\Phi(t) \qquad 2 \leq i,j \leq d$$

(ii) $\Phi(t)$ is an $\mathcal{F}_t$-adapted and continuous increasing process such that

$$\{\text{points of increase of } \Phi(\cdot)\} \subset \{t: X^1(t) = 0\}$$

(iii) $X(t)$ is an $\mathcal{F}_t$-adapted $\bar{D}$-valued continuous process satisfying the SDE

$$\begin{aligned} dX(t) &= \sigma(X(t))\, dB + b(X(t))\, dt + c(X(t))\, dt \\ &\quad + \tau(X(t))\, dM + \beta(X(t))\, d\Phi + \gamma(X(t))\, d\Phi \end{aligned} \tag{3.1}$$

$$X(0) = x$$

Moreover, the law uniqueness holds in the sense that the probability measure P^x induced by $X(\cdot)$ on the space of continuous paths is uniquely determined for each starting point x, and the family $\{P^x\}$ defines a diffusion process on $\bar{D}$.

The diffusion process in Proposition 3.1 is what we call the $(\mathcal{L}_1, L_1)$-diffusion in this chapter.

REMARKS

1. In (3.1) and in most of the sequel, τ and β are regarded as $d \times d$-matrix-and d-vector-valued functions, respectively, with the convention that $\tau^{i1} = \tau^{1i} = \beta^1 = 0$ for $1 \leq i \leq d$.

2. Equation (3.1) is an abbreviation for the following componentwise and integral expression:

$$X^i(t) = x^i + \sum_{j=1}^{d} \int_0^t \sigma^{ij}(X(s))\, dB^i + \int_0^t b^i(X(s))\, ds$$

$$+ \int_0^t c^i(X(s))\, ds + \sum_{j=1}^{d} \int_0^t \tau^{ij}(X(s))\, dM^j$$

$$+ \int_0^t \beta^i(X(s))\, d\Phi + \int_0^t \gamma^i(X(s))\, d\Phi \qquad (3.2)$$

Moreover, in the strict expression of the SDE (3.1) it is better to add the factor $1_D(X(t))$ in front of dB and dt, but this is not necessary in the present case because the Lebesgue measure of $\{t: X(t) \in \partial D\}$ vanishes (a.s.).

We make a comment about the scaling relation in the $(L_\varepsilon, L_\varepsilon)$-diffusion $X_\varepsilon(t)$. Assume for a moment that $c = 0$. Then it can be easily proved that for each $\varepsilon > 0$ the process $\varepsilon X_1(t/\varepsilon^2)$ is the $(L_\varepsilon, \tilde{L}_\varepsilon)$-diffusion where

$$\tilde{L}_\varepsilon = \varepsilon \sum_{i,j=2}^{d} \alpha^{ij}(\tfrac{x}{\varepsilon})\partial_i\partial_j + \sum_{i=1}^{d} [\beta(\tfrac{x}{\varepsilon}) + \gamma(\tfrac{x}{\varepsilon})]\partial_i$$

and therefore the term of second-order derivatives in the boundary condition will vanish in the limit $\varepsilon \downarrow 0$. The boundary condition $L_\varepsilon u = 0$ describing $X_\varepsilon(t)$ was modified so that the second-order derivatives does not vanish even in the limit. Thus the scaling relation that the process $\varepsilon X_1(\cdot/\varepsilon^2)$ is equivalent in law to $X_\varepsilon(\cdot)$ holds only in case 2 with $c = 0$.

3.2 Skorohod Equation

The SDE (3.1) in one dimension yields a very simple form and is known as the Skorohod equation (e.g., see Refs. 13 and 14; see Ref. 15 for a multidimensional analog). We explain it in a form convenient for our later use.

PROPOSITION 3.2 (Skorohod Equation). For a given real-valued continuous process $U(t)$ with $U(0) \geq 0$, the equation

$$X(t) = U(t) + \Phi(t)$$

has a unique pair of $X(t)$ and $\Phi(t)$ satisfying the following three conditions.

(i) $X(t)$ is a continuous process with $X(t) \geq 0$.

(ii) $\Phi(t)$ is an increasing continuous process with $\Phi(0) = 0$.

(iii) {Points of increase of $\Phi(\cdot)$} $\subset \{t : X(t) = 0\}$.

Moreover, we have

$$\Phi(t) - \Phi(s) \leq \max_{s\leq t_1\leq t_2\leq t} |U(t_2) - U(t_1)| \qquad 0 \leq s \leq t \tag{3.3}$$

The proof is easy (see Ref. 14; (3.3) follows from the fact that $\Phi(t) = 0$ for $0 \leq t \leq S$ and $= -\min\{U(s) : S \leq s \leq t\}$ for $t > S$ where $S = \inf\{t > 0 : U(t) < 0\}$.

DEFINITION If $\Phi(t)$ satisfies conditions (i) and (ii) of Proposition 3.2, then $\Phi(t)$ is said to be associated with $X(t)$.

3.3 Some Known Results on Homogenization

A result of Bensoussan, Lions, and Papanicollaou [4] is that the L_ε-diffusion in $\mathbb{R}^d$ converges in the law sense to the L_0-diffusion in $\mathbb{R}^d$ as $\varepsilon \downarrow 0$, where L_0 is given by (2.8). In connection with this result, we make a few remarks.

REMARKS

3. The elliptic differential operator L_0 is nondegenerate.

4. Under requirements 1 and 2, the centering condition (Assumption 1) is satisfied if and only if each component process $X^k(t)$ of the $L^{(0)}$-diffusion $X(t)$ is recurrent in the sense that $X^k(t)$ hits any state (real number) with probability 1.

Here is a proof of the "only if" part. The component process $X^k(t)$ satisfies the SDE: $dX^k(t) = \sigma^k\, dB + b^k\, dt$, where $\sigma^k = \sigma^k(X(t)) = (\sigma^{k1}(X(t)),\ldots,\sigma^{kd}(X(t))$ and $b^k = b^k(X(t))$.

Assuming that the centering condition (Assumption 1) is satisfied and taking a periodic solution of $L^{(0)}\chi^k = -b^k$, we set $\hat{X}^k(t) = X^k(t) + \chi^k(X(t))$. Then by Itô's formula we can write $\hat{X}^k(t) = X^k(0) + \chi^k(X(0)) + M^k(t)$, where $M^k(t)$ is a martingale with increasing process $A^k(t) = 2\int_0^t a_0^{kk}(X(s))\,ds$, $a_0^{kk}(x)$ being the (k,k)-entry of the matrix $a_0(x) = (I + \nabla\chi)a(x)^t(I + \nabla\chi)$. Let u be a periodic solution of $L^{(0)}u = 2[a_0^{kk}(x) - \bar{a}_0^{kk}]$. Then

$$u(X(t)) - u(X(0)) = \int_0^t \nabla u\, dB + \int_0^t L^{(0)}u(X(s))\, ds$$

$$= \int_0^t \nabla u\, dB + 2\int_0^t [a_0^{kk}(X(s)) - \bar{a}_0^{kk}]\, ds$$

and hence noting that $\bar{a}_0^{kk} > 0$ (Remark 3) we have

$$A^k(t) = t[2\bar{a}_0^{kk} + \frac{u(X(t)) - u(X(0))}{t} - \frac{1}{t}\int_0^t \nabla u\, dB]$$

$$\to \infty, \qquad \text{as } t \to \infty \text{ (a.s.)}$$

Therefore, $M^k(t)$ is recurrent, and hence $\hat{X}^k(t)$ and $X^k(t)$ are also recurrent. We omit the proof of the "if" part, which is not hard either.

4. SOME PRELIMINARY LEMMAS

All the assumptions on the coefficients of the differential operators are the same as stated in Sec. 2. Let $L = L_1$ and consider the $(L^{(0)},L)$-diffusion Y(t). By Proposition 3.1 this process can be constructed by solving the SDE:

$$dY(t) = \sigma(Y(t))\, dB + b(Y(t))\, dt + \tau(Y(t))\, dM + \beta(Y(t))\, d\psi + \gamma(Y(t))\, d\psi \qquad (4.1)$$

By the periodicity of the coefficients, the process Y(t) induces a diffusion process $Y_0(t)$ on $[0,\infty) \times \underline{T}^{d-1}$; it is defined by $Y_0^1(t) = Y^1(t)$ and $Y_0^i(t) = Y^i(t)$ (mod 1) with $0 \le Y_0^i(t) < 1$ for $2 \le i \le d$. Also, by Remark 4 of Sec. 3 the hitting time

$$T_a = \inf\{t > 0: Y_0^1(t) = a\} \qquad (4.2)$$

is finite a.s. for any $a \geq 0$. This fact, together with the uniform ellipticity of $L^{(0)}$, implies the following:

LEMMA 4.1 $Y_0(t)$ is recurrent in the sense that it hits any open set ($\neq \emptyset$) a.s.

Let $Y(0) = x$ and set $e(x) = 1 - E\{\exp[-\psi(\infty)]\}$. Then it can be proved that the function $e(x)$ is excessive for $Y(t)$; by the periodicity it is also excessive for the recurrent diffusion process $Y_0(t)$ and hence $e(x) \equiv$ const [since the continuity of $e(x)$ can also be proved]. Therefore, $e(x) = E[e(Y(t))]$, from which it follows that $E\{\exp[-\psi(\infty)]\} = E\{\exp[-(\psi(\infty) - \psi(t))]\}$ and hence $\psi(\infty) = \infty$ a.s.

LEMMA 4.2 The process $Y_1(t)$ defined by

$$Y_1(t) = Y_0(\psi^{-1}(t)) \qquad \psi^{-1}(t) = \max\{s : \psi(s) = t\}$$

is a right-continuous conservative strong Markov process on $\underline{T}^{d-1}$ with generator

$$LH = \sum_{i,j=2}^{d} \alpha^{ij}(x)\partial_i\partial_j + \partial_{\beta+\gamma}H$$

Proof. Since ψ is an additive functional of $Y(t)$, the strongly Markovian property of $Y_1(t)$ is a consequence of the general theory. The part concerning the generator can be seen as follows. If φ is a smooth function on ∂D with bounded partial derivatives of order ≤ 2 and if $u = H\varphi$ is the solution of (2.14), then $du(Y(t)) = (\nabla u)\sigma\, dB + Lu\, d\psi$ by an application of Itô's formula. Therefore,

$$u(Y(t)) - u(Y(0)) - \int_0^t (LH\varphi)(Y(s))\, d\psi$$

is a martingale. The time substitution $t \to \psi^{-1}(t)$ then yields

$$\varphi(Y_1(t)) - \varphi(Y_1(0)) - \int_0^t (LH\varphi)(Y_1(s))\, ds = \text{martingale}$$

provided that φ is periodic. This means that $Y_1(t)$ has the generator LH for smooth functions.

LEMMA 4.3 There exists a unique invariant probability measure m_1 of the Markov process $Y_1(t)$.

Proof. This result is not new, of course; but we give a construction of m_1 which will be useful for proving the next lemma. Let $a > 0$ be fixed and, assuming that $Y_0(0)$ is on the boundary, define the F_t-stopping times T_n, S_{n+1}, $n \geq 0$ by

$$T_0 = 0 \qquad S_n = \inf \{t > T_{n-1} : Y_0^1(t) = a\}$$

$$T_n = \inf \{t > S_n : Y_0^1(t) = 0\} \qquad n \geq 1$$

Then $\{Y_0(T_n), n \geq 1\}$ is a Markov process on $\underline{T}^{d-1}$ with transition function satisfying Doeblin's condition. Therefore, there is a unique invariant probability measure μ of this process; moreover, if $\mu_n(x,\cdot)$ denotes the probability distribution of $Y_0(T_n)$ starting at x, then

$$\text{Total variation of } \mu_n(x,\cdot) - \mu \to 0 \qquad n \to \infty \quad (4.3)$$

the convergence being uniform in x and exponentially fast. Now define a probability measure m_1 on $\underline{T}^{d-1}$ by

$$\int \varphi \, dm_1 = cE_\mu[\int_0^{\psi(T)} \varphi(Y_1(s)) \, ds]$$

$$= cE_\mu[\int_0^T \varphi(Y_0(s)) \, d\psi] \qquad T = T_1, \quad \varphi \in C(\underline{T}^{d-1})$$

where the suffix μ indicates that the initial distribution of $Y_0(\cdot)$ is μ and c is the normalizing constant. For proving that m_1 is an invariant measure of $Y_1(t)$, it is enough to show that

$$\int \varphi_t \, dm_1 = cE_\mu[\int_0^{\psi(T)} \varphi(Y_1(s)) \, ds] \qquad (4.4)$$

where $\varphi_t(x) = E[\varphi(Y_1(t))]$ with $Y_0(0) = x$; but (4.4) can be proved by making use of the strong Markov property of $Y_1(t)$ and the fact that $\psi(T)$ is a stopping time for the process $Y_1(\cdot)$, or more precisely, an F_t^*-stopping time where F_t^* denotes the sub-σ-field of F consisting of those A's for which $A \cap \{\psi^{-1}(t) < s\} \in F_s$ (for all $s \geq 0$); also, the fact that $Y_1(\psi(T)) = Y_0(T)$ has distribution μ is used. Finally, to prove the uniqueness of m_1, let m_2 be another invariant probability measure of $Y_1(t)$ and let $Y_0(0)$ be m_2-distributed. Then, by making use of (4.3) we can prove that

$$\frac{1}{n}\int_0^{T_n} \varphi(Y_1(s))\, ds = \frac{1}{n}\sum_{k=1}^{n}\int_{T_{k-1}}^{T_k} \varphi(Y_1(s))\, ds \underset{n\to\infty}{\to} c^{-1}\int \varphi\, dm_1$$

in probability. But this fact together with its special case $T_n/n \to c^{-1}$ (obtained by setting $\varphi = 1$) implies that

$$\frac{1}{t}\int_0^t \varphi(Y_1(s))\, ds \to \int \varphi\, dm_1 \qquad t \to \infty \text{ (in probability)}$$

Therefore, we have

$$\int \varphi\, dm_2 = E_{m_2}[\varphi(Y_1(t))] = E_{m_2}[\tfrac{1}{t}\int_0^t \varphi(Y_1(s))\, ds] \to \int \varphi\, dm_1$$

proving that $m_1 = m_2$.

LEMMA 4.4 Let u be a bounded and smooth function on $\overline{D}$, periodic in $x^2, \ldots, x^d$ and satisfying $L^{(0)}u = -b^1$ in D. Then

$$1 + \int_{\underline{T}^{d-1}} Lu\, dm_1 > 0$$

Proof. Since the function $h(x) = x^1 + u(x)$ satisfies $L^{(0)}h = 0$ in D and $Lh = 1 + Lu$ on ∂D, we have

$$dh(Y(t)) = (\nabla h)\sigma\, dB + (1 + Lu)\, d\psi$$

Therefore, taking $a > 2\|u\|_\infty$ and then defining T_a, T, and μ as in the proof of Lemma 4.3, we have

$$\begin{aligned} 0 < E_\mu[h(Y(T_a)) - h(Y(0))] &= E_\mu[\int_0^{T_a} (1 + Lu)(Y(s))\, d\psi] \\ &= E_\mu[\int_0^{T} (1 + Lu)(Y_0(s))\, d\psi] \\ &= c^{-1}\int (1 + Lu)\, dm_1 \end{aligned}$$

proving the lemma.

LEMMA 4.5 Both γ_0^1 and $\tilde{\gamma}^1$ are strictly positive.

Proof. $\tilde{\gamma}^1 > 0$ follows immediately from Lemma 4.4 by taking $L = \partial_\gamma$ and $u = \chi^1$. To prove $\gamma_0^1 > 0$, let ξ^1 be the bounded solution (periodic in $x^2,\ldots,x^d$) of $\mathcal{L}^{(0)}\xi^1 = -b^1$ in D with $\xi^1 = 0$ on ∂D, and set $h(x) = x^1 + \xi^1(x)$. Then, $\mathcal{L}^{(0)}h = 0$ in D, $h = 0$ on ∂D, and $h > 0$ on $[a,\infty) \times \mathbb{R}^{d-1}$ provided that $a > 2\|\xi^1\|_\infty$. Therefore, $h > 0$ in D and hence $\partial h/\partial x^1 = 1 + \partial\xi^1/\partial x^1 \geq 0$ on ∂D. On the other hand, by Lemma 4.4,

$$\int_{\underline{T}^{d-1}}\left(1 + \frac{\partial\xi^1}{\partial x^1}\right)dm_1 = \int_{\underline{T}^{d-1}} (1 + L\xi^1)\, dm_1 > 0$$

and hence $1 + \partial\xi^1/\partial x^1$ does not vanish identically. Thus

$$\gamma_0^1 = \int_{\underline{T}^{d-1}} \left(1 + \frac{\partial\xi^1}{\partial x^1}\right) dm_0 > 0$$

5. PROOF OF THEOREM 2.1

Let $X_\varepsilon(t)$ be the $(\mathcal{L}_\varepsilon, L_\varepsilon)$-diffusion obtained by solving the SDE

$$dX_\varepsilon(t) = \sigma_\varepsilon\, dB + \frac{1}{\varepsilon}b_\varepsilon\, dt + c_\varepsilon\, dt + \tau_\varepsilon\, dM_\varepsilon + \frac{1}{\varepsilon}\beta_\varepsilon\, d\Phi_\varepsilon + \gamma_\varepsilon\, d\Phi_\varepsilon \qquad (5.1a)$$

$$X_\varepsilon(0) = x \qquad (5.1b)$$

where the suffix ε in the coefficients means that the argument is $X_\varepsilon(t)/\varepsilon$, e.g., $\sigma_\varepsilon = \sigma(X_\varepsilon(t)/\varepsilon)$, etc. Writing down the first component we have

$$X_\varepsilon^1(t) = x^1 + \int_0^t \sigma_\varepsilon^1\, dB + \frac{1}{\varepsilon}\int_0^t b_\varepsilon^1\, ds + \int_0^t c_\varepsilon^1\, ds + \Phi_\varepsilon(t)$$
$$[\sigma^1 = (\sigma^{11},\ldots,\sigma^{1d})]$$

in which $\Phi_\varepsilon(t)$ can be considered to be associated with $X_\varepsilon^1(t)$. Therefore, by Proposition 3.2 we have the following:

LEMMA 5.1 For any constant $t_0 > 0$,

$$E[\Phi_\varepsilon(t)] \leq \frac{\text{const}}{\varepsilon} \qquad 0 \leq t \leq t_0$$

where const may depend on t_0 but not on ε.

Let ξ be the same as in (2.10) and set

$$\hat{X}_\varepsilon(t) = X_\varepsilon(t) + \varepsilon\xi\left(\frac{X_\varepsilon(t)}{\varepsilon}\right) \tag{5.2}$$

Since $L^{(0)}\xi^k = -b^k$ in D and $L^{(0)}\xi^k = -\beta^k$ on ∂D, an application of Itô's formula yields

$$d\hat{X}_\varepsilon(t) = (I + \nabla\xi)_\varepsilon(\sigma_\varepsilon\, dB + c_\varepsilon\, dt + \tau_\varepsilon\, dM_\varepsilon) + \hat{\gamma}_\varepsilon\, d\Phi_\varepsilon \tag{5.3}$$

where $\hat{\gamma} = (I + \nabla\xi)\gamma$. Next, for each k ($1 \leq k \leq d$) let η^k be a solution of $L^{(0)}\eta^k = 0$ in D with the boundary condition $L^{(0)}\eta^k = \hat{\gamma}^k - \gamma_0^k$ on ∂D, where γ_0^k is the kth component of γ_0 of (2.12), and set $\eta = {}^t(\eta^1,\ldots,\eta^d)$. Such a solution can be obtained by

$$\eta^k = H\varphi^k \qquad \varphi^k(x) = \int_0^\infty E[\gamma_0^k - \hat{\gamma}^k(Z(t))]\, dt$$

where Z(t) is the $L^{(0)}$-diffusion on ∂D starting at $x \in \partial D$. Then by Itô's formula

$$d\eta\left(\frac{X_\varepsilon(t)}{\varepsilon}\right) = \frac{1}{\varepsilon}(\nabla\eta)_\varepsilon(\sigma_\varepsilon\, dB + c_\varepsilon\, dt + \tau_\varepsilon\, dM_\varepsilon + \gamma_\varepsilon\, d\Phi_\varepsilon) + \frac{1}{\varepsilon^2}(\hat{\gamma} - \gamma_0)_\varepsilon\, d\Phi_\varepsilon$$

and hence

$$\hat{\gamma}_\varepsilon\, d\Phi_\varepsilon = \varepsilon^2\, d\eta_\varepsilon - \varepsilon(\nabla\eta)_\varepsilon(\sigma_\varepsilon\, dB + c_\varepsilon\, dt + \tau_\varepsilon\, dM_\varepsilon + \gamma_\varepsilon\, d\Phi_\varepsilon) + \gamma_0\, d\Phi_\varepsilon$$

Inserting this into (5.3) we have

$$X_\varepsilon(t) = \hat{U}_\varepsilon(t) + \hat{M}_\varepsilon(t) + \hat{C}_\varepsilon(t) + \tilde{M}_\varepsilon(t) + \hat{\Phi}_\varepsilon(t) \tag{5.4}$$

where

$$\hat{U}_\varepsilon(t) = x + \varepsilon\xi\left(\frac{x}{\varepsilon}\right) - \varepsilon\xi\,\frac{X_\varepsilon(t)}{\varepsilon} + \varepsilon^2\left[\eta\left(\frac{X_\varepsilon(t)}{\varepsilon}\right) - \eta\left(\frac{x}{\varepsilon}\right)\right]$$

$$\hat{M}_\varepsilon(t) = \int_0^t (I + \nabla\xi - \varepsilon\nabla\eta)_\varepsilon\sigma_\varepsilon\, dB$$

$$\hat{C}_\varepsilon(t) = \int_0^t (I + \nabla\xi - \varepsilon\nabla\eta)_\varepsilon c_\varepsilon\, ds$$

$$\tilde{M}_\varepsilon(t) = \int_0^t (I + \nabla\xi - \varepsilon\nabla\eta)_\varepsilon\tau_\varepsilon\, dM_\varepsilon$$

$$\hat{\Phi}_\varepsilon(t) = \int_0^t [\gamma_0 - \varepsilon(\nabla\eta)\gamma]_\varepsilon\, d\Phi_\varepsilon$$

The first component of (5.4) is

$$X_\varepsilon^1(t) = \hat{U}_\varepsilon^1(t) + \hat{M}_\varepsilon^1(t) + \hat{C}_\varepsilon^1(t) + \tilde{M}_\varepsilon^1(t) + \hat{\Phi}_\varepsilon^1(t) \tag{5.5}$$

For a continuous process $X = \{X(t)\}$ we use the notation

$$\begin{aligned} \mathrm{Osc}(X,\delta) &= \mathrm{Osc}(X,\delta,t_0) \\ &= \max\{|X(t) - X(s)| : |t-s| \le \delta,\ 0 \le t,\ s \le t_0\} \end{aligned}$$

$t_0 > 0$ being arbitrarily fixed. We remark that (5.5) can be considered as the Skorohod equation with respect to $X^1(t)$ and $\hat{\Phi}_\varepsilon^1(t)$. In fact, $\hat{\Phi}_\varepsilon^1(t) = \int_0^t [\gamma_0^1 - \varepsilon(\nabla\eta^1)_\varepsilon\gamma_\varepsilon]\, d\Phi_\varepsilon$ is associated with $X_\varepsilon^1(t)$ (for small $\varepsilon > 0$), because $\gamma_0^1 - \varepsilon(\nabla\eta^1)_\varepsilon\gamma_\varepsilon > 0$ for all sufficiently small ε by Lemma 4.5 and because $\Phi_\varepsilon(t)$ is associated with $X_\varepsilon^1(t)$. Thus by Proposition 3.2 we can control $\hat{\Phi}_\varepsilon^1(t)$ [and hence $\Phi_\varepsilon(t)$] by making use of $\hat{U}_\varepsilon^1(t)$, $\hat{M}_\varepsilon^1(t)$, $\hat{C}_\varepsilon^1(t)$, and $\tilde{M}_\varepsilon^1(t)$. The control of $\hat{U}_\varepsilon^1(t)$, $\hat{M}_\varepsilon^1(t)$, and $\hat{C}_\varepsilon^1(t)$ as $\varepsilon \downarrow 0$ is straightforward; that is,

$$\lim_{\delta\downarrow 0}\lim_{\varepsilon\downarrow 0} P\{\mathrm{Osc}(X,\delta) > \lambda\} = 0 \qquad \lambda > 0$$

for X = any one of $\hat{U}_\varepsilon^1$, $\hat{M}_\varepsilon^1$, $\hat{C}_\varepsilon^1$. As for $\tilde{M}_\varepsilon^1(t)$, it is enough to notice that $\tilde{M}^1(t)$ is a martingale with increasing process

$$\varepsilon^2 \int_0^t \sum_j \Big|\sum_i \partial_i\eta^1\tau^{ij}\Big|_\varepsilon^2\, d\Phi_\varepsilon \le \text{const}\ \varepsilon \qquad \text{(by Lemma 5.1)}$$

Consequently, an application of the estimate (3.3) yields

$$\begin{aligned} \mathrm{Osc}(\hat{\Phi}_\varepsilon^1,\delta) \le \mathrm{Osc}(\hat{U}_\varepsilon^1,\delta) + \mathrm{Osc}(\hat{M}_\varepsilon^1,\delta) + \mathrm{Osc}(\hat{C}_\varepsilon^1,\delta) \\ + \mathrm{Osc}(\tilde{M}_\varepsilon^1,\delta) \end{aligned}$$

and hence the following lemma.

LEMMA 5.2 For any $t_0 > 0$:

(i) $E[\Phi_\varepsilon(t_0)]$ is bounded in ε.

(ii) $\lim_{\delta\downarrow 0}\overline{\lim_{\varepsilon\downarrow 0}} P[\mathrm{Osc}(\Phi_\varepsilon,\delta,t_0) > \lambda] = 0$ for any $\lambda > 0$.

REMARK 1. $E[|\Phi_\varepsilon(t_0)|^2]$ is also bounded in ε.

We now set

$$U_\varepsilon(t) = \varepsilon\xi\left(\frac{x}{\varepsilon}\right) - \varepsilon\xi\,\frac{X_\varepsilon(t)}{\varepsilon} + \varepsilon^2\left[\eta\left(\frac{X_\varepsilon(t)}{\varepsilon}\right) - \eta\left(\frac{x}{\varepsilon}\right)\right]$$

$$+ \int_0^t (\nabla\xi_0 - \varepsilon\nabla\eta)_\varepsilon \sigma_\varepsilon \, dB + \int_0^t (\nabla\xi_0 - \varepsilon\nabla\eta)_\varepsilon c_\varepsilon \, ds$$
$$- \int_0^t \varepsilon(\nabla\eta)_\varepsilon \tau_\varepsilon \, dM_\varepsilon - \int_0^t \varepsilon(\nabla\eta)_\varepsilon \gamma_\varepsilon \, d\Phi_\varepsilon$$

$$M_\varepsilon^D(t) = \int_0^t (I + \nabla\chi)_\varepsilon \sigma_\varepsilon \, dB \qquad C_\varepsilon(t) = \int_0^t (I + \nabla\chi)_\varepsilon c_\varepsilon \, ds$$
$$M_\varepsilon^\partial(t) = \int_0^t (I + \nabla\xi)_\varepsilon \tau_\varepsilon \, dM_\varepsilon$$

where $\xi_0 = \xi - \chi$, χ being given by (2.5). Then (5.4) can be written as follows:

$$X_\varepsilon(t) = x + U_\varepsilon(t) + M_\varepsilon^D(t) + C_\varepsilon(t) + M_\varepsilon^\partial(t) + \gamma_0\Phi_\varepsilon(t) \tag{5.6}$$

By virtue of Lemma 5.2 we have the following:

LEMMA 5.3 If X is one of M_ε^D, C_ε, M_ε^∂, Φ_ε, then for any $\lambda > 0$,

$$\lim_{\delta\downarrow 0} \overline{\lim_{\varepsilon\downarrow 0}} P[\mathrm{Osc}\,(X,\delta,t_0) > \lambda] = 0$$

To estimate $|U_\varepsilon(t)|$ we need the following lemma.

LEMMA 5.4 Let u be a smooth and bounded function on $\overline{D}$ periodic in $x^2, \ldots, x^d$ such $u(x^1, \ldots, x^d) \to 0$ as $x^1 \to \infty$ uniformly in $(x^2, \ldots, x^d)$. Then for any $t_0 > 0$

$$\lim_{\varepsilon\downarrow 0} E[\max_{0\le t\le t_0} |\int_0^t u\left(\frac{X_\varepsilon(s)}{\varepsilon}\right) ds|] = 0 \tag{5.7}$$

Proof. For any $\delta > 0$ we can write $u = f + g$ where f vanishes on $\{x : x^1 \ge N\}$ for some $N > 0$ and $\|g\|_\infty < \delta$, both f and g being smooth and periodic in $x^2,\ldots,x^d$. Let X(t) be the $L^{(0)}$-diffusion on $\mathbb{R}^d$ and set $T = \inf\{t > 0 : X^1(t) = 0\}$. Making use of the fact that $f \equiv 0$ on $\{x : x^1 \ge N\}$, it is not hard to prove that

$$v(x) = -E[\int_0^T f(X(t))\,dt] \qquad X(0) = x$$

is bounded on $\overline{D}$; also, from a result on elliptic partial differential equations v is a smooth function on $\overline{D}$ satisfying $L^{(0)}v = f$ in D and $v = 0$ on ∂D. Therefore, an application of Itô's formula yields

$$dv\left(\frac{X_\varepsilon(t)}{\varepsilon}\right) = \varepsilon^{-1}(\nabla v)_\varepsilon(\sigma_\varepsilon\, dB + c_\varepsilon\, dt + \gamma_\varepsilon\, d\Phi_\varepsilon) + \varepsilon^{-2} f_\varepsilon\, dt$$

and hence

$$\int_0^t f_\varepsilon\, ds = \varepsilon^2[v\left(\frac{X_\varepsilon(t)}{\varepsilon}\right) - v(\tfrac{x}{\varepsilon})] + \varepsilon \int_0^t (\nabla v)_\varepsilon(\sigma_\varepsilon\, dB + c_\varepsilon\, ds + \gamma_\varepsilon\, d\Phi_\varepsilon)$$

Since the Brownian integral term is a continuous martingale,

$$E\left[\max_{0\le t\le t_0} \left|\int_0^t (\nabla v)_\varepsilon \sigma_\varepsilon\, dB\right|\right] \le E\left[\max_{0\le t\le t_0} \left|\int_0^t (\nabla v)_\varepsilon \sigma_\varepsilon\, dB\right|^2\right]^{1/2} \le \left[4E\left|\int_0^{t_0} (\nabla v)_\varepsilon \sigma_\varepsilon\, dB\right|^2\right]^{1/2}$$

which is bounded from above by a constant independent of ε. As for the term containing $d\Phi_\varepsilon$ we can apply Lemma 5.2 and thus we have

$$\lim_{\varepsilon\downarrow 0} E\left[\max_{0\le t\le t_0} \left|\int_0^t f\left(\frac{X_\varepsilon(s)}{\varepsilon}\right) ds\right|\right] = 0 \tag{5.8}$$

Since $\|g\|_\infty < \delta$ and $\delta > 0$ was arbitrary, we obtain (5.7).

LEMMA 5.5 For any $t_0 > 0$,

(i) $\lim_{\varepsilon\downarrow 0} E\left[\max_{0\le t\le t_0} |U_\varepsilon(t)|\right] = 0$

(ii) $\lim_{\delta\downarrow 0} \overline{\lim_{\varepsilon\downarrow 0}} P[\mathrm{Osc}\,(X_\varepsilon,\delta,t_0) > \lambda] = 0, \qquad \lambda > 0$

Proof. (ii) is immediate from (i) and Lemma 5.3. (i) follows from Lemma 5.4 once we prove that for each k $(1 \le k \le d)$,

$$\lim_{x^1\to\infty} \nabla\xi_0^k(x^1,x^2,\dots,x^d) = 0 \quad \text{uniformly in } (x^2,\dots,x^d) \tag{5.9}$$

To prove this let $X_0(t)$ be the $L^{(0)}$-diffusion on $\mathbb{R} \times \underline{T}^{d-1}$ starting at $(1,x^2,\dots,x^d)$ and set

$$H(x^2,\dots,x^d;A) = P[X_0(T) \in \{0\} \times A] \qquad A \in B(\underline{T}^{d-1})$$

where $T = \inf\{t > 0 : X_0^1(t) = 0\}$. Then $H(x^2,\ldots,x^d;A)$ is a transition function on $\underline{T}^{d-1}$ satisfying Doeblin's condition and hence

$$\xi^k(n,x^2,\ldots,x^d) = H^n(\chi_0^k - \chi^k) \to \text{const} \qquad n \to \infty \tag{5.10}$$

uniformly (exponentially fast) in $(x^2,\ldots,x^d)$, where H^n is the nth iterate of the operator

$$H : \psi(x^2,\ldots,x^d) \to (H\psi)(x^2,\ldots,x^d) = \int_{\underline{T}^{d-1}} \psi(y) \times H(x^2,\ldots,x^d;dy)$$

The equality in (5.10) follows from the periodicity of the diffusion coefficients and the fact that ξ^k is a solution of

$$L^{(0)}\xi^k = 0 \text{ in } D \tag{5.11}$$

with $\xi^k = \chi_0^k - \chi^k$ on ∂D. Now, (5.10) and (5.11) combined with Schauder's estimate (e.g., see Ref. 16, p. 85) implies (5.9).

LEMMA 5.6 If u is a smooth periodic function in $\mathbb{R}^d$ such that the integral of u over $\underline{T}^d$ with respect to the measure m vanishes, then for any $t_0 > 0$,

$$\lim_{\varepsilon \downarrow 0} E\left[\max_{0 \le t \le t_0} \left| \int_0^t u\left(\frac{X_\varepsilon(s)}{\varepsilon}\right) ds \right| \right] = 0$$

Proof. Let V_0 be a periodic solution of $L^{(0)}v_0 = u$ in $\mathbb{R}^d$ and set $v = v_0 - H[v_0]$ on $\overline{D}$, where $[v_0]$ is the restriction of v_0 on ∂D. Then an application of Itô's formula yields

$$dv\left(\frac{X_\varepsilon(t)}{\varepsilon}\right) = \varepsilon^{-1}(\nabla v)_\varepsilon(\sigma_\varepsilon\, dB + c_\varepsilon\, dt + \tau_\varepsilon\, dM_\varepsilon + \gamma_\varepsilon\, d\Phi_\varepsilon) + \varepsilon^{-2}[(L^{(0)}v)_\varepsilon\, dt + (L^{(0)}v)_\varepsilon\, d\Phi_\varepsilon]$$

Since the second term of the right-hand side equals $\varepsilon^{-2}u_\varepsilon\, dt$, the lemma follows as in the case of (5.8).

LEMMA 5.7 If φ is a smooth periodic function on ∂D such that the integral of φ over $\underline{T}^{d-1}$ with respect to the measure m_0 vanishes, then for any $t_0 > 0$,

$$\lim_{\varepsilon \downarrow 0} E\left[\max_{0 \le t \le t_0} \left| \int_0^t \varphi\left(\frac{X_\varepsilon(s)}{\varepsilon}\right) d\Phi_\varepsilon \right|\right] = 0$$

Proof. Let u be a periodic solution of $L^{(0)}u = \varphi$ on ∂D and set $v = Hu$ on $\bar{D}$. Then

$$dv\left(\frac{X_\varepsilon(t)}{\varepsilon}\right) = \varepsilon^{-1}(\nabla v)_\varepsilon(\sigma_\varepsilon\, dB + c_\varepsilon\, dt + \tau_\varepsilon\, dM_\varepsilon + \gamma_\varepsilon\, d\Phi_\varepsilon)$$
$$+ \varepsilon^{-2}\varphi\left(\frac{X_\varepsilon(t)}{\varepsilon}\right) d\Phi_\varepsilon$$

and the lemma follows as in (5.8).

The proof of Theorem 2.1 will now be given as follows. Let W be the space of continuous paths w from $[0,\infty)$ to the product space

$$\bar{D} \times \mathbb{R}^d \times \mathbb{R}^d \times \mathbb{R}^d \times \mathbb{R}^d \times [0,\infty) \tag{5.12}$$

Denote by $\mathcal{B}_t$, $t \ge 0$, the corrdinate σ-fields $\sigma\{w(s): 0 \le s \le t\}$ on W and set $\mathcal{B} = \vee_t \mathcal{B}_t$. Given the $(\mathcal{L}_\varepsilon, L_\varepsilon)$-diffusion $X_\varepsilon(t)$ satisfying the SDE (5.1), we define $\tilde{P}_\varepsilon$ as the probability measure in $(W,\mathcal{B})$ induced by the process

$$(X_\varepsilon(t),\ U_\varepsilon(t),\ M_\varepsilon^D(t),\ C_\varepsilon(t),\ M_\varepsilon^\partial(t),\ \Phi_\varepsilon(t))$$

with state space (5.12). We write $w = (w_0, w_1, \ldots, w_5)$ for a generic element of W and also w_k^i for the ith component of w_k $(0 \le k \le 4)$. Then the following equations (5.13) and (5.14) hold $\tilde{P}_\varepsilon$ a.s.

$$w_0(t) = x + w_1(t) + \cdots + w_4(t) + \gamma_0 w_5(t) \qquad t > 0 \quad \text{[by (4.6)]} \tag{5.13}$$

$$w_1(0) = w_2(0) = w_3(0) = w_4(0) = 0 \qquad w_4^1(t) = 0 \ (t \ge 0)$$
$$w_5(t) \text{ is increasing with } w_5(0) = 0 \tag{5.14}$$

Moreover, each of the following is a $\tilde{P}_\varepsilon$-martingale with respect to $\mathcal{B}_t$.

$$w_2^i(t) \qquad 1 \le i \le d$$

$$w_4^i(t) \qquad 2 \le i \le d$$

$$w_2^i(t)w_2^j(t) - 2\int_0^t a_0^{ij}\left(\frac{w_0(s)}{\varepsilon}\right) ds \qquad 1 \le i,\ j \le d \quad (5.15)$$

$$w_2^i(t)w_4^j(t) \qquad 1 \le i \le d,\ 2 \le j \le d$$

$$w_4^i(t)w_4^j(t) - 2\int_0^t \alpha_0^{ij}\left(\frac{w_0(s)}{\varepsilon}\right) dw_5(s) \qquad 2 \le i,\ j \le d$$

Here $a_0^{ij}(x)$ and $\alpha_0^{ij}(x)$ are the (i,j)-entries of the matrices defined respectively by

$$a_0(x) = (I + \nabla\chi)a(x)^t(I + \nabla\chi) \qquad x \in \mathbb{R}^d \qquad (5.16)$$

$$\alpha_0(x) = (I + \nabla\chi_0)\alpha(x)^t(I + \nabla\chi_0) \qquad x \in \partial D \qquad (5.17)$$

Denoting by $\tilde{E}_\varepsilon$ the expectation with respect to $\tilde{P}_\varepsilon$, we have

$$\lim_{\varepsilon\downarrow 0} \tilde{E}_\varepsilon\left[\max_{0\le t\le t_0} |w_1(t)|\right] = 0 \qquad \text{[by Lemma 5.5(i)]} \quad (5.18)$$

$$\lim_{\varepsilon\downarrow 0} \tilde{E}_\varepsilon\left[\max_{0\le t\le t_0} \left|\int_0^t a_0^{ij}\left(\frac{w_0(s)}{\varepsilon}\right) ds - a_0^{ij}t\right|\right] = 0 \qquad (5.19)$$

$$\lim_{\varepsilon\downarrow 0} \tilde{E}_\varepsilon\left[\max_{0\le t\le t_0} |w_3(t) - c_0t|\right] = 0 \qquad \text{(by Lemma 5.6)}$$

$$\lim_{\varepsilon\downarrow 0} \tilde{E}_\varepsilon\left[\max_{0\le t\le t_0} \left|\int_0^t \alpha_0^{ij}\left(\frac{w_0(s)}{\varepsilon}\right) dw_5(s) - \alpha_0^{ij}w_5(t)\right|\right] = 0 \qquad \text{(by Lemma 5.7)} \quad (5.20)$$

From Lemmas 5.3 and 5.5 it follows that for any sequence $\varepsilon_1' > \varepsilon_2' > \cdots$ tending to 0 there exists a subsequence $\varepsilon_1 > \varepsilon_2 > \cdots$ of $\{\varepsilon_k'\}$ such that $\tilde{P}_\varepsilon$ converges (weakly) to some probability measure on $(W,\mathcal{B})$ as $\varepsilon \downarrow 0$ via $\{\varepsilon_k\}$ (e.g., see Ref. 17). In what follows $\tilde{P}$ denotes any limiting probability measure of $\tilde{P}_\varepsilon$ as $\varepsilon \downarrow 0$ via some sequence $\varepsilon_0 > \varepsilon_1 > \cdots$. Then

$$w_0(t) = x + w_2(t) + w_3(t) + w_4(t) + \gamma_0 w_5(t) \qquad t \ge 0 \quad (5.21)$$

holds P a.s. Moreover, each of the following is a

$\tilde{P}$-martingale with respect to $\mathcal{B}_t$.

$$w_2^i(t) \qquad 1 \leq i \leq d \tag{5.22a}$$

$$w_4^i(t) \qquad 2 \leq i \leq d \tag{5.22b}$$

$$w_2^i(t)w_2^j(t) - 2a_0^{ij}t \qquad 1 \leq i,j \leq d \tag{5.22c}$$

$$w_2^i(t)w_4^j(t) \qquad 1 \leq i \leq d, \quad 2 \leq j \leq d \tag{5.22d}$$

$$w_4^i(t)w_4^j(t) - 2\alpha_0^{ij}w_5(t) \qquad 2 \leq i,j \leq d \tag{5.22e}$$

For example, the martingale property of (5.22e) can be proved as follows. For $0 \leq s < t$ and for any bounded continuous function $F(w)$ on W which is $\mathcal{B}_s$-measurable, we have

$$\tilde{E}_\varepsilon\left\{F(w)\left[w_4^i(t)w_4^j(t) - 2\int_0^t \alpha_0^{ij}\left(\frac{w_0(u)}{\varepsilon}\right) dw_5(u)\right]\right\}$$

$$= \tilde{E}_\varepsilon\left\{F(w)\left[w_4^i(s)w_4^j(s) - 2\int_0^s \alpha_0^{ij}\left(\frac{w_0(s)}{\varepsilon}\right) dw_5(u)\right]\right\}$$

Letting $\varepsilon \downarrow 0$ via $\varepsilon_1 > \varepsilon_2 > \cdots$ in the above and making use of (5.20) (note also Remark 1), we have

$$\tilde{E}\{F(w)[w_4^i(t)w_4^j(t) - 2\alpha_0^{ij}w_5(t)]\}$$

$$= \tilde{E}\{F(w)[w_4^i(s)w_4^j(s) - 2\alpha_0^{ij}w_5(s)]\}$$

which proves that (5.22e) is a $\tilde{P}$-martingale. Also we note that

$$w_3(t) = c_0 t \tag{5.23a}$$

$$w_5(t) \text{ is associated with } w_0(t) \tag{5.23b}$$

for almost all w with respect to $\tilde{P}$. In fact, (5.23a) follows from (5.19), and (5.23b) follows from the fact that the set W_0 of w for which (5.23b) holds is a closed subset of W with $\tilde{P}_\varepsilon(W_0) = 1$.

Finally from what we have proved, (5.21) can be regarded as an SDE for $w_0(t)$. By the result of S. Watanabe [11] this SDE determines uniquely the probability law of the process $\{w_0(t),\tilde{P}\}$, which is nothing but the (L_0,L_0)-diffusion. Thus we have proved that the $(L_\varepsilon, L_\varepsilon)$-diffusion converges in law to the $(L_0, L_0$-diffusion (in the sense of Remark 2 of Sec. 2) as $\varepsilon \downarrow 0$.

6. PROOF OF THEOREM 2.2

Since the proof of Theorem 2.2 is similar to that of Theorem 2.1, we only give the outline. In case 2 the $(L_\varepsilon, L_\varepsilon)$-diffusion $X_\varepsilon(t)$ can be obtained by solving the SDE

$$dX_\varepsilon(t) = \sigma_\varepsilon\, dB + \frac{1}{\varepsilon} b_\varepsilon\, dt + c_\varepsilon\, dt + \gamma_\varepsilon\, d\Phi_\varepsilon \qquad X_\varepsilon(0) = x \tag{6.1}$$

Taking the function χ of (2.5) and setting $\tilde{X}_\varepsilon(t) = X_\varepsilon(t) + \varepsilon\chi(x_\varepsilon(t)/\varepsilon)$, we have

$$d\tilde{X}_\varepsilon(t) = (I + \nabla\chi)_\varepsilon(\sigma_\varepsilon\, db + c_\varepsilon\, dt) + (I + \nabla\chi)_\varepsilon \gamma_\varepsilon\, d\Phi_\varepsilon$$

The first component process can be written as

$$d\tilde{X}^1_\varepsilon(t) = \tilde{\sigma}^1_\varepsilon\, dB + \tilde{c}_\varepsilon\, dt + (1 + \partial_\gamma \chi^1)_\varepsilon\, d\Phi_\varepsilon \tag{6.2}$$

where $\tilde{\sigma}$ is the first row vector of $(I + \nabla\chi)\sigma$ and $\tilde{c}$ is the first component of $(I + \nabla\chi)c$. Since the integral of $1 + \partial_\gamma\chi^1 - \tilde{\gamma}^1$ over $\underline{T}^{d-1}$ with respect to $\tilde{m}$ vanishes, there exists a (smooth and bounded) solution η of

$$L^{(0)}\eta = 0 \qquad \text{in } D$$

$$\partial_\gamma\eta = 1 + \partial_\gamma\chi^1 - \tilde{\gamma}^1 \qquad \text{on } \partial D$$

Taking such a solution η we have from Itô's formula

$$d\left[\varepsilon\eta\left(\frac{X_\varepsilon(t)}{\varepsilon}\right)\right] = (\nabla\eta)_\varepsilon(\sigma_\varepsilon\, dB + c_\varepsilon\, dt) + (1 + \partial_\gamma\chi^1 - \tilde{\gamma}^1)_\varepsilon\, d\Phi_\varepsilon$$

and hence

$$\int_0^t (1 + \partial_\gamma\chi^1 - \tilde{\gamma}^1)_\varepsilon\, d\Phi_\varepsilon$$
$$= \varepsilon\left[\eta\left(\frac{X_\varepsilon(t)}{\varepsilon}\right) - \eta\left(\frac{x}{\varepsilon}\right)\right] - \int_0^t (\nabla\eta)_\varepsilon(\sigma_\varepsilon\, dB + c_\varepsilon\, ds)$$

Inserting this into (6.2) we have

$$X^1_\varepsilon(t) = \tilde{U}_\varepsilon(t) + \tilde{\gamma}^1\Phi_\varepsilon(t) \tag{6.3}$$

where

$$\tilde{U}_\varepsilon(t) = x^1 - \varepsilon\left[\chi^1\left(\frac{X_\varepsilon(t)}{\varepsilon}\right) - \chi^1\left(\frac{x}{\varepsilon}\right)\right] + \varepsilon\left[\eta\left(\frac{X_\varepsilon(t)}{\varepsilon}\right) - \eta\left(\frac{x}{\varepsilon}\right)\right]$$

$$+ \int_0^t \tilde{\sigma}_\varepsilon \, dB + \int_0^t \tilde{c}_\varepsilon \, ds - \int_0^t (\nabla\eta)_\varepsilon(\sigma_\varepsilon \, dB + c_\varepsilon \, ds)$$

Since $\tilde{\gamma}^1 > 0$ by Lemma 4.5, (6.3) can be regarded as an equation of Skorohod type and hence applying (3.3) we obtain the following lemma.

LEMMA 6.1 For any fixed $t_0 > 0$,

(i) $\lim_{\delta\downarrow 0} \overline{\lim_{\varepsilon\downarrow 0}} P[\mathrm{Osc}(\Phi_\varepsilon,\delta,t_0) > \lambda] = 0$ for any $\lambda > 0$.

(ii) $E[\Phi_\varepsilon(t_0)^p]$ is bounded in ε for each $p \geq 1$.

LEMMA 6.2 If u is a smooth periodic function in $\mathbb{R}^d$ such that the integral of u over $\underline{T}^d$ with respect to m vanishes, then

$$\lim_{\varepsilon\downarrow 0} E\left[\max_{0\leq t\leq t_0} \left|\int_0^t u\left(\frac{X_\varepsilon(s)}{\varepsilon}\right) ds\right|\right] = 0$$

The proof is similar to that of Lemma 5.6.

LEMMA 6.3 Let φ be a smooth periodic function on ∂D such that the integral of φ over $\underline{T}^{d-1}$ with respect to $\tilde{m}$ vanishes, let u be a (smooth and bounded) solution of $L^{(0)}u = 0$ in D and $\partial_\gamma u = \varphi$ on ∂D, and set

$$A_\varepsilon(t) = \int_0^t (\nabla u)_\varepsilon c_\varepsilon \, ds + \int_0^t \varphi\left(\frac{X_\varepsilon(s)}{\varepsilon}\right) d\Phi_\varepsilon \tag{6.4}$$

Then for any $t_0 > 0$,

$$\lim_{\varepsilon\downarrow 0} E\left[\max_{0\leq t\leq t_0} |A_\varepsilon(t)|\right] = 0 \tag{6.5}$$

Proof. By Itô's formula

$$\varepsilon u\left(\frac{X_\varepsilon(t)}{\varepsilon}\right) - \varepsilon u\left(\frac{x}{\varepsilon}\right) = M_\varepsilon^0(t) + A_\varepsilon(t)$$

$$M_\varepsilon^0(t) = \int_0^t (\nabla u)_\varepsilon \sigma_\varepsilon \, dB$$

and hence

$$E[|M_\varepsilon^0(t) + A_\varepsilon(t)|^2] \leq 4\varepsilon^2 \|u\|_\infty$$

Denote by W the space of continuous paths $w = (w', w'')$ from $[0,\infty)$ to $\mathbb{R}^2$ and write $w(t) = (w'(t), w''(t))$ for the value of $w = (w', w'')$ at time t. Let Q_ε be the probability measure on W induced by the $\mathbb{R}^2$-valued process $(M_\varepsilon^0(t), A_\varepsilon(t))$. Then, by virtue of Lemma 6.1, for any sequence $\varepsilon_1 > \varepsilon_2 > \cdots$ tending to 0 we can choose a subsequence $\{\varepsilon_k'\}$ of $\{\varepsilon_k\}$ such that Q_ε converges (weakly) to some limit probability measure Q_0 as $\varepsilon \downarrow 0$ via $\{\varepsilon_k'\}$. It then follows that $w'(t)$ is a Q_0-martingale, $w''(t)$ is a process with bounded variation, and $w'(t) + w''(t) = 0$ (Q_0 a.s.). Therefore $w'(t) = w''(t) = 0$ (Q_0 a.s.), and hence

$$\lim_{\varepsilon\downarrow 0} E[M_\varepsilon^0(t)^2] = \lim_{\varepsilon\downarrow 0} E[A_\varepsilon(t)^2] = 0$$

from which (6.5) follows.

LEMMA 6.4 If φ is the same as in Lemma 6.3, then

$$\lim_{\varepsilon\downarrow 0} E\left[\max_{0\leq t\leq t_0} \left|\int_0^t \varphi\left(\frac{X_\varepsilon(s)}{\varepsilon}\right) d\Phi_\varepsilon\right|\right] = 0$$

Proof. By a similar method as employed in proving Lemma 5.5(i) [especially (5.9)], we can prove that $\lim_{\varepsilon\downarrow 0} E \max_{0\leq t\leq t_0} |A_\varepsilon^0(t)|] = 0$, where $A_\varepsilon^0(t)$ denotes the first term of the right-hand side of (6.4). Therefore, Lemma 6.4 follows from Lemma 6.3.

Finally, we write

$$X_\varepsilon(t) = x + U_\varepsilon(t) + M_\varepsilon(t) + C_\varepsilon(t) + \tilde{\gamma}\Phi_\varepsilon(t)$$

where

$$U_\varepsilon(t) = \varepsilon\left[\chi\left(\frac{x}{\varepsilon}\right) - \chi\left(\frac{X_\varepsilon(t)}{\varepsilon}\right)\right] + \int_0^t \{(I + \nabla\chi)\gamma - \tilde{\gamma}\}_\varepsilon \, d\Phi_\varepsilon$$

$$M_\varepsilon(t) = \int_0^t (I + \nabla\chi)_\varepsilon \sigma_\varepsilon \, dB \qquad C_\varepsilon(t) = \int_0^t (I + \nabla\chi)_\varepsilon c_\varepsilon \, ds$$

Then, recalling the notations of (2.6), (2.7), (2.15), and (5.16) we have

$$E\left[\max_{0\le t\le t_0} |U_\varepsilon(t)|\right] \to 0 \qquad (6.6)$$

(by Lemma 6.4)

$$E\left[\max_{0\le t\le t_0} \left|\int_0^t a_0^{ij}\left(\frac{X_\varepsilon(s)}{\varepsilon}\right) ds - a_0^{ij} t\right|\right] \to 0 \qquad (6.7)$$

(by Lemma 6.2)

$$E\left[\max_{0\le t\le t_0} |C_\varepsilon(t) - c_0 t|\right] \to 0 \qquad (6.8)$$

(by Lemma 6.2)

as $\varepsilon \downarrow 0$. Therefore, the proof of Theorem 2.2 can be completed as in the case of Theorem 2.1.

REFERENCES

1. E. di Giorgi and S. Spagnolo, Sulla convergenza degli integrali dell'energia per operatori ellitici del secoundo ordine, Boll. UMI 8:391-411 (1973).
2. E. Sanchez Palencia, Comportement local et macroscopique d'un type de milieux physiques hétérogènes, Int. J. Eng. Sci. 12:331-351 (1974).
3. I. Babuska, Solution of interface problems by homogenization, I, II, III, SIAM J. Math. Anal. 7:603-634, 635-645 (1976); 8:923-937 (1977).
4. A. Bensoussan, J. L. Lions, and G. C. Papanicolaou, Asymptotic Analysis for Periodic Structure, North-Holland, Amsterdam, 1978.
5. A. Bensoussan, J. L. Lions, and G. C. Papanicolaou, Boundary layers and homogenization of transport processes, Publ. RIMS, Kyoto Univ. 15:53-157 (1979).
6. G. C. Papanicolaou, D. W. Stroock, and S. R. S. Varadhan, Martingale approach to some limit theorems, in Statistical Mechanics and Dynamical Systems, Univ. Conf. Turbulence, (D. Ruelle, ed.), Duke Univ. Math. Ser., Vol. 3

Durham, N. C., 1977.

7. M. Fukushima, A generalized stochastic calculus in homogenization, in Quantum Fields-Algebra, Processes (L. Streit, ed.), Springer-Verlag, Vienna, 1980.

8. S. M. Kozlov, Averaging differential operators with almost periodic rapidly oscillating coefficients, Sov. Math. Dokl. 18(5):1323-1326 (1977).

9. S. M. Kozlov, Averaging random structures, Sov. Math. Dokl. 19(4):950-954 (1978).

10. G. C. Papanicolaou and S. R. S. Varadhan, Boundary value problems with rapidly oscillating random coefficients, Colloq. Math. Soc. J. Bolyai 27: Random Fields, Esztergom, Hungary, 1979.

11. S. Watanabe, On stochastic differential equations for multidimensional diffusion processes with boundary conditions, I, II, J. Math. Kyoto Univ. 11:169-180, 545-551 (1971).

12. N. Ikeda and S. Watanabe, Stochastic Differential Equations and Diffusion Processes, North-Holland/Kodansha, Amsterdam, 1981.

13. A. V. Skorohod, Stochastic equations for diffusion process in a bounded region 1, 2, Theory Prob. Appl. 6: 264-274 (1961): 7:3-23 (1962).

14. H. P. McKean, Stochastic Integrals, Academic Press, New York, 1969.

15. H. Tanaka, Stochastic differential equations with reflecting boundary condition in convex regions, Hir. Math. J. 9:163-177 (1979).

16. D. Gilbarg and N. S. Trudinger, Elliptic Partial Differential Equations of Second Order, Springer-Verlag, Berlin, 1977.

17. P. Billingsley, Convergence of Probability Measures Wiley, New York, 1968.

16

Diffusion Approximation of Some Stochastic Difference Equations

HISAO WATANABE/Faculty of Engineering, Kyushu University, Fukuoka, Japan

1. INTRODUCTION

In this chapter we are concerned with the diffusion approximations for a stochastic difference equation, generally expressed in the following manner:

$$x^{\varepsilon}_{n+1} - x^{\varepsilon}_{n} = \varepsilon F(n, x^{\varepsilon}_{n}, \omega) + \varepsilon^2 G(n, x^{\varepsilon}_{n}, \omega) \qquad n = 1, 2, \ldots \tag{1.1}$$

where $\{F(n,x,\omega)\}$ and $\{G(n,x,\omega)\}$ are certain random fields on a probability space $(\Omega, \mathcal{B}, P)$. Such models arise in population genetics, (e.g., Iizuka and Matsuda [3]) and in other fields (e.g., Kushner [6]). The motivation of this chapter is in Ref. 3. The fields naturally have the discrete structure.

We show that (1.1) can be considered as the discrete analog of Khasminskii [5], Borodin [2], Papanicolaou and Kohler [8], Blankenship and Papanicolaou [1], and Kesten and Papanicolaou [4]. Although our discussion is expository and not essentially new, it is important for applications to describe the discrete version. The second term of the right-hand side of (1.1) is not difficult to handle. So we suppose that $G \equiv 0$.

We treat the case in which the driving force comes from a Markov chain. In the continuous parameter case, it corresponds to Blankenship and Papanicolaou [1], but we consider slightly more general situations. The results of Khasminskii [5], Borodin [2], and Kesten and Papanicolaou [4] can also be transferred to the discrete case. We are interested in the martingale method, as found in Papanicolaou, Stroock, and Varadhan [9]. So we will consider the discrete version of Kesten and Papanicolaou [4]. Such results are discussed separately [11].

2. ASSUMPTIONS AND RESULT

We assume that the random fields are derived from a certain Markov chain $\{y_j\}$, namely, $F(n,x,\omega) = F(x,y_n(\omega))$. $\{y_j\}$ is the Markov chain taking values in a compact metric space S and defined on a probability measure space $(\Omega,\mathcal{B},P)$. We denote by $\Pi(x,dy)$ the transition probability of $\{y_j\}$. We put for $f \in B(S)$, $\Pi f(x) = \int_S f(y)\Pi(x,dy)$, where B(S) is the set of bounded measurable functions on S. Suppose that for any $f \in B(S)$, $\Pi f(x) \in C(S)$, where C(S) is the set of continuous functions of C(S). Then, by means of Grothendieck's theorem (see Meyer [7]) we can see that Πf is a compact operator on C(S). We assume that Π is conservative and that $\{\Pi^n(x,dy);\ x \in S\}$ are mutually absolutely continuous.

Then there exists a unique probability measure $\bar{\Pi}(\cdot)$ on S and $0 < \rho < 1$ such that

$$|\Pi^n f(x) - \bar{\Pi} f| < K\rho^n \|f\|_\infty \qquad x \in S \tag{2.1}$$

for any $f \in B(S)$ and $n \geq 0$, where we put $\bar{\Pi} f = \int_S f(y)\bar{\Pi}(dy)$, and K is a constant independent of f(x), n, and $x \in S$ (see theorem 1 in Watanabe [10]).

Next, we assume that for each component of $F(x,y) = (F_1(x,y), \ldots, F_d(x,y))$,

$$\int_S F_k(x,y)\Pi(dy) = 0 \qquad x \in R^d,\ k = 1, 2, \ldots, d \quad (2.2)$$

and

$$\frac{\partial^\beta}{\partial x_1^{\beta_1} \cdots \partial x_d^{\beta_d}} F(x,y) \in C(R^d \times S) \qquad |\beta| \le 2$$

$$|\beta| = \beta_1 + \cdots + \beta_d,\ \beta = (\beta_1, \ldots, \beta_d)$$

By (2.1) we can see that

$$\sum_{n=0}^{\infty} [\Pi^n(x,dy) - \Pi(dy)] \equiv \psi(x,dy)$$

is absolutely and boundedly convergent. If the integral

$$\psi_k^{(1)}(x,y) = \int_S F_k(x,z)\psi(y,dz) \quad (2.3)$$

is convergent, then (2.3) gives the unique solution of the following equations:

$$\int_S (\Pi - I)(y,dz)\psi_k^{(1)}(x,z) = -F_k(x,y) \quad (2.4)$$

For $f \in C^1(R^d)$, set

$$\psi_f^{(1)}(x,y) = \sum_{k=1}^{d} \frac{\partial f}{\partial x_k}(x)\psi_k^{(1)}(x,y)$$

Let $\varphi_M(x)$ be a $C^\infty(R^d)$ function such that $0 \le \varphi_M \le 1$:

$$\varphi_M(x) = \begin{cases} 1 & \text{if } |x| \le \frac{M}{2} \\ 0 & \text{if } |x| \ge M \end{cases}$$

and such that gradient of $\varphi_M(x)$ is bounded uniformly in x and $M \ge 1$. We define

$$F^M(x,y) = \varphi_M(x)F(x,y)$$

Define $f \in C^\infty(R^d)$ if the following integral converges (we assume it):

$$A_f^M(x,y) = \frac{1}{2} \sum_{k,\ell=1}^{d} F_k^M(x,y)F_\ell^M(x,y) \frac{\partial^2 f}{\partial x_k\, \partial x_\ell}(x)$$

$$+ \sum_{k=1}^{d} F_k^M(x,y) \int_S \frac{\partial \psi_f^{(1),M}(x,z)}{\partial x_k} \Pi(y,dz)$$

where $\varphi_f^{(1),M}(x,y) = \varphi_M(x)\psi_f^{(1)}(x,y)$. Also define

$$\bar{L}^M f(x) = \int_S \Pi(dy) A_f^M(x,y)$$

Furthermore, we define $\psi_f^{(2),M}$ as the solution of the equation

$$(\Pi - I)\psi_f^{(2),M}(x,y) = -A_f^M(x,y) + \bar{L}^M f(x) \tag{2.5}$$

Because $\int_S \Pi(dy)[-A_f^M(x,y) + \bar{L}^M f(x)] = 0$ such $\varphi_f^{(2),M}$ exists. It holds that

$$\lim_{M\to\infty} \bar{L}^M f(x) = \bar{L} f(x) = \int_S \Pi(dy) A_f(x,y)$$

where

$$A_f(x,y) = \frac{1}{2} \sum_{k=1}^{d} \sum_{\ell=1}^{d} F_k(x,y) F_\ell(x,y) \frac{\partial^2 f}{\partial x_k \, \partial x_\ell}(x)$$

$$+ \sum_{\ell=1}^{d} F_\ell(x,y) \int_S \frac{\partial \psi_f^{(1)}}{\partial x_\ell}(x,z)\Pi(y,dz)$$

If we extend the Markov chain $\{y_j\}$ by considering the reversed chain, we have

$$\int_S F_k(x,y)F_\ell(x,y)\Pi(dy) + 2\int_S F_k(x,y)[\int_{S\times S} \int F_\ell(x,u)$$

$$\times \psi(z,du)\Pi(y,dz)] \Pi(dy)$$

$$= \sum_{n=-\infty}^{\infty} E(F_k(x,y_0)F_\ell(x,y_n))$$

$$= \lim_{N\to\infty} \frac{1}{N} \sum_{n,m=0}^{N} E(F_k(x,y_n)F_\ell(x,y_m)) \equiv a_{k\ell}(x)$$

which shows the nonnegative definiteness of $\{a_{k\ell}(x)\}$. Here we assume that $\{y_0\}$ has the stationary invariant measure $\Pi(dy)$.

Let

$$b_\ell(x) = \sum_{k=1}^{d} \int_S \Pi(dy)F_k(x,y) \int_{S\times S} \int \frac{\partial F_\ell}{\partial x_k}(x,u)\psi(z,du)\Pi(y,dz)$$
$$= \sum_{k=1}^{\infty} \sum_{n=1}^{\infty} E\{F_k(x,y_0) \frac{\partial F_\ell}{\partial x_k}(x,y_n)\}$$

Then, $\bar{L}f$ can be written in the following manner:

$$\bar{L}f(x) = \frac{1}{2} \sum_{k=1}^{d} \sum_{\ell=1}^{d} a_{k\ell}(x) \frac{\partial^2 f}{\partial x_k \, \partial x_\ell} + \sum_{\ell=1}^{d} b_\ell(x) \frac{\partial f}{\partial x_\ell}$$

Assume that the probability measure R on $(C[0,\infty),F)$ such that $f(x_t) - f(x_0) = \int_0^t f(x(s))\, ds$ is a martingale with respect to $(C[0,\infty),F,R)$ is unique, where F is the topological σ-field. Also, we assume that the families R^M of measures which are the solutions of martingale problems associated with the generator L^M are tight.

Define $x^\varepsilon(t)$, for $\{x^\varepsilon_{j}\}$ defined in (1.1), as follows:

$$x^\varepsilon(t) = x_j^\varepsilon + \frac{t - j\varepsilon^2}{\varepsilon^2}(x_{j+1}^\varepsilon - x_j^\varepsilon) \qquad (2.6)$$
$$\text{if } j\varepsilon^2 \leq t < (j+1)\varepsilon^2$$

Then the stochastic process $\{x^\varepsilon(t)\}$ induces a probability measure Q^ε on $C[0,\infty)$.

THEOREM 2.1 The probability measure Q^ε converges weakly as $\varepsilon \to 0$ to the measure Q which is generated by diffusion Markov process with the generator $\bar{L}$.

REMARKS

1. If $G \not\equiv 0$, we have only to add an extra term to $A_f(x,y)$, namely,

$$\sum_{k=1}^{d} G_k(x,y) \frac{\partial f}{\partial x_k}$$

2. As will be seen from the proof of Theorem 2.1, if we put $\tau^{\varepsilon,U}(\omega) = \inf \{x_t^\varepsilon(\omega) \notin U\}$ when $x_0^\varepsilon(\omega) = x \in U$ for an

open neighborhood of x, $x_t^{\varepsilon,U}(\omega) = x^{\varepsilon}_{t\wedge\tau^{\varepsilon,U}(\omega)}(\omega)$, and similarly we define τ^U, x_t^U for x_t, we have that the probability measure $Q^{\varepsilon,U}$ which is generated by the stochastic process $\{x_t^{\varepsilon,U}(\omega)\}$ converges weakly to the probability measure Q^U which is generated by the stochastic process $\{x_t^U(\omega)\}$. Therefore, even if x_t^{ε} have boundary conditions, the minimal process x_t^{ε} up to attaining the boundary converges weakly to the minimal diffusion processes x_t.

3. PROOF OF THEOREM 2.1

Define $X_n^{\varepsilon,M}(\omega)$, successively,

$$x_{n+1}^{\varepsilon,M}(\omega) - x_n^{\varepsilon,M}(\omega) = \varepsilon F^M(x_n^{\varepsilon,M}(\omega), y_n(\omega))$$

We put

$$f^{\varepsilon,M}(x,y) = f(x) + \varepsilon\psi_f^{(1),M}(x,y) + \varepsilon^2\psi_f^{(2),M}(x,y) \quad \text{where } f \in C^\infty(R^d) \tag{3.1}$$

We consider

$$= f^{\varepsilon,M}(x^{\varepsilon,M}_{[t/\varepsilon^2]}, y_{[t/\varepsilon^2]}) - f^{\varepsilon,M}(x^{\varepsilon,M}_{[s/\varepsilon^2]}, y_{[s/\varepsilon^2]})$$
$$-\sum_{j=[s/\varepsilon^2]}^{[t/\varepsilon^2]-1} [E(f^{\varepsilon,M}(x^{\varepsilon,M}_{j+1}, y_{j+1})) \mid (x_j, y_j)] - f^{\varepsilon,M}(x_j^{\varepsilon,M}, y_j^2) \tag{3.2}$$

By Taylor's expansion, we have

$$\begin{aligned} f^{\varepsilon,M}(x^{\varepsilon,M}_{j+1}, y_{j+1}) &= f^{\varepsilon,M}(x^{\varepsilon,M}_{j+1}) + \varepsilon\psi_f^{(1),M}(x^{\varepsilon,M}_{j+1}, y_{j+1}) \\ &\quad + \varepsilon^2\psi_f^{(2),M}(x^{\varepsilon,M}_{j+1}, y_{j+1}) \qquad (3.3) \\ &= f(x_j) + \sum_{k=1}^{d} \varepsilon F_k^M(x_j^{\varepsilon,M}, y_j)\frac{\partial f}{\partial x_k}(x_j^{\varepsilon,M}) \\ &\quad + \frac{1}{2}\sum_{k=1}^{d}\sum_{\ell=1}^{d} \varepsilon^2 F_k^M(x_j^{\varepsilon,M}, y_j) F_\ell^M(x_j^{\varepsilon,M}, y_j) \\ &\quad \times \frac{\partial^2 f}{\partial x_k\, \partial x_\ell}(x_j^{\varepsilon,M}) \end{aligned}$$

$$+ \frac{1}{2} \sum_{k=1}^{d} \sum_{\ell=1}^{d} \sum_{m=1}^{d} \varepsilon^3 F_k^M(x_j^{\varepsilon,M}, y_j) F_\ell^M(x_j^{\varepsilon,M}, y_j) F_m^M(x_j^{\varepsilon,M}, y_j)$$

$$\times \int_0^1 (1 - u)^2 \frac{\partial^3 f}{\partial x_k\, \partial x_\ell\, \partial x_m}[x_j^{\varepsilon,M} + u(x_{j+1}^{\varepsilon,M} - x_j^{\varepsilon,M})]\, du$$

$$+ \varepsilon \psi_f^{(1),M}(x_j^{\varepsilon,M}, y_{j+1}) + \varepsilon^2 \sum_{k=1}^{d} F_k^M(x_j^{\varepsilon,M}, y_j)$$

$$\times \frac{\partial \psi_f^{\varepsilon,M}}{\partial x_k}(x_j^{\varepsilon,M}, y_{j+1}) + \varepsilon^3 \sum_{k=1}^{d} \sum_{\ell=1}^{d} F_k^M(x_j^{\varepsilon,M} y_j)$$

$$\times F_\ell^M(x_j^{\varepsilon,M}, y_j) \int_0^1 (1 - u) \frac{\partial^2 \psi_f^{(1),M}}{\partial x_k\, \partial x_\ell}(x_j^{\varepsilon,M} + u(x_{j+1}^{\varepsilon,M}$$

$$- x_j^{\varepsilon,M}), y_{j+1})\, du + \varepsilon^3 \psi_f^{(2),M}(x_j^{\varepsilon,M}, y_{j+1})$$

$$+ \varepsilon^3 \sum_{k=1}^{d} F_k^M(x_j^{\varepsilon,M}, y_j) \int_0^1 \frac{\partial \psi_f^{(2),M}}{\partial x_k}(x_j^{\varepsilon,M} + u(x_{j+1}^{\varepsilon,M}$$

$$- x_j^{\varepsilon,M}), y_{j+1})\, du$$

By (2.4), (3.2), and (3.3), we have

$$= f(x_{[t/\varepsilon^2]}^{\varepsilon,M}) - f(x_{[s/\varepsilon^2]}^{\varepsilon,M}) + \varepsilon[\psi_f^{(1),M}(x_{[t/\varepsilon^2]}^{\varepsilon,M}, y_{[t/\varepsilon^2]})$$

$$- \psi_f^{(1),M}(x_{[s/\varepsilon^2]}^{\varepsilon,M}, y_{[s/\varepsilon^2]})] + \varepsilon^2[\psi_f^{(2),M}(x_{[t/\varepsilon^2]}^{\varepsilon,M},$$

$$y_{[t/\varepsilon^2]}) - \psi_f^{(2),M}(x_{[s/\varepsilon^2]}^{\varepsilon,M}, y_{[s/\varepsilon^2]})]$$

$$- \varepsilon \sum_{j=[s/\varepsilon^2]}^{[t/\varepsilon^2]-1} \left\{ \sum_{k=1}^{d} F_k^M(x_j^{\varepsilon,M}, y_j) \frac{\partial f}{\partial x_k}(x_j^{\varepsilon,M}) \right.$$

$$\left. + E[\psi_f^{(1),M}(x_j^{\varepsilon,M}, y_{j+1}) \mid (x_j^{\varepsilon,M}, y_j)] - \psi_f^{(1),M}(x_j, y_j) \right\}$$

$$- \varepsilon^2 \sum_{j=[s/\varepsilon^2]}^{[t/\varepsilon^2]-1} \left\{ \frac{1}{2} \sum_{k=1}^{d} \sum_{\ell=1}^{d} F_k^M(x_j^{\varepsilon,M}, y_j) F_\ell^M(x_j^{\varepsilon,M}, y_j) \right.$$

$$\times \frac{\partial^2 f}{\partial x_k\, \partial x_\ell}(x_j^{\varepsilon,M}) + \sum_{k=1}^{d} F_k^M(x_j^{\varepsilon,M}, y_j)$$

$$\times E\left[\frac{\partial \psi_f^{(1),M}}{\partial x_k}(x_j^{\varepsilon,M}, y_{j+1}) \mid (x_j^{\varepsilon,M}, y_j)\right] + E(\psi_f^{(2),M}(x_j^{\varepsilon,M}, y_{j+1}) \mid$$

$$(x_j^{\varepsilon,M}, y_j)) - \psi_f^{(2),M}(x_j^{\varepsilon,M}, y_j)\Big\}$$

$$- \varepsilon^3 \sum_{j=[s/\varepsilon^2]}^{[t/\varepsilon^2]-1} \left[\frac{1}{2} \sum_{k=1}^{d} \sum_{\ell=1}^{d} \sum_{m=1}^{d} F_k^M(x_j^{\varepsilon,M}, y_j) F_\ell^M(x_j, y_j) F_m^M(x_j^{\varepsilon,M}, y_j)\right.$$

$$\times \int_0^1 (1-u)^2 \frac{\partial^3 f}{\partial x_k\ \partial x_\ell\ \partial x_m}\ x_j^{\varepsilon,M} + u(x_{j+1}^{\varepsilon,M} - x_j^{\varepsilon,M})\ du$$

$$+ \sum_{k=1}^{d} \sum_{\ell=1}^{d} F_k^M(x_j^{\varepsilon,M}, y_j) F_\ell^M(x_j^{\varepsilon,M}, y_j)$$

$$\times \int_0^1 (1-u) \frac{\partial^2 \psi_f^{(1),M}}{\partial x_k\ \partial x_\ell}(x_j^{\varepsilon,M} + u(x_{j+1}^{\varepsilon,M} - x_j^{\varepsilon,M}), y_j)\ du$$

$$+ \sum_{\ell=1}^{d} F_k^M(x_j^{\varepsilon,M}, y_j) \int_0^1 \frac{\partial \psi_f^{(2),M}}{\partial x_k}(x_j^{\varepsilon,M} + u(x_{j+1}^{\varepsilon,M}$$

$$\left. - x_j^{\varepsilon,M}), y_j)\ du\right]$$

$$= f(x_{[t/\varepsilon^2]}^{\varepsilon,M}) - f(x_{[s/\varepsilon^2]}^{\varepsilon,M}) + \varepsilon\left[\psi_f^{(1),M}(x_{[t/\varepsilon^2]}^{\varepsilon,M}, y_{[t/\varepsilon^2]})\right.$$

$$\left. - \psi_f^{(1),M}(x_{[s/\varepsilon^2]}^{\varepsilon,M}, y_{[/\varepsilon^2]})\right]$$

$$+ \varepsilon^2\left[\psi_f^{(2),M}(x_{[t/\varepsilon^2]}^{\varepsilon,M}, y_{[t/\varepsilon^2]}) - \psi_f^{(2),M}(x_{[s/\varepsilon^2]}, y_{[s/\varepsilon^2]})\right]$$

$$= \varepsilon^2 \sum_{j=0}^{[t/\varepsilon^2]-1} \overline{L}^M f(x_j^{\varepsilon,M})$$

$$- \varepsilon^3 K(x_j^{\varepsilon,M}, x_{j+1}^{\varepsilon,M}, y_j, y_{j+1})$$

Let

$$M_f^{\varepsilon,M}(n) = f^{\varepsilon,M}(x_n^{\varepsilon,M}, y_n) - \sum_{j=0}^{n-1} [E(f^{\varepsilon,M}(x_{j+1}^{\varepsilon,M}, y_{j+1}) \mid$$

$$(x_j, y_j)) - f^{\varepsilon,M}(x_j, y_j)]$$

Since $M_f^{\varepsilon,M}(n)$ is a martingale with respect to $Q^{\varepsilon,M}$, where $Q^{\varepsilon,M}$ is the measure on $C[0,\infty)$ generated by the stochastic process $\{x^{\varepsilon,M}(t)\}$ which is defined as in (2.6) by $\{x_n^{\varepsilon,M}\}$, we have for any $\Phi_s \in L^2(Q^{\varepsilon,M})$ which are measurable with respect to $\{x_n^{\varepsilon,M};\ n \leq [x/\varepsilon^2]\}$:

$$0 = E^{\varepsilon,M}\{(M_f^{\varepsilon,M}([t/\varepsilon^2]) - M_f^{\varepsilon,M}([s/\varepsilon^2])\Phi_s\}$$

$$= E^{\varepsilon,M}\left[f(x_{[t/\varepsilon^2]}^{\varepsilon,M}) - f(x_{[s/\varepsilon^2]}^{\varepsilon,M}) - \varepsilon^2 \sum_{j=[s/\varepsilon^2]}^{[t/\varepsilon^2]-1} {}^{M}f(x_j^{\varepsilon,M})\ \Phi_s^{\varepsilon,M}\right]$$

$$- \varepsilon E^{\varepsilon,M}\Big[(\psi_f^{(1),M}(x_{[t/\varepsilon^2]}^{\varepsilon,M}, y_{[/\varepsilon^2]})$$

$$- \psi_f^{(1),M}(x_{[s/\varepsilon^2]}^{\varepsilon,M}, y_{[s/\varepsilon^2]}))\Phi_s\Big]$$

$$- \varepsilon^2 E^{\varepsilon,M}\Big[(\psi_f^{(2),M}(x_{[t/\varepsilon^2]}^{\varepsilon,M}, y_{[t/\varepsilon^2]})$$

$$- \psi_f^{(2),M}(x_{[s/\varepsilon^2]}^{\varepsilon,M}, y_{[s/\varepsilon^2]}))\Phi_s\Big]$$

$$- \varepsilon^3 E^{\varepsilon,M}\left[K(x_j^{\varepsilon,M}, x_{j+1}^{\varepsilon,M}, y_j, y_{j+1})\Phi_s\right]$$

Suppose that the family of measures $Q^{\varepsilon,M}$ for each fixed $M > 0$ is tight. Since $f \in C^\infty(R^d)$ with compact supports, by means of lemma 3.4 in Papanicolaou, Stroock and Varadhan (Ref. 9, p. 83), for any subsequence $P^{\varepsilon_n,M}$ of $\{P^{\varepsilon,M}\}$ which converges weakly to the measure $P^{\{\varepsilon_n\},M}$, we have

$$\lim_{n\to\infty} E^{\varepsilon_n,M}\Big[(f(X_{[t/\varepsilon_n^2]}^{\varepsilon_n,M}) - f(X_{[x/\varepsilon_n^2]}^{\varepsilon_n,M})$$

$$- \varepsilon_n^2 \sum_{j=[s/\varepsilon_n^2]}^{[t/\varepsilon_n^2]-1} \bar{L}^M f(X_1^{\varepsilon_n,M}))\Phi_s\Big]$$

$$= E^{\{\varepsilon_n\},M}\left[(f(X_t) - f(X_s) - \int_s^t \bar{L}^M f(X_u)\, du)\Phi_s\right]$$

which turns out to be zero.

Therefore, we see that

$$M_f^M(t) = f(x_t) - f(x_0) - \int_0^t \bar{L}^M f(x_u)\, du$$

is a Martingale for any measure Q^M which are the weak limit points of the family $\{Q^{\varepsilon,M}\}$ of measures. It reamins to prove that the family $\{Q^{\varepsilon,M}\}$ of measures is tight.

In (3.1), we take the ith coordinate function as $f(\cdot)$,

$1 \leq i \leq d$; then we have

$$(x^{\varepsilon,M}_{[t/\varepsilon^2]} - x^{\varepsilon,M}_{[s/\varepsilon^2]})_i = M^{\varepsilon,M}_{x_i}([\tfrac{t}{\varepsilon^2}]) - M^{\varepsilon,M}_{x_i}([\tfrac{s}{\varepsilon^2}]) \tag{3.4}$$
$$- \varepsilon[\psi^{\varepsilon,M}_i(x^{\varepsilon,M}_{[t/\varepsilon^2]}, y_{[t/\varepsilon^2]}) - \psi^{(1),M}_i(x^{\varepsilon,M}_{[s/\varepsilon^2]}, y_{[s/\varepsilon^2]})]$$
$$- \varepsilon^2[\psi^{(2),M}_{x_i}(x^{\varepsilon,M}_{[t/\varepsilon^2]}, y_{[t/\varepsilon^2]}) - \psi^{(2),M}_{x_i}(x^{\varepsilon,M}_{[s/\varepsilon^2]}, y_{[s/\varepsilon^2]})]$$
$$- \varepsilon^2 \sum_{j=[s/\varepsilon^2]}^{[t/\varepsilon^2]-1} \int_{S\times S} \int \Pi(dy) \sum_{k=1}^{d} F^M_k(x^{\varepsilon,M}_j, y_j)$$
$$\times \frac{\partial\psi^{(1),M}_i}{\partial x_k}(x^{\varepsilon,M}_j, z)\Pi(y,dz) - \varepsilon^3 K(x^{\varepsilon,M}_j, x^{\varepsilon,M}_{j+1}, y_j, y_{j+1})$$

The second, third, and fourth terms on the right-hand side of (3.4) are uniformly small as ε tends to zero.

Let

$$(\tilde{x}_t)_i = M^{\varepsilon,M}_{x_i}\left(\left[\frac{t}{\varepsilon^2}\right]\right)$$
$$- \varepsilon^2 \sum_{j=0}^{[t/\varepsilon^2]-1} \int_{S\times S} \int \Pi(dy) \sum_{k=1}^{d} F^M_k(x^{\varepsilon,M}_j, y_j)$$
$$\times \frac{\partial\psi^{(1),M}_i}{\partial x_k}(x,z)\Pi(y,dz)$$

Then we have, for $F^{\varepsilon,M}_s$ the σ-field generated by $\{x^{\varepsilon,M}(t); t \leq s\}$,

$$E(|(\tilde{x}_t - \tilde{x}_s)_i|^2 F^{\varepsilon,M}_s) \leq 2E\left(\left|M^{\varepsilon,M}_{x_i}\left(\left[\frac{t}{\varepsilon^2}\right]\right) - M^{\varepsilon,M}_{x_i}\left(\left[\frac{s}{\varepsilon^2}\right]\right|^2 F^{\varepsilon,M}_s\right)$$
$$+ 2E|\Sigma_{j=[s/\varepsilon^2]}^{[t/\varepsilon^2]-1} \varepsilon^2 \int_{S\times S} \int \Pi_0(dy)$$
$$\times \sum_{k=1}^{d} F^M_k(x^{\varepsilon,M}_j, y_j) \frac{\partial\psi^{(1),M}_i}{\partial x_k}(x^{\varepsilon,M}_j, z)$$
$$\times \Pi(y,dz)|^2 | F^{\varepsilon,M}_s) \tag{3.5}$$

The second term on the right-hand side in (3.5) is majorized by const $|t - s|^2$.

On the other hand, it holds that

$$E\{|M^{\varepsilon,M}_{x_i}([\tfrac{t}{\varepsilon^2}]) - M^{\varepsilon,M}_{x_j}([\tfrac{s}{\varepsilon^2}])|^2 | F^{\varepsilon,M}_s\}$$

$$= \sum_{j=[s/\varepsilon^2]}^{[t/\varepsilon^2]-1} \{E[f^{\varepsilon,M}(x^{\varepsilon,M}_{j+1}, y_{j+1})^2 | (x^{\varepsilon,M}_j, y_j)]$$

$$- E[f^{\varepsilon,M}(x_{j+1}, y_{j+1}) | (x^{\varepsilon,M}_j, y_j)]^2\}$$

$$= \sum_{j=[s/\varepsilon^2]}^{[t/\varepsilon^2]-1} \varepsilon^2\{E[(\psi^{(1),M}_i + \varepsilon\psi^{(2),M}_{x_i})^2 | (x^{\varepsilon,M}_j, y_j)]$$

$$- [E(\psi^{(1),M}_i + \varepsilon\psi^{(2),M}_{x_i} | (x^{\varepsilon,M}_j, y_j))]^2\}$$

Since each term in Σ is bounded, we have

$$E(|M^{\varepsilon,M}_{x_i}([\tfrac{t}{\varepsilon^2}]) - M^{\varepsilon,M}_{x_i}([\tfrac{s}{\varepsilon^2}])|^2 | F^{\varepsilon,M}_s) \le \text{const } |t - s|$$

Therefore, we can see that

$$E[x^{\varepsilon,M}_{[t/\varepsilon^2]} - x^{\varepsilon,M}_{[s/\varepsilon^2]}|^2 | F^{\varepsilon,M}_s] \le \text{const } |t - s|$$

and also, we have that

$$E[|x^{\varepsilon,M}_{[t/\varepsilon^2]}|^2] \le \text{const}$$

Thus we can deduce the tightness of the measure $\{Q^{\varepsilon,M}\}$. By the same argument as in Kesten and Papanicolaou (Ref. 4, pp. 119-120), we can obtain the conclusion of theorem.

4. GENERALIZATIONS

Let us consider more general situations. The driving Markov chain $\{y_j\}$ also depends on ε. We denote $\{y_j\}$ by $\{y^\varepsilon_j\}$ to emphasize this, and we assume that all conditions in Sec. 2 for $\{y_j\}$ are satisfied for $\{y^\varepsilon_j\}$. First, we consider the case that

$$\sup_{z\in S}\ \sup_{0<\varepsilon} |\psi_\varepsilon|(z,S) < \infty$$

where $|\psi_\varepsilon|$ means the total variation of a signed measure $\psi_\varepsilon(z,\cdot)$. Furthermore, we suppose that for any bounded continuous function $f(\cdot)$,

$$\lim_{\varepsilon\to 0} \int_S \psi_\varepsilon(z,du)f(u) = \int_S \psi_0(z,du)f(u)$$

$$\lim_{\varepsilon\to 0} \int_S \Pi_\varepsilon(x,dy)f(y) = \int_S \Pi_0(x,dy)f(y)$$

and

$$\lim_{\varepsilon\to 0} \int_S \Pi_\varepsilon(dy)f(y) = \int_S \Pi_0(dy)f(y)$$

where $\psi_0(z,du)$ is a signed measure with finite total variations. $\Pi_0(x,dy)$ and $\Pi_0(dy)$ are probability measures.

Define

$$a_{k\ell}^{0,M}(x) = \int_S F_k^M(x,y)F_\ell^M(x,y)\Pi_0(dy) + 2\int_S F_k^M(x,y)[\int_{S\times S}\int F_\ell^M(x,u)\psi_0(z,du)\Pi_0(y,du) \times \Pi_0(dy)]$$

$$b_\ell^{0,M}(x) = \Sigma_{k=1}^d \int_s \Pi_0(dy)F_k^M(x,y)\int_{S\times S}\int \frac{\partial F_k^M}{\partial x_k}(x,u) \times \psi_0(z,du)\Pi_0(y,dz)$$

and

$$\bar{L}^{0,M}f(x) = \frac{1}{2}\sum_{k=1}^d\sum_{\ell=1}^d a_{k\ell}^{0,M}(x)\frac{\partial^2 f}{\partial x_k\,\partial x_\ell} + \sum_{\ell=1}^d b_\ell^{0,M}(x)\frac{\partial f}{\partial x_\ell}$$

Under the assumption of the uniqueness of the martingale problem to the generations L^0 and the tightness of the family of Markov diffusion measures $R^{0,M}$ corresponding to the generation L^0, we have the following theorem.

THEOREM 4.1 Let $\{x_n^\varepsilon\}$ be the solution of the stochastic difference equation $x_{n+1}^\varepsilon = x_n^\varepsilon + \varepsilon F(x_n^\varepsilon, y_n^\varepsilon)$, $n = 0, 1, 2, \ldots$. The stochastic processes $\{x^\varepsilon(t)\}$ defined by (2.6) converge weakly to the diffusion Markov process $\{x(t)\}$ with the generator L^0, where $L^{0,\infty} = L^0$.

Second, we consider the case that

$$\sup_{0<\varepsilon} |\psi_\varepsilon|(z,S) = \infty \qquad \text{for } z \in S$$

However, we assume that there exists the positive function $\Gamma(\varepsilon) \to \infty$ $(\varepsilon \to 0)$ such that $\lim_{\varepsilon\to 0} \varepsilon^2\Gamma(\varepsilon) = 0$, and for all bounded continuous functions it holds that

$$\lim_{\varepsilon\to 0} \int_S \frac{\psi_\varepsilon(z,du)}{\Gamma(\varepsilon)} f(u) = \int_S f(u)\psi_0(z,du)$$

uniformly with respect to z, where $\psi_0(z,du)$ is a signed measure with finite total variation. Also, assume that there exist probability measures $\Pi_0(y,du)$, $\Pi_0(du)$ such that

$$\lim_{\varepsilon\to 0} \int_S \Pi_\varepsilon(y,du)f(u) = \int_S \Pi_0(y,du)f(u)$$

and

$$\lim_{\varepsilon\to 0} \int_S \Pi_\varepsilon(du)f(u) = \int_S \Pi_0(du)f(u)$$

for all bounded continuous functions and for all $y \in S$. If we replace ε^2 by $\varepsilon^2\Gamma(\varepsilon)$ in the course of proof of Theorem 2.1, under assumptions similar to those of Theorem 4.1, we have the following theorem.

THEOREM 4.2 Let

$$x^\varepsilon(t) = x_j^\varepsilon + \frac{t - j\varepsilon^2\Gamma(\varepsilon)}{\varepsilon^2\Gamma(\varepsilon)}(x_{j+1}^\varepsilon - x_j^\varepsilon)$$

$$\text{if } j\varepsilon^2\Gamma(\varepsilon) \leq t < (j+1)\varepsilon^2\Gamma(\varepsilon).$$

Then the stochastic process $\{x^\varepsilon(t)\}$ converges weakly as $\varepsilon \to 0$ to the diffusion Markov processes with the generator

$$\bar{L}^0 = \frac{1}{2} \sum_{k,\ell=1}^{d} a_{k\ell}^0(x)\frac{\partial^2}{\partial x_k\, \partial x_\ell} + \sum_{\ell=1}^{d} b_\ell(x)\frac{\partial}{\partial x_\ell}$$

where

$$a_{k\ell}^0(x) = 2\int_S F_k(x,y) \int_{S\times S}\int F_\ell(x,u)\psi_0(z,du)\Pi_0(y,du)\Pi_0(dy)$$

and

$$b_\ell^0(x) = \sum_{k=1}^{d} \int_S \Pi_0(dy)F_k(x,y) \int_{S\times S} \int \frac{\partial F_\ell}{\partial x_k}(x,u)\psi_0(z,du)$$

$$\times\ \Pi_0(y,du)$$

Now, consider infinite dimensional cases. Let $F(X,Y)$ be an infinite dimensional vector function,

$$F(X,Y) = (F_1(X,Y),\ F_2(X,Y),\dots) \in R^\infty, \text{ where}$$
$$X = (x_1,x_2,\dots) \in R^\infty$$

$Y \in S$, and S is a compact metric space. In R^∞, we introduce the following metric:

$$\rho(X^1,X^2) = \sum_{n=1}^{\infty} \frac{1}{2^n} \frac{|x_n^1 - x_n^2|}{1 + |x_n^1 - x_n^2|}$$
$$\text{where } X^i = (x_1^i,x_2^i,\dots) \text{ for } i = 1,\ 2$$

Define successively

$$X_\varepsilon^{n+1} = X_\varepsilon^n + \varepsilon F(X_\varepsilon^n,Y^n)$$

where $\{Y^n\}$ is the Markov chain satisfying the condition of Sec. 2. Then unlike the continuous parameter case, we can define unique $\{X^n\}$ without ambiguity. Assume that the conditions for $F(X,Y)$ in Sec. 2 are satisfied for $d = \infty$, $F_k(x,y)$ are uniformly bounded with respect to k, $X \in K$ and $y \in S$, and

$$\sum_{k=1}^{\infty} F_k(X,Y)\frac{\partial F_\ell}{\partial X}(X,Z)$$

are uniformly, boundedly convergent for any $\ell = 1,\ 2,\ \dots$ $X \in K$ and $Y \in S$, where K is a compact subset of R^∞. Let

$$X_\varepsilon(t) = X_\varepsilon^j + \frac{t - j\varepsilon^2}{\varepsilon^2}(X_\varepsilon^{j+1} - X_\varepsilon^j)$$
$$\text{if } j\varepsilon^2 \le t < (j+1)\varepsilon^2$$

Define $\tau_k^\varepsilon = \inf\{t;\ X_\varepsilon(t) \notin K\}$ and $X_\varepsilon^K(t) = X_\varepsilon(t \wedge \tau_k^\varepsilon)$. Let $P^{\varepsilon,K}$ be a probability measure determined by stochastic process $X_\epsilon^K(t)$. Then for any weak limits of the family of measures $P^{\epsilon,K}$, by the same method as Theorem 2.1, we can see that

$$f(X_{t\wedge\tau_K}) - f(X_0) - \int_0^{t\wedge\tau_K} \bar{L}\, f(x)\, du$$

is a martingale for $f \in C_0^\infty$ such that

$$\sup_{x\in K} \sum_{|\alpha|\le 3} \left|\frac{\partial^\alpha f}{\partial x^\alpha}(X)\right| < \infty \qquad \text{for } \alpha = (\alpha_1,\alpha_2,\alpha_3,\ldots)$$

$|\alpha| = \alpha_1 + \alpha_2 + \alpha_3 + \cdots$, where

$$\bar{L} = \frac{1}{2}\sum_{k,\ell=1}^{\infty} a_{k,\ell} \frac{\partial^2 f}{\partial x_k\, \partial x_\ell} + \sum_{\ell=1}^{\infty} b_\ell \frac{\partial f}{\partial x_\ell}$$

$$a_{k,\ell} = \int_S F_k(X,Y)F_\ell(X,Y)\Pi(dY)$$

$$+ 2\int_S F_k(X,Y) \int_{S\times S} F_\ell(X,Y)\psi(Z,dU)\Pi(Y,dZ)\Pi(dY)$$

and

$$b_\ell = \sum_{k=1}^{\infty} \int_S \Pi(dY)F_k(X,Y) \int_{S\times S}\int \frac{\partial F_\ell}{\partial x_k}(X,U)\psi(Z,dU)\Pi(Y,dZ)$$

Under the assumption of the uniqueness of the Martingale problem for the generator $\bar{L}$, we can see that the stochastic process $X_\epsilon^K(t)$ converges weakly to the diffusion Markov process $X^K(t)$ with generator $\bar{L}$.

5. EXAMPLE

Let $\{y_j\}$ be a Markov chain with two states $\{-1,+1\}$. We assume that $\{y_j\}$ has the following stationary transition probabilities:

$$[\Pi(i,j)] = \begin{bmatrix} 1-\gamma & \gamma \\ \gamma & 1-\gamma \end{bmatrix}$$

where $0 < \gamma < 1$.

Then, we have $\Pi(1) = \Pi(-1) = 1/2$ and

$$\begin{matrix} \psi(1,1) & \psi(1,-1) \\ \psi(-1,1) & \psi(-1,-1) \end{matrix} = \frac{1}{4\gamma} \begin{matrix} 1 & -1 \\ -1 & 1 \end{matrix}$$

Let $F(x,y) = yg(x)$, where $g(x) = x(1 - x)$ $0 \leq x \leq 1$, $y = -1$ or 1. Therefore, we have

$$a(x) = g^2(x) + 2[\tfrac{1}{2} g(x), -\tfrac{1}{2} g(x)] \begin{bmatrix} 1-\gamma & \gamma \\ \gamma & 1-\gamma \end{bmatrix}$$

$$\times \begin{bmatrix} \frac{1}{4\gamma} & -\frac{1}{4\gamma} \\ -\frac{1}{4\gamma} & \frac{1}{4\gamma} \end{bmatrix} \begin{bmatrix} g(x) \\ -g(x) \end{bmatrix}$$

$$= g^2(x) + 2(\frac{1}{2\gamma} - 1)g^2(x) = 2(\frac{1}{2\gamma} - \frac{1}{2})g^2(x)$$

$$b(x) = [\tfrac{1}{2} g(x), -\tfrac{1}{2} g(x)] \begin{bmatrix} 1-\gamma & \gamma \\ \gamma & 1-\gamma \end{bmatrix} \begin{bmatrix} \frac{1}{4\gamma} & -\frac{1}{4\gamma} \\ -\frac{1}{4\gamma} & \frac{1}{4\gamma} \end{bmatrix} \begin{bmatrix} g'(x) \\ -g'(x) \end{bmatrix}$$

$$= (\frac{1}{2\gamma} - 1)g(x)g'(x)$$

Let us define successively,

$$x^{\varepsilon}_{j+1} = x^{\varepsilon}_j + \varepsilon y_j g(x^{\varepsilon}_j) \qquad j = 1, 2, \ldots$$

$$x^{\varepsilon}_0 = x \qquad 0 < x < 1$$

For this model, the conditions of theorem are satisfied. Therefore, $x^{\varepsilon}(t)$ defined by (2.6) converges weakly as $\varepsilon \to 0$ to the diffusion process with generator

$$\frac{1}{2} a(x)\frac{d^2f}{dx^2} + b(x)\frac{df}{dx}$$

when γ depends on ε and satisfies $\varepsilon^2/\gamma(\varepsilon) \to 0$ as $\varepsilon \to 0$. If we take as $\Gamma(\varepsilon) = [1/\gamma(\varepsilon)]$, then we have

$$\frac{\psi_\varepsilon(i,j)}{\Gamma(\varepsilon)} \to \frac{1}{4}\begin{bmatrix} 1 & -1 \\ -1 & 1 \end{bmatrix}$$

$$[\Pi_\varepsilon(i,j)] \to \begin{bmatrix} 1 & 0 \\ 0 & 1 \end{bmatrix}$$

Therefore, we have

$$a^0(x) = 2\left[\frac{1}{2}\,g(x), -\frac{1}{2}\,g(x)\right]\begin{bmatrix}1 & 0\\ 0 & 1\end{bmatrix}\begin{bmatrix}\frac{1}{4} & -\frac{1}{4}\\ -\frac{1}{4} & \frac{1}{4}\end{bmatrix}\begin{bmatrix}g(x)\\ g(x)\end{bmatrix}$$

$$= g^2(x)$$

$$b^0(x) = \left[\frac{1}{2}\,g(x), -\frac{1}{2}\,g(x)\right]\begin{bmatrix}1 & 0\\ 0 & 1\end{bmatrix}\begin{bmatrix}\frac{1}{4} & -\frac{1}{4}\\ -\frac{1}{4} & \frac{1}{4}\end{bmatrix}\begin{bmatrix}g'(x)\\ -g'(x)\end{bmatrix}$$

$$= \frac{1}{2}\,g(x)g'(x)$$

Therefore, $x^\varepsilon(t)$ defined as in Theorem 4.2 converges weakly as $\varepsilon \to 0$ to the diffusion process $x(t)$ with generator

$$\frac{1}{2}\,g^2(x)\frac{d^2}{dx^2} + \frac{1}{2}\,g(x)g'(x)\frac{d}{dx}$$

and $x(t)$ is the solution process of the stochastic differential equation

$$dx(t) = g(x(t))\;dB(t) + \frac{1}{2}\,g(x(t))g'(x(t))\;dt$$

$$= g(x(t))\circ dB(t)$$

where $\circ$ indicates the Stratonovich integral, and the limiting process $x(t)$ is the solution process of the Itô-type stochastic differential equation $dx(t) = g(x(t))\;dB(t)$ in the case $\gamma = 1/2$.

REFERENCES

1. G. Blankenship and G. C. Papanicolaou, Stability and control of stochastic systems with wide band noise disturbances, SIAM J. Appl. Math. 34:437-476 (1978).
2. A. N. Borodin, A limit theorem for solutions of differential equations with random right hand side, Theory Prob. Appl. 22 (1977).
3. M. Iizuka and H. Matsuda, Weak convergence of discrete time non-Markovian processes related to section models in population genetics, J. Math. Biology, 15(1982), 107-127.
4. H. Kesten and G. C. Papanicolaou, A limit theorem for

turbulent diffusion, Commun. Math. Phys. 65:97-128 (1978).

5. R. Z. Khasminskii, A limit theorem for solutions of differential equations with a random right hand side, Theory Prob. Appl. 11:390-406 (1966).
6. H. J. Kushner, A martingale method for the convergence of a sequence of processes to a jump diffusion process, Z. Wahrscheinlichkeitsth. Verwend. Geb. 53:207-219 (1980).
7. P. A. Meyer, Les résolventes fortement Fellérienes d'après Mokobodzki, in Séminaire de probabilités II, Springer-Verlag, Berlin, 171-174, 1968.
8. G. C. Papanicolaou and W. Kohler, Asymptotic theory of mixing stochastic ordinary differential equations, Commun. Pure Appl. Math. 27:641-668, (1974).
9. G. C. Papanicolaou, D. Stroock and S. R. S. Varadhan, Martingale approach to some limit theorems, in Statistical Mechanics and Dynamical Systems, Duke Univ. Conf. Turbulence (M. Reed, ed.), Duke Univ. Math. Ser., Vol. 3, Durham, N.C., 1977.
10. H. Watanabe, Potential operator of a recurrent strong Feller process in the strict sense and boundary value problem, J. Math. Soc. Jap. 16:83-95 (1964).
11. H. Watanabe, Diffusion approximations of some stochastic difference equations II, Hiroshima Math. J. 13:689-708 (1983).

Index

Milton Keynes UK
Ingram Content Group UK Ltd.
UKHW021902071024
449327UK00021B/1605

9 780367 451783